Springer Series on

Atoms+Plasmas

15

W0037441

Springer Series on

Atoms+Plasmas

Editors: G. Ecker P. Lambropoulos I. I. Sobel'man H. Walther

Managing Editor: H. K. V. Lotsch

1 **Polarized Electrons**
2nd Edition
By J. Kessler

2 **Multiphoton Processes**
Editors: P. Lambropoulos and S. J. Smith

3 **Atomic Many-Body Theory**
2nd Edition
By I. Lindgren and J. Morrison

4 **Elementary Processes in Hydrogen-Helium Plasmas**
Cross Sections and Reaction Rate Coefficients
By R. K. Janev, W. D. Langer, K. Evans, Jr. and D. E. Post, Jr.

5 **Pulsed Electrical Discharge in Vacuum**
By G. A. Mesyats and D. I. Proskurovsky

6 **Atomic and Molecular Spectroscopy** 2nd Edition
Basic Aspects and Practical Applications
By S. Svanberg

7 **Interference of Atomic States**
By E. B. Alexandrov, M. P. Chaika and G. I. Khvostenko

8 **Plasma Physics** 2nd Edition
Basic Theory with Fusion Applications
By K. Nishikawa and M. Wakatani

9 **Plasma Spectroscopy**
The Influence of Microwave and Laser Fields
By E. Oks

10 **Film Deposition by Plasma Techniques**
By M. Konuma

11 **Resonance Phenomena in Electron-Atom Collisions**
By V. I. Lengyel, V. T. Navrotsky and E. P. Sabad

12 **Atomic Spectra and Radiative Transitions** 2nd Edition
By I. I. Sobel'man

13 **Multiphoton Processes in Atoms**
By N. B. Delone and V. P. Krainov

14 **Atoms in Plasmas**
By V. S. Lisitsa

15 **Excitation of Atoms and Broadening of Spectral Lines**
By I. I. Sobel'man, L. Vainshtein and E. Yukov

16 **Reference Data on Multicharged Ions**
By V. G. Pal'chikov and V. P. Shevelko

17 **Lectures on Non-linear Plasma Kinetics**
By V. N. Tsytovich

Igor I. Sobel'man Leonid A. Vainshtein
Evgenii A. Yukov

Excitation of Atoms and Broadening of Spectral Lines

Second Edition
With 21 Figures

Springer

Professor Dr. Igor I. Sobel'man
Professor Leonid A. Vainshtein
Dr. Evgenii A. Yukov

P. N. Lebedev Physical Institute, Russian Academy of Sciences,
Leninsky Prospect 53, 117924 Moscow, Russia

Series Editors:

Professor Dr. Günter Ecker

Ruhr-Universität Bochum,
Lehrstuhl Theoretische Physik I,
Universitätsstraße 150,
D-44801 Bochum, Germany

Professor Peter Lambropoulos, Ph. D.

Max-Planck-Institut für Quantenoptik
D-85748 Garching, Germany, and
Foundation for Research and Technology –
Hellas (FO.R.T.H.),
Institute of Electronic Structure & Laser (IESL)
University of Crete,
PO Box 1527, Heraklion, Crete 71110, Greece

Professor Igor I. Sobel'man

Lebedev Physical Institute,
Russian Academy of Sciences,
Leninsky Prospekt 53, 117924 Moscow, Russia

Professor Dr. Herbert Walther

Universität München, Sektion Physik,
Am Coulombwall 1,
D-85748 Garching/München, Germany

Managing Editor:

Dr. Helmut K.V. Lotsch

Library of Congress Cataloging-in-Publication Data.
Sobel'man, I. I. (Igor'Il'ich), 1927 –
[Vvedenie v teoriiu atomnykh spectrov. English]
Excitation of atoms and broadening of spectral lines / Igor I. Sobel'man, Leonid A. Vainshtein, Evgenii A. Yukov. –
2nd ed. p.cm. – (Springer series on atoms + plasma ; 15)
Includes bibliographical references and index.
ISBN 978-3-540-58686-9 ISBN 978-3-642-57825-0 (eBook)
DOI 10.1007/978-3-642-57825-0
1. Cross sections (Nuclear physics) 2. Spectral line broadening. 3. Nuclear excitation. 4. Atomic spectroscopy.
I. Vaĭnshteĭn, L. A. (Leonid Abramovich), 1928– . II. IUkov, E. A. (Evgeniĭ Aleksandrovich), 1945– . III. Title.
IV. Series.
QC794.6.C7S6213 1995
539.7'54–dc20 95-211 CIP

The first edition appeared as Springer Series in Chemical Physics, Vol. 7

ISBN 978-3-540-58686-9

This work is subject to copyright. All rights are reserved, whether the whole or part of the material is concerned, specifically the rights of translation, reprinting, reuse of illustrations, recitation, broadcasting, reproduction on microfilms or in any other way, and storage in data banks. Duplication of this publication or parts thereof is permitted only under the provisions of the German Copyright Law of September 9, 1965, in its current version, and permission for use must always be obtained from Springer-Verlag. Violations are liable for prosecution under the German Copyright Law.

© Springer-Verlag Berlin Heidelberg 1981, 1995

Originally published by Physica-Verlag Heidelberg New York in 1981, 1995

The use of general descriptive names, registered names, trademarks, etc. in this publication does not imply, even in the absence of a specific statement, that such names are exempt from the relevant protective laws and regulations and therefore free for general use.

Typesetting: Macmillan India Ltd., Bangalore, India
Production: PRODUserv Springer Produktions-Gesellschaft, Berlin
SPIN 10089022 54/3020 - 5 4 3 2 1 0 - Printed on acid-free paper

Preface to the Second Edition

In this new edition – as in the first one of 1981 – the main goal is to present the theory of elementary processes responsible for the excitation and formation of atomic spectra. No attempt has been made to give a systematic and detailed treatment of the general theory of atomic collisions. Instead, stress is placed on efficient and comparatively simple approximate methods for the calculation of cross sections and rate coefficients. New results of importance and new understanding of many problems have appeared in recent years and require a significant revision of the original book.

The formulas for cross sections given in Chaps. 2 and 3 are derived and represented using the unified approach which is symmetrical with respect to orbital and spin quantum numbers. This allows us to give the final results in a more general form which is simpler and more convenient for computer programming.

Chapter 4 in which the collisions between heavy particles are considered, is slightly simplified and corrected. The references to recent sources and reviews on the subject are given.

The approximation of cross sections and rate coefficients by analytic formulas (Sect. 5.1) is considerably changed in this edition. All the numerical data given in Chap. 6, except Tables 6.2 and 6.3, are new. They have been calculated using the updated version of the computer code ATOM, which has substantially been extended and corrected during recent years.

Chapter 7 devoted to the broadening of spectral lines is slightly revised. In addition, references are added which communicate the progress in the field.

The authors are very grateful to Dr. Helmut Lotsch of Springer Verlag for encouragement and advices.

Moscow, April 1995

I. Sobelman
L. Vainshtein
E. Yukov

Preface to the First Edition

New applications of atomic spectroscopy in laser physics, laser spectroscopy, laser frequency and wavelength measurements, plasma physics, astrophysics, and some other related problems have been developed very intensively in the last years. As a result, the approximate methods of calculation of the transition probabilities and cross sections necessary for all these applications have become of vastly increased importance. At the same time, some new problems have arisen in the theory of spectral line broadening such as the shape of nonlinear resonances in the spectra of gas lasers, interference effects, and some other problems connected with various spectroscopic methods of plasma diagnostics.

This book is devoted to the systematic treatment of the theory of the elementary processes responsible for the excitation of atomic spectra and the theory of spectral line broadening. The choice of problems is significantly different from that traditional for books on the theory of atomic collisions. The main goal of the book is to present the most efficient and useful of comparatively simple approximate methods for the calculation and estimation of cross sections. Numerous tables containing the results of approximate cross section calculations for the most important elementary processes are included in the book. Comprehensive presentation of the theory of atomic collisions is out of the scope of this book and can be found elsewhere. However, the fundamentals of the general theory of collisions which are necessary for formulation of approximate methods are given in Chapter 2.

In dealing with the theory of spectral line broadening special attention is paid to the general approach to the problem based on the method of density matrix and quantum kinetic equation. This approach is of interest for some modern applications of broadening theory e.g. such as high-resolution laser spectroscopy.

We consider this book as complementary to I. Sobelman's *Atomic Spectra and Radiative Transitions* (first volume of present book-series). We wish to express our sincere gratitude to I.L. Beigman, who helped us to prepare Sects. 3.5, 5.4, to L.P. Presnyakov, who assisted us in Sects. 3.4, 4.1–3. We are grateful also to Prof H.-W. Drawin, Prof. H.R. Griem, Dr. V.S. Lisitsa and Prof. J.-P. Toennies, who have read the manuscript, for many valuable comments.

In conclusion we are particularly grateful to H.W. Drawin, H. Griem and J.P. Toennies for the final reading of the manuscript and their helpful suggestions.

Moscow, April 1980

I. Sobelman
L. Vainshtein
E. Yukov

Contents

1 Elementary Processes Giving Rise to Spectra

In this chapter, the most important elementary processes responsible for the excitation and ionization of atoms and ions are listed, and their relative role in various plasmas is briefly discussed. Throughout this book, $\mathscr{Z}$ is the nuclear charge, Z is the ion charge, and z is the charge of atomic core, i.e., the charge of the atom or ion without the optical electron. For a neutral atom $z = 1$; for a singly ionized atom, $z = 2$, and so on. An ion with charge $Z = z - 1$ will be denoted X_z. We shall usually use the word "atom" to mean both the neutral atom with $z = 1$ and ions with $z > 1$. As a rule plasma is supposed to be optically thin.

1.1 Cross Sections and Rate Coefficients

The intensity of the spectral line corresponding to the transition from the level k to the level i in an atom X_z is given by

$$I_{ki} = N_k^{(z)} A_{ki}^{(z)} \hbar\omega_{ki} \; [\text{erg cm}^{-3} s^{-1}] \,, \tag{1.1.1}$$

where $A_{ki}^{(z)}, \omega_{ki}$, are the probability and frequency of the transition, respectively, $N_k^{(z)}$ is the number of atoms X_z in the level k per unit volume, and $\hbar$ is Planck's constant divided by 2π.

Usually electrons play the main role in excitation and ionization processes in plasmas. Interaction with heavy particles (protons and ions) is important only for transitions between levels with very small energy splitting and also in some cases of transient plasma. The processes of main interest are the following:
Excitation and deexcitation:

$$X_z + e \rightleftarrows X_z^* + e \,, \tag{1.1.2}$$

where the asterisk denotes the excited state.
Ionization and three-body recombination:

$$X_z + e \rightleftarrows X_{z+1} + e + e \,. \tag{1.1.3}$$

Radiative ionization and recombination:

$$X_z + \hbar\omega \rightleftarrows X_{z+1} + e \,. \tag{1.1.4}$$

Dielectronic recombination and autoionization:

$$X_{z+1} + e \rightleftarrows X_z^{**} \rightarrow X_z^* + \hbar\omega \,. \tag{1.1.5}$$

Free-free emission and absorption:

$$X_z + e \rightleftarrows X_z + e + \hbar\omega\,. \tag{1.1.6}$$

Emission and absorption in spectral line:

$$X_z^* \rightleftarrows X_z + \hbar\omega\,. \tag{1.1.7}$$

Each of the processes (1.1.2–7) is a pair of direct and reverse reactions. The probabilities (cross sections) of such processes are connected by simple formulas (see Sect. 1.2).

All these processes can be separated into two groups, radiative [processes (1.1.4, 6, 7)] and nonradiative [processes (1.1.2, 3)]. In the case of dielectronic recombination (1.1.5) there are two stages: firstly, the nonradiative attachment of an electron and the excitation of the ion takes place; then there are two possibilities, nonradiative decay (autoionization), or radiative transition into a "stable" state below the ionization limit. This latter process provides the net recombination.

The number of radiative transitions (1.1.7) per second for one atom X_z^* is the transition probability $A_{ki}[s^{-1}]$, which does not depend on the plasma characteristics.[1] In the case of processes (1.1.2–6) the number of excitations suffered by one atom or ion per second is proportional to the electron density N_e and the velocity of the electron, v,

$$W = N_e v\sigma \quad [s^{-1}]\,, \tag{1.1.8}$$

where σ [cm^2] is the excitation cross section.

If the electrons in a plasma have an energy distribution $\mathcal{F}(\mathcal{E})$, i.e., the number density of electrons in the energy interval $(\mathcal{E},\ \mathcal{E} + d\mathcal{E})$ is

$$dN_e = N_e \mathcal{F}(\mathcal{E})\,d\mathcal{E}\,,$$

(1.1.8) has to be rewritten in the form

$$W = N_e \langle v\sigma\rangle\,,$$

$$\langle v\sigma\rangle = \int_{\Delta E}^{\infty} v\sigma(\mathcal{E})\mathcal{F}(\mathcal{E})\,d\mathcal{E}\,. \tag{1.1.9}$$

Here ΔE is the threshold energy of excitation, and $\langle v\sigma\rangle$ is called the excitation rate coefficient (for one atom X_z and unit electron density).

The same definition is used for cross sections of other processes, in which there are two particles in the initial and final states. Deexcitation (1.1.2), and radiative ionization and recombination (1.1.4) are processes of this kind. For example, the probability of radiative ionization is

$$W_\nu = \int_{E_z/\hbar}^{\infty} N_\omega c\sigma(\omega)\,d\omega\,,$$

[1] The radiative transition probabilities were discussed in detail in [1.1].

where N_ω is the photon density at frequency ω, and E_z is the threshold ionization energy.

If there are three particles in the final state, the energy can be arbitrarily distributed between them. For transitions with a definite energy for one of the particles, the cross section can be defined in a similar way, but it should have the dimensions of $cm^2\ erg^{-1}$ instead of cm^2. The cross section of free-free emission (1.1.6) and the "differential cross section" of ionization (1.1.3) are examples.

The total ionization cross section i.e., differential cross-section integrated over the energy of the ejected electron, is of most interest in applications. This cross section is defined as

$$\sigma_i(\mathscr{E}) = \int_0^{\mathscr{E}-E_z} \sigma_i(\mathscr{E};\mathscr{E}',\mathscr{E}'')\,d\mathscr{E}'\,, \qquad (1.1.10)$$
$$\mathscr{E}' + \mathscr{E}'' = \mathscr{E} - E_z\,,$$

where $\mathscr{E}'$ and $\mathscr{E}''$ are the energies of the ejected and scattered electrons. One can see that the dimensions of the ionization cross section (1.1.10) are cm^2.

In some cases there are three particles in the initial state of the reaction: the ion X_z and two electrons (three-body recombination) or the ion, an electron, and a photon (free–free absorbtion). Here the transition probability is proportional to the fluxes of both particles. For example, for three-body recombination we have

$$W_r = N_e^2 \kappa_r$$
$$\kappa_r = \int\int v_1 v_2 \mathscr{F}(\mathscr{E}_1)\mathscr{F}(\mathscr{E}_2)\sigma_r(\mathscr{E}_1,\mathscr{E}_2;\mathscr{E})\,d\mathscr{E}_1\,d\mathscr{E}_2 \qquad (1.1.11)$$

The dimensions of σ_r in this case are [cm^4 s]. The case of dielectronic recombination will be discussed in Sect. 5.2.

1.2 Populations of Atomic Levels in a Plasma; Rates of Direct and Reverse Processes

In this section we discuss briefly the simplest cases of level populations distribution in a plasma, in order to show in which way the populations are linked with the cross sections of elementary processes. Our consideration is confined to the case of homogeneous uniform and stationary plasmas. A more detailed treatment of level populations in a plasma is given in Sect. 5.4 (see also the review article by Drawin in [1.2], where transient phenomena are also discussed).

1.2.1 Thermodynamic Equilibrium

If some volume of plasma is in thermodynamic equilibrium, the following distributions are valid.

i) *Maxwell distribution of energies of free electrons:*

$$dN_e = N_e \mathcal{F}_M(\mathcal{E})\,d\mathcal{E}\,,$$
$$\mathcal{F}_M(\mathcal{E}) = 2\pi^{-1/2}T^{-3/2}\mathcal{E}^{1/2}\exp(-\mathcal{E}/T)\,, \tag{1.2.1}$$

where T is the electron temperature in energy units (1 eV = 11605 K = 8066 cm^{-1}).

ii) *Boltzmann distribution of atoms over energy levels* E_k

$$\frac{N_k}{N_0} = \frac{g_k}{g_0}\exp(-\beta_{k0})\,, \quad \beta_{k0} = \frac{E_k - E_0}{T}\,, \tag{1.2.2}$$

where g_k and g_0 are statistical weights of the levels k and 0.

iii) *Saha distribution of atoms over degrees of ionization:*

$$\frac{N^{(z+1)}}{N^{(z)}} = \frac{g_{z+1}}{g_z}S\exp(-\beta_z)\,, \quad \beta_z = \frac{E_z}{T}\,, \tag{1.2.3}$$

$$S = 2\left(\frac{mT}{2\pi\hbar^2}\right)^{3/2} N_e^{-1} = \frac{z^3}{4\pi^{3/2}}\cdot\frac{\Theta^{3/2}}{a_0^3 N_e}\,, \tag{1.2.4}$$

where E_z is the ionization energy of the atom X_z, $\Theta = T/z^2\mathrm{Ry}$ is the temperature in Rydberg units (1 Ry=13.6 eV=157894 K), $a_0 = \hbar^2/me^2 = 0.529\cdot 10^{-8}$cm is the Bohr radius, and g_z is the partition function for atom X_z:

$$g_z = \textstyle\sum_k g_k^{(z)}\exp(-\beta_{k0})\,.$$

In fact $\beta_{k0} \gg 1$ for all levels with principal quantum number different from that of the ground state.

The three distributions (1.2.1–3) are wholly determined by the plasma density N_e and temperature T. The factor S in (1.2.3) corresponds to the statistical weight of free electrons; one can see that usually $S \gg 1$. In cool plasmas only neutral atoms X_1 are present. With increasing temperature, the density of ions $N^{(z)}(z > 1)$ increases up to a maximum value and then decreases due to the further ionization $X_z \to X_{z+1}$. So atoms X_z predominate in a plasma only in the temperature interval $T_{z-1} \lesssim T \lesssim T_z$, where T_z is the temperature at which $N^{(z+1)} = N^{(z)}$. From (1.2.3) we obtain

$$\frac{E_z}{T} = \ln\left(\frac{g_{z+1}}{g_z}\cdot S\right) \gg 1\,. \tag{1.2.5}$$

For $N_e = 10^{13} - 10^{17}$, $E_z/T \simeq 20$.

Thus the density of atoms X_z is large enough only when $T \ll E_z$. Since for most levels the excitation energies E_{k0} are of the order of E_z, according to (1.2.2) populations of excited levels are very low compared to the density of atoms in the ground state.

For highly charged atoms (in fact for $z \geq 4$) important exclusion from this rule apply to the levels $n_0 l$ with the same principal quantum number n_0 as the ground state. The spacing of the levels $n_0 l_0$ and $n_0 l$ is of the order of $\Delta E_{l_0 l} \sim E_z/z$, and for high z, $\beta_{l_0 l} < 1$.

1.2.2 Rates of Direct and Reverse Processes

At thermodynamic equilibrium, detailed balance holds for any pair of collisional processes that are inverse to each other.

Let us consider a pair of atomic levels denoted by j and k. According to the principle of detailed balance the rate of collisional excitation of the j–k transition is equal to the rate of collisional deexcitation,

$$N_j N_e \langle v\sigma_{jk}\rangle = N_k N_e \langle v\sigma_{kj}\rangle \,. \tag{1.2.6}$$

Using (1.2.2) we obtain

$$g_j \langle v\sigma_{jk}\rangle = g_k \langle v\sigma_{kj}\rangle \exp(-\beta), \quad \beta = \Delta E/T \tag{1.2.7}$$

In the case of ionization, the same argument gives

$$g_z \langle v\sigma_i\rangle = 2\left(\frac{mT}{2\pi\hbar^2}\right)^{3/2} g_{z+1}\kappa_r \exp(-\beta_z)\,. \tag{1.2.8}$$

Here $\kappa_r = \langle\!\langle v_1 v_2 \sigma_r \rangle\!\rangle$ is the rate coefficient of three-body recombination.

Equations (1.2.7, 8) do not include the level populations and consequently they do not depend on the existence of thermodynamic equilibrium. It is only necessary that the energies of the electrons have a Maxwellian distribution with temperature T.

We shall now derive the formula connecting the cross sections of excitation and deexcitation. Equation (1.2.7) can be rewritten in the form

$$g_j \int_{\Delta E}^{\infty} \left(\frac{2\mathscr{E}}{m}\right)^{1/2} \sigma_{jk}(\mathscr{E})\,\mathscr{E}^{1/2} \exp\left(-\frac{\mathscr{E}}{T}\right) d\mathscr{E} = g_k \int_0^{\infty} \left(\frac{2\mathscr{E}'}{m}\right)^{1/2} \sigma_{kj}(\mathscr{E}')\,\mathscr{E}'^{1/2}$$
$$\exp\left(-\frac{\mathscr{E}'}{T}\right) \times \exp\left(-\frac{\Delta E}{T}\right) d\mathscr{E}'\,,$$

or

$$g_j \int_0^{\infty} (\mathscr{E} + \Delta E)\sigma_{jk}(\mathscr{E} + \Delta E) \exp\left(-\frac{\mathscr{E}}{T}\right) d\mathscr{E} = g_k \int_0^{\infty} \mathscr{E}\sigma_{kj}(\mathscr{E}) \exp\left(-\frac{\mathscr{E}}{T}\right) d\mathscr{E}\,.$$

This equation has to be true for any value of T. Hence it follows that

$$g_j(\mathscr{E} + \Delta E)\sigma_{jk}(\mathscr{E} + \Delta E) = g_k \mathscr{E}\,\sigma(\mathscr{E})\,. \tag{1.2.9}$$

This equation is usually called the Klein–Rosseland formula. In the particular case $\mathscr{E} \gg \Delta E$, it reduces to the simple equality $g_j \sigma_{jk} = g_k \sigma_{kj}$. Equation (1.2.9) does not depend on any particular characteristics of the plasma; thus it provides the general relation between excitation and deexcitation cross sections. A similar relation can be obtained for the recombination cross section and the differential cross section of ionization.

To connect the rates of radiative ionization σ_{iv} and recombination σ_{rv}, it is convenient to consider the atom to be in total thermodynamic equilibrium with the plasma and blackbody radiation. Then we obtain, recalling that c denotes the speed of light in vacuum,

$$g_z \langle c\sigma_{iv} \rangle = 2\left(\frac{mT}{2\pi\hbar^2}\right)^{3/2} g_{z+1}\kappa_v \,. \tag{1.2.10}$$

Here $\langle c\sigma_{iv}\rangle$ is the average over frequencies of the radiation field, and $\kappa_v = \langle v\sigma_{rv}\rangle$ is the average over electron energies. By the method similar to that used in deriving the formula (1.2.9), we find

$$\begin{aligned} &g_z \cdot q^2\sigma_{iv}(\omega) = g_{z+1} \cdot k^2 \cdot \sigma_{rv}(\mathscr{E}) \,, \\ &\hbar\omega = \mathscr{E} + E_z \,, \end{aligned} \tag{1.2.11}$$

where q and k are the wave numbers of the photon and electron respectively. Equation (1.2.11) can be rewritten in the form

$$g_z\sigma_{iv}(\omega) = \frac{2mc^2\mathscr{E}}{\hbar^2\omega^2} g_{z+1}\sigma_{rv}(\mathscr{E}) \,. \tag{1.2.12}$$

Formulas (1.2.11, 12) are known as the Milne formulas.

1.2.3 The Simplest Model

Thermodynamic equilibrium takes place for sufficiently high densities N_e. For moderate or low densities, the thermodynamic distributions (1.2.2, 3) do not hold; nor does the relation (1.2.6). Under these conditions the level population depends on the balance of all processes of excitation, radiation and so on. The general solution of the level-populations problem is very difficult because one has to consider an infinite set of equations including all levels and processes.

In the general case, the level populations differ greatly from those in thermodynamic equilibrium. However the distribution of the energies of the free electrons is as a rule almost Maxwellian. We shall suppose below that the Maxwellian distribution (1.2.1) as well as the formulas (1.2.7, 8) hold. To provide a qualitative description of the transition from thermodynamic distribution at high density to other distributions at intermediate and low density, we consider the simplest model including collisional excitation $j \to k$ and deexcitation $k \to j$, the radiative decay of the upper level k to the lower level j, and also the radiative decay of the upper level k to some other levels (i.e., we assume $A_k \geq A_{kj}$).

The equation of balance will have the form

$$N_j N_e \langle v\sigma_{jk}\rangle = N_k N_e \langle v\sigma_{kj}\rangle + N_k A_k \,,$$

where $A_k = \sum_n A_{kn}$ is the total probability of radiative decay from level 2. Using (1.2.7) we obtain

$$\frac{N_k}{N_j} = \frac{g_k}{g_j} \cdot \frac{\exp(-\beta_{kj})}{1+R}, \quad R = \frac{A_k}{N_e\langle v\sigma_{kj}\rangle}. \tag{1.2.13}$$

Equation (1.2.13) gives the level population in the framework of the two-level model for an arbitrary electron density N_e. The factor R describes the deviation from the thermodynamical limit. It can be shown that $\langle v\sigma_{kj}\rangle$ does not depend strongly on T so the main temperature dependence is given by the factor $\exp(-\beta_{kj})$, which is the same as in the Boltzmann formula.

If the level k is the ionization limit we obtain the equation for ionization equilibrium. In the case of ionization equilibrium one has to consider three recombination processes: three-body (which is proportional to N_e^2), radiative, and dielectronic (both of which are proportional to N_e). We thus obtain the equation

$$\frac{N^{(z+1)}}{N^{(z)}} = \frac{g_{z+1}}{g_z} \cdot S \cdot \frac{\exp(-\beta_z)}{1+R_z}, \tag{1.2.14}$$

$$R_z = \frac{\kappa_v + \kappa_d}{N_e\kappa_r}, \tag{1.2.15}$$

where κ_r, κ_v and κ_d are the rate coefficients of three-body, radiative, and dielectronic recombination.

If $R \gtrsim 1$ in (1.2.13) or (1.2.14) the ratios N_k/N_j and $N^{(z+1)}/N^{(z)}$ are strongly dependent on the cross sections of atomic translations in contrast to the case of thermodynamical equilibrium.

1.2.4 Coronal Limit

Let us now consider the case which is the opposite limit to the thermodynamical distributions, namely the low-density limit. This condition holds very well in the solar corona, where the electron density ($N_e \backsim 10^8 - 10^9\ \mathrm{cm}^{-3}$) is much less than in any laboratory plasma. For this reason the low-density limit is called the coronal limit.

In the limit $N_e \to 0$, using (1.2.7) again, we obtain

$$\frac{N_k}{N_j} = N_e\frac{\langle v\sigma_{jk}\rangle}{A_k}. \tag{1.2.16}$$

The intensity of the spectral line due to the k–n transition is

$$I_{kn} = N_j N_e \langle v\sigma_{jk}\rangle \hbar\omega \frac{A_{kn}}{A_k}. \tag{1.2.17}$$

The ratio A_{kn}/A_k is called the branching ratio. In the absence of branching, when only one radiative transition is possible, $A_{kn} = A_k$, and the intensity does not depend on the transition probability. This is an important feature of the coronal limit.

According to (1.2.13) the coronal limit is applicable at electron densities

$$N_e \ll N_e^*, \quad N_e^* = \frac{A_k}{\langle v\sigma_{kj}\rangle}. \tag{1.2.18}$$

As will be shown below $A_k \propto \Delta E^2 \propto z^4, \langle v\sigma\rangle \propto \Delta E^{-3/2} \propto z^{-3}$, so that $N_e^* \propto \Delta E^{7/2} \propto z^7$. At $z = 1$ $N_e^* \backsim 10^{16}\mathrm{cm}^{-3}$, at $z \backsim 10$ $N_e^* \backsim 10^{23}\mathrm{cm}^{-3}$. In other words in a high-temperature plasma with highly charged atoms, the coronal limit usually holds at least for the levels with allowed radiative decay to the ground state. In laboratory plasmas for densities greater than 10^{14} cm^3 the metastable levels can be collisionally coupled with other levels. In the solar corona ($N_e \backsim 10^9$ cm^{-3}) even for metastable levels the coronal limit usually holds.

Similarly, from (1.2.14, 15) we obtain the equation for ionization equilibrium in the coronal limit:

$$\frac{N^{(z+1)}}{N^{(z)}} = \frac{\langle v\sigma_i\rangle}{\kappa_v + \kappa_d}. \tag{1.2.19}$$

The ionization degree in this case does not depend on N_e, in contrast to the case of thermodynamical distribution.

2 Theory of Atomic Collisions

In this chapter the fundamentals of the general theory of electron collisions with atoms and ions are considered. The treatment begins with the simplest case of scattering in a central field. The quasi-classical (impact parameter) approximation for scattering phases $\eta(\rho)$ and their connection with quantum phases η_λ are obtained. Formulas expressing the cross sections in terms of unitary S matrices are given.

The general system of integrodifferential equations for the problem of electron collision with a complex atom is dealt with (Sect. 2.2). These equations are reduced to the integral radial equations with the polarization potential, which simplifies the formulation of different approximate methods of calculation.

In the last section of the chapter, the first-order approximation is considered. The final formulas are given which are used further in approximate calculations.

In this book the main interest lies in formulation and analysis of the approximate calculation methods for excitation and ionization cross sections. For this reason, only those questions of the collision theory are touched upon which are quite necessary for this purpose. Comprehensive representation of the general collision theory can be found in modern textbooks such as [2.1–3].

2.1 Fundamentals of Scattering Theory

2.1.1 Elastic Scattering in a Central Field

The scattering of particles is usually described by the ratio of the number of particles scattered in an element of solid angle dO per second to the flux density of incident particles, i.e., to the number of particles incident on 1 cm^2 per second. This ratio $d\sigma$ is measured in cm^2, and is identified as the differential scattering cross section.

Let the particles fall on the scattering center along the z axis with velocity v. The free motion of particles is described by the wave function $\psi = v^{-1/2}\exp(ikz)$ where $k = p/\hbar = mv/\hbar$. The wave function is normalized so that the flux density of particles is equal to $v|\psi^2| = 1$. The diverging spherical wave $f(\vartheta)\exp(ikr)/r$ corresponds to scattered particles far away from the scattering centre. Thus at large distance, ψ can be written in the form

$$\psi \simeq v^{-1/2}\exp(ikz) + v^{-1/2}f(\vartheta)\, r^{-1}\exp(ikzr)\,. \tag{2.1.1}$$

The number of particles scattered per second into a solid angle dO equals

$$v\left|\frac{f(\vartheta)}{\sqrt{v}}\frac{\exp(ikr)}{r}\right|^2 r^2 dO = |f(\vartheta)|^2 dO\,.$$

Since the flux density of incident particles is 1, we have

$$d\sigma = |f(\vartheta)|^2 dO\,. \tag{2.1.2}$$

Therefore to calculate $d\sigma$ it is necessary to find the function $f(\vartheta)$, which is determined by the asymptotic wave function (2.1.1). The function $f(\vartheta)$ is called the scattering amplitude.

The Schrödinger equation for a particle in a centrally symmetric field has the solution $R_l(r)\mathrm{P}_l(\cos\vartheta)$, where P_l is the Legendre polynomial and R_l is the solution of the radial equation

$$\frac{1}{r}\frac{d^2}{dr^2}(rR_l) - \left\{\frac{l(l+1)}{r^2} + \frac{2m}{\hbar^2}[\mathscr{E} - U(r)]\right\}R_l = 0\,, \tag{2.1.3}$$

where $\mathscr{E} = \hbar^2k^2/2m$ is the electron energy

$$rR_l \simeq k^{-1/2}\sin(kr - \pi l/2 + \eta_l), \quad (r \to \infty)\,. \tag{2.1.4}$$

The phases η_l in the asymptotic expression for R_l are defined by the potential $U(r)$ in the whole range $0 \le r < \infty$. To determine these phases it is necessary to find the solution of (2.1.3) for all values of r.

The wave function ψ can be expanded in terms of the functions R_lP_l in a so-called partial wave expansion:

$$\begin{aligned}\psi &= \sum_l A_lR_l(r)\mathrm{P}_l(\cos\vartheta) \approx \sum_l A_lk^{-1/2}r^{-1}\sin(kr - \pi l/2 + \eta_l)\mathrm{P}_l(\cos\vartheta)\\ &= \frac{\exp(ikr)}{2ir}\sum_l\frac{A_l}{\sqrt{k}}\mathrm{P}_l(\cos\vartheta)\exp[\mathrm{i}(\eta_l - l\pi/2)]\\ &\quad - \frac{\exp(-ikr)}{2ir}\sum_l\frac{A_l}{\sqrt{k}}\mathrm{P}_l(\cos\vartheta)\exp[-\mathrm{i}(\eta_l - l\pi/2)]\,.\end{aligned} \tag{2.1.5}$$

The analogous expansion of the plane wave in (2.1.1) is

$$\exp(ikz) = \sum_l \mathrm{i}^l(2l+1)\mathrm{P}_l(\cos\vartheta)j_l(kr)\,, \tag{2.1.6}$$

where

$$j_l(x) = \sqrt{\frac{\pi}{2x}}J_{l+1/2}(x) \underset{x\to\infty}{\simeq} \frac{1}{x}\sin\left(x - \frac{l\pi}{2}\right). \tag{2.1.7}$$

Substituting (2.1.5–7) in (2.1.1) we find

$$A_l = \sqrt{\frac{m}{\hbar}}\frac{1}{k}\mathrm{i}^l(2l+1)\exp(\mathrm{i}\eta_l)\,,$$

and therefore

$$\psi \simeq \left(\frac{m}{\hbar k}\right)^{1/2} \frac{1}{2ikr} \sum_l \mathrm{i}^l (2l+1)\,\mathrm{P}_l(\cos\vartheta)$$
$$\times\, [-\exp(-ikr + \mathrm{i}l\pi/2) + \exp(2\mathrm{i}\eta_l)\exp(ikr - \mathrm{i}l\pi/2)]\,, \tag{2.1.8}$$

$$f(\vartheta) = (2\mathrm{i}k)^{-1} \sum_l (2l+1)\,[\exp(2\mathrm{i}\eta_l) - 1]\,\mathrm{P}_l(\cos\vartheta)\,. \tag{2.1.9}$$

We shall now substitute (2.1.9) in (2.1.2) and integrate over the angles dO. Since

$$\int \mathrm{P}_l(\cos\vartheta)\,\mathrm{P}_{l'}(\cos\vartheta)\sin\vartheta\, d\vartheta = \delta_{ll'} \frac{2}{2l+1}\,, \tag{2.1.10}$$

for the total cross section of elastic scattering we have

$$\sigma = 4\pi k^{-2} \sum_l (2l+1)\sin^2\eta_l\,. \tag{2.1.11}$$

Comparing (2.1.9) and (2.1.11), it is easy to see that cross section of elastic scattering can be related to the amplitude of forward scattering $f(0)$ by

$$\sigma = 4\pi k^{-1}\,\mathrm{Im}\,\{f(0)\} = -2\pi\mathrm{i}k^{-1}[f(0) - f^*(0)]\,. \tag{2.1.12}$$

This relation is called the optical theorem. It is a general relation which is valid also for noncentral fields.

2.1.2 Wave Functions ψ_k^+, ψ_k^-

We shall introduce the wave function ψ_k^+, which describes the plane wave $\exp(\mathrm{i}\boldsymbol{k}\cdot\boldsymbol{r})$ and the scattered spherical wave. In contrast to (2.1.1) we normalize this function according to

$$\psi_k^+ \underset{r\to\infty}{\simeq} \exp(\mathrm{i}\boldsymbol{k}\cdot\boldsymbol{r}) + f(\vartheta)\exp(\mathrm{i}kr)/r\,. \tag{2.1.13}$$

For an arbitrary direction of the wave vector $\boldsymbol{k}$, $\cos\vartheta = (\boldsymbol{k}\cdot\boldsymbol{r})/kr$. Using (2.1.8) we obtain

$$\psi_k^+ = \frac{1}{\sqrt{k}} \sum_l \mathrm{i}^l (2l+1)\exp(\mathrm{i}\eta_l)\,\mathrm{P}_l(\cos\vartheta)\,R_l(r)$$
$$\simeq \frac{1}{2\mathrm{i}kr} \sum_l \mathrm{i}^l (2l+1)\,\mathrm{P}_l(\cos\vartheta)\,\{-\exp[-\mathrm{i}(kr - l\pi/2)]$$
$$+\, \exp(2\mathrm{i}\eta_l)\exp[\mathrm{i}(kr - l\pi/2)]\}\,. \tag{2.1.14}$$

Replacing $\exp(i\eta_l)$ in ψ_k^+ by $\exp(-i\eta_l)$ we define the function ψ_k^-,

$$\begin{aligned}\psi_k^- &= \frac{1}{\sqrt{k}}\sum_l i^l(2l+1)\exp(-i\eta_l)\mathrm{P}_l(\cos\vartheta)R_l(r)\\ &\simeq \frac{1}{2ikr}\sum_l i^l(2l+1)\mathrm{P}_l(\cos\vartheta)\{-\exp(-2i\eta_l)\exp[-i(kr-l\pi/2)]\\ &\quad+\exp[i(kr-l\pi/2)]\}\,. \end{aligned} \tag{2.1.15}$$

One can see that

$$\psi_k^- \simeq \exp(i\boldsymbol{k}\cdot\boldsymbol{r}) + f(\vartheta)\exp(-ikr)/r\,, \tag{2.1.13$'$}$$

$$\psi_k^- = (\psi_{-k}^+)^*\,. \tag{2.1.16}$$

The functions ψ_k^+ and ψ_k^- are the solutions of the integral equation

$$\psi_k^\pm = \exp(i\boldsymbol{k}\cdot\boldsymbol{r}) + \frac{2m}{\hbar^2}\int G_{\pm k}(\boldsymbol{r},\boldsymbol{r}')\,U(\boldsymbol{r}')\,\psi_k^\pm(\boldsymbol{r}')\,d\boldsymbol{r}'\,, \tag{2.1.17}$$

where $G_{\pm k}(\boldsymbol{r},\boldsymbol{r}')$ is the Green's function of the free electron, defined as the solution

$$G_{\pm k}(\boldsymbol{r},\boldsymbol{r}') = -\frac{1}{4\pi}\frac{\exp(\pm ik|\boldsymbol{r}-\boldsymbol{r}'|)}{|\boldsymbol{r}-\boldsymbol{r}'|} \tag{2.1.18a}$$

of the equation

$$(\nabla^2+k^2)\,G_{\pm k}(\boldsymbol{r},\boldsymbol{r}') = \delta(\boldsymbol{r}-\boldsymbol{r}')\,. \tag{2.1.18b}$$

At $r \gg r'$, we find

$$G_k(\boldsymbol{r},\boldsymbol{r}') = -\frac{1}{4\pi r}\exp(ikr - i\boldsymbol{k}'\cdot\boldsymbol{r}'),\ \boldsymbol{k}' = k\frac{\boldsymbol{r}}{r}\,; \tag{2.1.19}$$

and since ϑ is the angle between $\boldsymbol{k}$, $\boldsymbol{r}$, which is the same as that between $\boldsymbol{k}$, $\boldsymbol{k}'$, we obtain

$$f(\vartheta) = -\frac{m}{2\pi\hbar^2}\int\exp(-i\boldsymbol{k}'\cdot\boldsymbol{r})\,U(\boldsymbol{r})\,\psi_k^+(\boldsymbol{r})\,d\boldsymbol{r}\,. \tag{2.1.20}$$

Equations (2.1.17, 20) are useful for obtaining a solution of the problem by the method of successive approximation due to Born. In the integral equation (2.1.17) one can replace the unknown $\psi_k^\pm$ on the right-hand side by some approximation, and obtain a better approximation on the left-hand side. For the first approximation one can use $\exp(i\boldsymbol{k}\cdot\boldsymbol{r})$ in place of $\psi_k^\pm$ on the right-hand side; this gives

$$f^B(\vartheta) = -\frac{m}{2\pi\hbar^2}\int U(\boldsymbol{r})\exp[i(\boldsymbol{k}-\boldsymbol{k}')\cdot\boldsymbol{r}]\,d\boldsymbol{r}\,. \tag{2.1.21}$$

This is the first Born approximation for the scattering amplitude.

2.1.3 Quasi-Classical Approximation

As noted above, to find precise scattering phases η_l, a numerical solution of the radial equations (2.1.3) is necessary. However, the problem is considerably simplified in the quasi-classical approximation. In this approximation, the function R_l for a particle with angular momentum l in a centrally symmetric field $U(r)$ has the form

$$R_l \simeq \frac{1}{\sqrt{k}\,r} \sin\left(\frac{1}{\hbar}\int_{r_1}^{r} P_r\,dr + \frac{\pi}{4}\right), \tag{2.1.22}$$

where

$$P_r^2 = 2m[\mathcal{E} - U(r)] - \hbar^2(l+1/2)^2 r^{-2}\,. \tag{2.1.23}$$

For a free particle, this becomes

$$R_l \simeq \frac{1}{\sqrt{k}\,r} \sin\left[\frac{1}{\hbar}\int_{r_0}^{r} \sqrt{2m\mathcal{E} - \hbar^2(l+1/2)^2 r^{-2}}\,dr + \frac{\pi}{4}\right]. \tag{2.1.24}$$

The turning points r_1, r_0 are the zeros of the expression under the radical.

From (2.1.22–24) one can see that the presence of the scattering potential results in phase shift η_l in the argument of the sine, where

$$\begin{aligned}\eta_l =& \int_{r_1}^{\infty} \sqrt{2m\hbar^{-2}[\mathcal{E} - U(r)] - (l+1/2)^2 r^{-2}}\,dr \\ &- \int_{r_0}^{\infty} \sqrt{2m\hbar^{-2}\mathcal{E} - (l+1/2)^2 r^{-2}}\,dr\end{aligned} \tag{2.1.25}$$

This phase may be identified as the scattering phase.

It can be shown that the quasi-classical approximation is applicable when a large number of partial waves ψ_l make a substantial contribution in elastic scattering cross section. In that case, the main contribution to the sum over l is made by the terms with large values of l. At large l the lower integration limits in (2.1.25) should also be large

$$r_1 \backsim \frac{\hbar l}{\sqrt{2m(\mathcal{E} - U)}}, \quad r_0 \backsim \frac{\hbar l}{\sqrt{2m\mathcal{E}}}\,.$$

If $|U(r)|$ decreases with increasing r so quickly that in the whole range of r the following condition is fulfilled:

$$U(r) \ll \mathcal{E}\,, \tag{2.1.26}$$

then $r_1 \approx r_0 \approx l/k$, where $k = mv/\hbar$, and

$$\eta_l = -\int_{l/k}^{\infty} \frac{m}{\hbar^2} \frac{U(r)}{\sqrt{k^2 - (l+1/2)^2 r^{-2}}}\,dr\,. \tag{2.1.27}$$

In the quasi-classical approximation, the angular momentum of the particle is $mv\rho$, where ρ is impact parameter. Therefore $\hbar\sqrt{l(l+1)} \approx \hbar l \approx mv\rho$, and

$$l = k\rho \,. \tag{2.1.28}$$

If we assume that the particle moves in a straight line with constant velocity, then from (2.1.27, 28) we obtain

$$\eta_l = -\frac{1}{2}\eta(\rho)\,, \quad \eta(\rho) = \frac{1}{\hbar}\int_{-\infty}^{\infty} U(r)\,dt\,, \qquad r^2 = \rho^2 + v^2t^2\,. \tag{2.1.29}$$

In the case of elastic scattering by an atom in the state a one has to substitute the diagonal matrix element of the interaction $U_{aa}(r)$ which is the first-order correction to the energy of the state a due to the interaction with the scattered particle. Consequently,

$$\eta(\rho) = \frac{1}{\hbar}\int_{-\infty}^{\infty} \Delta E_a(t)\,dt\,. \tag{2.1.29'}$$

In other words, in the quasi-classical approximation, the value of $-2\eta_l$ is equal to the phase shift due to the shift of the atomic level during the collision. It is easy to prove that for the field $U(r) = \hbar C/r^n$, (2.1.29) gives the same result as the more accurate (2.1.27):

$$\eta(\rho) = -2\,\eta_l = \frac{\alpha_n C}{v\rho^{n-1}}\,, \quad \alpha_n = \frac{\Gamma\left(\frac{1}{2}\right)\Gamma\left(\frac{n-1}{2}\right)}{\Gamma\left(\frac{n}{2}\right)}\,. \tag{2.1.30}$$

If we replace in the formula (2.1.11) for the elastic cross section the sum with respect to l by the integral over ρ, according to

$$\sum_l (2l+1) \simeq k^2 \int \rho\, d\rho\,,$$

we obtain

$$\sigma = 4\pi \int_0^{\infty} [1 - \cos\,\eta(\rho)]\,\rho\, d\rho\,. \tag{2.1.31}$$

As has already been mentioned above, the quasi-classical approximation is valid when partial waves with large values of l give substantial contributions to the cross section. It means that collisions with the values of impact parameter ρ

$$\rho \gg \frac{\hbar}{mv} = \lambda\!\!\!^{-} \tag{2.1.32}$$

are most important. Here $\lambda\!\!\!^{-}$ is the de Broglie wavelength of the electron.

2.1.4 Inelastic Scattering

In the general case, when both elastic and inelastic scattering occurs, the wave function must contain, besides the incoming plane wave, a large number of outgoing waves, corresponding to different types of scattering, or different scattering channels. For purely elastic scattering the intensities of incoming and outgoing partial waves (l waves) are the same. For inelastic scattering, the intensity of an outgoing wave corresponding to elastic scattering must be smaller than that of an incoming one. Taking this into account, the wave function ψ describing elastic scattering at large r can be written in the form

$$\begin{aligned}\psi \simeq \left(\frac{m}{\hbar k}\right)^{1/2} \frac{1}{2ikr}\sum_l i^l(2l+1)\,\mathrm{P}_l(\cos\vartheta)\,\{-\exp[-\mathrm{i}(kr-l\pi/2)] \\ +\exp(-2\beta_l+2\mathrm{i}\eta_l)\exp[\mathrm{i}(kr-l\pi/2)]\,,\end{aligned} \tag{2.1.33}$$

where $\beta_l \geq 0$. Hence we obtain for the scattering amplitude

$$f(\vartheta) = \frac{1}{2ik}\sum_l(2l+1)\,\mathrm{P}_l(\cos\vartheta)\,[\exp(-2\beta_l+2\mathrm{i}\eta_l)-1]\,. \tag{2.1.34}$$

This formula differs from (2.1.9) only by the replacement of the real phase η by the complex one $\eta_l + \mathrm{i}\beta_l$. From (2.1.2) it follows that

$$\sigma_{\text{elastic}} = \frac{\pi}{k^2}\sum_l(2l+1)\,|\,1-\exp(-2\beta_l+2\mathrm{i}\eta_l)|^2\,. \tag{2.1.35}$$

Using the expression (2.1.33) we can also derive a formula for the cross section for inelastic collisions, which is defined by a difference of fluxes corresponding to incoming and outgoing waves:

$$\sigma_{\text{inelastic}} = \frac{\pi}{k^2}\sum_l(2l+1)[1-\exp(-4\beta_l)]\,. \tag{2.1.36}$$

The total cross section is

$$\begin{aligned}\sigma &= \sigma_{\text{elastic}} + \sigma_{\text{inelastic}} \\ &= \frac{2\pi}{k^2}\sum_l(2l+1)[1-\exp(-2\beta_l)\cos 2\eta_l]\,.\end{aligned} \tag{2.1.37}$$

For $\beta_l = 0$, $\sigma = \sigma_{\text{elastic}}$; for $\beta_l = \infty$, the term $\exp(-2\beta_l) = 0$, and so $\sigma_{l.\,\text{elastic}} = \sigma_{l.\,\text{inelastic}} = \pi(2l+1)/k^2$. It is easy to see that $\pi(2l+1)/k^2$ is the number of particles with angular momentum l incident on the scattering center per second when the flux density is equal to unity.

From (2.1.35–37), the following limits of variation of the cross sections can be obtained:

$$0 \le \sigma_{l.\text{elastic}} \le \frac{4\pi}{k^2}(2l+1)\,,$$
$$0 \le \sigma_{l.\text{inelastic}} \le \frac{\pi}{k^2}(2l+1)\,, \qquad (2.1.38)$$
$$0 \le \sigma_l \le \frac{4\pi}{k^2}(2l+1)\,.$$

We note that inelastic scattering is always followed by some elastic scattering. If $\exp(-2\beta_l) \neq 1$, then at any value of η_l including $\eta_l = 0$, $\sigma_{\text{elastic}} \neq 0$.

The cross sections of elastic and inelastic scattering can be related to the elements of a matrix, which is usually called the scattering matrix, or S matrix. Let us denote the initial state of the scattering system by a and write

$$\exp(-2\beta_l - 2\mathrm{i}\eta_l) = S_{aa}^{(l)}\,. \qquad (2.1.39)$$

Then

$$\sigma_{\text{elastic}} = \frac{\pi}{k^2}\sum_l (2l+1)\,|\,1 - S_{aa}^{(l)}\,|^2\,, \qquad (2.1.40)$$

$$\sigma_{\text{inelastic}} = \frac{\pi}{k^2}\sum_l (2l+1)(1 - |\,S_{aa}^{(l)}\,|^2)\,, \qquad (2.1.41)$$

$$\sigma = \sigma_{\text{elastic}} + \sigma_{\text{inelastic}} = \frac{2\pi}{k^2}\sum_l (2l+1)\text{Re}\,\{1 - S_{aa}^{(l)}\}\,. \qquad (2.1.42)$$

Inelastic scattering is connected with transitions of the scattering system from the state a to all energetically accessible states b (i.e., to the states with energy which does not exceed the sum of the initial energy E_a of the atom and the kinetic energy of the incident particle $\mathscr{E}$). We denote by σ_{ab} the cross section of the transition $a \to b$. In Sect. 2.2, it will be shown that the cross section σ_{ab} can be related to a nondiagonal element of S matrix:

$$\sigma_{ab} = \frac{\pi}{k^2}\sum_l (2l+1)\,|\,S_{ab}^{(l)}\,|^2\,. \qquad (2.1.43)$$

The conservation of the total particle flux implies the inequalities (2.1.38), and is equivalent to the requirement of S matrix unitarity,

$$\sum_b |\,S_{ab}^{(l)}\,|^2 = 1\,, \; \sum_{b \neq a} |\,S_{ab}^{(l)}\,|^2 = 1 - |\,S_{aa}^{(l)}\,|^2\,. \qquad (2.1.44)$$

According to these equations, the modulus of the matrix elements $S_{ab}^{(l)}$ cannot exceed unity. The inequalities (2.1.38) arise immediately from this condition, and also the inequality

$$\sigma_{ab}^{(l)} \le \sigma_{\text{inelastic}}^{(l)} \le \frac{\pi}{k^2}(2l+1)\,. \qquad (2.1.45)$$

2.2 Theory of Electron–Atom Collisions

In this section we consider the general problem of scattering of electrons on atoms and ions. In contrast to the previous discussion, we have to deal with a complex system composed of an N-electron atom and an incident electron. Their interaction depends, in particular, on the atomic state. Besides, it is necessary to consider exchange effects, so that the system has to be described by an antisymmetrical wave function. After collision, the atom may be in any of the excited states or in the initial state. Hence, general equations describing collision process are similar to multiconfigurational Hartree–Fock equations.

We now make some simplifying assumptions. We shall deal mainly with inelastic collisions resulting in the excitation of an atom. Therefore, we shall not analyze any specific characteristics of the ionization process [2.4, 5]. Moreover we shall restrict ourselves to such transitions when quantum numbers of only one electron are changed. We shall call this electron "optical" in what follows, and usually suppose it to be in the outer shell of the atom. The atomic core without the optical electron will be called the parent ion. We shall describe the atom by a wave function which is constructed from one-electron functions in accordance with the rules of angular-momenta coupling.

When we say "atom" we cover both neutral atoms and positive ions. We denote the nuclear charge by $\mathscr{Z}$ and the charge of the atomic core by z:

$$z = \mathscr{Z} - N + 1 \qquad (2.2.1)$$

where N is total number of bound electrons. z can be called asymptotic charge; the optical and incident electrons move in fields which are asymptotically equal to $-z/r$ and $-(z-1)/r$, respectively. z coincides with the spectroscopic symbol of an ion: HeI, LiII etc.

[A] means the isoelectronic ion sequence of an atom A.

We use atomic units with the Rydberg unit for energy:

$$\mathrm{Ry} = 1/2 \text{ a.u.} = 13.60 \text{ eV}$$

In these units $v = k = p = \sqrt{\mathscr{E}}$. Cross sections are in the units πa_0^2.

2.2.1 General Formulas for Cross Sections

We shall denote by $\Psi^{a_0 M_0 k_0 m_0^s}$ the completely antisymmetric wave function of the system. The upper indices describe the state of the system before the collision, namely the atom in the state $a_0 M_0$, and the incident electron with wave vector k_0 and spin projection m_0^s. M_0 denotes the magnetic quantum number of an atom (or z projection of the total angular momentum of an atom). We shall expand this function in terms of atomic eigenfunctions:

$$\Psi^{a_0 M_0 k_0 m_0^s} = \hat{A} \sum_{aMm^s} \Psi_{aM}(\xi_1 \ldots \xi_N)\ \psi^{aMm^s}(\xi), \hat{A} = \sum_j \frac{(-1)^{N+1-j}}{\sqrt{N+1}} P_{\xi\xi_j} \qquad (2.2.2)$$

where $\hat{A}$ is antisymmetrization operator, $P_{\xi\xi_j}$ is the permutation operator $\xi \rightleftarrows \xi_j$, ξ_j being a set of space and spin variables.

In accordance with the general definition, the differential cross section of transition $a_0M_0m_0^s \to aMm^s$ equals the flux of electrons with the spin projection m^s in the solid angle dO, provided that the atom occurs in the state aM, and the incident electron flux equals unity. The velocity of the scattered electron $v = k$ is in that case given by the relation

$$k_0^2 + E_{a_0} = k^2 + E_a \, . \tag{2.2.3}$$

Using the antisymmetrical properties of the function Ψ, we obtain

$$d\sigma = v(N+1)\int |\hat{A}\Psi_{aM}\psi_{aMm^s}|^2 d\xi r^2 dO \quad (r \to \infty)\, , \tag{2.2.4}$$

where $d\xi$ denotes the integration over all variables besides $\boldsymbol{r}$. The asymptotic form of the integrand in (2.2.4) depends on whether the state aM belongs to a discrete or to a continuous spectrum. If we restrict ourselves to the simpler case of a discrete spectrum (for the other case, see [2.4]), only the term with $P_{\xi\xi}$ remains in (2.2.2) as $r \to \infty$, because Ψ_{aM} decays exponentially. Using the asymptotic expression for ψ_{aMm^s} similar to (2.1.1),

$$\begin{aligned}\psi_{aMm^s} \simeq \frac{1}{\sqrt{k}}[&\exp(\mathrm{i}\boldsymbol{k}\cdot\boldsymbol{r})\,\delta(aMm^s, a_0M_0m_0^s)\\ &+ f_{aMm^s}(\vartheta,\varphi)\exp(\mathrm{i}kr)/r]\chi_{m^s}\, ,\end{aligned} \tag{2.2.5}$$

where χ_{m^s} is the spin function, we obtain

$$d\sigma(a_0M_0m_0^s, aMm^s) = |f_{aMm^s}(\vartheta,\varphi)|^2 dO\, . \tag{2.2.6}$$

We shall now separate the radial and angular variables. This can be done by expanding the wave function of an outer electron in partial waves. To simplify the formulas we take the z axis along the direction of the vector $\boldsymbol{k}_0$. Then (2.2.2) can be rewritten in the form

$$\psi^{a_0M_0k_0m_0^s} = \hat{A}\sum_{\lambda_0\gamma}\beta_{\lambda_0}\Psi_{aM}(\xi_1\ldots\xi_N)\frac{1}{r}F_\gamma^{\gamma_0}(r)Y_{\lambda m}(\vartheta,\varphi)\,\chi_{m^s}\, , \tag{2.2.7}$$

$$\gamma = aM\lambda mm^s\, , \quad \gamma_0 = a_0M_0\lambda_0\,0m_0^s\, . \tag{2.2.8}$$

Here λ and m are, respectively, the angular momentum and its z projection of the outer electron (we reserve the letter l for an optical electron of an atom); $Y_{\lambda m}(\vartheta,\varphi)$ are the spherical harmonics. The radial functions $F_\gamma^{\gamma_0}(r)$ in the expansion (2.2.7) are the solutions of a scattering problem. They fulfil the conditions:

$$\begin{aligned}&F_\gamma^{\gamma_0}(0) = 0\, ,\\ &F_\gamma^{\gamma_0} \simeq \frac{1}{\sqrt{k}}\left\{\delta_{\gamma\gamma_0}\sin\left(k_0r - \frac{\lambda\pi}{2}\right) + T_{\gamma\gamma_0}\exp\left[\mathrm{i}\left(kr - \frac{\lambda\pi}{2}\right)\right]\right\}.\end{aligned} \tag{2.2.9}$$

The asymptotic form in (2.2.9) and below is written for simplicity for the case of neutral atom. For scattering by positive ions it should be replaced by the expression (2.2.40) given below.

Comparing (2.2.7–9) with (2.2.2 and 5), we obtain

$$\beta_{\lambda_0} = \mathrm{i}^{\lambda_0} \frac{\sqrt{4\pi(2\lambda_0+1)}}{k_0}\,, \tag{2.2.10}$$

$$f_{aMm^s}(\vartheta,\varphi) = \sum_{\lambda_0\lambda_m} \mathrm{i}^{\lambda_0-\lambda} \frac{\sqrt{4\pi(2\lambda_0+1)}}{k_0} T_{\gamma\gamma_0} Y_{\lambda m}(\vartheta,\varphi)\,. \tag{2.2.11}$$

Thus, if we know the radial functions $F_{\gamma}^{\gamma_0}(r)$, and hence the matrix $T_{\gamma\gamma_0}$, we can determine $f_{aMm^s}(\vartheta,\varphi)$, and then the scattering cross section according to formula (2.2.6).

The representation γ used up to now is in fact useful only for perturbation-theory calculations. To derive general equations of the Hartree–Fock type it is necessary to use the total angular momentum representation. In the *LS* coupling scheme, the system is described by the set of quantum numbers

$$\Gamma = a\lambda\frac{1}{2}L_{\mathrm{T}}S_{\mathrm{T}}\,,\quad a = \alpha LS\,, \tag{2.2.12}$$

where αLS is the set of atomic quantum numbers, and $L_{\mathrm{T}}, S_{\mathrm{T}}$ are the total orbital and spin momenta of the system consisting of the atom plus outer electron. The formulas that follow are independent of the particular coupling scheme, i.e., the definite form of Γ.

We shall denote the symmetric transformation matrix $\gamma \rightleftarrows \Gamma$ by $(\gamma|\Gamma)$. Then the total wave function of the system is

$$\begin{aligned}\Psi^{a_0M_0k_0m_s} &= \hat{A}\sum_{\lambda_0\Gamma} \beta_{\lambda_0}(\gamma|\Gamma)\,\Phi_\Gamma \frac{1}{r} F_\Gamma^{\Gamma_0}(r)\,,\\ \Phi_\Gamma &= \sum_{\gamma/\Gamma}(\Gamma|\gamma)\Psi_{aM}\, Y_{\lambda m}\chi_{m^s}\,.\end{aligned} \tag{2.2.13}$$

Using the unitarity of the transformation matrix, we obtain

$$F_\Gamma^{\Gamma_0}(r) \simeq \frac{1}{\sqrt{k}}\left\{\delta_{\Gamma\Gamma_0}\sin\left(k_0r-\frac{\lambda_0\pi}{2}\right) + T_{\Gamma\Gamma_0}\exp\left[\mathrm{i}\left(kr-\frac{\lambda\pi}{2}\right)\right]\right\}, \tag{2.2.14}$$

$$T_{\gamma\gamma_0} = \sum_{\Gamma_0/\gamma_0,\Gamma/\gamma} (\gamma_0|\Gamma_0)(\Gamma|\gamma)\,T_{\Gamma\Gamma_0}\,.$$

In this formula the summation is over those quantum numbers from the set Γ that do not occur in the γ set. We shall denote them by Γ/γ. In the *LS* coupling case, $\Gamma/\gamma = L_{\mathrm{T}}S_{\mathrm{T}}$. From (2.2.14), (2.2.6), and (2.2.11) we obtain the following expression for the differential cross section:

$$\begin{aligned}&d\sigma(a_0M_0m_0^s,\, aMm^s)\\ &\quad= \frac{4\pi}{k_0^2}\left|\sum_{\Gamma_0/a_0,\Gamma/a,m} \mathrm{i}^{\lambda_0-\lambda}\sqrt{2\lambda_0+1}\,(\gamma_0|\Gamma_0)(\Gamma|\gamma)\,T_{\Gamma\Gamma_0}Y_{\lambda m}(\vartheta,\varphi)\right|^2 dO\,.\end{aligned} \tag{2.2.15}$$

Usually one has to deal with collisions of unpolarized electrons with randomly oriented atoms. The orientation of an excited atom, described by the quantum number M, may be important because it determines the polarization of subsequent radiation. In order to obtain the corresponding cross section it is necessary to average (2.2.15) over $M_0 m_0^s$, and sum over m^s. It is convenient also to expand the product $Y_{\lambda m} Y^*_{\lambda' m'}$ in (2.2.15) in spherical functions $Y_{\kappa\mu}$ of the same angles. It is easy to show that only $\mu = 0$ terms give nonzero contributions, i.e., the cross section is independent of φ, as would be expected. We shall write the final result in the form

$$d\sigma(a_0, aM) = \frac{1}{2g_0 k_0^2} \sum_{\kappa} B_\kappa \mathrm{P}_\kappa(\cos\vartheta)\, dO \,. \tag{2.2.16}$$

In the case of LS coupling,

$$\begin{aligned} B_\kappa = & \frac{2S_\mathrm{T}+1}{2(2S_0+1)} \sum \mathrm{i}^{\lambda_0 - \lambda_0' + \lambda' - \lambda - 2m} [\lambda_0 \lambda_0' \lambda \lambda'][\kappa L_\mathrm{T} L_\mathrm{T}']^2 \\ & \times \begin{pmatrix} \lambda & \lambda' & \kappa \\ 0 & 0 & 0 \end{pmatrix} \begin{pmatrix} \lambda & \lambda' & \kappa \\ -m & m & 0 \end{pmatrix} \begin{pmatrix} L_0 & \lambda_0 & L_\mathrm{T} \\ M_0 & 0 & -M \end{pmatrix} \begin{pmatrix} L_0 & \lambda_0' & L_\mathrm{T}' \\ M_0 & 0 & -M_0 \end{pmatrix} \\ & \times \begin{pmatrix} L & \lambda & L_\mathrm{T} \\ M & m & -M_0 \end{pmatrix} \begin{pmatrix} L & \lambda' & L_\mathrm{T}' \\ M & m & -M_0 \end{pmatrix} T_{\Gamma\Gamma_0} T^*_{\Gamma'\Gamma_0'} \end{aligned} \tag{2.2.17}$$

the summation extending over $L_\mathrm{T}, L_\mathrm{T}', \lambda_0, \lambda_0', \lambda, \lambda', m, M_0$. Here and below we use the following designations

$$\begin{aligned} &[j_1 j_2 \ldots] = (2j_1+1)^{1/2}(2j_2+1)^{1/2} \ldots . \\ &\delta(ab\ldots, a'b'\ldots) = \delta_{aa'}\delta_{bb'}\ldots \end{aligned} \tag{2.2.18}$$

(The properties of Wigner's $3j$ symbols $\begin{pmatrix} j_1 & j_2 & j_3 \\ m_1 & m_2 & m_3 \end{pmatrix}$ were described in [Ref. 2.6, Sect. 4.2]).

We shall now consider the total cross section. Equation (2.2.16) after integration over angles gives

$$\sigma(a_0, aM) = \frac{2\pi}{g_0 k_0^2} B_0 \,. \tag{2.2.19}$$

Summing with respect to M, i.e. over the final orientations of the atom, we obtain

$$\begin{aligned} &\sigma(a_0, a) = \sum_{\lambda_0 \lambda} \sigma(a_0 \lambda_0, a\lambda) \,, \\ &\sigma(a_0\lambda_0, a\lambda) = \pi a_0^2 \frac{2}{k_0^2} \sum_{\Gamma_0/a_0\lambda_0, \Gamma/a\lambda} \frac{g_\Gamma}{g_0} |T_{\Gamma\Gamma_0}|^2 \,, \end{aligned} \tag{2.2.20}$$

where g_0 is the statistical weight of the state a_0 of the atom, and g_Γ is the statistical weight of the state Γ of the system.

Equation (2.2.20) provides a simple relation between the cross section σ and the matrix T for an arbitrary coupling scheme. The partial cross sections $\sigma(a_0\lambda_0, a\lambda)$ are introduced for convenience in the further discussion. In the case of LS coupling,

$$\sigma(a_0\lambda_0,\, a\lambda) = \pi a_0^2 \frac{2}{k_0^2} \sum_{L_\mathrm{T}S_\mathrm{T}} \frac{[S_\mathrm{T}L_\mathrm{T}]^2}{[S_0L_0]^2} |T_{\Gamma\Gamma_0}|^2 \tag{2.2.21}$$

$$\Gamma = a\lambda s L_\mathrm{T} S_\mathrm{T}\,, \quad a = \alpha LS, \quad s = \frac{1}{2}\,.$$

For transitions between J levels one should derive the formula using Jj (or $J\lambda$) coupling: $\Gamma = aJ, \lambda\frac{1}{2}(j)J_\mathrm{T}$. If the atom is described in LS coupling, and magnetic electron–atom interaction is neglected, the dependence on J is purely kinematic and can be expressed explicitly. For the matrix elements $T_{\Gamma\Gamma_0}$ we can return to LS coupling and obtain

$$\sigma(a_0J_0\lambda_0,\, aJ\lambda) = \pi a_0^2$$

$$\times \frac{2}{k_0^2} \sum_{j_0 j J_\mathrm{T}} [J_\mathrm{T} j_0 j J]^2 \left| \sum_{S_\mathrm{T}L_\mathrm{T}} [S_\mathrm{T}L_\mathrm{T}]^2 \begin{Bmatrix} S_\mathrm{T} & s & S_0 \\ L_\mathrm{T} & \lambda_0 & L_0 \\ J_\mathrm{T} & j_0 & J_0 \end{Bmatrix} \begin{Bmatrix} S_\mathrm{T} & s & S \\ L_\mathrm{T} & \lambda & L \\ J_\mathrm{T} & j & J \end{Bmatrix} \cdot T_{\Gamma\Gamma_0} \right|^2, \tag{2.2.22a}$$

where the $9j$ symbols $\begin{Bmatrix} j_2 & j_1 & j_0 \\ l_2 & l_1 & l_0 \\ k_2 & k_1 & k_0 \end{Bmatrix}$ were defined in [Ref. 1.1, Sect. 4.2], summation and transformation formulas are given in Sect. 6.3, too. In particular, (6.3.5) provides a possibility to rearrange the columns (or rows) in the sums of $9j$ symbols. Using this relation we can change the summation over angular momenta $j_0 j J_T$ to the sum over multipole indices kqv and write (2.2.22a) in a more convenient and simpler form:

$$\sigma(a_0J_0\lambda_0, aJ\lambda) = \pi a_0^2 \frac{2}{k_0^2} \sum_{q\kappa v} [q\kappa v J]^2 \begin{Bmatrix} S_0 & S & q \\ L_0 & L & \kappa \\ J_0 & J & v \end{Bmatrix}^2 \tag{2.2.22b}$$

$$\times \left| \sum_{L_\mathrm{T}S_\mathrm{T}} (-1)^{L_\mathrm{T}+S_\mathrm{T}} [S_\mathrm{T}L_\mathrm{T}]^2 \begin{Bmatrix} S_0 & S & q \\ s & s & S_\mathrm{T} \end{Bmatrix} \begin{Bmatrix} L_0 & L & \kappa \\ \lambda & \lambda_0 & S_\mathrm{T} \end{Bmatrix} \cdot T_{\Gamma\Gamma_0} \right|^2$$

We introduce now M-factors [2.12]:

$$M_{q\kappa v}(SLJ) \equiv M_{q\kappa v}(S_0L_0J_0, SLJ) = [J_0 J v] \begin{Bmatrix} S_0 & S & q \\ L_0 & L & \kappa \\ J_0 & J & v \end{Bmatrix}. \tag{2.2.23}$$

Utilizing the relations for $9j$ symbols with one zero we can write (2.2.22b) in a compacter and symmetrical form:

$$\sigma(a_0J_0\lambda_0, aJ\lambda) = \pi a_0^2 \frac{2}{k_0^2(2J_0+1)} \sum_{q\kappa v} (q\,\kappa)^4 M^2_{q\,\kappa\,v}(SLJ) \times \left| \sum_{L_T S_T} [S_T L_T] M_{qq0}(Ss S_T) M_{\kappa\kappa 0}(L\lambda L_T) \cdot T_{\Gamma\Gamma_0} \right|^2 . \tag{2.2.22c}$$

2.2.2 S-matrix and Collision Strength

In scattering theory, one usually uses an S matrix linked with our T matrix by the simple relation

$$S_{\Gamma\Gamma_0} = \delta_{\Gamma\Gamma_0} - 2\mathrm{i}T_{\Gamma\Gamma_0} , \qquad \sigma(a_0\lambda_0, a\lambda) = \pi a_0^2 \frac{1}{2k_0^2} \sum \frac{g_\Gamma}{g_0} |S_{\Gamma\Gamma_0} - \delta_{\Gamma\Gamma_0}|^2 . \tag{2.2.24}$$

The asymptotic form of $F_\Gamma^{\Gamma_0}(r)$ is represented in this case by

$$F_\Gamma^{\Gamma_0}(r) \simeq k^{-1/2} \left\{ \delta_{\Gamma\Gamma_0} \exp\left[-\mathrm{i}\left(k_0 r - \frac{\lambda_0\pi}{2}\right)\right] - S_{\Gamma\Gamma_0} \exp\left[\mathrm{i}\left(kr - \frac{\lambda\pi}{2}\right)\right]\right\} . \tag{2.2.25}$$

The S matrix is symmetric, and satisfies the unitarity condition

$$\sum_\Gamma |S_{\Gamma\Gamma_0}|^2 = 1 . \tag{2.2.26}$$

The definition of cross section directly in terms of the unitary matrix S (or T) often proves to be inconvenient because the approximate matrix may no longer be unitary. The most important example shows up the 1st-order perturbation theory when $S_{\Gamma\Gamma_0} \sim \langle\Gamma|U|\Gamma_0\rangle$ and not limited by any condition. Moreover, even the individual terms $|S_{\Gamma\Gamma_0}|$ can exceed unity. Therefore, the number of scattered particles can exceed the number of incident ones. To avoid this defect, the K matrix is often used [2.8]. It is connected with the S matrix by a nonlinear relation,

$$S = \frac{I + \mathrm{i}K}{I - \mathrm{i}K} , \tag{2.2.27}$$

where I is the unit matrix: $I_{\Gamma\Gamma_0} = \delta_{\Gamma\Gamma_0}$. This K-matrix representation is used in particular in the quantum defect method [2.9]. Discussion of various matrices of the collision theory can be found in [2.10].

The matrix K is symmetric and Hermitian, but nonunitary. The corresponding radial functions are real and have the asymptotic form

$$F_\Gamma^{\Gamma_0}(r) \simeq k^{-1/2} \left[\delta_{\Gamma\Gamma_0} \sin\left(k_0 r - \frac{\lambda_0\pi}{2}\right) + K_{\Gamma\Gamma_0} \cos\left(kr - \frac{\lambda\pi}{2}\right)\right] .$$

These functions can be represented as linear combinations of functions (2.2.14). No matter what kind of approximation is used for the calculation of the K matrix, the S matrix derived from (2.2.27) is unitary, and the corresponding cross sections, being approximate, are however in accordance with the particle-number conservation condition.

Sometimes instead of the transition cross section σ_{a_0a}, the dimensionless quantity, collision strength, is used

$$\Omega_{a_0a} = \frac{2g_0k_0^2}{\pi}\sigma_{a_0a}\,, \quad \Omega_{a_0a}(k_0,k) = \Omega_{aa_0}(k,k_0)\,. \tag{2.2.28}$$

The collision strength is symmetric with respect to direct and reverse processes, and is additive with respect to atomic level structure. In case of transitions with no spin change at large energies, $\sigma \propto \mathscr{E}^{-1}$, i.e., $\Omega = \text{const}$, or increases slowly (logarithmically).

As mentioned above, see (2.1.38), the total partial cross sections of inelastic processes obey definite inequalities. Since the cross section of a specific transition cannot exceed the total inelastic cross section, these inequalities can be written in the form

$$\sum_{\lambda} \Omega(a_0\lambda_0, a\lambda) \le 2\lambda_0 + 1\,. \tag{2.2.29}$$

2.2.3 Radial Equations

The functions $F_{\Gamma}^{\Gamma_0}(r)$ are the solutions of radial equations, which can be derived by means of a variational principle similarly to the derivation of the Hartree–Fock equation for discrete spectrum states. Although the analogy with the Hartree–Fock equation is rather close there are certain differences, which we shall briefly discuss below.

First, in collision theory the total wave function Ψ of the system must contain many different channels, i.e. it is multiconfigurational. Consequently the outer electron is described by a set of functions $F_{\Gamma}^{\Gamma_0}(r)$, satisfying a (generally infinite) set of integrodifferential equations.

On the other hand, the self-consistent field of the free electron is zero. This makes it possible to determine the atomic wave functions independent of the outer electron. In other words we can consider the atomic wave functions as known. Normally, no orthogonality conditions with atomic functions are imposed on the functions of an outer electron. The equations are derived regarding possible nonorthogonality. This leads naturally to a wider class of trial functions, but the equations become more complicated. However if certain additional assumptions are made, the equations become substantially simpler, and resemble the usual multiconfigurational Hartree–Fock equations.

Finally, we note that the energy of the system is also considered as known, in contrast to the discrete-spectrum case, where it has to be determined by solving an eigenvalue problem.

We shall not describe here the rather cumbersome derivation of equations [2.7], and give only the final results. The set of integrodifferential equations to be solved can be written in the form

$$(\mathscr{L}_\Gamma + k^2)F_\Gamma = \sum_{\Gamma' \neq \Gamma}{}' U_{\Gamma\Gamma'}(r)F_{\Gamma'} \,. \tag{2.2.30}$$

The operator $\mathscr{L}_\Gamma$ is the usual Hartree–Fock operator,

$$\mathscr{L}_\Gamma = \frac{d^2}{dr^2} - \frac{\lambda(\lambda+1)}{r^2} - U_\Gamma(r),\; U_\Gamma(r) = U_\Gamma^{\mathrm{c}} + U_{\Gamma\Gamma} \,, \tag{2.2.31}$$

where U_Γ^{c} describes the interaction of the outer electron with the atomic core, and $U_{\Gamma\Gamma}$, the interaction with the optical electron. The potentials $U_{\Gamma\Gamma'}$ (including the case $\Gamma' = \Gamma$) are integral operators which are expressed in terms of radial integrals:

$$U_{\Gamma\Gamma'}(r)F_{\Gamma'} = \sum_{\kappa'} \alpha^{\kappa'}_{\Gamma\Gamma'}\, y^{\kappa'}_{ll'}\, F_{\Gamma'} - \sum_{\kappa''} \beta^{\kappa''}_{\Gamma\Gamma'}\, y^{\kappa''}_{l\lambda'}\, P_{l'} \,, \tag{2.2.32}$$

$$y^{\kappa}_{ll'} = [ll'\lambda\lambda'] \begin{pmatrix} l & \kappa & l' \\ 0 & 0 & 0 \end{pmatrix} \begin{pmatrix} \lambda & \kappa & \lambda' \\ 0 & 0 & 0 \end{pmatrix} 2\int_0^\infty \frac{r_<^\kappa}{r_>^{\kappa+1}} P_l(r_1)P_{l'}(r_1)\,dr_1 \,,$$

$$\begin{aligned} y^{\kappa}_{l\lambda'} = [ll'\lambda\lambda'] \begin{pmatrix} l & \kappa & \lambda' \\ 0 & 0 & 0 \end{pmatrix} \begin{pmatrix} \lambda & \kappa & l' \\ 0 & 0 & 0 \end{pmatrix} 2\int_0^\infty \frac{r_<^\kappa}{r_>^{\kappa+1}} (1 - c_{ll'}\delta_{\kappa 0}\, r_>) \\ \times P_l(r_1)F_{\Gamma'}(r_1)\,dr_1 \,, \end{aligned} \tag{2.2.33}$$

$$c_{ll'} = \frac{1}{2}(-\varepsilon_l + k'^2) = \frac{1}{2}(-\varepsilon_{l'} + k^2) \,. \tag{2.2.34}$$

Here $P_l(r)$ and ε_l are the radial function and energy parameter of the optical electron. In (2.2.34) we neglect the difference between the energy parameter and the level energy.

The factors α and β depend only on the angular-momentum quantum numbers. If the atom is described in a fractional parentage scheme, $a = \gamma_{\mathrm{p}} L_{\mathrm{p}} S_{\mathrm{p}}\, nlsLS$, then

$$\begin{aligned} \alpha^{\kappa'}_{\Gamma\Gamma'} &= \delta(S,S')\,(-1)^{L_{\mathrm{T}}+L+L'+L_{\mathrm{p}}}[LL'] \begin{Bmatrix} L & L' & \kappa' \\ \lambda' & \lambda & L_{\mathrm{T}} \end{Bmatrix} \begin{Bmatrix} L & L' & \kappa' \\ l' & l & L_{\mathrm{p}} \end{Bmatrix} , \\ \beta^{\kappa''}_{\Gamma\Gamma'} &= (-1)^{l+l'+1-S-S'}[SS'LL'] \begin{Bmatrix} S & s & S_{\mathrm{p}} \\ S' & s & S_{\mathrm{T}} \end{Bmatrix} \begin{Bmatrix} L & \lambda & L_{\mathrm{T}} \\ L_{\mathrm{p}} & l' & L' \\ l & \lambda' & \kappa'' \end{Bmatrix} \end{aligned} \tag{2.2.35}$$

where $[j_1 j_2 \ldots]$ is defined by (2.2.18).

Eqs. (2.2.32, 35) can be written in more symmetrical form which is convenient for general derivations and for computer codes. We use the factors M defined by (2.2.23). Then

$$U_{\Gamma\Gamma'}(r)F_{\Gamma'} = \sum_{\kappa q}(-1)^{l+\lambda}\alpha_{\kappa q}\left[2\delta(q)y'_\kappa F_{\Gamma'} - [q]^2 y''_\kappa P_{l'}\right] ,$$

$$y'_\kappa = y^\kappa_{ll'} , \quad y''_\kappa = \sum_{\kappa''}[kk'']^2(-1)^{\kappa+\kappa''}\left\{\begin{matrix} k & l & l' \\ k'' & \lambda & \lambda' \end{matrix}\right\} y^{\kappa''}_{l\lambda'} , \tag{2.2.32a}$$

$$\alpha_{\kappa q} = (-1)^{\kappa+q}\frac{[kqL_pS_p]}{[L_\mathrm{T}S_\mathrm{T}]}M_{0qq}(S_psS)M_{0\kappa\kappa}(L_plL)M_{qq0}(SsS_\mathrm{T})M_{\kappa\kappa 0}(L\lambda L_\mathrm{T}) . \tag{2.2.35a}$$

These formulas are symmetrical on orbital and spin variables. If the "spin part" of the interaction is 1 and $s=\frac{1}{2}$, the "radial spin factors" are $y'_q = (\frac{1}{2}\|1^q\|\frac{1}{2}) = 2\delta(q,0)$, and $y''_q = 2q+1$. The factor $(-1)^{1+\lambda}$ is connected to the $3j$-symbols in (2.2.33) and (2.3.10–11) in place of $(l\|C^\kappa\|l')$.

If there are m equivalent optical electrons in the state nl^m, the factors α and β in (2.2.35) must be replaced by

$$\sum_{L_pS_p} G^{LS}_{L_pS_p}\sqrt{m}\,\alpha^{\kappa'}_{\Gamma\Gamma'} , \quad \sum_{L_pS_p} G^{LS}_{L_pS_p}\sqrt{m}\,\beta^{\kappa''}_{\Gamma\Gamma'} , \tag{2.2.36}$$

where $G^{LS}_{L_pS_p}$ are fractional parentage coefficients. (See [Ref. 2.6, Sect. 5.1.5]. The tables of fractional parentage coefficients for electronic configurations p^3, p^4 and p^5 are given in Sect. 6.3). For transitions between terms of one nl^m configuration, the α, β are related to the coefficients ($\|U^\kappa\|$) and ($\|V^\kappa\|$) [Ref. 2.6, Sect. 5.4.1].

The potential U^c_Γ can be expressed in terms of radial integrals in a similar way. Very often, however, simpler approximations are used for U^c_Γ since for inelastic collisions it is not very important.

In the above formulas, the summation limits over κ are defined by triangular and parity conditions, namely

$$\kappa = \kappa_\mathrm{min}, \kappa_\mathrm{min}+2, \dots \kappa_\mathrm{max} ,$$

$$\kappa'_\mathrm{min} = \max(|l-l'|, |\lambda-\lambda'|), \kappa'_\mathrm{max} = \min(l+l', \lambda+\lambda') , \tag{2.2.37}$$

$$\kappa''_\mathrm{min} = \max(|l-\lambda'|, |\lambda-l'|), \kappa''_\mathrm{max} = \min(l+\lambda', \lambda+l') .$$

The radial equations should be supplemented by boundary conditions. When $r=0$, all $F^{\Gamma_0}_\Gamma(0)=0$. The asymptotic form of these functions depends on the sign of k^2:

$$k^2 \geq 0, F_\Gamma \underset{r\to\infty}{\simeq} k^{-1/2}\left\{\delta_{\Gamma\Gamma 0}\sin\left(kr - \frac{\lambda\pi}{2}\right) + T_{\Gamma\Gamma_0}\exp\left[\mathrm{i}\left(kr-\frac{\lambda\pi}{2}\right)\right]\right\}; \tag{2.2.38}$$

$$k^2 < 0, F_\Gamma \underset{r\to\infty}{\to} 0 . \tag{2.2.39}$$

If the target particle is a positive ion $X_z(z > 1)$, it is necessary to change the argument in the asymptotic form of $F_\Gamma^{\Gamma_0}$ according to

$$\left(kr - \frac{\lambda\pi}{2}\right) \to \left[kr - \frac{\lambda\pi}{2} + \frac{z-1}{k}\ln 2kr + \arg\Gamma\left(\lambda + 1 - \mathrm{i}\frac{z-1}{k}\right)\right] \quad (2.2.40)$$

where $\Gamma(x)$ is the gamma function.

The value of k^2 in the above formulas is determined by energy conservation. There is no scattered wave for energetically inaccessible final levels ($k^2 < 0$). The inclusion of these states in a general system of equations corresponds to the inclusion of polarization terms in perturbation theory.

2.2.4 Integral Radial Equations

To analyze the equations of collision theory, and in some cases for their numerical solution, it is useful to pass from differential equations to integral equations. They can be derived by means of the formal solution of (2.2.30) using the Green's function that satisfies the equation

$$(\mathscr{L}_\Gamma + k^2)\, G_\Gamma(r, r') = \delta(r - r')\,. \quad (2.2.41)$$

The Green's function can be expressed in terms of two linearly independent solutions of the corresponding homogeneous equation:

$$G_\Gamma(r, r') = -\bar{F}_\Gamma(r_<)\,\bar{\bar{F}}_\Gamma(r_>)\,, \quad (2.2.42)$$

$$\begin{aligned} &(\mathscr{L}_\Gamma + k^2)\bar{F}_\Gamma = (\mathscr{L}_\Gamma + k^2)\bar{\bar{F}}_\Gamma = 0\,,\\ &\bar{F}_\Gamma(0) = 0, \quad \bar{\bar{F}}_\Gamma(r \to 0) = ar^{-\lambda} \end{aligned} \quad (2.2.43)$$

$$\left.\begin{aligned} \bar{F}_\Gamma &\underset{r\to\infty}{\simeq} k^{-1/2}\exp(\mathrm{i}\eta)\sin\left(kr - \frac{\lambda\pi}{2} + \eta\right)\\ \bar{\bar{F}}_\Gamma &\underset{r\to\infty}{\simeq} k^{-1/2}\exp\left[\mathrm{i}\left(kr - \frac{\lambda\pi}{2}\right)\right] \end{aligned}\right\} k^2 \geq 0 \quad (2.2.44)$$

$$\left.\begin{aligned} \bar{F}_\Gamma &\underset{r\to\infty}{\simeq} \frac{1}{2}q^{-1/2}\exp(qr)\\ \bar{\bar{F}}_\Gamma &\underset{r\to\infty}{\simeq} q^{-1/2}\exp(-qr) \end{aligned}\right\} k^2 < 0, \quad q = \mathrm{i}k \quad (2.2.45)$$

With the Green's function the integral equations for the functions F_Γ are written in the form[1]

$$F_\Gamma(r) = \delta_{\Gamma\Gamma_0}\bar{F}_{\Gamma_0}(r) + \int_0^\infty G_\Gamma(r,r') \sum_{\Gamma'\neq\Gamma}' U_{\Gamma\Gamma'}(r')F_{\Gamma'}(r')\, dr' . \qquad (2.2.46)$$

Substituting (2.2.42) and (2.2.44) in (2.2.46), and comparing with (2.2.38), we obtain

$$T_{\Gamma\Gamma_0} = \delta_{\Gamma\Gamma_0} \exp(\mathrm{i}\eta)\sin\eta - \int_0^\infty \bar{F}_\Gamma \sum_{\Gamma'\neq\Gamma}' U_{\Gamma\Gamma'}F_{\Gamma'}\, dr . \qquad (2.2.47)$$

The operator $\mathscr{L}$ in (2.2.43) as defined in (2.2.31) describes the motion of the particle in the field U_Γ. Therefore the solution $\bar{F}_\Gamma$ of (2.2.43) is usually called a distorted wave. Other representations are also possible. In particular it is possible to transmit the term with U_Γ from $\mathscr{L}_\Gamma F_\Gamma$ to the right-hand side of (2.2.30), i.e., take the free-motion operator as the basis. This representation is called the Born representation. We shall give here the explicit formulas for $\bar{F}$ and $\bar{\bar{F}}$ in the Born representation, which are needed below:

$$\begin{aligned} &\bar{F}_\Gamma = \sqrt{k}\, r\, j_\lambda(kr), \quad \bar{\bar{F}} = \mathrm{i}\sqrt{k}\, r\, h_\lambda^{(1)}(kr), \quad (k^2 > 0)\,, \\ &\bar{F}_\Gamma = \sqrt{q}\, r\, i_\lambda(qr), \quad \bar{\bar{F}} = \frac{2}{\pi}\sqrt{q}\, r\, k_\lambda(qr), \quad (k^2 = -q^2 < 0)\,. \end{aligned} \qquad (2.2.48)$$

Here j_λ and $h_\lambda^{(1)}$ are spherical Bessel and Hankel functions, and i_λ, k_λ are the same functions for an imaginary argument.[2]

2.2.5 Polarization Potential

As shown in the preceding sections, the problem of the calculation of cross sections amounts to solving an infinite set of integrodifferential or integral equations. Using the iterative procedure for the solution of integral equations, we obtain another formulation of the problem with a clearer physical interpretation.

---

[1] See [2.11], where the Green's function for homogeneous boundary conditions is given. It can be shown that the same formulas hold for inhomogeneous conditions of the type (2.2.38) if $\Gamma \neq \Gamma_0$. For $\Gamma = \Gamma_0$, two solutions of the homogeneous equation, one of which satisfies the condition at $r = 0$, and the other satisfies the condition (2.2.38) at $r \to \infty$, are linearly dependent. In this case the second solution should satisfy some other condition, for example, (2.2.35) without the sine term. This is the reason for the additional term with $\Gamma = \Gamma_0$ in the right-hand side of (2.2.46).

[2] The spherical functions j_λ, $h_\lambda^{(1)}$, i_λ, k_λ, n_λ are linked with the usual Bessel, Hankel and Neumann functions J_λ, $H_\lambda^{(1)}$, I_λ, K_λ and N_λ by the relation

$$z_\lambda(x) = \sqrt{\frac{\pi}{2x}} Z_{\lambda+\frac{1}{2}}(x) .$$

We shall proceed from the set of integral equations (2.2.46) and take the free term as zeroth approximation:

$$F_{\Gamma}^{(0)} = \delta_{\Gamma\Gamma_0} \bar{F}_{\Gamma_0} \,. \tag{2.2.49}$$

Then for the first approximation we get

$$\begin{aligned} F_{\Gamma_0}^{(1)} &= \bar{F}_{\Gamma_0} \,, \\ F_{\Gamma}^{(1)} &= \int_0^\infty G_{\Gamma}(r,r') U_{\Gamma\Gamma_0}(r') \bar{F}_{\Gamma_0}(r') \, dr' \,, \quad (\Gamma \neq \Gamma_0) \,. \end{aligned} \tag{2.2.50}$$

Extending the iterative procedure to increasingly higher approximations, one can obtain

$$\begin{aligned} F_{\Gamma_0} &= \bar{F}_{\Gamma_0} + \int_0^\infty G_{\Gamma_0}(r,r') V_{\Gamma_0\Gamma_0}(r') \bar{F}_{\Gamma_0}(r') \, dr' \,, \\ F_{\Gamma} &= \int_0^\infty G_{\Gamma}(r,r') [U_{\Gamma\Gamma_0}(r') + V_{\Gamma\Gamma_0}(r')] \bar{F}_{\Gamma_0}(r') \, dr' \,, \quad (\Gamma \neq \Gamma_0) \,; \end{aligned} \tag{2.2.51}$$

and for the T matrix,

$$\begin{aligned} T_{\Gamma_0\Gamma_0} &= \exp(i\eta_0) \sin \eta_0 - \int_0^\infty \bar{F}_{\Gamma_0} V_{\Gamma_0\Gamma_0} \bar{F}_{\Gamma_0} \, dr \\ T_{\Gamma\Gamma_0} &= - \int_0^\infty \bar{F}_{\Gamma} (U_{\Gamma\Gamma_0} + V_{\Gamma\Gamma_0}) \bar{F}_{\Gamma_0} \, dr \,, \quad (\Gamma \neq \Gamma_0) \,. \end{aligned} \tag{2.2.52}$$

The quantity $V_{\Gamma\Gamma_0}$ is called the polarization potential. This is an integral operator of the type

$$V(r)\varphi(r) = \int_0^\infty V(r,r') \varphi(r') \, dr' \,, \tag{2.2.53}$$

and is represented by the series

$$\begin{aligned} &V_{\Gamma\Gamma_0} = \sum_{n=2}^{\infty} V_{\Gamma\Gamma_0}^{(n)} \,, \quad V_{\Gamma\Gamma_0}^{(n)} = \sum_{\Gamma_1 \cdots \Gamma_{n-1}}' U_{\Gamma\Gamma_1 \cdots \Gamma_{n-1}\Gamma_0} \,, \\ &U_{\Gamma\Gamma_1 \cdots \Gamma_{n-1}\Gamma_0}(r,r') = \int dr_1 \ldots dr_{n-2} \, U_{\Gamma\Gamma_1}(r) \, G_{\Gamma_1}(r,r_1) \, U_{\Gamma_1\Gamma_2}(r_1) \ldots \\ &\qquad \times G_{\Gamma_{n-1}}(r_{n-2},r') \, U_{\Gamma_{n-1}\Gamma_0}(r') \,. \end{aligned} \tag{2.2.54}$$

The last formulas are applicable both for $\Gamma \neq \Gamma_0$ and $\Gamma = \Gamma_0$. In the sum over $\Gamma_1 \ldots \Gamma_{n-1}$ it is necessary to omit all terms that include one or more diagonal factors $U_{\Gamma_k\Gamma_k}$.

Thus the solution of the collision theory equations is expressed in the closed from (2.2.51). The correction for the first-order matrix $T_{\Gamma\Gamma_0}^{(1)}$ is determined through the value $V_{\Gamma\Gamma_0}$. From the second formula (2.2.52) it is seen that $V_{\Gamma\Gamma_0}$ is the correction for the Hartree–Fock potential $U_{\Gamma\Gamma_0}$, which is where the term "polarization potential" stems from.

Strictly speaking, only a formal solution has been obtained, since a lengthy solution of an infinite set of equations has been replaced by a no less cumbersome calculation of an infinite series (2.2.54). Besides, the question of the convergence of the series is still not clear. However, if the series converges, the use of a polarization potential to get the approximate solutions has a number of obvious advantages. In particular, in some cases it is far easier to formulate an approximate expression for the potential than for the wave function.

2.3 First-Order Approximation

2.3.1 General Formulas

In the previous section the formal solution of integral equations and the equation (2.2.52) for the T matrix were obtained. Here we consider in more detail the first-order approximation, and derive explicit expressions for the T matrix and cross sections in terms of radial integrals and some angular factors. The latter are expressed in terms of $3nj$ symbols. The equations derived here are used in Chap. 3 for a discussion of some approximate methods.

The first-order approximation is readily obtained from a general solution (2.2.52), if one omits the terms with polarization potential:

$$T^{\mathrm{I}}_{\Gamma_0\Gamma_0} = \exp(i\eta_0)\sin\eta_0, \quad T^{\mathrm{I}}_{\Gamma\Gamma_0} = -\int\limits_0^\infty \bar{F}_\Gamma U_{\Gamma\Gamma_0}\bar{F}_{\Gamma_0}\,dr\,. \tag{2.3.1}$$

This expression corresponds to replacement of the full set of equations by pair of independent equations,

$$(\mathscr{L}_{\Gamma_0} + k_0^2)F_{\Gamma_0} = 0\,, \quad (\mathscr{L}_\Gamma + k^2)F_\Gamma = U_{\Gamma\Gamma_0}F_{\Gamma_0}\,. \tag{2.3.2}$$

From comparison of the boundary condition (2.2.38) for $\Gamma_0 = \Gamma$ with (2.2.44) it is clear that $F_{\Gamma_0} = \bar{F}_{\Gamma_0}$.

Below we shall restrict ourselves to the discussion of inelastic collisions (i.e., $\Gamma \neq \Gamma_0$), and as usual, we shall not allow for the relativistic interaction of the outer electron with the atom. We begin with the case of LS coupling with the states of the type

$$a = \gamma_p S_p L_p nlSLJ\,,$$

where $\gamma_p L_p S_p$ define the state of the atomic core. On substitution of eqs (2.3.1), (2.2.32a, 35a) in (2.2.22c) we obtain

$$\sigma(a_0\lambda_0, a\lambda) = \pi a_0^2 \frac{2}{k_0^2(2J_0+1)} \sum_{q\kappa v}\Big| M_{q\kappa v}(SLJ) \sum_{L_\mathrm{T}S_\mathrm{T}} M_{qq0}(SsS_\mathrm{T}) M_{\kappa\kappa 0}(L\lambda L_\mathrm{T})[q\kappa]^2$$
$$\sum_{q'\kappa'} M_{0q'q'}(S_p sS) M_{0\kappa'\kappa'}(L_p lL)[L_p S_p q'\kappa'] M_{\kappa'\kappa' 0}(L\lambda L_\mathrm{T})$$
$$\times M_{q'q'0}(SsS_\mathrm{T})(-1)^{q'+\kappa'+l+\lambda}[2\delta(q')R'_{\kappa'} - [q']^2 R''_{\kappa'}]\Big|^2$$

and after summation over $L_T S_T$:

$$\sigma(a_0\lambda_0, a\lambda) = \pi a_0^2 \frac{2}{k_0^2(2J_0+1)} \sum_{q\kappa v} |M_{q\kappa v}(SLJ) M_{0qq}(S_p sS) M_{0\kappa\kappa}(L_p\lambda L)$$

$$\times [L_p S_p] \cdot \frac{1}{[q\kappa]} (-1)^{q+\kappa+l+\lambda} \left[2\delta(q) R'_\kappa - [q]^2 R''_\kappa\right]|^2 , \quad (2.3.3)$$

$$R'_\kappa = \int_0^\infty \bar{F}_{\Gamma_0}\, y'_\kappa \bar{F}_\Gamma dr , \quad R''_\kappa = \int_0^\infty \bar{F}_{\Gamma_0}\, y''_\kappa P_l\, dr .$$

$\bar{F}_\Gamma$ and R_κ depend, in general, on all quantum numbers Γ. However the case when they are independent of $L_T S_T$ is most interesting, and we restrict ourselves by this case. It is necessary to sum over $L_T S_T$.

Now we can separate the radial and angular factors, and we write σ in the form

$$\sigma(a_0, a) = \sum_\kappa \left[Q'_\kappa(a_0 J_0, aJ)\sigma'_\kappa(l_0, l) + Q''_\kappa(a_0 J_0, aJ)\sigma''_\kappa(l_0, l)\right]$$

$$Q_\kappa^{(p)} = \frac{2l_0+1}{g(a_0)} \sum_{qv} b_{q\kappa v}^2 C^{(p)}(q) , \quad C'(q) = 2\delta(q) , \quad C''(q) = \frac{1}{2}[q]^2 , \quad (2.3.4)$$

where $C^{(p)}$ are C' or C'', σ'_κ and σ''_κ are defined by (2.3.9) below. The amplitude angular factor is

$$b_{q\kappa v} = [L_p S_p] M_{q\kappa v}(SLJ) M_{0qq}(S_p sS) M_{0\kappa\kappa}(L_p\lambda L) . \quad (2.3.5)$$

Here, b and Q', Q'' depend only on the angular-momentum quantum numbers of the states a_0, a. Q'_κ coincides with the angular factor Q_κ for the probability of 2κ-pole radiative transitions [Ref. 2.6, Sect. 9.3.6)]. We use both notations Q and Q'. The radial factors $\sigma'_\kappa(l_0, l)$, $\sigma''_\kappa(l_0, l)$ are equal to the cross sections of one-electron transition $n_0 l_0 - nl$. σ' includes the direct and mixed parts, and σ'' is the exchange part.

The interval in which κ can vary in both the terms of (2.3.4) is the same, $|l_0 - l| \le \kappa \le l_0 + l$. However, the physical meaning of κ for the two terms is different. In the first one it is the index of multipole interaction. Therefore, only those σ'_κ do not vanish for which the value of κ has the same parity as $|l_0 - l|$. In the second term the quantity κ is simply the summation index, see (2.2.32a), and σ''_κ with any value of κ may exist. The set of formulas for $\sigma, Q \dots$ is given in Sect. 2.3.2.

For transitions between the mixed states $\bar{a}_0, \bar{a}$, where

$$\Psi(\bar{a}) = \sum_{a/\bar{a}} (\bar{a}|a)\Psi(a) , \quad (2.3.6)$$

the cross section is given by (2.3.4) with

$$b_{q\kappa v}(\bar{a}_0, \bar{a}) = \sum_{a_0/\bar{a}_0,\, a/\bar{a}} (\bar{a}_0|a_0) b_{q\kappa v}(a_0, a)(a|\bar{a}) . \quad (2.3.7)$$

where $b_{q\kappa v}(a_0, a)$ has been defined in (2.3.5). For intermediate coupling $a/\bar{a} = LS$. Thus, only the angular part depends on the coupling scheme. The radial part can be calculated independently. We see from (2.3.5) that angular factors for direct and mixed terms are the same (that permits to include the mixed term into the radial part σ'_κ). The presence of $\delta(q)$ provides $\delta(S_0, S)$ both in direct and mixed terms; therefore, the contribution of the direct term to the probability of intercombination transitions is $\sim (v/c)^2$. The case of configuration mixing is more complicated since the radial part (one electron cross section) also depends on the configuration. The total cross section cannot be written for this case in the form (2.3.4).

The following approximations are most often used:
(i) *Born–Oppenheimer Approximation*: The functions $\bar{F}_\Gamma$ in (2.3.3) are the free-motion radial functions, i.e., solutions of (2.2.43) at $U_\Gamma = 0$. If exchange terms are neglected in Eqs. (2.3.3–5) we obtain the usual Born approximation.
(ii) *Coulomb–Born Approximation* (with or without exchange): The F_Γ are the Coulomb radial functions for the potential $[U_\Gamma = -(z-1)]/r$. For neutral atoms $(z = 1)$, approximations (i) and (ii) coincide.
(iii) *Distorted Wave Approximation*: F_Γ are the solutions of (2.2.43) taking U_Γ as the Hartree–Fock atomic potential. (In some cases Thomas–Fermi or other approximate expressions for U_Γ are used.)

The approximations and also some modifications are discussed in Chap. 3. Approximations (i) and (ii) are the most important for applications.

2.3.2 List of Formulas for σ and Q-Factors

For arbitrary coupling scheme,

$$\begin{aligned}\sigma_{\bar{a}_0\bar{a}} &= \sigma'_{\bar{a}_0\bar{a}} + \sigma''_{\bar{a}_0\bar{a}} \\ &= \sum_\kappa [Q'_\kappa(\bar{a}_0, \bar{a})\, \sigma'_\kappa(l_0, l) + Q''_\kappa(\bar{a}_0, \bar{a})\, \sigma''_\kappa(l_0, l)]\,,\end{aligned} \tag{2.3.8}$$

where $\sigma'_\kappa(l_0, l)$ and $\sigma''_\kappa(l_0, l)$ are one-electron cross sections, depending on quantum numbers $n_0 l_0$, nl only:

$$\begin{aligned}\sigma'_\kappa(l_0, l) &= \pi a_0^2 \frac{4}{k_0^2 [l_0 \kappa]^2} \sum_{\lambda_0 \lambda} R'_\kappa (R'_\kappa - R''_\kappa) \\ \sigma''_\kappa(l_0, l) &= \pi a_0^2 \frac{4}{k_0^2 [l_0 \kappa]^2} \sum_{\lambda_0 \lambda} (R''_\kappa)^2\end{aligned} \tag{2.3.9}$$

$R' \equiv R^{\mathrm{d}}$ and R'' are direct and exchange radial integrals:

$$\begin{aligned}R'_\kappa &= [l_0\, l \lambda_0\, \lambda] \begin{pmatrix} l_0\, \kappa\, l \\ 0\,0\,0 \end{pmatrix} \begin{pmatrix} \lambda_0\, \kappa\, \lambda \\ 0\,0\,0 \end{pmatrix} \\ &\times 2 \int_0^\infty \int_0^\infty \bar{F}_{\Gamma_0}(r') P_{l_0}(r'') \frac{r_<^\kappa}{r_>^{\kappa+1}} P_l(r'') \bar{F}_\Gamma(r')\, dr''\, dr'\end{aligned} \tag{2.3.10}$$

$$R''_{\kappa} = \sum_{\kappa''} (-1)^{\kappa+\kappa''} [\kappa\kappa'']^2 \begin{Bmatrix} \kappa\, l_0\, l \\ \kappa''\, \lambda_0\, \lambda \end{Bmatrix} R^{\mathrm{e}}_{\kappa''}$$

$$R^{\mathrm{e}}_{\kappa''} = [l_0\, l\lambda_0\, \lambda] \begin{pmatrix} l_0\, \kappa\, \lambda \\ 0\,0\,0 \end{pmatrix} \begin{pmatrix} \lambda_0\, \kappa\, l \\ 0\,0\,0 \end{pmatrix}$$

$$2 \int_0^{\infty}\int_0^{\infty} \bar{F}_{\Gamma_0}(r') P_{l_0}(r'') \frac{r_<^{\kappa}}{r_>^{\kappa+1}} (1 - c r_> \delta(\kappa'', 0)) \bar{F}_{\Gamma}(r'') P_l \, dr'' \, dr' \tag{2.3.11}$$

$$c = \frac{1}{2}(-\varepsilon_{l_0} + k^2) = \frac{1}{2}(-\varepsilon_l + k_0^2) .$$

From the $3j$ and $6j$ symbols in (2.3.10, 11) it follows

$$\kappa_{\min} \le \kappa \le \kappa_{\max} , \quad \kappa''_{\min} \le \kappa'' \le \kappa''_{\max} ,$$
$$\kappa_{\min} = \max(|l - l_0|, |\lambda - \lambda_0|), \quad \kappa_{\max} = \min(l + l_0, \lambda + \lambda_0) , \tag{2.3.12}$$
$$\kappa''_{\min} = \max(|\lambda - l_0|, |l - \lambda_0|), \quad \kappa''_{\max} = \min(\lambda + l_0, l + \lambda_0) .$$

In the sum of $Q'_{\kappa}\sigma'_{\kappa}$ all κ have the parity of $\Delta l = l - l_0$:

$$\kappa = \kappa_{\min} + 2m , \quad m = 0, 1, 2 \ldots . \tag{2.3.13a}$$

In the sum of $Q''_{\kappa}\sigma''_{\kappa}$ κ can be of either parity, but all κ'' have the parity of $l - \lambda_0$:

$$\kappa = \kappa_{\min} + m , \quad \kappa'' = \kappa''_{\min} + 2m , \quad m = 0, 1, 2 \ldots . \tag{2.3.13b}$$

The angular factors Q'_{κ}, Q''_{κ} for the transition $\bar{a}_0$–$\bar{a}$ in intermediate coupling are equal to

$$Q^{(\mathrm{p})}_{\kappa}(\bar{a}_0, \bar{a}) = \frac{2l_0 + 1}{g(\bar{a}_0)} \sum_{qv} b^2_{q\kappa v}(\bar{a}_0, \bar{a}) C^{(\mathrm{p})}(q) ,$$

$$b_{q\kappa v}(\bar{a}_0, \bar{a}) = \sum_{L_0 S_0 LS} (\bar{a}_0 | a_0) \, b_{q\kappa v}(a_0, a) \, (a | \bar{a}) , \tag{2.3.14}$$

where a_0, a are the states in the LS coupling, $g(\bar{a}_0)$ is the statistical weight of the state $\bar{a}_0$, and the shorthand superscript (p) stands for prime and double primes, respectively, i.e.,

$$C' = 2\delta(q, 0), \quad C'' = \frac{1}{2}[q]^2 . \tag{2.3.15}$$

The (amplitude) angular factor b in the LS coupling for the transition

$$a_0 = \gamma_p L_p S_p n_0 l_0 s L_0 S_0 J_0 \rightarrow a = \gamma_p L_p S_p \, n l s L S J \quad \left(s = \frac{1}{2}\right)$$

is equal to

$$b_{q\kappa v}(a_0, a) = [L_pS_p]M_{q\kappa v}(SLJ)M_{0qq}(S_psS)M_{0\kappa\kappa}(L_plL)$$
$$= [J_0Jv]\begin{Bmatrix} S_0 & S & q \\ L_0 & L & \kappa \\ J_0 & J & v \end{Bmatrix} \cdot [S_0\, SL_0\, L]\begin{Bmatrix} S_0 & S & q \\ s & s & S_p \end{Bmatrix} \tag{2.3.16}$$
$$\times \begin{Bmatrix} L_0 & L & \kappa \\ l & l_0 & L_p \end{Bmatrix} (-1)^{S+s+S_p+L+l_0+L_p+\kappa+q} .$$

We give now the formulas for the Q factors in the case of LS coupling for the same transition (2.3.16). Formulas for other types of transitions will be given in Sect. 6.2. For simplicity we show as arguments only the last quantum numbers for a_0 and a. From (2.3.14–16) we obtain

$$Q'_\kappa(J_0, J) = \delta(S_0, S)\,[l_0JL_0L]^2 \begin{Bmatrix} J_0 & J & \kappa \\ L & L_0 & S \end{Bmatrix}^2 \begin{Bmatrix} L_0 & L & \kappa \\ l & l_0 & L_p \end{Bmatrix}^2$$
$$Q''_\kappa(J_0,J) = \frac{1}{2}\sum_{qv}[l_0JS_0SL_0Lqv]^2 \begin{Bmatrix} S_0 & S & q \\ L_0 & L & \kappa \\ J_0 & J & v \end{Bmatrix}^2 \begin{Bmatrix} S_0 & S & q \\ s & s & S_p \end{Bmatrix}^2 \begin{Bmatrix} L_0 & L & \kappa \\ l & l_0 & L_p \end{Bmatrix}^2 \tag{2.3.17}$$

For transitions between the terms $L_0\, S_0$–LS we should sum (2.3.17) with respect to J, and average it with respect to J_0:

$$Q'_\kappa(L_0\, S_0,\, LS) = \sum_{JJ_0} \frac{2J_0 + 1}{(2L_0 + 1)(2S_0 + 1)} Q'_\kappa(J_0, J)\, , \tag{2.3.18}$$

and similarly for Q''_κ. As a result we obtain considerably simpler equations:

$$Q'_\kappa(L_0\, S_0,\, LS) = Q'_\kappa(J_0, LS) = \delta_{S_0 S}\, Q_\kappa(L_0, L)\, ,$$
$$Q''_\kappa(L_0\, S_0,\, LS) = Q''_\kappa(J_0, LS) = \frac{2S+1}{2(2S_p+1)}\, Q_\kappa(L_0, L)\, , \tag{2.3.19}$$

where

$$Q_\kappa(L_0,\, L) = [l_0L]^2 \begin{Bmatrix} \kappa & L_0 & L \\ L_p & l & l_0 \end{Bmatrix}^2 \tag{2.3.20}$$

Thus the cross section for transition between terms $L_0\, S_0$–LS can be written in the form

$$\sigma_{a_0a} = \sum_\kappa Q_\kappa(L_0, L)\left[\delta_{S_0 S}\sigma'_\kappa(l_0, l) + \frac{2S+1}{2(2S_p+1)}\sigma''_\kappa(l_0, l)\right] . \tag{2.3.21}$$

Summing over L we have

$$Q(L_0, L_pS_pl) = \sum_L Q_\kappa(L_0, L_pS_plL) = 1\, ,$$
$$Q(L_pS_pl_0, L_pS_pl) = 1\, . \tag{2.3.22}$$

For a more general case $l_0^m - l_0^{m-1}l$, by summing over S, L_p, S_p, we obtain

$$Q_\kappa(l_0^m, l_0^{m-1}l) = Q_\kappa(l_0^m L_0 S_0, l_0^{m-1}l) \\ = Q_\kappa''(l_0^m, l_0^{m-1}l) = Q_\kappa''(l_0^m L_0 S_0, l_0^{m-1}l) = m\,, \tag{2.3.23}$$

and therefore

$$\sigma(l_0^m, l_0^{m-1}l) = m \sum_k [\sigma_\kappa'(l_0, l) + \sigma_\kappa''(l_0, l)]\,. \tag{2.3.24}$$

One can see from (2.3.24) that $\sum_\kappa \sigma_\kappa'(l_0, l)$ and $\sum_\kappa \sigma_\kappa''(l_0, l)$ correspond to cross sections of one-electron transitions. However, the interpretation of particular terms in the sum requires special consideration. According to (2.3.9, 10),

$$\sum_\kappa \sigma_\kappa' = \sum_\kappa \sigma_\kappa^{\mathrm{d}} + \sum_{\kappa\kappa''} \sigma_{\kappa\kappa''}^{\mathrm{de}}\,. \tag{2.3.25}$$

Here $\sigma_\kappa^{\mathrm{d}}$ is the cross section for direct 2κ-pole interaction; $\sigma_{\kappa\kappa''}^{\mathrm{de}}$ is a mixed term consisting of direct 2κ-pole and exchange $2\kappa''$-pole interactions. In contrast, σ_κ'' contains the sum of exchange terms of all multipole orders κ''. The sum $\sum_\kappa \sigma_\kappa''$ can be rewritten in the form

$$\sum_\kappa \sigma_\kappa'' = \sum_{\kappa''} \sigma_{\kappa''}^{\mathrm{e}}\,, \quad \sigma_{\kappa''}^{\mathrm{e}} = \pi a_0^2 \frac{4}{k_0^2[l_0\kappa'']^2} \sum_{\lambda_0\lambda} (R_{\kappa''}^{\mathrm{e}})^2 \tag{2.3.26}$$

where $\sigma_{\kappa''}^{\mathrm{e}}$ is $2\kappa''$-pole exchange cross section, and $R_{\kappa''}^{\mathrm{e}}$ is defined by (2.3.11). It should be noted that an explicit summation over κ and the transformation to the form $\sum_{\kappa''} \sigma_{\kappa''}^{\mathrm{e}}$ are possible only for the total cross section of a transition between configurations, when Q_κ'' does not depend on κ.

For some problems the partial cross sections $\sigma(\Gamma_0, \Gamma)$ with definite values of $L_\mathrm{T}S_\mathrm{T}$ are necessary. For transitions between terms LS we can use (2.2.21) without the sum over $L_\mathrm{T}S_\mathrm{T}$. The derivation similar to that used for the total cross section gives

$$\sigma(\Gamma_0, \Gamma) = \pi a_0^2 \frac{2}{k_0^2} \frac{1}{[S_0L_0]^2} \{ \sum_{q\kappa} M_{qq0}(SsS_\mathrm{T}) M_{\kappa\kappa0}(L\lambda L_\mathrm{T}) \\ M_{0qq}(S_psS) M_{0\kappa\kappa}(L_p\lambda L)\, [L_pS_pq\kappa]\, (-1)^{q+\kappa+l+\lambda}\, [2\delta(q)R_\kappa' \; - [q]^2 R_\kappa'']\}^2$$

or

$$\sigma(\Gamma_0, \Gamma) = \pi a_0^2 \frac{2}{k_0^2} \frac{[S_\mathrm{T}L_\mathrm{T}]^2}{[S_0L_0]^2} \left[\sum_\kappa A_\kappa \left(R_\kappa' - B_{S_0S} R_\kappa'' \right) \right]^2 \tag{2.3.27}$$

$$\Gamma = a\lambda s L_\mathrm{T} S_\mathrm{T}$$

where R' and R'' are defined by (2.3.10, 11). The factors A and B depend only on angular momenta. For the transition l_0^m–$l_0^{m-1}\,l$ we obtain

$$a_0 = l_0^m L_0 S_0\,, \quad a = l_0^{m-1}[L_p S_p] l L S$$

$$A_\kappa = (-1)^{L_T+L_0+L+L_p}[\kappa L_0 L] \begin{Bmatrix} \kappa & L_0 & L \\ L_p & l & l_0 \end{Bmatrix} \begin{Bmatrix} \kappa & L_0 & L \\ L_T & \lambda & \lambda_0 \end{Bmatrix} G^{L_0 S_0}_{L_p S_p} \sqrt{m} \qquad (2.3.28)$$

$$B_{S_0 S} = (-1)^{1-S-S_0}[S S_0] \begin{Bmatrix} \frac{1}{2} & S_0 & S_T \\ \frac{1}{2} & S & S_p \end{Bmatrix}$$

In particular for $S_p = 0$, $B_{S_0 S} = (-1)^{S_T+1}$; while for $S_0 = 0$,

$$B_{S_0 S} = (-1)^S \frac{1}{2}(2S+1)^{1/2} \delta_{S_T\,1/2}\,.$$

3 Approximate Methods for Calculating Cross Sections

Various approximate methods for calculating the cross sections of excitation and ionization by electron impact are considered.

Firstly, the Born approximation and its modifications taking into account the Coulomb field, exchange interaction, and normalization of the cross sections are described. The range of applicability and accuracy of these methods are discussed.

In Sects. 3.3 and 3.4 some other more complicated methods are briefly discussed, in particular, the second Born approximation, the method of polarization potential, and the close-coupling method. For more detailed discussions of these and some other approximate methods see [3.1–3].

Special consideration is given to the case of transitions between highly excited levels with $n \gg 1$. The Born approximation as well as the quasi-classical approximation are reduced to comparatively simple formulas. The results of numerical calculations are also given.

3.1 Born Approximation

3.1.1 Collisions of Fast Electrons with Atoms; Multipole Expansion

In those cases when the interaction responsible for scattering can be considered as a perturbation and the exchange can be neglected, it proves to be possible to obtain simple general formulas for the cross sections without using a partial wave expansion.

We consider first the scattering of a charged particle of arbitrary mass on an atom. We denote the reduced mass of the system by μ. In accordance with the well-known perturbation theory, the formula for the probability of an atomic transition between discrete states a_0M_0, aM, accompanied by a change in the perturbing particle wave vector $\boldsymbol{k}_0 - \boldsymbol{k}$ is

$$dW_{a_0M_0\boldsymbol{k}_0,\, aM\boldsymbol{k}} = \frac{\pi}{2}\, |U_{a_0M_0\boldsymbol{k}_0,\, aM\boldsymbol{k}}|^2\, \delta(E_0 - E)\, d\boldsymbol{k}\,, \tag{3.1.1}$$

where

$$\begin{aligned} &U_{a_0M_0\boldsymbol{k}_0,\, aM\boldsymbol{k}} = \int d\boldsymbol{r}\psi^*_{\boldsymbol{k}}(\boldsymbol{r}) U_{a_0M_0,\, aM}(\boldsymbol{r})\psi_{\boldsymbol{k}_0}(\boldsymbol{r})\,, \\ &U_{a_0M_0,\, aM}(\boldsymbol{r}) = \langle a_0M_0| \sum_i \frac{2}{|\boldsymbol{r}-\boldsymbol{r}_i|} - \frac{2\mathscr{Z}}{r}\, |aM\rangle\,, \\ &E_0 = E_{a_0} + \frac{k_0^2}{\mu}\,, \qquad E = E_a + \frac{k^2}{\mu}\,. \end{aligned} \tag{3.1.2}$$

The factor 2 in the expression for U is due to the use of the Rydberg unit of energy (Ry); $\psi_{\boldsymbol{k}_0}$ and $\psi_{\boldsymbol{k}}$ are the free-motion wave functions of the perturbing particle. The final state function is normalized to $\delta(\boldsymbol{k}-\boldsymbol{k}')$, or $\psi_{\boldsymbol{k}} = (2\pi)^{-3/2}\exp(\mathrm{i}\boldsymbol{k}\cdot\boldsymbol{r})$. We normalize the wave function of the initial state to unit flux density, $\psi_{\boldsymbol{k}_0} = v_0^{-1/2}\exp(\mathrm{i}\boldsymbol{k}_0\cdot\boldsymbol{r})$. Then the differential cross section $d\sigma$ coincides with the transition probability dW. Integrating (3.1.1) over k, we obtain the Born cross section

$$d\sigma^{\mathrm{B}}_{a_0M_0,\,aM} = \frac{\mu^2}{16\pi^2}\cdot\frac{k}{k_0}\,|\int \exp[-\mathrm{i}(\boldsymbol{k}_0-\boldsymbol{k})\cdot\boldsymbol{r}]U_{a_0M_0,\,aM}(\boldsymbol{r})\,d\boldsymbol{r}|^2\,dO\,,$$
$$k^2/\mu = E_{a_0} - E_a + k_0^2/\mu\,. \tag{3.1.3}$$

If we are not interested in the orientation of the atom after scattering, we can sum $d\sigma$ over M and average over M_0. Then

$$d\sigma^{\mathrm{B}}_{a_0a} = \frac{\mu^2}{16\pi^2}\cdot\frac{k}{k_0g_0}\cdot\sum_{M_0M}|\int \exp[-\mathrm{i}(\boldsymbol{k}_0-\boldsymbol{k})\cdot\boldsymbol{r}]\,U_{a_0M_0,\,aM}(\boldsymbol{r})\,d\boldsymbol{r}|^2\,dO, \tag{3.1.4}$$

where g_0 is the statistical weight of the state a_0.

Equation (3.1.4) is called the Born formula. The case $a = a_0$, $k = k_0$ corresponds to elastic scattering; the case $a \neq a_0$, $k \neq k_0$, to inelastic scattering.

The Born approximation is a good approximation for calculating electron–atom scattering, if the electron velocity is large compared with the velocities of the atomic electrons. In the following part of this section, we treat collisions with electrons only, and so we can suppose $\mu = 1$.

After integration of (3.1.4) over $d\boldsymbol{r}$ by means of the relation

$$\int\frac{\exp(-\mathrm{i}\boldsymbol{q}\cdot\boldsymbol{r})}{|\boldsymbol{r}-\boldsymbol{r}_i|}d\boldsymbol{r} = \frac{4\pi}{q^2}\exp(-\mathrm{i}\boldsymbol{q}\cdot\boldsymbol{r}_i)\,, \tag{3.1.5}$$

we obtain

$$d\sigma^{\mathrm{B}}_{a_0a} = \frac{8\pi}{k_0^2g_0}\cdot\sum_{M_0M}|F_{a_0M_0,\,aM}(q) - \mathscr{Z}\delta_{a_0M_0,\,aM}|^2\,\frac{dq}{q^3} \tag{3.1.6}$$

$$F_{a_0M_0,\,aM}(q) = \langle\, aM|\sum_j \exp(-\mathrm{i}\boldsymbol{q}\cdot\boldsymbol{r}_j)\,|a_0M_0\,\rangle$$
$$\boldsymbol{q} = \boldsymbol{k}_0 - \boldsymbol{k}\,,\quad q^2 = k_0^2 + k^2 - 2k_0k\cos\vartheta \tag{3.1.7}$$

We now separate the radial and angular variables. We expand $\exp(-\mathrm{i}\boldsymbol{q}\cdot\boldsymbol{r}_j)$ in (3.1.7) in spherical harmonics, see (2.1.6),

$$\exp(-\mathrm{i}\boldsymbol{q}\cdot\boldsymbol{r}) = 4\pi\sum_{\kappa\mu}\mathrm{i}^{-\kappa}\,Y^*_{\kappa\mu}(\vartheta_q,\varphi_q)\,Y_{\kappa\mu}(\vartheta,\varphi)j_\kappa(qr)\,. \tag{3.1.8}$$

Substituting (3.1.7) and (3.1.8) in (3.1.6) and using the Wigner–Eckart theorem [Ref. 3.4, Sect. 4.3],

$$\langle \gamma JM|T_{\kappa\mu}|\gamma_0 J_0 M_0 \rangle = (-1)^{J-M}(\gamma J\|T_\kappa\|\gamma_0 J_0)\begin{pmatrix} J & \kappa & J_0 \\ -M & \mu & M_0 \end{pmatrix}, \tag{3.1.9}$$

and the summation rules for $3j$ symbols and $Y_{\kappa\mu}$, we obtain

$$d\sigma^{\mathrm{B}}_{a_0a} = \frac{8\pi}{k_0^2 g_0} \,|\,(a\|T_\kappa\|a_0) - \mathscr{Z}\delta_{a_0a}\,|^2 \,\frac{dq}{q^3}\,, \tag{3.1.10}$$

where $T_{\kappa\mu}$ is the operator

$$T_{\kappa\mu} = \sum_j \left(\frac{4\pi}{2\kappa+1}\right)^{1/2} Y_{\kappa\mu}(\vartheta_j, \varphi_j)\, j_\kappa(qr_j)\,. \tag{3.1.11}$$

We shall begin with inelastic scattering $a_0 \neq a$. In that case the matrix element in (3.1.10) is nonzero only for one-electron transitions and (3.1.10) becomes simplified to

$$d\sigma^{\mathrm{B}}_{a_0a} = \frac{8\pi}{k_0^2(2l_0+1)} \sum_\kappa Q_\kappa(a_0, a)\,[R_\kappa(q)]^2\,\frac{dq}{q^3}\,. \tag{3.1.12}$$

The total cross section for transition $a_0 - a$ is

$$\sigma^{\mathrm{B}}_{a_0a} = \frac{8\pi}{k_0^2(2l_0+1)} \sum_\kappa Q_\kappa(a_0, a) \int\limits_{k_0-k}^{k_0+k} [R_\kappa(q)]^2\,\frac{dq}{q^3}\,. \tag{3.1.13}$$

Here $R_\kappa(q)$ is the radial integral

$$R_\kappa(q) = [\kappa l_0 l]\begin{pmatrix} \kappa & l_0 & l \\ 0 & 0 & 0 \end{pmatrix} \int\limits_0^\infty P_l(r)P_{l_0}(r)[\,j_\kappa(qr) - \delta_{\kappa 0}]\,dr\,, \tag{3.1.14}$$

which is similar to the corresponding integral in the multipole transition probability formula, but contains $j_\kappa(qr)$ in place of r^κ. We note that

$$j_\kappa(qr) \underset{q\to 0}{\to} \frac{q^\kappa}{(2\kappa+1)!!}\, r^\kappa\,. \tag{3.1.15}$$

The factor $Q_\kappa(a_0, a)$ in (3.1.13) depends on the angular momenta of the states a_0, a. The same factor is present in the expression for the line strength of electric 2κ-pole radiation [Ref. 3.4, Sects. 9.3, 6]. We note also that the Q_κ in (3.1.13) and Q'_κ in (2.3.8) are the same. Generally speaking, the radial functions $P_l(r)$ of the optical electron of the atom depend on the whole set of quantum numbers a. However below we assume for simplicity that the radial functions depend only on quantum numbers n, l, and consequently we shall denote them as $P_l(r)$ or $P_{nl}(r)$.

We shall write (3.1.13) in the form

$$\sigma^{\mathrm{B}}_{a_0a} = \sum_\kappa Q_\kappa(a_0, a)\,\sigma^{\mathrm{B}}_\kappa(l_0, l)\,, \tag{3.1.16}$$

$$\kappa = \kappa_{\min}, \kappa_{\min} + 2, \ldots\ l_0 + l\,; \quad \kappa_{\min} = |l_0 - l|$$

$$\sigma_\kappa^{\mathrm{B}}(l_0, l) = \pi a_0^2 \cdot \frac{8}{k_0^2(2l_0+1)} \int\limits_{k_0-k}^{k_0+k} [R_\kappa(q)]^2 \frac{dq}{q^3}\,. \tag{3.1.17}$$

Equation (3.1.16) is called the multipole expansion, and $\sigma_\kappa(l_0, l)$ is the 2κ-pole one-electron cross section for the transition $n_0 l_0 - nl$. For $a_0 = l_0^m L_0 S_0 J_0$, $a = l_0^{m-1}[L_p S_p]\, lLSJ$,

$$Q_\kappa(a_0,\ a) = \delta_{S_0 S}[l_0 J L_0 L]^2 \begin{Bmatrix} \kappa & J_0 & J \\ S & L & L_0 \end{Bmatrix}^2 \begin{Bmatrix} \kappa & L_0 & L \\ L_p & l & l_0 \end{Bmatrix}^2 \left(G_{L_p S_p}^{L_0 S_0}\right)^2 m\,, \tag{3.1.18}$$

where $G_{L_p S_p}^{L_0 S_0}$ is the fractional parentage coefficient. A fuller list of formulas for Q_κ is given in Sect. 2.3.2 and 6.2.

The possible values of κ in (3.1.16) are determined by the $3j$ symbol in (3.1.14) and correspond to a 2κ-pole electron–atom interaction (i.e., they do not pertain to any partial wave). All the κ are of the same parity according to the change of atomic parity during the transition $l_0 - l$. In many cases it is sufficient to consider only minimum multiplicity $\kappa = \kappa_{\min}$, since σ_κ decreases rapidly with increasing κ.

We should note that the decreasing of σ_κ with increasing κ is not connected with any small parameter. In this point, the electron–atom interaction radically differs from the interaction of the atom with an electromagnetic field. In the latter case, the higher multipoles contain the factor $(ze^2/\hbar c)^{2\kappa+1} = (z/137)^{2\kappa+1}$, and each successive term is smaller by a factor of about $10^{-4}\,z^2$. In the case of electron–atom collisions such small parameter does not exist. Numerical calculations show that in the Born approximation the multipole cross section $\sigma_{\kappa+2}$ is usually 5–10 times smaller than σ_κ.

From (3.1.14, 18) it follows that the selection rules when $\kappa \neq 0$ are the same as for electric multipole transitions. For example, in LS coupling

$$\begin{aligned} \text{at } \kappa = 1 : &\ \Delta l = \pm 1;\ \Delta L = 0, \pm 1;\ \Delta J = 0, \pm 1\,; \\ &\ L_0 + L \geq 1;\ J_0 + J \geq 1;\ \Delta S = 0\,; \end{aligned} \tag{3.1.19}$$

$$\begin{aligned} \text{at } \kappa = 2 : &\ \Delta l = 0, \pm 2;\ \Delta L = 0, \pm 1, \pm 2;\ \Delta J = 0, \pm 1, \pm 2\,; \\ &\ l_0 + l \geq 2;\ L_0 + L \geq 2;\ J_0 + J \geq 2\,; \Delta S = 0\,. \end{aligned} \tag{3.1.20}$$

At $\kappa = 0$, in contrast to radiative transitions, $\sigma_0 \neq 0$.

By means of (3.1.13, 14, 18), the relation between the cross sections of direct and reverse processes can be obtained. Exchanging in these formulas the initial and final states of the atom, and also k_0 and k, we have

$$g_0 k_0^2 \sigma^{\mathrm{B}}(a_0 k_0,\ ak) = g k^2 \sigma^{\mathrm{B}}(ak,\ a_0 k_0)\,. \tag{3.1.21}$$

It is easy to see that this relation coincides with (1.2.8) which was obtained from the principle of detailed balance.

For the elastic scattering on the atom in the state a containing closed shells and the open shell l^m, the differential cross section is equal

$$d\sigma_a^B = \frac{8\pi}{k^2(2l+1)}\left\{\left[\int_0^\infty \frac{\sin qr}{qr}\,\rho(r)\,dr - \mathscr{Z}\right]^2 + \sum_\kappa Q_\kappa(a)[R_\kappa(q)]^2\right\}\frac{dq}{q^3} \tag{3.1.22}$$

where

$$\rho(r) = \sum N_{n_jl_j}\,P^2_{n_jl_j}(r), \quad \int_0^\infty \rho(r)\,dr = N$$

$$Q_\kappa(a) = \frac{2l+1}{2L+1}\,|\,(l^mSL\|U^\kappa\|l^mSL)|^2\,.$$

The formulas for the reduced matrix element of U^κ are given in Sect. 6.2. The sum over κ is limited by the conditions $\kappa \leqslant 2l$, $\kappa \leqslant 2L$, κ being even. If $l \geqslant 1$, $L > 0$, the term with $\kappa = 2$ can be expressed through the atomic quadrupole moment.

3.1.2 Bethe Formula

By expanding $j_\kappa(qr)$ in the radial integral (3.1.14) in a power series, and neglecting all terms beyond the first nonvanishing term, we obtain

$$R_\kappa(q) \propto q^\kappa, \quad R_0(q) \propto q^2\,. \tag{3.1.23}$$

It is clear that at $\kappa = 1$ (i.e., $\Delta l = \pm 1$), the main contribution in the integral (3.1.13) is made just by the range of small q where this expansion is justified. Therefore we can get a simple approximation for the cross section. One should however substitute in the upper limit of the integral (3.1.13) some limited value q_0, since at large q the function $j_\kappa(qr)$ oscillates and this range can be neglected. The formula obtained in such a way contains

$$Q_1 l_{\max}\left|\int_0^\infty P_{l_0}P_l\,r\,dr\right|^2 \propto \frac{f_{a_0a}}{\Delta E}\,,$$

where f_{a_0a} is the oscillator strength for the transition $a_0 - a$[1]; $\Delta E = E_a - E_{a_0}$, the transition energy. As a result, we obtain the Bethe formula (which can be called the "dipole approximation"),

$$\sigma_{a_0a} \simeq \pi a_0^2 \cdot \frac{8}{k_0^2\Delta E}\cdot f_{a_0a}\ln\frac{q_0}{k_0-k}\,. \tag{3.1.24}$$

[1] About calculation of the oscillator strengths see [Ref. 3.4, Sects. 9.4, 7].

According to this formula, the cross section depends on q_0 only logarithmically. To estimate q_0 we should note that $R_\kappa(q)$ is determined by the range of $r \lesssim r_0$ with $r_0 \simeq |E_{a_0}|^{-1/2}$. Therefore

$$q_0 = \min(k_0 + k, \sqrt{|E_{a_0}|}) \tag{3.1.25}$$

For large energies the logarithmic term in (3.1.24) is

$$\ln \frac{q_0}{k_0 - k} \simeq \ln \frac{2qk_0}{\Delta E} \tag{3.1.26}$$

The Bethe formula (3.1.24) is applicable to dipole transitions only (in other words, to optically allowed transitions). Due to its simplicity it is useful for estimating cross sections, although its accuracy in some cases is considerably less than that of Born formulas (3.1.13) or (3.1.16, 17).

One can see from (3.1.23) and (3.1.13) that at $\kappa \neq 1$, the approximation (3.1.23) is not adequate since the value of $q^{-3}R_\kappa(q)$ becomes large at large q. In other words, the approximate result is too sensitive to the value of q_0. A detailed discussion of the Bethe approximation and related questions can be found in [3.5].

3.1.3 Brief Description of Born Cross Sections

Excitation cross sections for neutral atoms calculated by the Born formulas (3.1.16, 17) have a number of characteristic features. Some of them can be obtained by analysis of general formulas; others arise from the results of numerical calculations (Chap. 6) [3.6, 7]. We stress that here and everywhere in this section we consider only transitions without change of spin.

At high energies $\mathscr{E} \gg \Delta E$

$$\sigma_\kappa^{\mathrm{B}} \propto \frac{\ln \mathscr{E}}{\mathscr{E}} \quad \kappa = 1, \quad \Delta l = \pm 1 , \tag{3.1.27}$$

$$\sigma_\kappa^{\mathrm{B}} \propto \frac{1}{\mathscr{E}} \quad \kappa \neq 1 . \tag{3.1.28}$$

Below we often use the scaled energy

$$u = \frac{\mathscr{E}_0 - \Delta E}{DE} = \frac{\mathscr{E}}{DE} = \frac{k^2}{DE} . \tag{3.1.29}$$

For various atoms and transitions the Born cross sections reveal a similar behavior if the electron energy is expressed in threshold units

$$DE = \Delta E . \tag{3.1.30a}$$

These units are widely utilized in many calculations. However, they are certainly not appropriate for transitions between closely spaced levels when

$\Delta E \ll |E_0|$, E_0 being energy of the initial state counted from the ionization limit. Here the choice

$$DE = |E_0| \tag{3.1.30b}$$

is more adequate. If the same energy scale is necessary for different transitions in the ion X_z the scale

$$DE = z^2\,\mathrm{Ry} \tag{3.1.30c}$$

is more appropriate.

As an illustration (Fig. 3.1 a–c) excitation cross sections of some transitions in H and Na with ΔE from 0.3 to 10 eV are depicted in the scales *a*, *b* and *c* from (3.1.30). For convenience of comparison, all cross sections are normalized to the same maximum value. Usually cross sections reach a maximum value at $u \backsim 1$. For optically forbidden transitions the maximum is moved to smaller u. One can see from the Fig. 3.1 that this similarity is better with the scale $DE = |E_0|$.

As a rule, for transitions between levels with $\Delta n \geq 1$, the maximum value $\sigma_{\mathrm{m}}^{\mathrm{B}} = \sigma^{\mathrm{B}}(u_{\mathrm{m}}) \propto \Delta E^{-2}$. For transitions $n_0 - n$ with $n \gg n_0$, the cross section is proportional to n^{-3}. For optically allowed transitions, both rules can be written in the form

$$\sigma_{a_0a}^{\mathrm{B}}(u_m) \backsim \pi a_0^2 \frac{f_{a_0a}}{(\Delta E)^2}\,, \tag{3.1.31}$$

where f_{a_0a} is the oscillator strength.

The descriptions given above have to be modified in the case of small $\Delta E \ll |E_0|$, where E_0 is the energy of the initial atomic state. For such transitions, threshold units are not useful, and one is most interested in energies $\mathcal{E} \gtrsim |E_0| \gg \Delta E$. At these energies the cross section is almost independent of ΔE. Only for $\kappa = 1$ is there the weak logarithmic dependence.

It should be noted that all these features can be seen even in the Bethe formulas (3.1.24, 26):

$$\sigma_{a_0a}^{\mathrm{B}} \backsim \pi a_0^2 \cdot \frac{f_{a_0a}}{(\Delta E)^2} \cdot \frac{\ln Cu}{u+1}\,.$$

At large ΔE, this gives (3.1.31), and at small ΔE and $\mathcal{E} \gg \Delta E, \sigma^{\mathrm{B}} \propto \ln \mathcal{E}/\mathcal{E}$.

It must be emphasized that the formulas above can be used only for preliminary estimates of cross sections. These properties also have to be taken into account when constructing semiempirical formulas.

We have not mentioned above the problems of applicability of the Born approximation, and its agreement with experimental data. At large energies the Born cross section and the experimental one should agree. In fact, experimental results are often normalized to the calculated Born cross section at large energies (however, there is a question what energy value is sufficiently large). As an illustration we give here two examples: excitation of optically allowed $1s$–$2p$ and forbidden $1s$–$2s$ transitions in H atom. (Figs. 3.2 and 3) The Born approximation

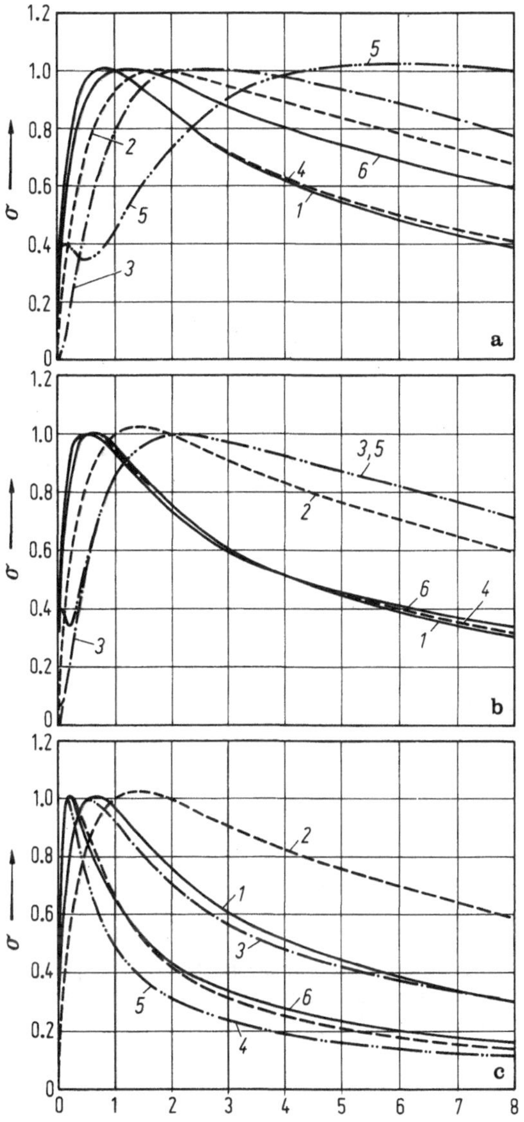

Fig. 3.1. Effective cross sections for transitions in atoms H (curves 1–5) and Na (curve 6): Transition $1s$–$2s$ (1); Transition $1s$–$2p$ (2); Transition $2s$–$5p$ (3); Transition $3s$–$4p$ (4); Transition $4s$–$5p$ (5); Transition $3s$–$3p$ in Na atom (6). (a) $DE = \Delta E$, (b) $DE = E_0$, (c) $DE = \mathrm{Ry}$

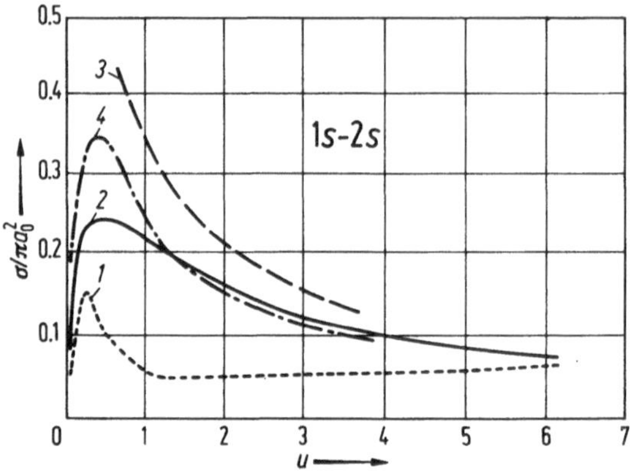

Fig. 3.2. Effective cross section for the transition $1s$–$2s$ of the H atom: (1) Experiment [3.6]; (2) Born approximation; (3) Distorted wave method without exchange; (4) Approximation of close coupling between three levels $1s$–$2s$–$2p$ with exchange

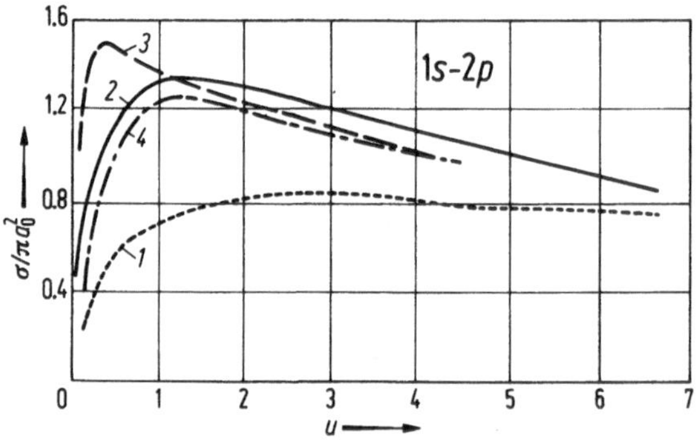

Fig. 3.3. Effective cross section for the transition $1s$–$2p$ of the H atom: (1) Experiment [3.6]; (2) Born approximation; (3) Distorted wave method without exchange; (4) Approximation of close coupling between three levels $1s$–$2s$–$2p$ with exchange

usually overestimates the cross sections up to a factor of two in the vicinity of the maximum, and the maximum is located too close to the threshold. Such a feature of the Born and the experimental cross sections is rather typical. There are two main reasons:

(i) The repulsion of optical and incident electrons is neglected in the Born approximation (atom polarization effect);

(ii) The Born cross section is increasing infinitely with increasing interaction $\langle a_0|U|a\rangle$, this results in S-matrix unitarity violation (normalization effect).

In the sections to follow this will be discussed in more detail. For a detailed comparison of calculated and experimental cross sections see [3.10].

3.1.4 Ionization and Three-Body Recombination

It is not difficult to generalize the Born formula to transitions for which the initial or final state of the atom belongs to the continuum. The transition of an atom from a discrete state to the continuum means ionization. The reverse process is called three-body recombination, since it implies the capture of an electron by an ion with simultaneous scattering of the second electron. The presence of the third particle is necessary to fulfill the conservation of energy and momentum.

The equation for the ionization cross section can be derived from (3.1.16, 17), if we assume that the state a belongs to the continuum: $a = a_\mathrm{i}\varepsilon lLSJ$, where a_i is the state of the ion, and ε is the energy of ejected electron. Usually we are not interested in the particular total angular momentum of the system: ion plus ejected electron, and the cross section should be summed over those quantum numbers. If the continuum radial function $P_{\varepsilon l}$ is normalized to $\delta(\varepsilon - \varepsilon')$, the differential ionization cross section is

$$d\sigma^\mathrm{i}(a_0,\, a_\mathrm{i}\; \varepsilon l) = \sum_\kappa Q_\kappa \, d\sigma^\mathrm{i}_\kappa(l_0,\, \varepsilon l)\,, \tag{3.1.32}$$

$$d\sigma^\mathrm{i}_\kappa(l_0,\, \varepsilon l) = \pi a_0^2 \cdot \frac{8}{k_0^2(2l_0+1)}\, d\varepsilon \int\limits_{k_0-k}^{k_0+k} [R_\kappa(q)]^2 \frac{dq}{q^3}\,. \tag{3.1.33}$$

The radial integral R_κ can be calculated according to (3.1.14) with the function $P_{\varepsilon l}$ in place of $P_l(r)$. For ionization from a shell l_0^m according to (3.1.18) we have, cf. (2.3.22),

$$\sum_{LSJ} Q_\kappa(a_0,\, a) = Q_\mathrm{i} = m\left(G^{L_0S_0}_{L_iS_i}\right)^2 \,. \tag{3.1.34}$$

Therefore Q_i does not depend on κ. If the state a_0 is described by definite genealogical scheme $a_0 = \gamma_p L_p S_p l_0 L_0 S_0 J_0$ then $Q_\mathrm{i} = 1$ (to be more exact, $\delta_{L_iL_p}\delta_{S_iS_p}$).

For applications, the total ionization cross section is required:

$$\sigma^\mathrm{i}_{a_0a_i} = Q_\mathrm{i} \sum_{l\kappa} \int\limits_0^{E_\mathrm{m}} \frac{d\sigma^\mathrm{i}(\varepsilon)}{d\varepsilon}\, d\varepsilon, \quad d\sigma^\mathrm{i}(\varepsilon) = d\sigma^\mathrm{i}(n_0l_0,\, l\varepsilon), \quad E_\mathrm{m} = \mathscr{E} - E_z \tag{3.1.35}$$

where E_z is the ionization energy of the atom X_z.

The cross section of the reverse process, namely, three-body recombination, is defined in a similar way, exchanging the initial and final states. The wave function of the electron in the continuum has to be normalized to unit flux instead of $\delta(\varepsilon - \varepsilon')$. Besides, the result has to be averaged over the directions of wave vector $\boldsymbol{k}$ of this electron, and over the states of the ion X_{z+1}. In this way we get the relation

$$2\pi^2 k_0^2 g_0 \frac{d\sigma^{\mathrm{i}}(a_0,\ a_{\mathrm{i}}\varepsilon l)}{d\varepsilon} = \varepsilon k^2 g_i \sigma^{\mathrm{r}}(a_{\mathrm{i}}\varepsilon l,\ a_0)\,. \tag{3.1.36}$$

As mentioned in Chap. 1, the dimensions of the differential ionization cross section $d\sigma^{\mathrm{i}}/d\varepsilon$ in the CGS system are $\mathrm{cm}^2\ \mathrm{erg}^{-1}$, while the total ionization cross section is measured in cm^2 and for three-body recombination σ^{r} has dimensions cm^4 s, so that $N_{\mathrm{e}}^2 vv'\sigma^{\mathrm{r}}$ has the dimensions of s^{-1}.

The structure of (3.1.32, 33) is similar to that of corresponding formulas for discrete spectra. The total ionization cross section (3.1.35) includes an additional summation over l, and integration over the energy of the ejected electron ε. This entails considerably more lengthy calculations as compared to the case of excitation cross section. In fact, up to ten terms in the sum over l should be taken into account. Several values of κ correspond generally to every value of l.

The first Born approximation is certainly applicable for large velocities of the incident electrons. However, in the case of ionization this condition should be discussed in more details. If the energy of the ejected electron is ε, the energy of the scattered one is $\mathscr{E}' = \mathscr{E} - E_z - \varepsilon$; it may be small if the ejected electron takes most of the energy $\mathscr{E} - E_z$. In other words, the range of small $\mathscr{E}'$ where the Born approximation does not work, can contribute to the integral (3.1.35) for any $\mathscr{E}$. Analysis and numerical calculations show that for large $\mathscr{E}$ and $\varepsilon \to \mathscr{E} - E_z$ $d\sigma^{\mathrm{B}}(\varepsilon)$ is small and the corresponding error is not considerable. If, however, an exchange is included, $d\sigma^{\mathrm{i}}(\varepsilon)$ become symmetrical with respect to ejected and scattered

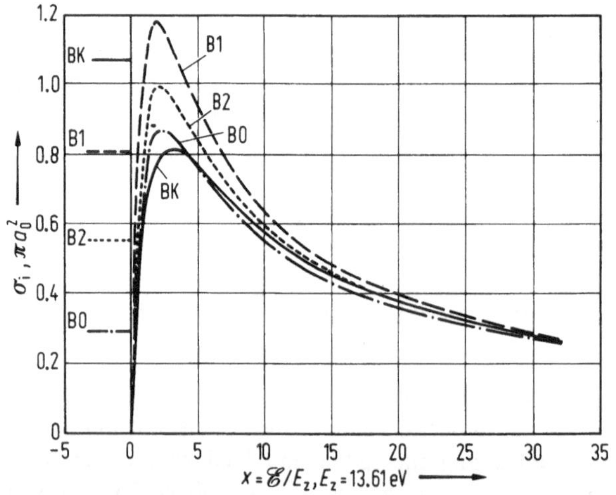

Fig. 3.4. Ionization cross section for 1s level of the hydrogen atom given in various approximations (Sect. 3.1.4)

electrons, i.e.,

$$d\sigma^{\mathrm{i}}(\varepsilon) = d\sigma^{\mathrm{i}}(\mathscr{E}'), \quad \mathscr{E}' = \mathscr{E} - E_z - \varepsilon \tag{3.1.37}$$

and the contribution of small $\mathscr{E}'$ may be considerable. On the other hand, from the (3.1.37) follows that the total cross section may be written [with an exact $d\sigma^{\mathrm{i}}(\varepsilon)$] as

$$\sigma^{\mathrm{i}} = \frac{1}{2}\int_0^{E_{\mathrm{m}}} \frac{d\sigma^{\mathrm{i}}(\varepsilon)}{d\varepsilon} d\varepsilon = \int_0^{E_{\mathrm{m}}/2} \frac{d\sigma^{\mathrm{i}}(\varepsilon)}{d\varepsilon} d\varepsilon \,.$$

For the Born approximation without exchange the two expressions are not equal and the second one is preferable since it excludes the range of slow scattered electron. In what follows we shall always use for the total ionization cross section the following definition

$$\sigma^{\mathrm{i}} = \int_0^{E_{\mathrm{m}}/2} \frac{d\sigma^{\mathrm{i}}(\varepsilon)}{d\varepsilon} d\varepsilon, \quad E_{\mathrm{m}} = \mathscr{E} - E_z \,. \tag{3.1.38}$$

A detailed and mathematically more accurate consideration of the symmetry of the ionization amplitude was done in [3.8, 9]. In particular, it was shown that general equations include the interference term with a phase choice which is rather uncertain. Usually this interference term disappears from the final expression for the total cross section.

In most cases (3.1.38) is in better agreement with the experimental data than (3.1.35). However, it should be noted that this agreement is partly occasional: one can see that (3.1.38) gives always smaller result than (3.1.35), but the Born approximation usually overestimates cross sections.

Generally speaking, the energy dependence of the ionization cross section is similar to that of the excitation cross section. We note that for ionization the two scales (3.1.30a, b) coincide: $DE = E_z$. In the sum over l there are terms with $\kappa = 1$ as well as $\kappa \neq 1$. At large energies they are $\propto \ln(u)/u$ and $1/u$, respectively. It is difficult to predict at what energy the logarithmic term will become important. Near the threshold $\sigma^{\mathrm{i}} \propto u^{3/2}$ in the Born approximation. A more accurate analysis which is beyond the scope of the Born approximation shows that, in fact, $\sigma^{\mathrm{i}} \propto u$ [3.9–11].

In Fig. 3.4 the ionization cross sections of the H atom in the ground state $1s$ are presented. The results ($B1$, $B2$) obtained using (3.1.35, 38) are shown together with the cross section (BK) recommended in the compilation [3.12]. The curve BO was obtained with exchange included by Ochkur method (see below).

3.2 Some Refinements of the Born Approximation

The Born method for calculating inelastic collision cross sections differs from all other approximate methods in its simplicity: it does not require a partialwave expansion. The calculation of cross sections in the Born approximation amounts to the numerical calculation of a few integrals.

At the same time, the Born method provides a sufficiently accurate qualitative description of cross sections, and often does not lead to large quantitative disagreements with experimental data. For many types of transitions from the ground state, an overestimate of the maximum cross section by a factor 1.5–2 is typical. For transitions between excited states, there is insufficiency of experimental data. However, for transitions between the excited levels of multiply charged ions the Born method should provide very reasonable accuracy of cross sections in the energy ranges that are of the most interest in practical applications.

In some cases, however, the Born approximation is totally inadequate. For example, it gives the zero result for intercombination transitions in LS coupling. It is therefore of interest to consider the possibility of refining the Born approximation by taking some evident physical effects into consideration. It should be emphasized that we mean only the possibility of including new types of transitions with the same typical accuracy, i.e., a factor of the order of 2 near the maximum cross section. We do not consider here any attempts to improve this accuracy.

The main physical effects mentioned above are the distortion of incident and scattered waves by the atomic field, the exchange, and normalization. All these refinements can be realized only in the partial wave representation (Sect. 2.2), so that the simpler q representation (Sect. 3.1) cannot be used. The following will be based on the first-order approximation for the solution of a general set of equations (Sect. 2.3). This approximation requires, however, some refinements and modifications to take into account properly the physical effects mentioned above. In the following sections we discuss these refinements.

3.2.1 Distortion of Incident and Scattered Waves; Excitation of Ions

The simplest refinement of the Born approximation is the distorted wave approximation which corresponds to the first-order approximation without exchange (Sect. 2.3; $R^e = 0$). In this case the functions $\bar{F}_\Gamma$ are the solutions of the radial equations for an electron in the attractive field U_Γ of the atom. At this point, however, we should distinguish between the effects of attraction of the electron by the short-range field of the neutral atom and the attraction by the long-range Coulomb field of the ion. In fact, the polarization of the atom due to repulsion of the outer and optical electrons overcomes the first effect, so that the average distance between electrons increases as compared to the case of free movement of the outer electron. For this reason, the Born approximation overestimates the cross section. Inclusion of the attractive field without regard for polarization effects would result in larger overestimation of the cross section at small energies. Allowance for the polarization effects (i.e., the repulsion of electrons) is possible only in second-order perturbation theory. From the above it is clear that the distorted-wave method is not appropriate for the calculation of excitation cross sections. This is confirmed by the results of numerical calculations.

In contrast to the attraction of electrons to neutral atoms, the Coulomb attraction to positive ions leads to an increase of the electron flux density in the vicinity of the ion owing to electrons with large initial values of the impact parameter. The effect of Coulomb attraction of electrons to the ion provides the qualitative change of the excitation cross section at threshold, namely,when $\mathscr{E}_0 = \Delta E, \sigma \neq 0$, in contrast to the case of the excitation of neutral atoms. For this reason one should not neglect the Coulomb interaction in calculations of ion excitation, at least in the vicinity of threshold.

The Coulomb–Born approximation refers to the method in which only the distortion by a Coulomb field with asymptotic value $z-1$ of the charge is taken into account. The cross section is defined by the formulas of the first-order approximation (Sect. 2.3) without exchange terms. In the integral R^{d} in (2.3.10) we have to substitute the Coulomb radial functions. We obtain

$$\sigma_{a_0 a} = \sum_\kappa Q_\kappa(a_0, a)\,\sigma_\kappa(l_0, l), \quad \sigma_\kappa(l_0, l) = \pi a_0^2 \frac{4}{k_0^2[l_0\kappa]^2} \sum_{\lambda_0\lambda} (R_\kappa^{\mathrm{d}})^2 \,, \tag{3.2.1}$$

$$\begin{aligned} R_\kappa^{\mathrm{d}} &= [l_0 l \lambda_0 \lambda] \begin{pmatrix} \kappa & l_0 & l \\ 0 & 0 & 0 \end{pmatrix} \begin{pmatrix} \kappa & \lambda_0 & \lambda \\ 0 & 0 & 0 \end{pmatrix} \\ &\times 2 \iint_0^\infty dr' dr'' F_{\lambda_0}(r') P_{l_0}(r'') \frac{r_<^\kappa}{r_>^{\kappa+1}} P_l(r'')\, F_\lambda(r') \,. \end{aligned} \tag{3.2.2}$$

where P_l are the radial functions of the optical electron of the ion, and F_λ are the functions of the outer electron which are defined by equation

$$\begin{aligned} &\left(\frac{d^2}{dr^2} - \frac{\lambda(\lambda+1)}{r^2} + \frac{2(z-1)}{r} + k^2 \right) F_\lambda = 0 \,, \\ &F_\lambda \underset{r\to\infty}{\simeq} k^{-1/2} \sin\left(kr - \frac{\lambda\pi}{2} + \frac{z-1}{k} \ln kr + \eta \right), \; F_\lambda(0) = 0 \,. \end{aligned} \tag{3.2.3}$$

The angular factors Q_κ are the same as in the Born approximation, and are defined by formulas of Sect. 2.3.

The sum over λ_0, λ in (3.2.1) converges very slowly. However, the value of R_κ^{d} tends to its Born limit at $\lambda_0, \lambda \gg 1$; thus the total cross section can be calculated by means of

$$\sigma_\kappa(l_0, l) = \sigma_\kappa^{\mathrm{B}}(l_0, l) + \frac{4\pi}{k_0^2[l_0\kappa]^2} \sum_{\lambda_0\lambda} [(R_\kappa^{\mathrm{d}})^2 - (R_\kappa^{\mathrm{B}})^2] \,. \tag{3.2.4}$$

Here R_κ^{B} is defined according to (3.2.2, 3) with $z=1$, and $\sigma_\kappa^{\mathrm{B}}$ is the Born cross section (3.1.17), which can be calculated without partial-wave expansion. The sum in (3.2.4) converges much better.

In spite of apparent similarity, the Coulomb–Born approximation and the distorted-wave method are radically different. At $z = 1$, the Coulomb–Born

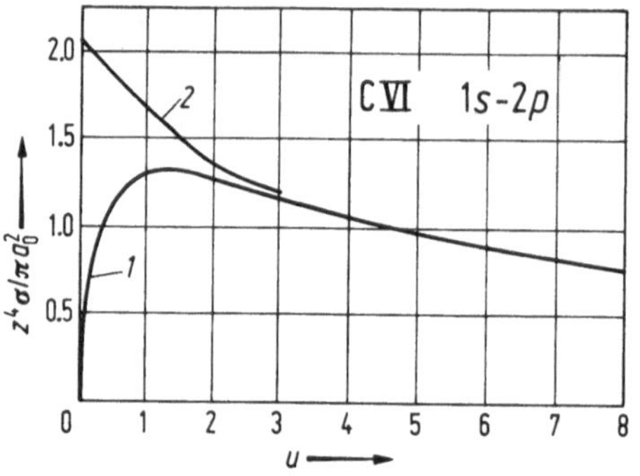

Fig. 3.5. Effective cross section for the transition $1s$–$2p$ in the ion C VI (C^{5+}): (1) Born approximation; (2) Coulomb–Born approximation; $u = (\mathscr{E} - \Delta E)/\Delta E$

approximation coincides with the Born approximation. By a simple substitution of the variables

$$zr \to r, \quad k/z \to k ,$$

one can show that the Coulomb–Born approximation is the first-order perturbation theory result with parameter $1/z$. Therefore the error of this approximation should decrease with increasing z.

In Fig. 3.5 the excitation cross sections are given for the hydrogenlike ion CVI calculated in the Born and Coulomb–Born approximations. One can see that the Coulomb–Born cross section, in contrast to the Born cross section, jumps at threshold to its maximum value. When the electron energy increases to $u \backsim 1-2$, the difference between Born and Coulomb–Born approximations vanishes. For a detailed discussion of electron impact excitation and ionization of positive ions see [3.13], and a recent review on experimental data [3.14].

3.2.2 Allowance for Exchange

a. Intercombination transitions ($\Delta S = 1$). The transitions with change of spin in the LS coupling approximation take place due to the exchange interaction. The corresponding generalization of the Born (or Coulomb–Born) approximation can be obtained from the first-order approximation formulas. We shall use the exchange term in (2.3.8), and substitute the free-electron (or Coulomb) wave functions into the radial integral $R^{e}_{\kappa''\kappa}$ of (2.3.11). This method is called the Born–Oppenheimer approximation. Numerical calculations imply, however, that the accuracy of the Born–Oppenheimer approximation is much poorer than that of the Born approximation for transitions with $\Delta S = 0$.

This failure is mostly due to nonorthogonality of the wave functions of the initial and final states of the system. One can demonstrate this nonorthogonality in the case of a hydrogenlike ion. The functions

$$\Phi_0(r_1, r_2) = P_{l_0}(r_1)F_{\lambda_0}(r_2), \quad \Phi(r_1, r_2) = P_l(r_2)F_\lambda(r_1)$$

are nonorthogonal because $P(r)$ is the eigenfunction for the field $-z/r$, and $F(r)$ is that for the field $-(z-1)/r$. The angular parts of the wave functions provide the orthogonality for the case $\lambda_0 \neq l$, $\lambda \neq l_0$.

The lack of orthogonality of the total wave functions in initial and final states implies that the Born–Oppenheimer approximation isn't the first-order approximation of a perturbation theory. To avoid this defect the method of orthogonalized functions was suggested [3.15]. In the first-order theory, the modification of the formulas of Sect. 2.3 consists simply in the substitution of the functions $\tilde{F}_\lambda$ instead of F_λ in the exchange radial integrals:

$$\begin{aligned} \tilde{F}_{\lambda_0} &= F_{\lambda_0} - \langle F_{\lambda_0} | P_l \rangle \cdot P_l \cdot \delta_{l\lambda_0}, \\ \tilde{F}_\lambda &= F_\lambda - \langle F_\lambda | P_{l_0} \rangle \cdot P_{l_0} \cdot \delta_{l_0\lambda}. \end{aligned} \tag{3.2.5}$$

Then using the formulas of Sect. 2.3 for transitions with $\Delta S = 1$, we obtain

$$\sigma_{a_0a} = \sum_\kappa Q''_\kappa(a_0, a)\,\sigma''_\kappa(l_0, l), \tag{3.2.6}$$

$$\sigma''_\kappa(l_0, l) = \pi a_0^2 \frac{4}{k_0^2[l_0\kappa]^2} \sum_{\lambda_0\lambda} (R''_\kappa)^2$$

$$\begin{aligned} R''_\kappa = &\sum_{\kappa''} (-1)^{\kappa+\kappa''} [\kappa\kappa'']^2 \begin{Bmatrix} \kappa & l_0 & l \\ \kappa'' & \lambda_0 & \lambda \end{Bmatrix} [l_0 l \lambda_0 \lambda] \begin{pmatrix} l_0 & \kappa'' & \lambda \\ 0 & 0 & 0 \end{pmatrix} \begin{pmatrix} \lambda_0 & \kappa'' & l \\ 0 & 0 & 0 \end{pmatrix} \cdot \\ &2\int_0^\infty\int_0^\infty \tilde{F}_{\Gamma_0}(r') P_{l_0}(r'') \frac{r_<^\kappa}{r_>^{\kappa+1}} [1 - c r_> \delta(\kappa'', 0)]\, \tilde{F}_\Gamma(r'') P_l(r')\, dr'' dr'. \end{aligned} \tag{3.2.7}$$

The angular factors Q''_κ are the same as defined in the Sect. 2.3. Numerical calculations show that the accuracy of the orthogonalized function method for intercombination transitions is about the same as that of the Born approximation for allowed transitions. The method is applicable to neutral atoms as well as to ions. Below we shall use the designations B, CB and CBE (or BE for $z = 1$) for the Born, Coulomb–Born and orthogonalized functions approximations. With increasing z the role of the orthogonalization is decreasing (P and F functions are in almost the same field) and CBE method gives the same results as the CB–Oppenheimer approximation. For $z = 1$ the contributions from the two terms in (3.2.5) are much larger than their difference. In that case the simple subtraction procedure defined by (3.2.5) may be inadequate; a somewhat better one is to omit the contributions from $\lambda_0 = l$ and $\lambda = l_0$.

Another modification of the Born–Oppenheimer approximation was proposed for neutral atoms by *Ochkur* [3.16]. The *Ochkur* approximation is often used in practical calculations due to its simplicity, although the range of its applicability is more limited. The basic assumption of this approximation is as follows: in the asymptotic expansion of the exchange-scattering amplitude in powers of $1/k$, only the first nonvanishing term has physical meaning at small or medium energies. If we neglect all the other terms, the exchange cross section can be calculated in the simpler q representation in a way similar to that for direct cross sections [cf. (3.1.16, 17)]. The result for the transition between the atomic terms L_0S_0 and LS with $\Delta S = 1$ is

$$\sigma_{a_0a} = \sum_\kappa Q''_\kappa(a_0, a)\,\tilde{\sigma}_\kappa(l_0, l)\,,$$

$$\tilde{\sigma}_\kappa(l_0, l) = \pi a_0^2 \cdot \frac{8}{k_0^2(2l_0+1)} \int_{k_0-k}^{k_0+k} [\tilde{R}_\kappa(q)]^2 \,\frac{dq}{q^3}\,, \tag{3.2.8}$$

$$\tilde{R}_\kappa(q) = \frac{q^2}{k_0^2}[\kappa l_0 l] \begin{pmatrix} \kappa & l_0 & l \\ 0 & 0 & 0 \end{pmatrix} \int_0^\infty P_{l_0}(r) P_l(r) j_\kappa(qr)\, dr\,. \tag{3.2.9}$$

In contrast to the orthogonalized functions and the Born–Oppenheimer approximations the Ochkur method does not require a partial wave representation – results can be given in a q-representation in similar way to the Born cross section. This much simplifies its applications. However until now, no generalization of this approach was proposed for ions. It is easy to see from (3.2.8) and (3.1.14) that for large energies $\mathscr{E}$ the cross section of the exchange transition decreases as $\mathscr{E}^{-3}$ as compared with $\mathscr{E}^{-1}$ or $\mathscr{E}^{-1}\ln\mathscr{E}$. Asymptotically (at large $\mathscr{E}$), the Born–Oppenheimer approximation and its two modifications considered above coincide. Usually, however, this asymptotic agreement is achieved only for very large values of $\mathscr{E}$, when the cross section is too small.

b. Allowed Transitions with $\Delta S = 0$. The transitions with $\Delta S = 0$ are connected mainly with direct interaction, and the correction due to exchange is usually comparatively small. In some cases (in particular, for ions) the exchange correction proves to be more important because of a considerable compensation of the direct and interference terms. For $\Delta l = \pm 1$, this compensation can occur at threshold when $\Delta E \simeq |E_0|$ (E_0 being the initial energy of the atom). The typical example (and important one) is the excitation of singlet states of [He] ions[1] from the ground state. For $|\Delta l| \neq 1$, the influence of exchange can be even more impor tant because the exchange term can exceed the direct term at threshold.

To calculate the cross section taking exchange into account, we should use (2.3.8) with both the direct and exchange terms. According to the above discussion we shall use in the radial integral R^{d}_κ the Coulomb wave functions defined by (3.2.3), and in the exchange radial integral $R^{\mathrm{e}}_{\kappa''\kappa'}$, we shall substitute the orthogonalized functions $\tilde{F}_\lambda$ defined by (3.2.5). Of course we cannot consider this method as quite consistent, but it provides good results for medium values

[1] Here and everywhere we denote an ion of the isoelectronic sequence of atom A by [A].

of z (for large z, $\tilde{F}_\lambda$ coincides with F_λ). The cross section is determined by

$$\sigma_{a_0a} = \sigma'_{a_0a} + \sigma''_{a_0a} = \sum_\kappa [Q'_\kappa(a_0, a)\sigma'_\kappa(l_0, l) + Q''_\kappa(a_0, a)\sigma''_\kappa(l_0, l)], \quad (3.2.10)$$

$$\begin{aligned} \sigma'_\kappa(l_0, l) &= \pi a_0^2 \frac{4}{k_0^2[l_0\kappa]^2} \sum_{\lambda_0\lambda} R'_\kappa(R'_\kappa - R''_\kappa) \\ \sigma''_\kappa(l_0, l) &= \pi a_0^2 \frac{4}{k_0^2[l_0\kappa]^2} \sum_{\lambda_0\lambda} (R''_\kappa)^2 \end{aligned} \quad (3.2.11)$$

where R^d_κ and R''_κ are defined by (3.2.2, 7), and formulas for Q'_κ and Q''_κ are given in Sect. 2.3.2.

As mentioned above, the exchange is important mainly due to compensation of the direct and interference terms in σ'. We shall consider as an example the transition between states LS (without consideration of the fine structure). According to (2.3.19), the cross section (3.2.10) can be written in the form

$$\begin{aligned} \sigma_{a_0a} &= \pi a_0^2 \cdot \frac{4}{k_0^2(2l_0+1)} \cdot \sum_{\lambda_0\lambda\kappa} Q_\kappa(L_0, L)\left(f^2 - fg + \frac{2S+1}{2(2S_p+1)}g^2\right), \\ f &= R^\mathrm{d}_\kappa, \quad g = \sum_{\kappa''} R^\mathrm{e}_{\kappa''\kappa}. \end{aligned} \quad (3.2.12)$$

Usually for $\Delta l = \pm 1$, $g \lesssim f$. The exchange becomes important if the exchange amplitude g is of the order of the direct amplitude f, and at the same time the spin factor in (3.2.12) is small. For example, for the transition $1s^2\,{}^1S - 1s\,2p\,{}^1P$ in a [He] ion at threshold $g \simeq 3/4f, (2S+1)/2(2S_p+1) = 1/4$. Therefore, the expression in brackets in (3.2.12) is equal to $0.4f^2$, compared with the value f^2 without exchange.

For $\kappa > 1$ (i.e., $\Delta l \neq \pm 1$), g can exceed f at threshold due to exchange interactions with $\kappa'' = 0$ and 1 in (3.2.11). For this reason the role of exchange for such transitions can be even more important than for $\Delta l = \pm 1$, and even for $\Delta E \ll |E_0|$ exchange can be important.

In any case for $\mathcal{E} \gg \Delta E$, the exchange can be neglected because of the decrease $\propto \mathcal{E}^{-3}$ in the exchange cross section.

In Fig. 3.6 the cross sections of excitation of $2\,{}^1P$ and $2\,{}^3P$ levels in O VII are given. At the threshold, the two cross sections are almost equal. This fact is confirmed by experimental observation of X-ray spectra of laser plasma and the solar corona. The calculation of $\sigma(2\,{}^1P)$ without exchange gives for the ratio $\sigma(2\,{}^1P)/\sigma(2\,{}^3P) > 2$, which contradicts the experimental data.

It is worth noting that if we sum the cross sections over S the expression in brackets of (3.2.12) would take the form

$$f^2 - gf + g^2$$

which is insensitive to exchange if only $g \lesssim f$ (Fig. 3.6). For example for [H] ions (only one value of S), the exchange can be neglected.

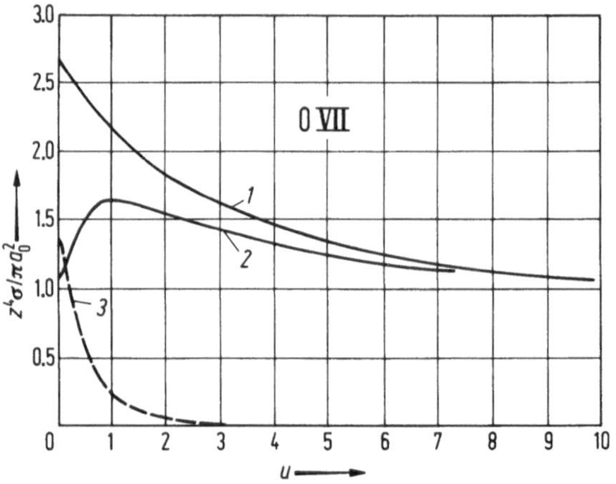

Fig. 3.6. Effective excitation cross-sections for heliumlike ion O VII: (1) Coulomb–Born approximation for the transition $1s^2\,{}^1S$–$1s2p\,{}^1P$; (2) Coulomb–Born approximation with exchange for the transition $1s^2\,{}^1S$–$1s2p\,{}^1P$; (3) Exchange excitation cross section for the transition $1s^2\,{}^1S - 1s2p\,{}^3P$; $u = (\mathscr{E} - \Delta E)/\Delta E$

3.2.3 Normalization

The partial excitation cross section must satisfy certain inequalities arising from the conservation of particle number similar to the inequalities (2.1.38). According to (2.2.24) and the unitarity condition of the S matrix (2.2.26), we have

$$\sigma(\Gamma_0, \Gamma) \le \frac{\pi(2S_T+1)(2L_T+1)}{2(2S_0+1)(2L_0+1)k_0^2} \equiv \sigma^{\lim}(\Gamma_0, \Gamma)\,,$$
$$\sigma(a_0\lambda_0, a\lambda) \le \frac{\pi}{k_0^2}(2\lambda_0+1)\,. \tag{3.2.13}$$

The Born approximation in the general case does not ensure the fulfillment of conditions (3.2.13). For sufficiently strong interaction, the partial cross section $\sigma(\Gamma_0, \Gamma)$ can exceed the limiting value $\sigma^{\lim}(\Gamma_0, \Gamma)$. As a consequence the total Born cross section in some cases is too large. Below, this situation will be referred to as "normalization failure." A Born cross section corrected with regard for the inequalities (3.2.13) by means of any procedure will be called a normalized cross section and denoted by $\bar{\sigma}$.[1]

We consider here the normalization method based on the use of the K matrix. According to the definition (2.2.27) of the K matrix, the matrix relation

$$S = \frac{I + \mathrm{i}K}{I - \mathrm{i}K} \tag{3.2.14}$$

[1] Particle number conservation (or probability conservation) requires the unitarity of scattering matrix which imposes stricter limitations than inequalities (3.2.13).

ensures the unitarity of the S matrix, independent of the approximation used in calculation of the K matrix. In (3.2.14) I denotes the unit matrix. For the purposes of normalization it is sufficient to take into account those matrix elements which include the initial state Γ_0, i.e., we assume

$$K = \begin{pmatrix} 0 & K_{\Gamma_0\Gamma_1} & K_{\Gamma_0\Gamma_2} & \cdots \\ K_{\Gamma_1\Gamma_0} & 0 & 0 & \cdots \\ K_{\Gamma_2\Gamma_0} & 0 & 0 & \cdots \\ \cdots & \cdots & \cdots & \cdots \end{pmatrix} . \tag{3.2.15}$$

The K matrix method has been applied by Seaton [3.17] to ensure the unitarity condition. However, in [3.17] the full K matrix has been used instead of (3.2.15) and in consequence some other effects were admixed (cf. discussion in Sect. 3.3.2). After substitution of definition (3.2.14) in (2.2.24) we obtain the following expressions for the normalized cross section:

$$\bar{\sigma}_{a_0a} = \sum_{\lambda_0\lambda L_T S_T} \bar{\sigma}(\Gamma_0, \Gamma) , \quad \bar{\sigma}(\Gamma_0, \Gamma) = \frac{\sigma(\Gamma_0, \Gamma)}{[1 + \sum_{a'\lambda'} K^2_{\Gamma'\Gamma_0}]^2} . \tag{3.2.16}$$

In (3.2.16), $\Gamma = \alpha LS\lambda\frac{1}{2}L_T S_T$. In the case of transitions between levels LSJ, $\Gamma = \alpha LSJ, \lambda\frac{1}{2}(j)J_T$ and the summation in (3.2.16) should be over $\lambda_0\lambda j_0 jJ_T$.

It is easy to show that in the first-order approximation,

$$K^2_{\Gamma'\Gamma_0} = \frac{1}{4}\frac{\sigma(\Gamma_0, \Gamma')}{\sigma^{\text{lim}}(\Gamma_0, \Gamma')} . \tag{3.2.17}$$

We note that in the denominator of the second of equations (3.2.16), we should not sum over $L_T S_T$. That means independent normalization of each $L_T S_T$ channel. We cannot perform the summation over $L_T S_T$ in (3.2.16) explicitly due to the nonlinear relation of $\bar{\sigma}(\Gamma_0, \Gamma)$ with $\sigma(\Gamma_0, \Gamma)$. For this reason, instead of the formulas of Sect. 2.3.2 we have to use a general expression (3.2.16) and (2.3.27, 28) for $\sigma(\Gamma_0, \Gamma)$:

$$\sigma(\Gamma_0, \Gamma) = \pi a_0^2 \frac{2}{k_0^2}\frac{[S_T L_T]^2}{[S_0 L_0]^2}\left[\sum_\kappa A_\kappa(R'_\kappa - B_{S_0 S} R''_\kappa)\right]^2 \tag{3.2.18}$$

The calculations with (3.2.16–18) are much more complicated than those with the usual formulas of the first-order approximation (Sect. 2.3.2). Recalling the approximate nature of the K matrix normalization procedure we simplify (3:2.16) by substituting in the denominator the average value of $K^2_{\Gamma'\Gamma_0}$:

$$\begin{aligned} K^2_{\lambda'\lambda_0} &= \sum_{L_T S_T} \frac{(2L_T + 1)(2S_T + 1)}{(2L_0 + 1)(2S_0 + 1)2(2\lambda_0 + 1)} K^2_{\Gamma'\Gamma_0} \\ &= \sum_{L_T S_T} \frac{k_0^2}{4\pi(2\lambda_0 + 1)} \sigma(\Gamma_0, \Gamma') = \frac{k_0^2}{4\pi(2\lambda_0 + 1)} \sigma(a_0\lambda_0, a'\lambda') . \end{aligned} \tag{3.2.19}$$

With this simplification we can perform the explicit summation over $L_T S_T$ in (3.2.16). The values of $\sigma(a_0\lambda_0,\ a\lambda)$ can be calculated by the formulas of Sect. 2.3.2:

$$\sigma(a_0\lambda_0,\ a'\lambda') = \sum_\kappa [Q'_\kappa \sigma'_\kappa(l_0\lambda_0,\ l'\lambda') + Q''_\kappa \sigma''_\kappa(l_0\lambda_0,\ l'\lambda')], \tag{3.2.20}$$

where $\sigma'_\kappa(l_0\lambda_0,\ l'\lambda')$ and $\sigma''_\kappa(l_0\lambda_0,\ l'\lambda')$ are partial cross sections from the sums (2.3.9).

From (2.3.22, 23) one can see that the sum over a' does not depend on the total momenta L_0, L'. Therefore, we can finally write (3.2.16) in the form

$$\bar{\sigma}(a_0, a) = \sum_{\lambda_0\lambda} \sigma(a_0\lambda_0, a\lambda)/D,$$

$$D = \left[1 + \frac{k_0^2}{4\pi(2\lambda_0+1)} \sum_{n'l'\lambda'\kappa} \sigma_\kappa(l_0\lambda_0, l\lambda)\right]^2 . \tag{3.2.21}$$

We also note that for a calculation of the exchange part of $\sum_\kappa$ it is better to use (2.3.26).

The sum over λ' in denominator of (3.2.21) contains a few terms $\backsim \Delta l = |l_0 - l|$. The sum over atomic states $n'l'$ is infinite. Usually however only one or two states $n'l'$ are important. It has to be emphasized that in many cases this level does not coincide with the final state of the transition. For example, in the case of the $3s$–$4p$ transition in the Na atom, the normalization is determined by the $3p$ level, i.e., by the resonance transition $3s - 3p$ in the sum over $n'l'$. In other words, normalization can be important even in the case of fulfillment of the conditions (3.2.13) for a given transition. In fact it is necessary to fulfill the condition (3.2.13) for all the transitions from a given initial state a_0.

One can infer from (3.2.21) that normalization by means of the K matrix provides a normalized cross section $\bar{\sigma} \ll \sigma^{\mathrm{lim}}$ when $\sigma \gg \sigma^{\mathrm{lim}}$.

The most consistent method of obtaining normalized cross sections is the accurate solution of the set of equations (2.2.30) with regard to close coupling of the states which are important for normalization (Sect. 3.3). The unitarity of the S matrix in this case is ensured if only the interaction matrix U is Hermitian. It is necessary, however, for the diagonal potentials $U_{\Gamma_0\Gamma_0}$ and $U_{\Gamma\Gamma}$ to be omitted in (2.2.30); these terms provide the distortion of the incident and scattered waves. The last effect is not related to the problem of normalization, and can be a source of inaccuracies in the cross sections (Sect. 3.2.2).

The results of numerical calculations by this method are in good agreement with those of the K matrix method. We note that when $\sigma \gg \sigma^{\mathrm{lim}}$, the close coupling method decreases the cross section $\bar{\sigma}$ compared with σ^{lim} even more than the K matrix method does.

In Figs. 3.7, 3.8, cross sections of the $3s - 3p$ and $3s$–$4p$ transitions in the Na atom are given.

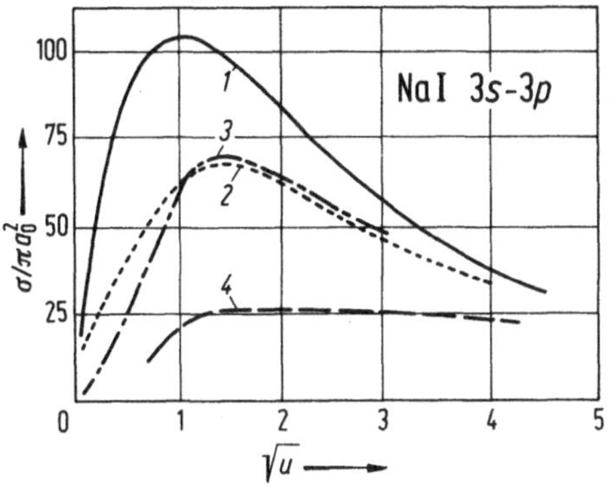

Fig. 3.7. Effective cross section for the transition $3s$–$3p$ of the Na atom: (1) Born approximation; (2) Cross section normalized with the use of the K matrix; (3) Close-coupling approximation; (4) Experiment [3.18]; $u = (\mathcal{E} - \Delta E)/\Delta E$

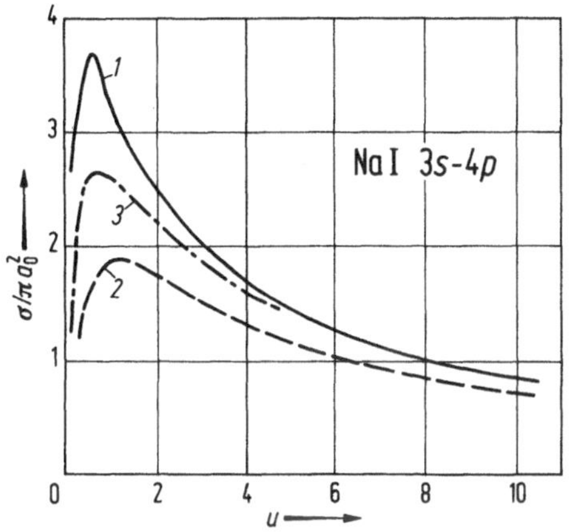

Fig. 3.8. Effective cross section for the transition $3s$–$4p$ of the Na atom: (1) Born approximation; (2) Cross section normalized with the use of K matrix; (3) Close-coupling method; $u = (\mathcal{E} - \Delta E)/\Delta E$

3.2.4 Concluding Remarks: Generalized Born Approximation

The Born approximation with the refinements discussed in the above sections provides excitation cross sections with accuracy sufficient for most applications. Usually the Born approximation gives a result which overestimates the cross section in the vicinity of the maximum by a factor of 1.5–2, and becomes more accurate with increasing energy. A typical example has been given above in Figs. 3.2 and 3.3 for excitation of resonance levels of the hydrogen atom.

According to recent measurements and also astrophysical data, similar accuracy is provided by the Coulomb–Born approximation in the case of excitation of ions. Moreover, the accuracy of Coulomb–Born approximation increases with increasing z.[1] To obtain sufficiently accurate results in some cases, the exchange has to be taken into account. The typical example is the excitation of [He] ions from the ground state.

The transitions with change of spin are fully connected with the exchange interaction. The method of orthogonalized functions outlined in Sect. 3.2.3 provides for these transitions in ions accuracy comparable with the characteristic accuracy of the Born approximation for transitions without spin change. However for neutral atoms, the accuracy of the orthogonalized-function method can be poorer.

At least for transitions between levels with small energy difference ΔE, the normalization procedure is necessary to provide similar accuracy.

The Born approximation taking into account the effects of the Coulomb field, exchange, and normalization will be called the generalized Born approximation (GBA). The range of applicability of the GBA is much broader than usual Born approximation. At large energies, $\mathcal{E} \gg \Delta E$ both methods give the same results. We note that the GBA does not include transitions through a virtual level (Sect. 3.3). The calculations in the framework of the GBA are lengthier than for the Born approximation. The computer program for these calculations isn't too complicated, however, and does not require much computer time. The results of such calculations will be presented in Chap. 6.

In this section we have discussed the calculation of excitation cross sections. In the case of the ionization cross section, only Coulomb attraction is important. Moreover the influence of the Coulomb field does not result in any qualitative effect similar to that in the case of excitation. More accurate analysis shows that the behavior of the ionization cross section near threshold is the same for ions and neutral atoms: $\sigma_i \propto (\mathcal{E}_0 - \Delta E)$.

3.3 More Accurate Methods of Calculation of Excitation Cross Sections

In two previous sections we discussed methods based on the first-order approximation. To obtain more accurate results one should take into consideration higher-order terms of the perturbation theory or find the exact (numerical) solution of the equation set (2.2.30) with a finite number of equations. Computations of this kind are very complicated and of little use for the systematic calculations of cross

[1] The important exception to this rule are the near-threshold resonances connected with the long-range Coulomb interaction (Sect. 3.4).

sections which are necessary for applications. Analysis of the results of some individual calculations reveals poor convergence of both methods with respect to the number of states that were included.

One of the reasons for such a poor convergence is based on the opposite signs of corrections due to different terms in the equation set or in the series. One possible example is the mutual compensation of the effects of electron attraction to a neutral atom and polarization of the atom in respect to inelastic scattering (Sect. 3.2.2).

One can distinguish two types of refinement of the results. Firstly, there are quantitative refinements in which the characteristic inaccuracy of the generalized Born approximation (GBA) is decreased or totally removed. The second type applies to cases where the GBA is for some reason inadequate, and the inaccuracy is much greater than that characteristic of the method. In the latter case the need for more accurate methods is evident. However instead of simply increasing the number of terms or equations it is advisable to add only those terms which overcome the deficiency of the GBA.

In the rest of this section we discuss some of the problems mentioned above. The treatment is rather illustrative in places, because a complete account of the problem is beyond the scope of this book. We do not discuss here such methods as the Glauber approximation, impulse approximation, free electron scattering model, and variational methods. For the first two of these methods we refer to [3.2, 3, 19]. The free electron scattering model is discussed in [3.20]. About variational methods see [3.21].

3.3.1 Transitions Via Virtual States

The Born approximation for excitation cross sections corresponds to the first term of a series in perturbation theory. The higher-order terms contain infinite sums over all possible virtual states. Generally we cannot select a few terms from such a sum because of its poor convergence (see above). There are however cases in which the first term of the series in perturbation theory is small or even vanishing for some reason. Then it is necessary to treat the first non-vanishing term. Two-electron excitation or ionization with a simultaneous excitation of the ion are examples of this type of transition. If the one-electron wave functions in the initial and final states are supposed to be orthogonal, the transition matrix element in the first Born approximation is zero. A non-vanishing cross section in the first order can be obtained only with slight nonorthogonality of the one-electron atomic wave functions before and after the transition. In second order, such a transition occurs through virtual levels corresponding to the excitation of one electron. The number of such virtual levels is not more than two. The orthogonality of wave functions in this case is of no importance, and the cross section in second order can prove to be considerably larger than in the first Born approximation.

A similar situation can arise for some one-electron transitions. We consider as an example quadrupole transitions (for example, $s-d$). In the Born approximation, the cross section of such transitions is much smaller than that of dipole transitions. In second order, the transition can be accomplished through the virtual level as a "sequence" of two dipole transitions. The matrix element of such a transition often exceeds the first-order matrix element of the quadrupole transition. Usually one should consider only one virtual level. For example, in the case of the $ns-n'd$ transition, this level is np, i.e., in second-order, we have the transition $ns-np-n'd$. The choice of virtual level is based on the high probability (around unity) of the excitation $ns-np$.

For calculation of cross sections in second order we can use (2.2.52–54):

$$T^{\mathrm{I+II}}_{\Gamma\Gamma_0} = -\int_0^\infty F_\lambda U_{\Gamma\Gamma_0} F_{\lambda_0} dr - \sum_{a_1\lambda_1} \int\int_0^\infty F_\lambda(r) U_{\Gamma\Gamma_1}(r) G_{\Gamma_1}(r,r') U_{\Gamma_1\Gamma_0}(r') F_{\lambda_0}(r')\, dr\, dr' , \tag{3.3.1}$$

where the orthogonalized functions $\tilde{F}_\lambda$ (3.2.5) should be substituted in the exchange integrals. The Green's function G_{Γ_1} for a neutral atom in accordance with (2.2.42) and (2.2.48) is

$$G_{\Gamma_1}(r,r') = -ik_1 rr' j_{\lambda_1}(k_1 r_<) h^{(1)}_{\lambda_1}(k_1 r_>) , \quad k_1^2 = \mathscr{E}_0 - E_{a_1 a_0} . \tag{3.3.2}$$

In the sum over a_1 we can consider only a few terms. Usually one state a_1 which is most strongly coupled with the initial state a_0, and the final state a is sufficient.

In Sect. 3.1 we expressed the Born cross section in the q representation without a partial-wave expansion. In second order we cannot derive an analogous expression. It is possible however to obtain the approximate formula in the q representation. We begin with the three-dimensional expression for the excitation cross section, cf. (3.1.4),

$$\sigma_{a_0 a} = \frac{1}{8\pi g_0 k_0^2} \int_{k_0-k}^{k_0+k} q\,dq \sum_{M_0 M} \Big| \sum_{a_1} \int [\delta_{a_1 a} + W_{aa_1}(\boldsymbol{r})] U_{a_1 a_0}(\boldsymbol{r}) \exp(-i\boldsymbol{q}\cdot\boldsymbol{r})\, d\boldsymbol{r} \Big|^2 , \tag{3.3.3}$$

where M_0, M are total angular momentum projections, $\boldsymbol{q} = \boldsymbol{k}_0 - \boldsymbol{k}$, and

$$U(\boldsymbol{r}) = \sum_j \left(\frac{2}{|\boldsymbol{r}-\boldsymbol{r}_j|} - \frac{2}{r} \right) , \quad W(\boldsymbol{r}) = \int U(\boldsymbol{r}-\boldsymbol{x}) G_1(\boldsymbol{x}) \exp(i\boldsymbol{k}\cdot\boldsymbol{x})\, d\boldsymbol{x} . \tag{3.3.4}$$

Here G_1 is the three-dimensional Green's function. Below we assume that G_1 is the Green's function of a free electron (even for ions):

$$G_1(\boldsymbol{x}) = -\frac{1}{4\pi x}\exp(\mathrm{i}k_1 x)\,, \quad k_1^2 + E_{a_1 a_0} = k^2 + E_{aa_0} = k_0^2 \tag{3.3.5}$$

In spite of this approximation, the calculation of W_{aa_1} is very complicated. Only in the case of $k \to 0$ can we obtain a simpler expression for W. We assume that the level a_1 is situated lower than the final level a, and consider at first the interelectron interaction

$$U = \frac{2}{|\boldsymbol{r} - \boldsymbol{r}_1|} \equiv \frac{2}{\rho}\,.$$

By substituting (3.3.5) in (3.3.4) and integrating over $\boldsymbol{x}$ we obtain for $k = 0$

$$W = \frac{2}{E_{aa_1}\rho}[1 - \exp(\mathrm{i}\sqrt{E_{aa_1}}\rho)]\,, \qquad k \to 0\,. \tag{3.3.6}$$

In the opposite case of large k one can show that W is imaginary and proportional to k^{-1}. The simplest interpolation formula which satisfies both this condition and (3.3.6) is

$$W(\rho) = 2\left(\frac{1 - \cos(\sqrt{E_{aa_1}}\rho)}{k_1^2\rho} - \mathrm{i}\frac{\sin(\sqrt{E_{aa_1}}\rho)}{k_1\sqrt{E_{aa_1}}\rho}\right)\,, \tag{3.3.6a}$$

$$k_1^2 = k^2 + E_{aa_1}\,, \quad \rho = |\boldsymbol{r} - \boldsymbol{r}_1|\,.$$

For a many-electron atom, taking into account the interaction with the nucleus, we must replace W by

$$\sum_j [W(\rho_j) - W(r)]\,. \tag{3.3.7}$$

On substituting (3.3.6a, 7) in (3.3.3) we obtain an approximate expression for the excitation cross section in the second Born approximation in the q representation.

For numerical calculations it is necessary to separate the radial and angular variables by means of multipole expansion of the potentials (3.3.4) and (3.3.6a). In spite of the more complicated expression of $W(\rho)$, as compared to $U(\rho)$, its expansion is not very difficult. If we confine ourselves to excitation transitions without change of the spin, we obtain

First order: $\quad n_0 l_0^m L_0 S - n_0 l_0^{m-1}[L_p S_p]\, nl\, LS$
Second order: $\quad n_0 l_0^m L_0 S - n_0 l_0^{m-1}[L_p S_p]\, n_1 l_1 L_1 S - n_0 l_0^{m-1}[L_p S_p]\, nl\, LS$

$$\sigma_{a_0 a}^{\mathrm{I+II}} = \pi a_0^2 \cdot \sum_\kappa \frac{2}{k_0^2} C_\kappa \int_{k_0-k}^{k_0+k} q\,dq \left| \int_0^\infty j_\kappa(qr)(A^{\mathrm{I}} + A^{\mathrm{II}})\, r^2 dr \right|^2\,, \tag{3.3.8}$$

$$A^{\mathrm{I}} = (l\|K_\kappa\|l_0)\,,$$

$$A^{\mathrm{II}} = \sum_{\kappa'\kappa''} (2\kappa + 1) \begin{pmatrix} \kappa' & \kappa'' & \kappa \\ 0 & 0 & 0 \end{pmatrix} \begin{Bmatrix} \kappa' & \kappa'' & \kappa \\ l_0 & l & l_1 \end{Bmatrix} (l\|W_{\kappa'}\|l_1)(l_1\|K_{\kappa''}\|l_0)\,, \tag{3.3.9}$$

where

$$C_\kappa = \frac{2L+1}{2\kappa+1} \begin{Bmatrix} L_0 & L & \kappa \\ l & l_0 & L_p \end{Bmatrix}^2 , \tag{3.3.10a}$$

$$(l\|T_\kappa\|l') = (-1)^l [ll'] \begin{pmatrix} l & l' & \kappa \\ 0 & 0 & 0 \end{pmatrix} \times \int_0^\infty P_l(r') P_{l'}(r') T_\kappa(r, r')\, dr' . \tag{3.3.10b}$$

The radial operators K_κ and W_κ are defined by the following formulas:

$$\begin{aligned} K_\kappa(r, r') &= 2\frac{r_<^\kappa}{r_>^{\kappa+1}} - \frac{2}{r}\delta_{\kappa o} , \\ W_\kappa(r, r') &= \frac{1}{k_1^2} K_\kappa(r, r') + 2\frac{\sqrt{E_{aa_1}}}{k_1^2}\omega'_\kappa - \frac{2i}{k_1}\omega''_\kappa , \\ \omega'_\kappa(r, r') &= (2\kappa+1) j_\kappa(x_<) n_\kappa(x_>) - \delta_{\kappa o} n_0(x) , \\ \omega''_\kappa(r, r') &= (2\kappa+1) j_\kappa(x) j_\kappa(x') - \delta_{\kappa o} j_0(x) , \\ x &= \sqrt{E_{aa_1}}\, r , \quad x' = \sqrt{E_{aa_1}}\, r' . \end{aligned} \tag{3.3.11}$$

Here j_κ and n_κ are the spherical Bessel and Neumann functions (see footnote at page 27). We note that the sum over κ' , κ'' in (3.3.9) includes only a few terms in accordance with the triangle conditions

$$|l - l_1| \le \kappa' \le l + l_1 , \quad |l_1 - l_0| \le \kappa'' \le l_1 + l_0 , \tag{3.3.12}$$

and parity conservation.

3.3.2 Use of the *K* matrix

Equations (3.3.8–12) for cross sections of transitions via virtual levels are approximate especially at large energies. Moreover they are not applicable for transitions with change of spin. For this reason we consider another approximate method based on the K matrix (2.2.27).

We shall use the K matrix elements, calculated in first order,

$$K_{\Gamma''\Gamma'} = -T^{\mathrm{I}}_{\Gamma''\Gamma'} = \int_0^\infty F_{\lambda''} U_{\Gamma''\Gamma'} F_{\lambda'}\, dr , \tag{3.3.13}$$

as well as the cross section calculated in accordance with (2.2.24), and the matrix relation (2.2.27). If we include in the K matrix the elements coupling the states $a_0\lambda_0$, $a_1\lambda_1$, $a\lambda$, the resulting S matrix will describe the direct transition a_0–a, the transition via a virtual level a_0–a_1–a, and also the normalization effects (Sect. 3.2.4). The latter property is an important advantage of the K matrix approximation.

With the three types of transitions mentioned, we obtain from (2.2.27)

$$S_{\Gamma\Gamma_0} = \frac{2\mathrm{i}}{D}[K_{\Gamma\Gamma_0} + \mathrm{i}\sum_{\lambda_1} K_{\Gamma\Gamma_1} K_{\Gamma_1\Gamma_0} + o(K^3)] \tag{3.3.14}$$

where $O(K^3)$ includes the terms of the fourth and higher order with respect to $K_{\Gamma'\Gamma''}$ and D is the determinant of the matrix $I - \mathrm{i}K$:

$$D = \|I - \mathrm{i}K\| = 1 + K_{\Gamma\Gamma_0}^2 + \sum_{\lambda_1} K_{\Gamma_1\Gamma_0}^2 + \sum_{\lambda_1} K_{\Gamma\Gamma}^2 + o(K^2) \tag{3.3.15}$$

If we neglect the terms $o(K^2)$ and $o(K^3)$ in (3.3.14, 15) we can obtain the following expression for the cross section:

$$\sigma_{a_0a} = \sum_{\lambda_0\lambda} D^{-1}\left[\sigma^{\mathrm{I}}(a_0\lambda_0, a\lambda) + \frac{4\pi}{k_0^2}\sum_{L_\mathrm{T}S_\mathrm{T}} \frac{(2L_\mathrm{T}+1)(2S_\mathrm{T}+1)}{(2L_0+1)(2S_0+1)}\left|\sum_{\lambda_1} K_{\Gamma\Gamma_1}K_{\Gamma_1\Gamma_0}\right|^2\right] . \tag{3.3.16}$$

We note that in contrast to (3.3.8) the mixing term in (3.3.16) is absent because we used here the high velocity limit for scattering amplitude.

3.3.3 Polarization Potential

The polarization potential $V_{\Gamma\Gamma_0}$ (2.2.54) is used mostly in elastic scattering problems since no adequate approximation for off-diagonal potentials $V_{\Gamma\Gamma_0}$ has been found up to now. The cross section for elastic scattering is determined by the diagonal potential $V_{\Gamma_0\Gamma_0}$. For simplicity, the zero indices will be omitted in this section.

Instead of the equation (2.2.52) for the T matrix we introduce the potential $V_{\Gamma\Gamma}$ in the left-hand side of the basic radial equation (2.2.41, 43) and confine ourselves to the second-order term in $V_{\Gamma\Gamma}$. Inclusion of $V_{\Gamma\Gamma}$ in the radial equation corresponds to partial allowance for the higher-order terms. We obtain the following equations for elastic scattering:

$$(\mathscr{L}_\Gamma - V_\Gamma + k^2)F_\Gamma = 0\,, \quad T_{\Gamma\Gamma} = \exp(\mathrm{i}\eta)\sin\eta\,,$$
$$F_\Gamma \underset{r\to\infty}{\simeq} k^{-1/2}\exp(\mathrm{i}\eta)\sin\left(kr - \frac{\lambda\pi}{2} + \eta\right)\,, \tag{3.3.17}$$

$$V_\Gamma F_\Gamma = -\sum_{\Gamma_1\neq\Gamma} U_{\Gamma\Gamma_1}\int_0^\infty G_{\Gamma_1}(r, r')\,U_{\Gamma_1\Gamma}(r')\,dr'. \tag{3.3.18}$$

We shall consider now the so-called adiabatic approximation, in which we neglect the exchange term in (3.3.18) and approximate the Green's function by a δ function:

$$G_{\Gamma_1} = -\frac{1}{\Delta E}\delta(r - r')\,, \quad \Delta E = E_{a_1} - E_a\,, \tag{3.3.19}$$

where E_a and E_{a_1} are atomic energies in initial and virtual states (in Ry units). In this approximation, the potential V_Γ is local. We average it over L_T as in Sect. 3.2.3. Then we obtain the simple expression

$$V_\Gamma = V_a(r) = -\sum_{\kappa a_1} \frac{s_\kappa(a, a_1)}{[l\kappa]^2 \Delta E} [y^\kappa_{ll_1}(r)]^2 ,$$

$$s_\kappa(a, a_1) = Q_\kappa(a, a_1)[ll_1]^2 \begin{pmatrix} l & l_1 & \kappa \\ 0 & 0 & 0 \end{pmatrix}^2 , \tag{3.3.20}$$

where Q_κ is the angular factor that was determined in Sect. 2.3.2.

From (3.3.20) one can derive the two limits: $r \to 0$, and $r \to \infty$. In the case of $r \to 0$, only the term with $\kappa = 0$ is nonzero, and therefore $l_1 = l$, $L_1 = L$, $Q(a, a_1) = 1/g_a$, and

$$V_a(0) = \sum_{n_1} \frac{4}{g_a \Delta E} \left[\int_0^\infty P_{n_1 l}(r) P_{nl}(r) \frac{dr}{r} \right]^2 . \tag{3.3.21}$$

The main term in the asymptotic region ($r \to \infty$) is that with $\kappa = 1$, i.e., $l_1 = l \pm 1$, and we have

$$V_a(r) = -\frac{b}{r^4}, \quad b = \sum_{a_1} \frac{4 f_{aa_1}}{(\Delta E)^2} \ (r \to \infty) , \tag{3.3.22}$$

where b is the polarizability of the atom in the state a, and f_{aa_1} is the oscillator strength.

In calculations a simple approximate expression for the polarization potential is often used:

$$V_a(r) = -\frac{b}{(r^2 + r_0^2)^2} , \tag{3.3.23}$$

where r_0 is the atomic radius. This approximation asymptotically coincides with (3.3.22), and is finite at $r \to 0$. A better estimate of r_0 can be obtained by application of (3.3.21).

The adiabatic approximation is applicable at small energies of the outer electron: $k^2 \ll \Delta E$. It is necessary also that $\sqrt{\Delta E} r_0 \ll 1$.

3.3.4 Close-Coupling Method

Usually by the close-coupling approximation is meant the exact solution of the set of equations

$$(\mathscr{L}_\Gamma + k^2) F_\Gamma = \sum_{\Gamma'} U_{\Gamma\Gamma'} F_{\Gamma'} , \quad \Gamma = a\lambda L_T S_T ,$$

$$\mathscr{L}_\Gamma = \frac{d^2}{dr^2} - \frac{\lambda(\lambda+1)}{r^2} - U^c_\Gamma(r), \tag{3.3.24}$$

where the number of the states a (or "channels") is limited by some condition. Below we shall use the Born representation when the term with diagonal potential $U_{\Gamma\Gamma}$ is included in the sum in the right-hand side of (3.3.24).

The simplest version of the close-coupling method is the two-level approximation,

$$\begin{aligned}(\mathscr{L}_{\Gamma_0} + k_0^2) F_{\Gamma_0} &= \sum_{\lambda} U_{\Gamma_0 \Gamma} F_\Gamma \,, \\ (\mathscr{L}_\Gamma + k^2) F_\Gamma &= \sum_{\lambda_0} U_{\Gamma\Gamma_0} F_{\Gamma_0} \,.\end{aligned} \tag{3.3.25}$$

The first-order approximation can be obtained from (3.3.25) if we neglect the right-hand side in the equation for F_{Γ_0}. In other words, we neglect the influence of the final channel on the initial one. The consideration of this influence is the main feature of the close-coupling method.

The equation set (3.3.24) or (3.3.25) is Hermitian and therefore its solution provides a unitary S matrix in the framework of included channels. That means we obtain a normalized cross section automatically.

The set Γ can include the energetically accessible states a (or open channels with $\Delta E < \mathscr{E}_0, k^2 > 0$) and also some energetically inaccessible states a (or closed channels with $\Delta E > \mathscr{E}_0, k^2 < 0$). In the system of equations (3.3.24), the open channels $a' \neq a$ describe the excitation through virtual levels a'. The inclusion of closed channels corresponds to allowance for the effects of atom polarization by the outer electron. Of course this separation of the role of open and closed channels is approximate but it proves to be useful for general discussion.

As mentioned in the beginning of Sect. 3.3, the convergence of the method with increasing number of channels is rather slow. Now we can explain this in the following way. The diagonal potentials $U_{\Gamma\Gamma}$ in the set (3.3.24) describe the attraction of the electron to the atom by the average field. This attraction determines the elastic scattering of the electron, and at the same time provides a decreasing of the distance between the atomic and outer electrons. The latter effect, however, is considerably reduced by the interelectron repulsion, i.e., by polarization of the atom. To allow for this polarization, we have to include in the equation set many other states mainly corresponding to closed channels.

It should be noted that the diagonal polarization potential $V_{\Gamma\Gamma}$ also describes the attraction of the electron to the center of the atom. The increasing of interelectron distance is described by the off-diagonal terms $V_{\Gamma\Gamma'}$.

The inclusion of energetically accessible states in the set of equations describes, as mentioned above, mainly the transitions through virtual states. It is clear that this effect is important only when we deal with the influence of the strong channel on the weak one. In particular, this effect is unimportant in the case of resonance-level excitation.

As an illustration of the above qualitative statements we consider the excitation of the $2s$ and $2p$ levels of atomic hydrogen. The results of some calculations by various approximate methods together with experimental data are shown in

Figs. 3.2 and 3.3. As mentioned above the Born cross sections exceed the experimental data in the vicinity of the maximum cross section by a factor of 1.5–2.

The inclusion of the electron-atom attraction (the distorted-wave approximation) provides a further increase of the cross section, and an increasing discrepancy with experiment. This effect is especially great in the case of the $1s - 2s$ transition. We note that the effect of attraction is considerably compensated by exchange (not shown in the figures). The close coupling three-state approximation $1s - 2s - 2p$ (with exchange) provides slightly better results than the Born approximation does in the case of the $1s - 2p$ transition. However this improvement is insignificant compared with the total inaccuracy. In the case of the $1s - 2s$ transition, the close-coupling method provides worse results than the Born approximation. The reason is the inclusion in the equations of the diagonal potentials without adequate regard for polarization effects. It is interesting that the results of the close-coupling approximation are near to those of the distorted-wave approximation with exchange.

The six-state close-coupling approximation ($1s - 2s - 2p - 3s - 3p - 3d$) was used only in the energy range between the $n = 2$ and $n = 3$ thresholds. Taking into account $n = 3$ states in that case, one can partially take into account polarization of the atom, which results in a considerable decrease of the cross section of the $1s - 2s$ transition.

The above-mentioned possibilities of the close-coupling approximation do not concern the near-threshold region. The influence of close coupling in this region can be very important, particularly in the case of degenerate l levels. Due to degeneracy, the cross section appears to be finite at the threshold and oscillating in its vicinity. The close-coupling approximation adequately describes these features. In this book, however, we do not consider such effects.

One can see from the above that the use of the close-coupling method requires an accurate choice of the set of levels and even potentials which have to be included in the equation set in accordance with a particular problem. For example, to obtain a normalized cross section it is sufficient to use the two-state approximation, and moreover, we should neglect the diagonal potentials apart from the Coulomb field. It is useful to omit the non-Coulomb diagonal potentials in all cases when the polarization effects are not taken into account.

The consideration of polarization requires the solution of a large equation set. For this reason in [3.22] the so-called pseudostate method was proposed. In this book we cannot describe this method in detail, and refer to the reviews [3.23, 24].

3.4 Excitation of Highly Charged Atoms

In this section we consider some special properties of the inelastic scattering of electrons on highly charged atoms. One of such features, namely the finite value of the cross section at the threshold, was mentioned in Sect. 3.2.1. Here we consider in more detail the pecularities connected, in particular, with thresholds of other channels. These pecularities are closely connected with dielectronic recombination which will be considered in the Sect. 5.2 on the basis of a different approach which is more convenient for applications. Here we are interested mostly in the general theory of inelastic scattering on highly charged atoms. The outline of this section is based on the works of Presnyakov and Urnov [3.25, 26] (see also [3.27]).

When we deal with highly charged atoms ($z \gg 1$) it is convenient to use the Coulomb system of units. In this system the unit of length is a_0/z, and that of energy is z^2Ry. Equations (2.2.30) in the Coulomb system are written in the form

$$(\mathscr{L}_\Gamma + k^2) F_\Gamma = \frac{1}{z} \sum_{\Gamma'} U_{\Gamma\Gamma'} F_{\Gamma'} ,$$
$$\mathscr{L} = \frac{d^2}{dr^2} - \frac{\lambda(\lambda+1)}{r^2} + \frac{2\alpha}{r} , \quad \alpha = 1 - 1/z , \tag{3.4.1}$$

where $U_{\Gamma\Gamma'}$ are defined by equations (2.2.32) and (3.3.35), but the terms $U_\Gamma^c + 2\alpha/r$ are included in $U_{\Gamma\Gamma}$. At $z \gg 1$, the potentials $U_{\Gamma\Gamma'}$ are almost independent of z. The boundary conditions for the functions F_Γ, in accordance with (2.2.14) and (2.2.40), are

$$F_\Gamma(0) = 0 , \quad F_\Gamma \underset{r\to\infty}{\simeq} k^{-1/2} \{\delta_{\Gamma\Gamma_0} \sin(k_0 r + d_0) + T_{\Gamma\Gamma_0} \exp[\mathrm{i}(kr + d)]\} ,$$
$$d = -\frac{\lambda\pi}{2} + \frac{\alpha}{k} \ln(2kr) + \arg\Gamma\left(\lambda + 1 - \mathrm{i}\frac{\alpha}{k}\right) , \tag{3.4.2}$$
$$k^2 = \mathscr{E}_0 - E_{aa_0} .$$

In place of (3.4.1, 2), a set of integral equations of the type (2.2.46) can be introduced. We shall write these equations separately for open channels, $\Gamma = \gamma\,(k_\gamma^2 > 0)$, and closed channels $\Gamma = c(k_c^2 < 0)$:

$$F_\gamma = \delta_{\gamma\Gamma_0} F_{\lambda_0} + \frac{1}{z} \sum_{\Gamma'} \hat{G}_\gamma U_{\gamma\Gamma'} F_{\Gamma'} ,$$
$$F_c = \frac{1}{z} \sum_{\Gamma'} \hat{G}_c U_{c\Gamma'} F_{\Gamma'} , \tag{3.4.3}$$

where F_{λ_0} is the regular solution of the homogeneous equation [see (3.2.3)]. The integral operator $\hat{G}$ is defined by the relation

$$\hat{G}\varphi(r) = \int_0^\infty G(r, r')\, \varphi(r')\, dr' , \tag{3.4.4}$$

where $G(r, r')$ is the radial Coulomb Green's function of (3.4.1).

In Sect. 3.4.2 the regularity of $\hat{G}_\gamma$ will be proved. This means that coupling of open channels at $z \gg 1$ is weak and perturbation theory is thus applicable for these channels. In contrast to the case of neutral atoms, now we have the parameter $1/z$, and for $z \gg 1$, first-order perturbation theory (i.e., the Coulomb–

Born approximation) should be a good approximation in the absence of closed channels. In contrast to $\hat{G}_\gamma$, the closed channel Green's function $\hat{G}_c$ contains the poles, and perturbation theory isn't applicable to these channels. This question will be considered in Sect. 3.4.2.

3.4.1 Coulomb Green's Function

The Green's function is determined by (2.2.42–45). We substitute $C^{-1}\Phi$ in place of $\bar{F}$, and write G_Γ in the form

$$G_\Gamma(r, r') = -\frac{1}{C}\Phi(\mu, k, r_<)\bar{\bar{F}}(\mu, k, r_>)\,,$$
$$\Phi(\mu, k, r) = r^{\mu+1/2} \quad \text{when } r \to 0\,, \tag{3.4.5}$$
$$\mu = \lambda + 1/2\,,$$

where $C = C(\mu, k)$ does not depend on r, and is defined by the equation

$$C(\mu, k) = \Phi\,\frac{d\bar{\bar{F}}}{dr} - \frac{d\Phi}{dr}\,\bar{\bar{F}}\,. \tag{3.4.6}$$

Now we introduce instead of the two linearly independent functions $\Phi(\mu, k, r)$ and $\bar{\bar{F}}(\mu, k, r)$ a new pair of functions $\Phi_\pm = \Phi(\pm\mu, k, r)$. Using (3.4.6) and an asymptotic expression (2.2.44) for $\bar{\bar{F}}$ we obtain an expression for G_Γ in terms of $\Phi_\pm$:

$$\begin{aligned} G_\Gamma(r, r') = &-\frac{2\mathrm{i}}{C(\mu, k)C(\mu, -k)}\Phi_+(r)\,\Phi_+(r') \\ &+\frac{1}{2\mu}\Phi_+(r_<)\left[\frac{C(-\mu, -k)}{C(-\mu, k)}\Phi_+(r_>) - \Phi_-(r_>)\right]. \end{aligned} \tag{3.4.7}$$

In the case of the Coulomb field $U_\Gamma = -2\alpha/r$, from (3.4.7) we have

$$G_\Gamma(r, r') = \cot(\pi\nu)\,\bar{F}_{\nu\mu}(r)\,\bar{F}_{\nu\mu}(r') + G_\Gamma^{(p)}(r, r')\,,$$
$$G_\Gamma^{(p)}(r, r') = \frac{1}{\sin(\pi\mu)}\bar{F}_{\nu\mu}(r_<)[\cos(2\pi\mu)\bar{F}_{\nu\mu}(r_>) - \bar{F}_{\nu,-\mu}(r_>)]\,, \tag{3.4.8}$$
$$\mu = \lambda + 1/2\,, \quad \nu = \mathrm{i}\alpha/k\,, \quad \mathrm{Im}\,\{k\} \geq 0\,.$$

The radial functions $\bar{F}_{\nu\mu}$ are related to the Coulomb functions F_λ from (3.2.3) for a complex value of $k = \mathrm{i}\alpha/\nu$:

$$\bar{F}_{\nu\mu}(r) = a_\nu F_\lambda(r)\,, \quad a_\nu = [\mathrm{i}\cot(\pi\nu) + 1]^{-1/2}\,. \tag{3.4.9}$$

In the particular case when k is real,

$$\bar{F}_{\nu\mu} = \left(\coth\frac{\pi\alpha}{k} + 1\right)^{-1/2} F_\lambda\,.$$

For imaginary k, when $\nu = n$ is a positive integer and $E = -\alpha^2/n^2$, we have

$$\bar{F}_{\nu\mu} = \left(\frac{\pi\nu^3}{2\alpha^2}\right)^{1/2} P_{n\lambda}, \quad \nu = n, \quad \mu = \lambda + 1/2 . \tag{3.4.9a}$$

The radial Coulomb Green's function in (3.4.8) includes two terms. The first term in the case of closed channel ($k^2 < 0$) is singular (over energy) because of the factor $\cot\pi\nu$. The poles at $\nu = n$ correspond to a series of resonances which converges to the new channel threshold ($k^2 \to -0$). The second term $G_\Gamma^{(p)}$ does not contain any singularities as a function of energy. At $k \to 0$, $G_\Gamma^{(p)}$ is independent of k and is a regular function of r, r'. It is important for applications that the singular part of G_Γ is factorized over its arguments r and r'.

3.4.2 Potential and Resonance Scattering

We rewrite the integral equation set (3.4.3) to show explicitly the singular part of the closed channel Green's functions:

$$\begin{aligned} F_\gamma &= \delta_{\gamma\Gamma_0} F_{\lambda_0} + \frac{1}{z}\sum_{\Gamma'} \hat{G}_\gamma U_{\gamma\Gamma'} F_{\Gamma'} , \quad \gamma = ak\lambda\tfrac{1}{2}L_T S_T , \\ F_c &= \bar{F}_c \Lambda_c + \frac{1}{z}\sum_{\Gamma'} \hat{G}_c^{(p)} U_{c\Gamma'} F_{\Gamma'} , \quad c = a_c\nu\lambda\tfrac{1}{2}L_T S_T . \end{aligned} \tag{3.4.10}$$

Here Λ_c does not depend on r and is defined by equation

$$z\Lambda_c = \cot(\pi\nu) \cdot \sum_{\Gamma'} \int_0^\infty \bar{F}_c U_{c\Gamma'} F_{\Gamma'} dr , \tag{3.4.11}$$

with the function $\bar{F}_c$ defined by (3.4.9).

For $\nu \neq n$ an iterative method can be applied to get the solution. For $\gamma \neq \Gamma_0$ we obtain, cf. (2.2.51),

$$F_\gamma = \frac{1}{z}\hat{G}_\gamma (U_{\gamma\Gamma_0} + V_{\gamma\Gamma_0}) F_{\lambda_0} + \frac{1}{z}\sum_c \hat{G}_\gamma (U_{\gamma c} + V_{\gamma c}) \cdot \bar{F}_c \Lambda_c , \tag{3.4.12}$$

and the set of algebraical equations for Λ_c,

$$\sum_{c'} [z \tan(\pi\nu)\, \delta_{cc'} - R_{cc'}] \Lambda_{c'} = R_{c\Gamma_0} , \tag{3.4.13}$$

where

$$R_{cc'} = \int_0^\infty \bar{F}_c (U_{cc'} + V_{cc'})\, \bar{F}_{c'}\, dr , \tag{3.4.14}$$

and $R_{c\Gamma_0}$ is defined by (3.4.14) with substitution of F_{λ_0} in place of $\bar{F}_{c'}$. Polarization potentials $V_{\Gamma\Gamma'}$ are defined by (2.2.53, 54) with $G_\Gamma = G_\gamma$ for open channels and $G_\Gamma = G_c^{(p)}$ for closed channels. In contrast to ordinary perturbation theory

$V_{\Gamma\Gamma'}$ does not contain any singularities. Equations (3.4.12–14) are applicable also for $\nu = n$, the singularities being seen explicitly.

From (3.4.12) we obtain

$$T_{\Gamma\Gamma_0} = T^{\rm pot}_{\Gamma\Gamma_0} + T^{\rm res}_{\Gamma\Gamma_0}, \tag{3.4.15}$$

$$\begin{aligned} T^{\rm pot}_{\Gamma\Gamma_0} &= -\frac{1}{z}\int_0^\infty F_\lambda(U_{\Gamma\Gamma_0} + V_{\Gamma\Gamma_0})F_{\lambda_0}\,dr\,, \\ T^{\rm res}_{\Gamma\Gamma_0} &= -\frac{1}{z}\sum_c A_c\int_0^\infty F_\lambda(U_{\Gamma c} + V_{\Gamma c})\,\bar{F}_c\,dr\,. \end{aligned} \tag{3.4.16}$$

The matrix $T^{\rm pot}_{\Gamma\Gamma_0}$ is similar to that determined by (2.2.52), and corresponds to usual "potential" scattering. The matrix $T^{\rm res}_{\Gamma\Gamma_0}$ describes the additional "resonance" scattering, which is related to the closed channels. Every closed channel $c(a_c, k_c^2 = \mathscr{E}_0 - E_{ca_0})$ is connected with a set of resonances in the vicinity of $\nu = i\alpha/k_c = n$ (i.e., at values of $\mathscr{E}_0$ near $E_{ca_0} - \alpha^2/n^2$). Below we assume that all the resonances are independent, and hence we neglect the off-diagonal integrals $R_{cc'}$, in (3.4.13). Then we obtain

$$A_c = \frac{R_{c\Gamma_0}}{z\tan(\pi\nu) - R_{cc}}, \tag{3.4.17}$$

$$T^{\rm res}_{\Gamma\Gamma_0} = -\frac{1}{z^2}\sum_c \frac{R_{\Gamma c}R_{c\Gamma_0}}{\tan(\pi\nu) - R_{cc}/z}. \tag{3.4.18}$$

The scattering amplitude $T^{\rm res}_{\Gamma\Gamma_0}$ reveals the resonances, their positions $\mathscr{E}_0(c, n)$ and widths δ being given by

$$\begin{aligned} &\mathscr{E}_0(c, n) = E_{ca_0} - \frac{\alpha^2}{\nu_c^2}, \quad \tan(\pi\nu_c) = \frac{1}{z}\,\mathrm{Re}\,\{R_{cc}\} \\ &\mathscr{E}_0(c, n) \simeq E_{ca_0} - \frac{\alpha^2}{n^2} + \frac{2\alpha^2}{\pi n^3 z}\,\mathrm{Re}\,\{R_{cc}\}, \quad \delta \simeq -\frac{4\alpha^2}{\pi n^3 z}\,\mathrm{Im}\,\{R_{cc}\}. \end{aligned} \tag{3.4.19}$$

Each closed channel c (i.e., energetically inaccessible state a_c) corresponds to an infinite set of resonances in the small energy range from threshold $\mathscr{E}_0 = E_{aa_0}$ up to $\mathscr{E}_0 = E_{ca_0}$. Hence $n \geq \alpha/\sqrt{E_{ca}}$, and usually $n \gg 1$. For large n, the radial functions $P_{n\lambda} \propto n^{-3/2}$ and in accordance with (3.4.9a) $\bar{F}_c$ does not depend on ν. We see that all the energy dependence of $T^{\rm res}_{\Gamma\Gamma_0}$ is concentrated in a factor $\tan(\pi\nu)$.

The excitation cross section for the transition $a_0 - a$ in accordance with (2.2.20) is

$$\sigma_{a_0 a} = \frac{\pi a_0^2}{z^2}\cdot\frac{2}{k_0^2}\sum_{\lambda_0\lambda L_T S_T}\frac{g_\Gamma}{g_0}\,|\,T^{\rm pot}_{\Gamma\Gamma_0} + T^{\rm res}_{\Gamma\Gamma_0}\,|^2, \tag{3.4.20}$$

and also contains the resonances at energies $\mathscr{E}_0 = \mathscr{E}_0(c, n)$. One can prove that in the case of nonoverlapping (independent) resonances when (3.4.17, 18) are applicable, the averaged cross section can be expressed in the form

$$\begin{aligned}\bar{\sigma}_{a_0a} &= \frac{1}{2\delta\mathscr{E}_0}\int\limits_{\mathscr{E}_0-\delta\mathscr{E}}^{\mathscr{E}_0+\delta\mathscr{E}} \sigma_{a_0a}(\mathscr{E}_0)\, d\mathscr{E}_0 \\ &= \sum_{\lambda_0\lambda L_T S_T} [\sigma^{\mathrm{pot}}(\Gamma_0, \Gamma) + \sum_c \bar{\sigma}^{\mathrm{res}}(\Gamma_0\, \Gamma; c)]\,,\end{aligned} \tag{3.4.21}$$

$$\bar{\sigma}^{\mathrm{res}}(\Gamma_0\Gamma; c) = \frac{\pi a_0^2}{z^4}\cdot\frac{2}{k_0^2}\cdot\frac{|R_{\Gamma c}R_{c\Gamma_0}|^2}{z\,\mathrm{Im}\,\{R_{cc}\}}\left[\frac{1-\Theta(\mathscr{E}_0 - E_{ca_0})}{2}\right], \tag{3.4.22}$$

where function $\Theta(x) = \pm 1$ for $x \gtrless 0$. In the averaging of $\sigma_{a_0a}(\mathscr{E}_0)$ in (3.4.21) we supposed that in the range $2\delta\mathscr{E}$ all the matrix elements $R_{c\Gamma}$ and R_{cc} are constant, and only the factor $\tan(\pi\nu)$ is energy dependent. We note that in this case all the interference terms are vanishing. This is relevant both to potential and resonance scattering and to different closed channels.

We shall consider now the first-order approximation for $T_{\Gamma\Gamma_0}$ and σ_{a_0a}. In this approximation we can neglect potentials $V_{\Gamma\Gamma_0}$ in (3.4.16), (3.4.18) everywhere besides the resonance denominator in (3.4.18), in which we have to keep the first nonvanishing term in the imaginary part of R_{cc}. From (3.4.14) we have

$$R_{cc} = \int\limits_0^\infty \bar{F}_c\left(U_{cc} + \frac{1}{z}\sum_{\Gamma_1} U_{c\Gamma_1}\hat{G}_{\Gamma_1}U_{\Gamma_1c} + \dots\right)\bar{F}_c\, dr\,.$$

In accordance with (3.4.9a) the functions $\bar{F}_c$ are real. To get the imaginary part of $\hat{G}_\Gamma(k^2 > 0)$ it is convenient to use the spectral representation of $\hat{G}_\Gamma$:

$$G_\Gamma(r, r') = \frac{2k}{\pi}\mathrm{P}\int\frac{F_{k'\lambda}(r)F_{k'\lambda}(r')}{k^2 - k'^2}dk' - \mathrm{i}F_{k\lambda}(r)F_{k\lambda}(r')\,, \tag{3.4.23}$$

where $F_{k\lambda} = F_\lambda$ are real functions and P denotes the principal value of the integral. If we determine Im $\{R_{cc}\}$ from (3.4.23) we can get the first-order approximation for $T^{\mathrm{res}}_{\Gamma\Gamma_0}$. Thus we obtain

$$\begin{aligned}T^{\mathrm{pot}}_{\Gamma\Gamma_0} &= -\frac{1}{z}R^{\mathrm{I}}_{\Gamma\Gamma_0}\,, \\ T^{\mathrm{res}}_{\Gamma\Gamma_0} &= -\frac{1}{z^2}\sum_c\frac{R^{\mathrm{I}}_{\Gamma c}\cdot R^{\mathrm{I}}_{c\Gamma_0}}{\tan(\pi\nu) - R^{\mathrm{I}}_{cc}/z + \mathrm{i}\sum_\gamma |R^{\mathrm{I}}_{\gamma c}|^2/z^2}\,,\end{aligned} \tag{3.4.24}$$

where

$$R^{\mathrm{I}}_{\Gamma'\Gamma''} = \int\limits_0^\infty F_{\lambda'}U_{\Gamma'\Gamma''}F_{\lambda''}\, dr\,, \tag{3.4.25}$$

whence for closed channels $\Gamma = c$ the functions $\bar{F}_c = F_{\nu\mu}$ should be substituted in place of F_λ.

On substituting (3.4.24) in (3.4.22) and (3.4.20) we obtain the finial expressions for potential and resonance parts of the cross section in the first order approximation:

$$\sigma^{\text{pot}}(\Gamma_0,\Gamma) = \frac{\pi a_0^2}{z^4}\cdot\frac{2}{k_0^2}\cdot\frac{g_\Gamma}{g_0}|R^{\text{I}}_{\Gamma\Gamma_0}|^2 \qquad (3.4.26)$$

$$\bar{\sigma}^{\text{res}}(\Gamma_0,\Gamma;c) = \frac{\pi a_0^2}{z^4}\cdot\frac{2}{k_0^2}\cdot\frac{g_\Gamma}{g_0}\cdot\frac{|R^{\text{I}}_{\Gamma c}|^2}{\sum\limits_\gamma |R^{\text{I}}_{\gamma c}|^2}\cdot|R^{\text{I}}_{c\Gamma_0}|^2\cdot\frac{1-\Theta(\mathscr{E}_0 - E_{ca_0})}{2}. \qquad (3.4.27)$$

Equation (3.4.26) coincides with the usual expression for the first-order partial cross section which was obtained in Sect. 2.3. It corresponds to the Coulomb–Born approximation since we used Coulomb wave functions F_λ. In accordance with (3.4.21), the total cross section is increased due to the resonance part $\bar{\sigma}^{\text{res}}$, connected with all closed channels.

In the vicinity of the threshold of the channel c when the closed channel becomes open (γ_c), the value of $R^{\text{I}}_{c\Gamma_0}$ is continuously transferred to $R^{\text{I}}_{\gamma_c\Gamma_0}$. That means

$$\bar{\sigma}^{\text{res}}(\Gamma_0\Gamma;c) = \sigma^{\text{pot}}(\Gamma_0,\gamma_c)\cdot\frac{|R^{\text{I}}_{\Gamma c}|^2}{\sum\limits_\gamma |R^{\text{I}}_{\gamma c}|^2},\quad \mathscr{E}_0\to E_{ca_0} \qquad (3.4.28)$$

near to the threshold of the channel c, and hence

$$\sum_\Gamma \bar{\sigma}^{\text{res}}(\Gamma_0\Gamma;c) = \sigma^{\text{pot}}(\Gamma_0,\gamma_c) \qquad (3.4.29)$$
$$\mathscr{E}_0 = E_{ca_0} - 0 \quad \mathscr{E}_0 = E_{ca_0} + 0$$

When energy reaches the threshold, the terms $\bar{\sigma}^{\text{res}}(\Gamma_0\Gamma;c)$ related to this channel disappear in (3.4.21). At the same time a new channel $a_0 - a_c$ is opened. In accordance with (3.4.29) the sum over a of the resonance parts of the total cross section is equal to the threshold value of the cross section of the new channel.

3.4.3 Discussion and Examples

In the previous section it was shown that the excitation cross section for the transition $a_0 - a$ includes besides the usual part $\sigma^{\text{pot}}_{a_0a}$ an additional part $\bar{\sigma}^{\text{res}}_{a_0a}$. This additional part is connected with the possibility of electron attachment into a quasi-stationary state a_cnl of the atom X_{z-1} with subsequent decay of this state into $X_z(a) + e$.

From (3.4.19), (3.4.18), and (3.4.24) it follows that the width of the resonance in first order is

$$\delta = \frac{4\alpha^2}{\pi n^3 z^2}\sum_\gamma |R^{\text{I}}_{\gamma c}|^2. \qquad (3.4.30)$$

Since δ is equal to the autoionization probability[1] W_a (to the open channels γ), the factor in (3.4.27, 28) is

$$\frac{|R^{\mathrm{I}}_{\Gamma c}|^2}{\sum\limits_{\gamma}|R^{\mathrm{I}}_{\gamma c}|^2} = \frac{W_a(c,\Gamma)}{W_a(c)} \,. \tag{3.4.31}$$

One can see that this factor is simply a branching ratio when more than one decay channel is possible.

The sum $W_a(c)=\sum_\gamma W_a(c,\gamma)$ besides the channel $\gamma = \Gamma$ includes the initial state Γ_0 and perhaps some other states Γ' which satisfy the energy conservation condition

$$E_{ca'} - \frac{\alpha}{n^2} > 0 \,. \tag{3.4.32}$$

For a given value of n, the most probable are the decays with minimum $E_{ca'}$ and the optically allowed transition $\lambda_c - \lambda_{\Gamma'} = \pm 1$. However at very small $E_{ca'}$, in accordance with (3.4.32) only very large n are possible and since $W_a \propto n^{-3}$, the value of W_a can be very small.

Besides the autoionization $c - \Gamma$ a radiative decay $c - c'$ is possible into a stationary state of the atom X_{z-1}. Usually in this case an attached electron does not change its quantum numbers: $a_c n\lambda - a_{c'} n\lambda$. If the value of n isn't too small we can use the approximation (3.4.28). If we add the radiative transition probability A and use (3.4.31) we obtain for the resonance part of the cross section

$$\bar{\sigma}^{\mathrm{res}}_{a_0 a} = \sum_{\lambda_0 \lambda L_T S_T c} \bar{\sigma}^{\mathrm{res}}(\Gamma_0 \Gamma; c) \,,$$

$$\bar{\sigma}^{\mathrm{res}}(\Gamma_0 \Gamma; c) = \sigma^{\mathrm{pot}}(\Gamma_0, \gamma_c)\frac{W_a(c,\Gamma)}{W_a(c) + A(c)} \,, \tag{3.4.33}$$

where $\sigma^{\mathrm{pot}}(\Gamma_0, \gamma_c)$ is the threshold value of the $a_0 - a_c$ partial cross section. Since $W_a \propto z^{-2} n^{-3}$ and the radiative transition probability $A \propto z^2 (137)^{-3}$ (in Coulomb units) is independent of n, we have

$$A/W_a \simeq 10^{-6} z^4 n^3 .$$

Usually radiative transitions can compete with autoionization at very large n only, i.e., in a narrow energy interval. Besides this small band (which widens with increasing z), the ratio $\bar{\sigma}^{\mathrm{res}}/\sigma$ does not depend on z.

Although the energy range $\mathscr{E}_0 = E_{aa_0} \div E_{ca_0}$ where resonance excitation can take place isn't large, in some cases its part in the total excitation rate proves to be significant. This is especially true for optically forbidden transitions $a_0 - a$ when there are higher levels a_c for which $\sigma_{a_0 a_c}$ is large. An illustrative example is shown in Fig. 3.9 for the transition $2s - 3s$ in the [Li] ion, O VI. At small energies the role of closed channels $np\,(n \geq 3)$ is much larger than the direct

[1] We recall that $\delta = W_a$ is in Coulomb units z^2Ry; in the Rydberg units, the values of δ and W_a are practically independent of z.

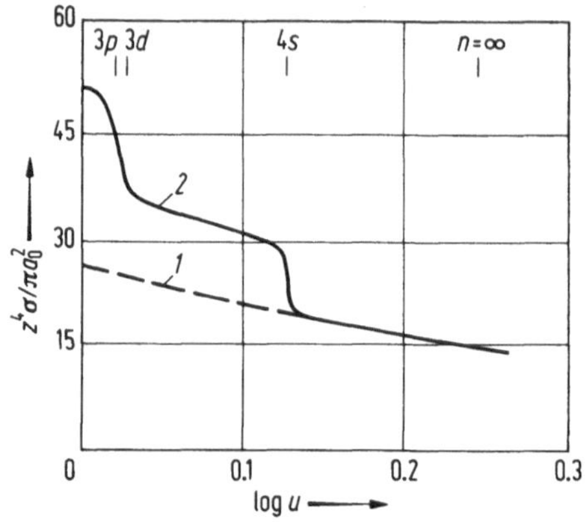

Fig. 3.9. Effective cross section for the transition $2s$–$3s$ of the lithiumlike ion O VI: (1) Coulomb-Born approximation; (2) Cross section with resonances taken into account; $u = (\mathscr{E} - \Delta E)/E$

excitation. We note however that the branching onto the $3s$ level decreases with increasing n.

It should be emphasized that resonance excitation is important for the total excitation rate only in those cases when the potential excitation cross section σ^{pot} is small for any reason.

We note in conclusion that resonance scattering is closely related to dielectronic recombination, which will be considered in detail in Sect. 5.2. In fact these two processes are alternative final channels after the attachment of an electron in a quasi-stationary state:

$$\mathrm{X}_z(a_0) + \mathrm{e} \to \mathrm{X}_{z-1}(a_c n\lambda) \begin{cases} \nearrow \mathrm{X}_z(a') + \mathrm{e} \\ \searrow \mathrm{X}_{z-1}(a'nl) + \hbar\omega \end{cases} \tag{3.4.34}$$

Among the resonance scattering channels, elastic scattering ($a' = a_0$) and inelastic scattering in different states $a' \neq a_0$ can be distinguished.

3.5 Transitions Between Highly Excited Levels

3.5.1 Born Approximation

For transitions between the highly excited levels n_0, $n \gg 1$, the cross sections summed over quantum numbers lm and averaged over $l_0 m_0$ are of the most interest. The cross section of the transition $n_0 \to n$ can be written in the form:

$$\sigma_{n_0 n} = \frac{\pi a_0^2}{z^4} \cdot \frac{8}{n_0^2 k_0^2} \int_{k_0-k}^{k_0+k} f(q) \frac{dq}{q^3} , \tag{3.5.1}$$

$$f(q) = \sum_{l_0 m_0 lm} |\langle n_0 l_0 m_0 | \exp(i\boldsymbol{q} \cdot \boldsymbol{r}) | nlm \rangle|^2 , \tag{3.5.2}$$

where k and q are expressed in units z/a_0. When n_0, n are not large, numerical summation in (3.5.2) is used. For highly excited levels n_0, $n \gg 1$, such summation is too cumbersome. In the case of the hydrogen atom, however, this summation can be carried out in an explicit form [3.28]. Moreover, when n_0, $n \gg 1$ and n_0, $n \gg |n - n_0|$, further simplification of the expression for the Born cross sections is possible. Formulas obtained for hydrogen may be applied to the transitions between highly excited levels of any atom, since for large n the system of levels is close to that of hydrogen.

We shall rewrite (3.5.2) in the following form:

$$f(q) = \lim_{\substack{E_0 \to E_n \\ E \to E_{n_0}}} (E_0 - E_{n_0})(E - E_n) \sum_{a'a''} \frac{\langle a'|\exp(-\mathrm{i}\boldsymbol{q}\cdot\boldsymbol{r})|a''\rangle\langle a''|\exp(\mathrm{i}\boldsymbol{q}\cdot\boldsymbol{r})|a'\rangle}{(E_0 - E_{n'})(E - E_{n''})} \tag{3.5.3}$$

where $a' = n'l'm'$, $a'' = n''l''m''$. It is not difficult to see that the terms with $n' \neq n_0$, $n'' \neq n$ added in (3.5.3) vanish when $E_0 \to E_{n_0}$, $E \to E_n$. By using the well-known spectral representation of the Green's function

$$G_E(\boldsymbol{r}, \boldsymbol{r}') = \sum_a \frac{\psi_a^*(\boldsymbol{r})\psi_a(\boldsymbol{r}')}{E - E_a}, \tag{3.5.4}$$

we obtain

$$f(q) = \lim_{\substack{E_0 \to E_{n_0} \\ E \to E_n}} (E_0 - E_{n_0})(E - E_n) \times \int\int d\boldsymbol{r}\, d\boldsymbol{r}'\, G_{E_0}(\boldsymbol{r}, \boldsymbol{r}') \exp[\mathrm{i}\boldsymbol{q}\cdot(\boldsymbol{r} - \boldsymbol{r}')] G_E(\boldsymbol{r}, \boldsymbol{r}'). \tag{3.5.5}$$

The Green's function for the Coulomb potential is [3.29]

$$G(\boldsymbol{r}, \boldsymbol{r}') = \frac{\Gamma(1-\nu)}{2\pi(x-y)} \cdot \nu \cdot \hat{L}[W_{\nu,\frac{1}{2}}(x/\nu) M_{\nu\frac{1}{2}}(y/\nu)],$$

$$x = r + r' + |\boldsymbol{r} - \boldsymbol{r}'|, \quad y = r + r' - |\boldsymbol{r} - \boldsymbol{r}'|, \quad \nu = \sqrt{z^2\mathrm{Ry}/(-E)},$$

$$\hat{L} = \frac{\partial}{\partial x} - \frac{\partial}{\partial y}, \tag{3.5.6}$$

where W and M are the Whittaker functions of the first and second kind respectively.

By means of (3.5.6) the calculation of $f(q)$ can be reduced to the integral

$$f(q) = \frac{1}{q}\int_0^q A(q')\, dq', \tag{3.5.7}$$

where the function $A(q)$ is expressed in terms of hypergeometric functions

$F(\alpha, \alpha', 1; \beta)$ and their derivatives [3.28]:

$$A(q) = \frac{1}{(n_0 n)^2} \mathrm{Re}\left\{ I'_\lambda(-n_0+1,\ -n+1) I'^*_\lambda(-n_0,\ -n) \right.$$
$$-I'_\lambda(-n_0+1, -n) I'^*_\lambda(-n_0, -n+1)$$
$$-\frac{1}{6}\frac{d^2}{dq^2}\left[I_\lambda(-n_0+1,\ -n+1) I^*_\lambda(-n_0,\ -n)\right.$$
$$\left.\left. -I_\lambda(-n_0+1,\ -n) I^*_\lambda(-n_0,\ -n+1)\right]\right\}, \tag{3.5.8}$$

where

$$I_\lambda(\alpha, \alpha') = \lambda\alpha + \alpha'^{-1}(\lambda - p_0)^{-\alpha}(\lambda - p)^{-\alpha'} F(\alpha, \alpha', 1;\ \beta),$$
$$\lambda = \frac{1}{2}(p_0 + p + iq),\quad p_0 = 1/n_0,\quad p = 1/n, \tag{3.5.9}$$
$$\beta = \frac{p_0 p}{(\lambda - p_0)(\lambda - p)},\quad I'_\lambda = \frac{d}{dq} I_\lambda .$$

Formulas (3.5.7) and (3.5.8) are exact. To obtain these expressions no simplifying assumptions have been made. Numerical calculations using (3.5.7, 8) have been made in [3.7].

We shall now consider the transitions between highly excited levels. By using the known asymptotic expression of the hypergeometric functions in terms of the Bessel functions J, formula (3.5.9) can be considerably simplified [3.28]. When $n_0 \gg 1$, $n \gg 1$, $\Delta n = |n - n_0| \ll n_0, n$,

$$A(q) = \frac{32}{3}\frac{(n_0 n)^3}{\Delta n (n_0 + n)^4} \exp\left(-q^2 \frac{(n_0 n)^2}{2(n_0 + n)}\right)[b_1(\varepsilon) + b_2(\varepsilon)], \tag{3.5.10}$$

$$b_1(\varepsilon) = \frac{\varepsilon^2 - 1}{\varepsilon^3}\frac{4 - \varepsilon^2}{\varepsilon^2} J_{\Delta n}(\Delta n \varepsilon) J'_{\Delta n}(\Delta n \varepsilon), \tag{3.5.11}$$

$$b_2(\varepsilon) = \Delta n \frac{(\varepsilon^2 - 1)^2}{\varepsilon^2}\left\{\frac{\varepsilon^2 - 1}{\varepsilon^2}\left(1 + \frac{12}{\varepsilon^2}\right) J^2_{\Delta n}(\Delta n \varepsilon) \right.$$
$$\left. - \left(1 - \frac{12}{\varepsilon^2}\right)[J'_{\Delta n}(\Delta n \varepsilon)]^2\right\}, \tag{3.5.12}$$

$$\varepsilon = \left[1 + q^2\left(\frac{n_0 n}{\Delta n}\right)^2\right]^{1/2}. \tag{3.5.13}$$

A correction factor

$$\exp[\Delta n^2/2(n_0 + n)][1 + (\Delta n)^2/4 n_0 n]^{-2}$$

is introduced in (3.5.10). This factor ensures that in the limit $q \to 0$, the quantity $f(q)$ from (3.5.7) tends towards the known classical expression for the oscillator strength [3.30]:

$$\lim_{q\to 0} f_{n_0 n}(q) = n_0^2 f_{n_0 n}$$

$$= \frac{32}{3}\left(\frac{n_0 n}{\Delta n(n_0+n)}\right)^3 \Delta n J_{\Delta n}(\Delta n) J'_{\Delta n}(\Delta n). \qquad (3.5.14)$$

As is known, formula (3.5.14) gives a good approximation even at small values of n_0, n. The correction factor in (3.5.10) is substantial just for small values of $\Delta n = n - n_0$, n_0, n. When $\Delta n/n \to 0$ it tends towards unity.

One can expect similar results for the cross sections also. As an illustration the results of accurate calculations based on the initial formulas of the Born approximation and the results obtained with the aid of approximate formulas (3.5.10–13) are shown in Fig. 3.10. It is evident from this figure that (3.5.10–13) provide a quite satisfactory approximation.

The results of numerical calculations using the approximate formula (3.5.10) are given in Table 3.1. For convenience of interpolation the principal dependence

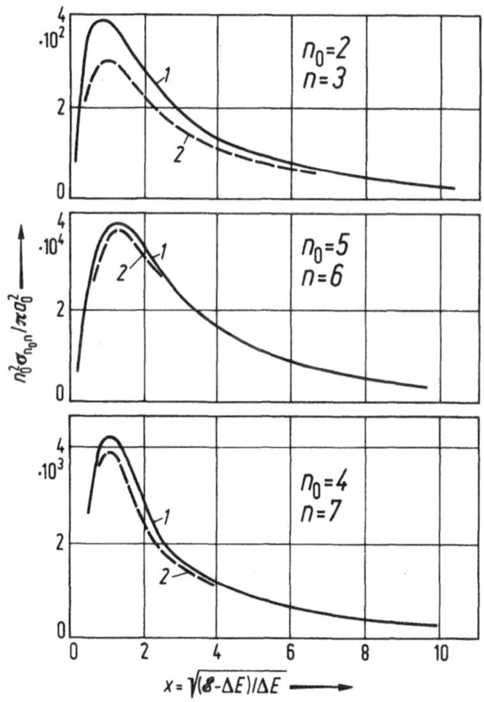

Fig. 3.10. Born effective cross sections for the transitions between highly excited levels: (1) Numerical calculations by means of exact formulas of the Born approximation; (2) Calculations with the approximate formula (3.5.10)

of the cross sections on n_0, n, Δn is given in explicit form

$$\sigma_{n_0 n} = \frac{\pi a_0^2}{z^4} \cdot \frac{1}{n_0^2} \cdot \left(\frac{n_0 n}{\Delta n}\right)^3 \cdot \frac{z^2 \mathrm{Ry}}{\mathscr{E}_0} \cdot \mathscr{F}_{n_0}(\Delta n, \mathscr{E}_0) . \tag{3.5.15}$$

The values of $\mathscr{F}_{n_0}$ are given in Table 3.1. The function $\mathscr{F}_{n_0}$ can be approximately expressed as follows:

$$\mathscr{F}_{n_0} \simeq \frac{8(n_0 n)^2}{(n_0+n)^4 \Delta n}\left(1-\frac{0.25}{\Delta n}\right) \ln\left(1+\frac{\mathscr{E}_0}{E_{n_0}}\right) + \frac{8 n_0^3}{(n_0+n)^3}\left(1-\frac{0.6}{\Delta n}\right)\left(\frac{4}{3}+\frac{(n_0+n)\Delta n}{n^2}\right) \frac{\mathscr{E}_0/E_{n_0}}{1+\mathscr{E}_0/E_{n_0}} . \tag{3.5.16}$$

Under the conditions $1 \ll \Delta n \ll n_0$, n, (3.5.15, 16) give the well-known classical expression for the cross section obtained by Stabler [3.31].

The rate coefficient $\langle v\sigma \rangle$ averaged over the Maxwellian distribution can be written in the form similar to that of (3.5.15):

$$\langle v \sigma_{n_0 n} \rangle = \frac{10^{-8}}{z^3} \cdot \frac{1}{n_0^2} \cdot \left(\frac{n_0 n}{\Delta n}\right)^3 \cdot \left(\frac{z^2 \mathrm{Ry}}{T}\right)^{1/2} \exp\left(-\frac{\Delta E}{T}\right) \Phi_{n_0}(\Delta n, T) [\mathrm{cm}^3 s^{-1}]. \tag{3.5.17}$$

Table 3.1. Born effective cross sections for transitions $n_0 - n$ between highly excited levels. The function $\mathscr{F}_{n_0}(\Delta n, \mathscr{E}_0)$ is given in tabular form. The cross sections are expressed in terms of $\mathscr{F}_{n_0}$ by means of the formula

$$\sigma_{n_0 n} = \frac{\pi a_0^2}{z^4} \frac{1}{n_0^2} \frac{(n_0 n)^3}{(\Delta n)^3} \frac{z^2 \mathrm{Ry}}{\mathscr{E}_0} \mathscr{F}_{n_0}(\Delta n, \mathscr{E}_0)$$

n_0	10			50			100		
$\sqrt{\mathscr{E}_0/E_{n_0}}$ \ Δn	1	2	4	1	2	4	1	2	4
0.20	—	—	—	—	—	—	0.01	—	—
0.28	—	—	—	0.03	0.01	—	0.04	0.03	0.01
0.40	—	—	—	0.10	0.07	0.03	0.11	0.09	0.07
0.57	0.18	0.05	—	0.24	0.22	0.16	0.24	0.24	0.20
0.80	0.40	0.36	0.19	0.43	0.46	0.40	0.43	0.47	0.44
1.13	0.64	0.64	0.50	0.66	0.71	0.69	0.66	0.71	0.73
1.60	0.90	0.86	0.70	0.90	0.92	0.91	0.91	0.93	0.95
2.26	1.16	1.05	0.84	1.16	1.11	1.06	1.16	1.12	1.10
3.20	1.42	1.21	0.94	1.42	1.28	1.17	1.42	1.29	1.21
4.53	1.68	1.37	1.03	1.69	1.44	1.27	1.69	1.45	1.48
9.05	2.21	1.66	1.18	2.22	1.74	1.44	2.22	1.75	1.56
1.81×10^1	2.74	1.96	1.33	2.75	2.04	1.59	2.75	2.05	1.64
3.62×10^1	3.27	2.25	1.48	3.28	2.33	1.75	3.28	2.34	1.79
7.24×10^1	3.80	2.53	1.63	3.81	2.63	1.90	3.81	2.63	1.95
1.45×10^2	4.33	2.83	1.78	4.34	2.92	2.06	4.34	2.93	2.10
2.90×10^2	4.86	3.11	1.92	4.87	3.21	2.21	4.88	3.22	2.26
5.79×10^2	5.38	3.40	2.07	5.40	3.50	2.37	5.40	3.51	2.42
1.16×10^3	5.87	3.66	2.20	5.89	3.77	2.52	5.89	3.78	2.56

The values of Φ_{n_0} are given in Table 3.2. Using (3.5.16) it is possible to obtain the approximate formula for the function Φ_{n_0},

$$\Phi_{n_0}(\Delta n, T) \simeq 2.18 \left\{ \frac{8(n_0 n)^2}{(n_0+n)^4 \Delta n} \left(1 - \frac{0.25}{\Delta n}\right) \varphi(x) \right.$$
$$\left. + \frac{8n_0^3}{(n_0+n)^3} \left(1 - \frac{0.6}{\Delta n}\right) \left(\frac{4}{3} + \frac{(n_0+n)\Delta n}{n^2}\right) [1 - x\varphi(x)] \right\} \tag{3.5.18}$$

where

$$\varphi(x) \simeq \ln\left(1 + \frac{1 + 1.4\gamma x}{\gamma x(1+1.4x)}\right),$$
$$x = E_{n_0}/T, \quad \gamma = \exp(0.5772) = 1.78\,. \tag{3.5.19}$$

It should be remembered that in the region of low energies $\mathcal{E}$, the Born approximation as a rule does not satisfy the normalization condition. The normalized cross sections can be calculated in the quasi-classical approximation (Sects. 3.5.2, 3).

Table 3.2. Rate coefficients $\langle v\sigma_{n_0 n}\rangle$ for transitions between highly excited levels in the Born approximation. The function $\Phi_{n_0}(\Delta n, T)$, which is given in tabular form, can be used to calculate the rate coefficients according to the formula

$$\langle v\sigma_{n_0 n}\rangle = \frac{10^{-8}}{z^3} \frac{1}{n_0^2} \left(\frac{n_0 n}{\Delta n}\right)^3 \left(\frac{z^2 \mathrm{Ry}}{T}\right)^{1/2} \exp\left(-\frac{\Delta E}{T}\right) \Phi_{n_0}(\Delta n, T)\ [\mathrm{cm^3 s^{-1}}]$$

100	n_0	10			50			100		
E_{n_0}/T	Δn	1	2	4	1	2	4	1	2	4
0.01		8.34	5.55	3.57	8.36	5.75	4.17	8.17	5.77	4.26
0.02		7.76	5.24	3.41	7.78	5.43	4.00	7.78	5.45	4.09
0.04		7.18	4.92	3.25	7.20	5.11	3.83	7.20	5.13	3.92
0.08		6.61	4.61	3.08	6.63	4.79	3.66	6.62	4.81	3.75
0.16		6.03	4.29	2.92	6.05	4.47	3.49	6.05	4.49	3.58
0.32		5.46	3.97	2.76	5.47	4.14	3.31	5.47	4.16	3.40
0.64		4.88	3.65	2.59	4.89	3.82	3.13	4.89	3.84	3.22
1.28		4.31	3.33	2.42	4.32	3.49	2.94	4.32	3.50	3.03
2.56		3.75	3.00	2.23	3.75	3.15	2.73	3.75	3.16	2.82
5.12		3.19	2.67	2.04	3.19	2.79	2.51	3.18	2.81	2.59
$1.02.10^1$		2.65	2.32	1.83	2.64	2.43	2.25	2.63	2.44	2.32
$2.05.10^1$		2.13	1.97	1.60	2.11	2.04	1.95	2.10	2.05	2.01
$4.10.10^1$		1.65	1.61	1.34	1.61	1.64	1.60	1.60	1.64	1.64
$8.19.10^1$		1.22	1.25	1.07	1.16	1.23	1.22	1.15	1.22	1.24
$1.64.10^2$		0.86	0.92	0.80	0.78	0.84	0.84	0.77	0.83	0.84
$3.28.10^2$		0.57	0.63	0.56	0.48	0.52	0.52	0.46	0.50	0.50
$6.55.10^2$		0.36	0.41	0.37	0.36	0.29	0.30	0.25	0.26	0.27

3.5.2 Transitions Between Highly Excited Levels in the Quasi-Classical Approximation

When the principal quantum number n is large, the motion of an atomic electron can be treated as quasi-classical motion. We shall consider the problem of the calculation of the probability for the transition $n_0 - n$ in a quasi-classical limit $\hbar \to 0$.[1] The quasi-classical theory may be developed in different ways. One may proceed from the wave packets which in the limit of $\hbar \to 0$ give the classical trajectories. We shall use below a somewhat different approach which enables one to treat the distortion of motion of the highly excited electron during the collision in the simplest way.

We proceed from the following representation of the atomic wave function:

$$\Psi_a \propto \exp\left(\frac{\mathrm{i}}{\hbar} S_a\right) , \tag{3.5.20}$$

where S_a is the classical action function. The probability of transition $a_0 - a$ is then written in the form

$$\begin{aligned} W_{a_0 a} &= \lim_{t\to\infty} |\langle \Psi_a^{(0)}(t)|\Psi_{a_0}(t)\rangle|^2 \\ &\propto \lim_{t\to\infty} \left|\left\langle \exp\left(\frac{\mathrm{i}}{\hbar} S_a^{(0)}\right) \right| \exp\left(\frac{\mathrm{i}}{\hbar} S_{a_0}\right)\right\rangle\right|^2 \end{aligned} \tag{3.5.21}$$

where the superscript 0 corresponds to an unperturbed atom. To determine the action function in the quasi-classical limit $\hbar \to 0$, one may use the classical Hamilton–Jacobi equation. The remainder of this section is based on [3.34], where this approach has been developed.[2]

We shall write the action function $S_{a_0}(t)$ in the form $S_{a_0}(t) = S_{a_0}^{(0)}(t) + \Delta S(t)$ where $\Delta S(t)$ is the increment of the action function due to the collision. When $t \to -\infty$, $\Delta S \to 0$. It is convenient to use as dynamical variables the phase variables u_j and their conjugated momenta I_j. The phase variables vary between 0 and 2π. The unperturbed action function is defined by [3.36]

$$S_a^{(0)} = \sum_j I_j u_j - E_a t . \tag{3.5.22}$$

From the Bohr quantization conditions it follows that

$$I_1 = n\hbar, \quad I_2 = l\hbar, \quad I_3 = m\hbar , \tag{3.5.23}$$

where n, l, and m are the principal, orbital, and magnetic quantum numbers;

---

[1] The purely classical impulse approximation is developed in [3.31–33].

[2] The correspondence principle method has been developed by Percival and Richards [3.35]. In this article one can also find a survey of classical, semiclassical and quantal methods for collisions between charged particles and highly excited atoms.

$a = nlm$. Using (3.5.22, 23) we obtain

$$W_{a_0 a} = \lim_{t\to\infty} \left| \frac{1}{(2\pi)^3} \int \int_0^{2\pi} \int du_1 du_2 du_3 \exp\left[-\frac{i}{\hbar}(S_a^{(0)} - S_{a_0}^{(0)}) + \frac{i}{\hbar}\Delta S \right] \right|^2$$

$$= \lim_{t\to\infty} \left| \frac{1}{(2\pi)^3} \int d\boldsymbol{u} \exp\left(-i\boldsymbol{k}\cdot\boldsymbol{u} + \frac{i}{\hbar}\Delta S \right) \right|^2 , \qquad (3.5.24)$$

where the vector $\boldsymbol{k}$ has the components $\Delta n = n - n_0$, $\Delta l = l - l_0$, $\Delta m = m - m_0$.

It is not difficult to show that the transition probability (3.5.24) satisfies the unitarity condition independently of the approximation used in calculating ΔS. Since the summation over the final states a is equivalent to summation over $\boldsymbol{k}$, we have

$$\sum_a W_{a_0 a} = \frac{1}{(2\pi)^3} \int d\boldsymbol{u} \int d\boldsymbol{u}' \delta(\boldsymbol{u} - \boldsymbol{u}') \exp\left[\frac{i}{\hbar}\Delta S(\boldsymbol{u}) - \frac{i}{\hbar}\Delta S(\boldsymbol{u}') \right] = 1 .$$

The Hamilton–Jacobi equation for the action function S is

$$\frac{\partial S}{\partial t} + H^{(0)}\left(u_j, \frac{\partial S}{\partial u_j} \right) + V\left(u_j, \frac{\partial S}{\partial u_j}, t \right) = 0 , \qquad (3.5.25)$$

where $H^{(0)}$ is the unperturbed Hamiltonian function of an atom and $V(u_j, \partial S/\partial u_j, t)$ is the perturbation induced by the incident particle. When $\Delta S \ll S^{(0)}$ in first-order classical perturbation theory, the following approximate equation for ΔS can be obtained:

$$\frac{\partial}{\partial t}\Delta S - \omega_0 \frac{\partial}{\partial u_1}\Delta S = -V\left(u_j, \frac{\partial S_{a_0}^{(0)}}{\partial u_j}, t \right) . \qquad (3.5.26)$$

We take into account that for the particular case of the Coulomb field $E_{a_0} = E_{n_0}$, and the unperturbed Hamiltonian function $H^{(0)}$ depends on phase variable u_1 only. The quantity $\omega_0 = \partial H^{(0)}/\partial I_1 = 2z^2 \mathrm{Ry}/\hbar n_0^3$ is the classical frequency for an electron with binding energy E_{n_0}. The term $V(u_j, \partial S^{(0)}/\partial u_j, t)$ in (3.5.26) includes the derivatives of unperturbed action function $\partial S^{(0)}/\partial u_j$.

The right-hand side of (3.5.26) can be expanded in a Fourier series,

$$\begin{aligned} V &= \sum_{\boldsymbol{q}} V_{\boldsymbol{q}}(t) \exp(i\boldsymbol{q}\cdot\boldsymbol{u}) \\ &= \sum_{q_1, q_2, q_3} V_{q_1 q_2 q_3}(t) \exp[i(q_1 u_1 + q_2 u_2 + q_3 u_3)] , \end{aligned} \qquad (3.5.27)$$

where the quantities q_1, q_2, q_3 are integers. With the aid of expansion (3.5.27), the solution of (3.5.26) can be written in the following form:

$$\begin{aligned} \Delta S(t) &= \sum_{\boldsymbol{q}} C_{\boldsymbol{q}}(t) \exp(i\boldsymbol{q}\cdot\boldsymbol{u} - iq_1\omega_0 t) \\ C_{\boldsymbol{q}}(t) &= \int_{-\infty}^{t} V_{\boldsymbol{q}}(t') \exp(-iq_1\omega_0 t') dt' . \end{aligned} \qquad (3.5.28)$$

Thus

$$W_{a_0 a} = \left| \frac{1}{(2\pi)^3} \int_0^{2\pi} d\boldsymbol{u} \exp\left(-\mathrm{i}\boldsymbol{k} \cdot \boldsymbol{u} + \frac{\mathrm{i}}{\hbar} \sum_{\boldsymbol{q}} C_{\boldsymbol{q}} e^{\mathrm{i}\boldsymbol{q} \cdot \boldsymbol{u}} \right) \right|^2 , \tag{3.5.29}$$

where

$$C_{\boldsymbol{q}} = C_{\boldsymbol{q}}(\infty) = \int_{-\infty}^{\infty} V_{\boldsymbol{q}}(t) \exp(-\mathrm{i}\omega_0 t)\, dt . \tag{3.5.30}$$

Formulas (3.5.29, 30) enable the probability of the transition $a_0 - a$ to be calculated in the quasi-classical limit within the framework of the classical perturbation theory $\Delta S \ll S^{(0)}$. The validity condition $\Delta S \ll S^{(0)}$ for (3.5.29) differs substantially from the validity condition for quantum perturbation theory $\Delta S \ll \hbar$. In fact, when n is large it follows from (3.5.22, 23) that $S^{(0)} \sim n\hbar \gg \hbar$. Therefore the classical perturbation theory is valid in the wider range $\Delta S \ll n\hbar$.

The first order of quantum perturbation theory is obtained from (3.5.29) by putting $\exp(\mathrm{i}\Delta S/\hbar) \simeq 1 + \mathrm{i}\Delta S/\hbar$. Then

$$W^{\mathrm{I}}_{a_0 a} = \frac{1}{\hbar^2} |C_{\boldsymbol{k}}|^2 = \frac{1}{\hbar^2} \left| \int_{-\infty}^{\infty} V_{\boldsymbol{k}}(t) \exp[\mathrm{i}(n - n_0)\omega_0 t]\, dt \right|^2 . \tag{3.5.31}$$

The calculation by means of (3.5.29) requires the use of a Fourier expansion of the classical interaction potential $V(u_j, \partial S^{(0)}/\partial u_j, t)$. Such calculations are extremely cumbersome. It is known, however, that in the quasi-classical limit the quantum matrix element of the time-dependent interaction $V(t)$ coincides with the Fourier component $V_{\boldsymbol{k}}(t) (k_1 = n - n_0,\ k_2 = l - l_0,\ k_3 = m - m_0)$:

$$\langle n_0 l_0 m_0 | V(t) | nlm \rangle \rightarrow V_{\boldsymbol{k}}(t) . \tag{3.5.32}$$

Hence (3.5.31) is equivalent to the first order of quantum perturbation theory for the probability of the transition $n_0 l_0 m_0 - nlm$ caused by a time-dependent potential $V(t)$. Thus the relation (3.5.32) enables one to calculate the increment of the action function ΔS and the transition probability $W_{a_0 a}$ using the quantum matrix elements. The summation over $\boldsymbol{q} = (q_1, q_2, q_3)$ in (3.5.29) means that the transitions via virtual levels $a' \equiv n'l'm'$ are taken into account.

To obtain the probability for the transition $n_0 - n$ one has to sum with respect to lm and to average over $l_0 m_0$:

$$W_{n_0 n} = \frac{1}{n_0^2} \sum_{l_0 m_0 l m} W_{n_0 l_0 m_0,\, nlm} . \tag{3.5.33}$$

For the quasi-classical probability (3.5.29) this is equivalent to summation with respect to $k_2 = l - l_0$, $k_3 = m - m_0$, and averaging over l_0, m_0.

Calculation of the probability $W_{n_0 n}$ can be simplified if one assumes that ΔS does not depend on u_2 and u_3, and extends the summation with respect to k_2 and k_3 from $-\infty$ to ∞. Then, using the average quantities C_q that do not depend on the angular momenta $q_2 = l_0 - l'$ and $q_3 = m_0 - m'$, one can write

$$W_{n_0n} = \left| \frac{1}{2\pi} \int_0^{2\pi} du \exp\left[-i(n - n_0)u + \frac{i}{\hbar} \sum_q C_q e^{iqu} \right] \right|^2 , \qquad (3.5.34)$$

where the notations $u = u_1$, $q = q_1$ are introduced.

It is convenient to define the average value C_q by means of the following relation:

$$C_q = C_{n_0n'} = \sqrt{\frac{1}{n_0^2} \sum_{l_0m_0l'm'} |C_{n_0l_0m_0\, n'l'm'}|^2}, \quad q = n' - n_0 ,$$

$$C_{n_0l_0m_0\, n'l'm'} = \int_{-\infty}^{\infty} \langle n_0l_0m_0|V(t)|n'l'm'\rangle \exp(-iq\omega_0 t)\, dt . \qquad (3.5.35)$$

This definition ensures an accurate result for the probability $W^{\mathrm{I}}_{n_0n}$ in first-order quantum perturbation theory. The effective cross section for the transition $n_0 - n$ is

$$\sigma_{n_0n} = 2\pi \int_0^{\infty} \rho\, d\rho\, W_{n_0n}(\rho) . \qquad (3.5.36)$$

When formula (3.5.34) for the probability W_{n_0n} is used we obtain the normalized cross section.

3.5.3 Transitions Between Adjacent Levels $\Delta n = 1$[1]

Now we shall consider the special case when only matrix elements V_q with $q = \Delta n = \pm 1$ exist. We shall assume that $C_{+1} = C_{-1} = C$. In this case the analytical expression for the transition probability can be obtained [3.39]:

$$W_{n_0n} = \left| J_{\Delta n}\left(\frac{2}{\hbar} C \right) \right|^2 \qquad (3.5.37)$$

where $J_{\Delta n}$ is the Bessel function, $\Delta n = |n - n_0|$.

We shall use the dipole approximation for the interaction potential

$$V(t) = e\boldsymbol{d} \cdot \boldsymbol{R}/R^3 , \qquad (3.5.38)$$

where $\boldsymbol{d}$ is the electric dipole moment of the atom. In order to eliminate the pole in the potential it is convenient to define $\boldsymbol{R}$ in the following way:

$$\boldsymbol{R} = \boldsymbol{r} + \frac{\boldsymbol{\rho}}{\rho} \left(\sqrt{\rho^2 + \rho_0^2} - \rho \right) ,$$

[1] The classical path first-order approximation with the cutoff procedure to ensure conservation of probability has been used for transition between adjacent levels in [3.37, 38]. The results obtained by this method are close to those given below.

where $\boldsymbol{r}$ is the radius vector of the outer electron, $\boldsymbol{\rho}$ is the vector of the impact parameter ρ which lies in a collision plane and is perpendicular to the velocity vector $\boldsymbol{v}$. The parameter $\rho_0 \sim n^2 a_0/z$ is introduced to avoid the pole in W at $\rho = 0$. The magnitude of ρ_0 has to be taken so that at high velocities of the outer electron the quasi-classical cross section coincides with the quantum Born cross sections.

When the energy of the outer electron $\mathscr{E} \gg |E_n|$ both in the case of neutral atom and in the case of an ion, the rectilinear trajectory may be used:

$$\boldsymbol{r}(t) = \boldsymbol{\rho} + \boldsymbol{v}t \,. \tag{3.5.39}$$

Defining ω_0 in (3.5.35) by means of the symmetrical relation

$$\omega_0 = \frac{n_0 + n}{(n_0 n)^2} \frac{z^2 \mathrm{Ry}}{\hbar}, \quad \hbar\omega_0(n - n_0) = E_n - E_{n_0} \,, \tag{3.5.40}$$

we obtain

$$C_q^2 = 4\left(\frac{\hbar}{mv}\right)^2 \frac{\alpha_{n_0 n}^2}{q^4(\rho_0^2 + \rho^2)} \xi(\beta) \,, \tag{3.5.41}$$

$$\alpha_{n_0 n}^2 = \frac{n_0^2 f_{n_0 n} q^3}{(n_0 + n) n_0 n \, z^2}, \quad q = \Delta n \,, \tag{3.5.42}$$

$$\begin{gathered} \xi(\beta) = \beta^2 [\mathrm{K}_0^2(\beta) + \mathrm{K}_1^2(\beta)] \,, \\ \beta = q \frac{n_0 + n}{(n_0 n)^2} \frac{z^2 \mathrm{Ry}}{\hbar v} \sqrt{\rho_0^2 + \rho^2} \,, \end{gathered} \tag{3.5.43}$$

where $\mathrm{K}_0(\beta)$ and $\mathrm{K}_1(\beta)$ are the modified Bessel functions, and $f_{n_0 n}$ is the oscillator strength for the transition $n_0 \to n$. In the Kramers approximation for $f_{n_0 n}$ we have

$$\alpha_{n_0 n}^2 = \frac{32}{3\pi\sqrt{3}} \cdot \frac{(n_0 n)^2}{(n_0 + n)^4 z^2}, \quad \alpha_{n_0 n} \simeq 1.40 \frac{n_0 n}{(n_0 + n)^2 z} \,. \tag{3.5.44}$$

The more accurate calculation using (3.5.14) gives for the case $n = n_0 + 1$ a numerical factor in (3.5.44) of 1.31. The function $\xi(\beta)$ can be approximately fitted to the simple formula

$$\xi(\beta) \simeq (1 + \pi\beta) \exp(-2\beta) \,. \tag{3.5.45}$$

The error does not exceed 10 percent.

Formula (3.5.37) and the dipole approximation are applicable for transitions with $\Delta n = 1$ only. When $\Delta n > 1$ the expression (3.5.37) takes into account only transitions via the virtual levels $n_0 \to n' \to n_0 + 1 \to n_0 + 2 \cdots$ but it ignores the direct transitions $n_0 \to n$. Therefore the asymptotic dependence of the cross section on the electron energy $\mathscr{E}$ will not be correct. Moreover, in the region of low energy of the outer electron $\mathscr{E} \ll z^2\mathrm{Ry}$, the higher multipole interactions

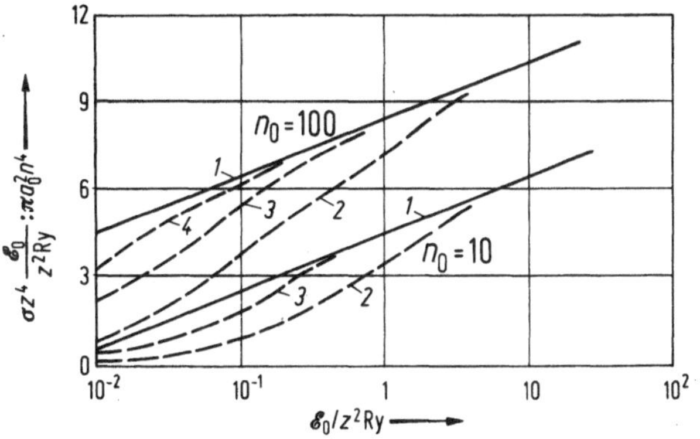

Fig. 3.11. Cross-sections for the transitions $n = n_0 + 1$ between highly excited levels: (1) Born approximation; (2) Quasi-classical cross sections for the transitions 10–11 and 100–101 of neutral hydrogen; (3) Quasi-classical cross sections for ions with $z = 4$; (4) Quasi-classical cross section for ion with $z = 8$

Table 3.3. Quasi-classical (normalized) cross sections for transitions $n_0 \to n_0 \pm 1$ in neutral hydrogen. Function $\mathscr{F}_{n0}(\mathscr{E}_0)$.

$\sigma_{n_0,n_0\pm 1} = \pi a_0^2 n_0(n_0 \pm 1)^3(\text{Ry}/\mathscr{E}_0)\mathscr{F}_{n0}(\mathscr{E}_0)$

$\mathscr{E}_0/\text{Ry}$ n_0:	10	20	30	40	50	100	150	200	250
0.01	0.016	0.049	0.090	0.139	0.186	0.414	0.617	0.791	0.942
0.02	0.052	0.144	0.233	0.324	0.413	0.790	1.07	1.29	1.47
0.04	0.138	0.325	0.500	0.655	0.791	1.29	1.61	1.85	2.03
0.08	0.338	0.664	0.921	1.13	1.30	1.85	2.19	2.42	2.60
0.16	0.670	1.13	1.44	1.67	1.85	2.42	2.75	2.98	3.16
0.32	1.14	1.68	2.00	2.24	2.42	2.98	3.30	3.53	3.70
0.64	1.66	2.21	2.54	2.77	2.95	3.50	3.82	4.04	4.21
1.28	2.12	2.67	3.00	3.22	3.40	3.94	4.25	4.48	4.65
2.56	2.51	3.06	3.37	3.60	3.77	4.31	4.62	4.84	5.02
5.12	2.85	3.39	3.70	3.92	4.10	4.63	4.94	5.17	5.34
10.24	3.15	3.69	4.00	4.22	4.40	4.93	5.24	5.42	5.63

must be taken into account [3.40]. In the case $\Delta n = 1$ the dipole approximation provides good results.

The results of numerical calculations by means of formulas (3.5.36, 37, 41) and the quantum Born approximation for the transitions 10–11 and 100–101 are compared in Fig. 3.11. In the region $\mathscr{E} \ll z^2\text{Ry}$, the Born approximation gives an overestimate of the cross section. This is because of the fact that when ρ is small the probability W^{I} calculated by first-order perturbation theory is far greater than unity. When formula (3.5.37) is used, the maximum value of $W(\rho)$ is approximately equal to 0.3. At high energies, the quasi-classical approximation and the Born approximation coincide.

Table 3.4. Quasi-classical cross sections for transitions $n_0 \rightarrow n_0 \pm 1$ in hydrogenlike ions. Function $\mathscr{F}_{n_0}(z, \mathscr{E}_0)$.

$$\sigma_{n_0,n_0\pm 1} = \frac{\pi a_0^2}{z^4} n_0(n_0 \pm 1)^3 (z^2 \mathrm{Ry}/\mathscr{E}_0) \mathscr{F}_{n_0}(z, \mathscr{E}_0)$$

$\mathscr{E}_0/z^2\mathrm{Ry}$ \ n_0	10			50			100			
z	2	4	∞	2	4	∞	2	4	8	∞
0.01	0.06	0.14	0.41	0.41	0.79	1.65	0.79	1.29	1.81	2.22
0.02	0.15	0.33	0.65	0.79	1.29	1.94	1.29	1.85	2.28	2.52
0.04	0.33	0.64	0.92	1.29	1.82	2.23	1.85	2.38	2.67	2.79
0.08	0.67	1.01	1.21	1.85	2.28	2.52	2.42	2.84	3.01	3.07
0.16	1.11	1.38	1.50	2.39	2.67	2.80	2.95	3.22	3.31	3.34
0.32	1.56	1.72	1.79	2.85	3.01	3.07	3.39	3.55	3.60	3.61
0.64	1.95	2.04	2.07	3.23	3.31	3.34	3.77	3.85	3.88	3.88
1.28	2.29	2.34	2.36	3.56	3.60	3.62	4.09	4.14	4.15	4.15
2.56	2.60	2.63	2.64	3.86	3.88	3.89	4.39	4.41	4.42	4.42
5.12	2.89	2.91	2.91	4.14	4.15	4.16	4.68	4.69	4.69	4.69
10.24	3.18	3.18	4.42	4.42	4.42	4.95	4.96	4.96	4.96	4.96

Table 3.5. Quasi-classical rate coefficients $\langle v\sigma_{n_0,n_0\pm 1}\rangle$ for neutral hydrogen. Function $\Phi_{n_0}(T)$.

$$\langle v\sigma_{n_0,n_0\pm 1}\rangle = 10^{-8} n_0(n_0 \pm 1)^3 (\mathrm{Ry}/T)^{1/2} \exp(-\Delta E/T) \Phi_{n_0}(T) \quad [\mathrm{cm^3 s^{-1}}]$$

T/Ry \ n_0	10	20	30	40	50	100	150	200	250
0.01	0.052	0.130	0.21	0.30	0.37	0.78	1.09	1.34	1.56
0.02	0.12	0.29	0.46	0.62	0.76	1.35	1.75	2.14	2.42
0.04	0.29	0.61	0.89	1.13	1.35	2.15	2.69	3.10	3.43
0.08	0.62	1.14	1.54	1.87	2.14	3.11	3.72	4.17	4.53
0.16	1.15	1.87	2.38	2.78	3.10	4.17	4.93	5.30	5.67
0.32	1.87	2.77	3.36	3.80	4.15	5.28	5.94	6.42	6.79
0.64	2.72	3.75	4.38	4.85	5.21	6.35	7.02	7.49	7.85
1.28	3.62	4.71	5.37	5.84	6.20	7.34	8.00	8.46	8.83

Table 3.6. Quasi-classical rate coefficients $\langle v\sigma_{n_0,n_0\pm 1}\rangle$ for hydrogenlike ions. Function $\Phi_{n_0}(z, T)$.

$$\langle v\sigma_{n_0,n_0\pm 1}\rangle = \frac{10^{-8}}{z^3} n_0(n_0 \pm 1)^3 (z^2\mathrm{Ry}/T)^{1/2} \exp(-\Delta E/T) \Phi_{n_0}(z, T) \quad [\mathrm{cm^3 s^{-1}}]$$

$T/z^2\mathrm{Ry}$ \ n_0	10			50			100			
z	2	4	∞	2	4	∞	2	4	8	∞
0.01	0.14	0.30	0.69	0.77	1.34	2.37	1.34	2.03	2.68	3.13
0.02	0.31	0.60	1.08	1.36	2.11	3.22	2.14	3.01	3.72	4.17
0.04	0.62	1.06	1.56	2.14	3.01	3.99	3.10	4.04	4.68	5.05
0.08	1.13	1.64	2.09	3.08	3.92	4.70	4.15	5.03	5.55	5.81
0.16	1.80	2.31	2.67	4.08	4.80	5.35	5.20	5.94	6.31	6.49
0.32	2.57	3.00	3.26	5.06	5.60	5.93	6.19	6.74	6.97	7.06
0.64	3.36	3.69	3.85	5.95	6.32	6.50	7.08	7.45	7.58	7.63
1.28	4.13	4.35	4.44	6.75	6.98	7.07	7.87	8.10	8.17	8.19

The results of numerical calculations by means of formulas (3.5.36, 37, 41) for transitions $\Delta n = 1$ in hydrogen and hydrogenlike ions are given in Tables 3.3 and 3.4. The rate coefficients averaged over the Maxwellian distribution are given in Tables 3.5 and 3.6. Similar to Tables 3.1 and 3.2 for the Born cross sections, Tables 3.3–6 give the values of the functions $\mathcal{F}_{n_0}$ and Φ_{n_0}. The cross sections σ and the rate coefficients $\langle v\sigma \rangle$ are determined by formulas (3.5.15) and (3.5.17).

In conclusion we note the following. In Coulomb units, $C_q \propto 1/z$. Therefore when $z \gg 1$ the normalization is not substantial and the cross sections are close to the values obtained in first-order quantum perturbation theory.

4 Collisions Between Heavy Particles

Collisions between heavy particles are treated here in the impact-parameter approximation. In the two-state approximation, a simple formula for the transition probability is obtained which is correct in both limiting cases of high and low velocities, and provides reasonable interpolation in the intermediate case. The simple formulas for the case of the multipole potentials and the numerical results for the most interesting cases of Coulomb and dipole–dipole interactions are given. Charge exchange cross sections are also estimated. A more detailed account of the physics of collisions between heavy particles can be found in [4.1–7].

4.1 Impact-Parameter Method

4.1.1 General Formulas

The collisions of atoms with heavy particles (e.g., atoms, ions) in a wide range of velocities can be described quasi-classically by considering the distance between the centers of the colliding particles as an explicit function of time $\boldsymbol{R}(t)$. If the colliding particles are neutral the trajectory $\boldsymbol{R}(t)$ is usually assumed to be rectilinear: $\boldsymbol{R}(t) = \boldsymbol{\rho} + \boldsymbol{v}t$, where $\boldsymbol{\rho}$ and $\boldsymbol{v}$ are the impact parameter and relative velocity. With such an approach the effective cross section for the transition of a system from one state to another is determined by the formula

$$\sigma = 2\pi \int_0^\infty W(\rho, v)\rho d\rho, \tag{4.1.1}$$

where $W(\rho, v)$ is the probability of the transition in a collision with impact parameter ρ and relative velocity v. The problem of calculating the transition probability $W(p, v)$ reduces to a solution of a system of equations for the time-dependent amplitudes of the state, which we write in the form

$$\mathrm{i}\hbar \frac{da_k}{dt} = \sum_m V_{km}(t) \exp(\mathrm{i}\omega_{km}\, t)\; a_m, \tag{4.1.2}$$

where $\hbar\omega_{km} = E_k - E_m$ is the energy difference between the unperturbed levels, and V_{km} are the matrix elements of the interaction, depending on the parameters $\boldsymbol{\rho}$ and $\boldsymbol{v}$.

In order to calculate the probability of the transition $0 \to n$ it is necessary to find the solution of the system (4.1.2) satisfying the initial conditions

$$a_k(-\infty) = \delta_{0k}. \tag{4.1.3}$$

The required probability is given by

$$W_{0n} = |a_n(\infty)|^2 . \tag{4.1.4}$$

The matrix V_{km} is Hermitian. Therefore the amplitudes of the states a_n and also the transition probabilities W_{0n} satisfy the normalization conditions

$$\sum_n |a_n(x)|^2 = 1, \sum_n W_{0n} = 1 . \tag{4.1.5}$$

By integrating the system (4.1.2) in first-order perturbation theory, i.e., assuming in the right-hand side of (4.1.3) $a_m(x) = \delta_{m0}$, it is not difficult to obtain the quasi-classical formula of the Born approximation,

$$W_{0n}^B = \frac{1}{\hbar^2}\left|\int_{-\infty}^{\infty} V_{0n}(t) \exp(\mathrm{i}\, \omega_{n0}\, t)\, dt\right|^2 , \tag{4.1.6}$$

In many cases the approximation (4.1.6) proves to be completely unsuitable. This approximation does not satisfy the normalization conditions (4.1.5), as a result of which W^{B} can exceed unity, which contradicts the physical meaning of this quantity. The approximation (4.1.6) often gives incorrect results at low velocities even when $W^{\mathrm{B}} \ll 1$. All these deficiencies are inherent also in the Born approximation for the collisions between electrons and atoms. However in the case of collisions between heavy particles the thermal velocities are considerably smaller than the thermal velocities of electrons and the errors of the Born approximation prove to be far greater than the errors in the case of electronic collisions. Therefore it is necessary to solve the system (4.1.2) without recourse to the series expansion. A general survey of the problem is given in [4.3–5].

We shall consider below only the cases when the main contribution to the transition probability is given by a multipole interaction at comparatively large distances, larger than atomic dimensions. In these cases the maximum values of the cross sections are of the order of atomic dimension squared or greater. To investigate the properties of the solutions of the system (4.1.2) we shall restrict ourselves to the consideration of a system of two equations, which is analogous to the two-state close coupling approximation in the general theory of inelastic collisions.

The exact analytical solution of the system of the two equations can be obtained only for some special potentials V_{km}. One of the well known example of such solution is the so-called Rozen–Zener model. At present many such models are being investigated [4.3–5]. These models however do not describe several potentials met in practice. For example, the exact solutions do not exist for the multipole potentials $V \propto R^{-n}$.

It is of interest therefore to use another approach based on an approximate solution of the system of two equations (4.1.2) not connected with some special form of the potentials. It is possible to construct an approximate solution correctly describing both limiting cases of high and low velocities: $v \to \infty$ and $v \to 0$. Examination of special examples, for which there exists an accurate solution of the two equations, shows that this approximation also describes sufficiently well the intermediate range.

4.1.2 Two-State Approximation

We shall consider the transition $0 \to 1$ and retain in the system (4.1.2) only the two equations corresponding to the levels 0, 1. We shall introduce the notation

$$V_0 = V_{00}/\hbar, \quad V_1 = V_{11}/\hbar, \quad V e^{i\varphi} = V_{01}/\hbar = V_{10}^*/\hbar\,,$$

$$\omega = (E_1 - E_0)/\hbar\,,$$

where V is a real quantity. The substitution

$$x = vt\,,$$

$$a_0 = b_0 \exp\left[-\frac{\mathrm{i}}{2}\left(\varphi + \frac{\omega}{v}x\right)\right] \tag{4.1.7}$$

$$a_1 = b_1 \exp\left[\frac{\mathrm{i}}{2}\left(\varphi + \frac{\omega}{v}x\right)\right]$$

leads to the system of two equations

$$\begin{aligned} \mathrm{i}\frac{db_0}{dx} &= \left(\frac{1}{v}V_0 - \frac{\omega}{2v}\right) b_0 + \frac{1}{v}Vb_1\,, \\ \mathrm{i}\frac{db_1}{dx} &= \frac{1}{v}Vb_0 + \left(\frac{1}{v}V_1 + \frac{\omega}{2v}\right) b_1\,, \end{aligned} \tag{4.1.8}$$

with the initial conditions

$$b_0 = \exp\left(\mathrm{i}\frac{\omega}{v}x\right), \quad b_1(-\infty) = 0\,. \tag{4.1.9}$$

The probability of the transition $0 \to 1$ is obviously given by $W = |b_1(\infty)|^2$.

If the states 0,1 are the unperturbed states of electrons in the system of two colliding particles, the transition is described by nondiagonal matrix element V of the interaction, and the diagonal matrix elements $V_0 = V_1 = 0$, then

$$\begin{aligned} \mathrm{i}\frac{db_0}{dx} &= \frac{1}{v}Vb_1 - \frac{\omega}{2v}b_0\,, \\ \mathrm{i}\frac{db_1}{dx} &= \frac{1}{v}Vb_0 - \frac{\omega}{2v}b_1\,. \end{aligned} \tag{4.1.10}$$

The exact solution of this very simplified system can be obtained only in few special cases.

(i) *The case of zero-energy defect* $\omega = 0$. The solution of the system (4.1.10) with the initial conditions (4.1.9) is

$$b_0 = \cos\left(\frac{1}{v}\int_{-\infty}^{x} V\,dx\right), \quad b_1 = -\mathrm{i}\,\sin\left(\frac{1}{v}\int_{-\infty}^{x} V\,dx\right) \tag{4.1.11}$$

and

$$W = \sin^2\left(\frac{1}{v}\int_{-\infty}^{\infty} V\,dx\right). \tag{4.1.12}$$

(ii) *Square well* $V(x) = V_C$ *at* $|x| < \rho_0$ *and* $V(x) = 0$ *at* $|x| > \rho_0$. The exact solution of the system (4.1.10) yields

$$W = \frac{4V_C^2}{4V_C^2 + \omega^2}\sin^2\left(\frac{\rho_0}{v}\sqrt{4V_C^2 + \omega^2}\right). \tag{4.1.13}$$

(iii) $V(x) = V\cosh^{-1}(\gamma x)$, γ being a constant. For the transition probability one obtains

$$W = \cosh^{-2}\left(\frac{\pi\omega}{2\gamma v}\right)\sin^2\left(\frac{\pi V_C}{\gamma v}\right). \tag{4.1.14}$$

Let us assume now that the states 0, 1 are the states of the electron of a system of two particles at the fixed internuclear distance R with energies $E_0(R), E_1(R)$ (quasimolecular complex). If the principal contribution to the transition probability $0 \to 1$ is given by the "point of intersection" of the terms $E_0(R) = E_1(R)$ the system (4.1.8) leads to the Landau–Zener formula [4.8].

$$W = 2\exp(-\delta)[1 - \exp(-\delta)] \tag{4.1.15}$$

$$\delta = \frac{2\pi V^2(x)}{v^2\left|\frac{dV_1}{dx} - \frac{dV_0}{dx}\right|_{x = x_0}}. \tag{4.1.16}$$

The point x_0 is given by

$$V_1(x_0) - V_0(x_0) + \omega = 0. \tag{4.1.17}$$

For more general assumptions about the form of the functions $V_0(x), V_1(x), V(x)$ and the magnitude of initial energy defect ω, the system of coupled equations (4.1.8) can be solved only approximately. Various approximate methods have been discussed in [4.6].

An approach to an approximate solution of the system (4.1.8) with a wide region of validity has been proposed in [4.9].

Introducing the new function $K(x)$ by the relation

$$K(x) = b_1(x)/b_0(x) \tag{4.1.18}$$

we can transform the system (4.1.8) into a nonlinear Riccatti equation for $K(x)$. Solution of this equation permits to determine $K(x)$. The probabilities are given by

$$b_0(x)^2 = \frac{1}{1+|K(x)|^2}, \quad b_1(x)^2 = \frac{|K(x)|^2}{1+|K(x)|^2} . \tag{4.1.19}$$

According to (4.1.19) the normalization (unitarity) conditions (4.1.5) are satisfied independently of the approximation used in the calculation of $K(x)$. The Born approximation for the transition probability can be deduced for small values of V/v.

An approximate solution of the equation for $K(x)$ leads to the following transition probability

$$W = \left| \frac{1}{v} \int_{-\infty}^{\infty} dx\, V \exp\left\{ \mathrm{i} \int_0^x dx_1 \sqrt{[\alpha(x_1)]^2 + 4\left[\frac{1}{v}V(x_1)\right]^2} \right\} \right|^2 ,$$

$$\alpha(x) = (\omega + V_1 - V_0)/v . \tag{4.1.20}$$

When $|V_1 - V_0| \ll \omega$, it can be assumed that

$$W = \left| \frac{1}{v} \int_{-\infty}^{\infty} dx\, V \exp\left\{ \frac{\mathrm{i}}{v} \int_0^x dx_1 [\omega^2 + 4V^2(x_1)]^{1/2} \right\} \right|^2 . \tag{4.1.21}$$

The interpolation formula (4.1.20) agrees with the accurate values of the transition probability in both the limiting cases $v \to 0$ and $v \to \infty$. Examination of various specific potentials reveals that (4.1.20) gives good results also in the intermediate range. For example, (4.1.20) yields exact results in the above-considered cases of square well (i) and zero-energy defect $\omega = 0$ (ii). In the case (iii) the expression for W, obtained by means of (4.1.20), coincides with (4.1.14) for $\pi\omega/2\gamma v < 1$.

For $\pi\omega/2\gamma v \gg 1$, when W is exponentially small, (4.1.20) gives the correct order of magnitude for the preexponential factor.

By means of (4.1.20, 21) it is possible to obtain the Bates results [4.10]. Instead of the Landau–Zener formula (4.1.15) the evaluation of the integral in (4.1.20) yields

$$W = \begin{cases} 2\delta & \text{when } \delta \ll 1 \\ \text{const.} \exp(-\delta) & \text{when } \delta \gg 1 . \end{cases}$$

For small δ this formula coincides with the Landau–Zener formula, and for large δ differs from it only by the constant pre-exponential factor of order unity. It can be shown that in the intermediate region $0.2 \leq \delta \leq 5$ the method considered above leads to the results which differ from the Landau–Zener fromula by no more than 20 percent.

The general formula (4.1.20) has been compared with the results of numerical solution of the system of two equations for different potentials. In all cases it gives good interpolation between two limiting cases $v \to 0$ and $v \to \infty$. It has also been shown that the cross section for the excitation of atomic hydrogen into the stationary states $2s$ and $2p$ in the collision between the two hydrogen atoms calculated by means of formula (4.1.20) differs from the cross section obtained by means of an accurate numerical solution by no more than several percent [4.3, 11, 12]. The difference of the results for the excitation by charged particles does not exceed 30 percent [4.13].

4.2 Transitions Caused by a Multipole Interaction

4.2.1 Two-State Approximation

We shall assume that one of the two conditions,

$$\omega \gg |V_1 - V_0|, \quad |V| \gg |V_1 - V_0| , \tag{4.2.1}$$

is fulfilled and take expression (4.1.21) for the transition probability. The second of the inequalities (4.2.1) can be fulfilled, for example, in those cases when the matrix element V decreases with increasing R more slowly than V_0, V_1, and the principal contribution to the transition probability is given by the interaction at large distances, which considerably exceed the dimensions of an atom.

The approximation (4.2.1) under the conditions (4.1.21) is of interest for many applications such as the transfer of excitation energy in the collisions of atoms with a small resonance defect, the excitation in a collision with an ion, and so on.

The matrix element V can be calculated accurately only in the case of collisions between two atoms of hydrogen or collisions of a hydrogen atom with a structureless charged particle. It is not difficult to show that in a collision of two atoms of hydrogen, as a result of which the states of both atoms change, the matrix element V has the form

$$V = \frac{\lambda}{R^n} - \exp(-b_1 R) \sum_{k=-n+1}^{n} \frac{C_k'}{R^k} - \exp(-b_2 R) \sum_{k=-n+1}^{n} \frac{C_k''}{R^k} , \tag{4.2.2}$$

where n is determined by the multipole order of the transition under consideration: if one of the atoms changes from the state a_1 to the state a_1' and the other changes from the state a_2 to the state a_2', electric multipole transitions $a_1 \to a_1', a_2 \to a_2'$, of the orders κ_1 and κ_2 being allowed, then $n = \kappa_1 + \kappa_2 + 1$. The constants b_1 and b_2 are expressed in terms of the ionization potentials I of the states a_1, a_1' and a_2, a_2' : $b_1 = \sqrt{2I_1} + \sqrt{2I_1'}$, $b_2 = \sqrt{2I_2} + \sqrt{2I_2'}$ (the quantities I are given in atomic energy units 2Ry.) The constants C_k' and C_k'' are expressed in terms of the quantum numbers of the levels under consideration, and besides C_k' and C_k'', and λ are linked in such a way that when $R \to 0, V(R) \to \text{const}$.

In the case of collisions between arbitrary atoms the matrix element V must have the same basic properties as (4.2.2). We shall therefore assume that

$$V(R) = \lambda/R^n - \exp(-bR) f(R) , \tag{4.2.3}$$

where the function $f(R)$ is a polynomial of the same sort as in (4.2.2) and satisfies the condition $V(R) \to \text{const}$ when $R \to 0$, and λ is the constant with dimensions s^{-1} cm^n. The trajectories of the colliding particles will be considered to be rectilinear: $R^2 = \rho^2 + v^2t^2 = \rho^2 + x^2$.

We shall substitute (4.2.3) into (4.1.21), replace the quantity $\sqrt{\omega^2 + 4V^2}$ by $\omega+2V$, which retains the basic analytical properties and limiting cases and which differs from the primary quantity by no more than the factor of $\sqrt{2}$, and integrate with respect to the complex variable z:

$$W = \mathrm{Re}\left\{\frac{1}{v}\int_{-\infty}^{\infty} V(\rho^2 + z^2)\exp\left[\mathrm{i}\frac{2}{v}\int_0^z V(\rho^2 + z'^2)\,dz' + \omega z\right] dz\right\} . \tag{4.2.4}$$

We shall assume further that the principal contribution to the transition probability is given by the distances $R > R_0 = 1/b$, for which $V \simeq \lambda/R^n$. To clarify the further approximation we shall assume at the beginning that $n = 2$ and in integrating use a contour, closed in the upper half-plane of z, with a cut along the imaginary axis from i ∞ to iρ. When calculating the integral we take the function exp (i$\omega z/v$), monotonic along the imaginary axis, outside the integrand sign at the point z_1, at which the derivative of the index of the exponent vanishes:

$$z_1 = \mathrm{i}\sqrt{\frac{\lambda}{\omega} + \rho^2} . \tag{4.2.5}$$

(The point z_1 should be found before the replacement of $\sqrt{\omega^2 + 4V^2}$ by $\omega+2V$.) In the case $n \neq 2$ the calculations can be carried out in a similar way. The difference is only that the lines of the cut do not lie along the imaginary axis but make an angle $\pi(1/2 - 1/n)$ with it. Then

$$\mathrm{Im}\{z_1\} = \sqrt{2^{2/n}\left(\frac{\lambda}{\omega}\right)^{2/n}\sin^2\frac{\pi}{2n} + \rho^2} . \tag{4.2.6}$$

As a result, for the transition probability we have

$$W = \exp\{-2[2^{2/n}\beta_n \sin^2(\pi/2n) + \chi^2]^{1/2}\} \times \sin^2\left[\frac{1}{v}\int_{-\infty}^{\infty} V(\rho,x)\,dx\right] , \tag{4.2.7}$$

where dimensionless constant β_n and parameter χ are

$$\beta_n = \frac{\lambda^{2/n}\omega^{2(n-1)/n}}{v^2} , \quad \chi = \frac{\rho\omega}{v} . \tag{4.2.8}$$

Formula (4.2.7) is obviously valid if the point z_1 lies outside the circle of radius R_0:

$$\left(\frac{2\lambda}{\omega}\right)^{2/n}\sin\frac{\pi}{2n} + \rho^2 > R_0^2 . \tag{4.2.9}$$

Thus, for all impact parameters satisfying this condition, the transition probability W is limited by the relationship

$$W < \exp\left(-2^{(n+1)/n}\sqrt{\beta_n}\,\sin\frac{\pi}{2n}\right) . \tag{4.2.10}$$

For sufficiently large λ and small ω, when

$$(2\lambda/\omega)^{1/n}\sin(\pi/2n) > R_0 , \tag{4.2.11}$$

formulas (4.2.7, 9) are valid for any ρ. It is clear from these formulas that with decreasing velocity the transition probability decreases exponentially. When $\omega = 0$ (zero energy defect), formula (4.2.7) coincides with the formula (4.1.12) obtained above for this case.

We shall now go on to calculate the preexponential factor in (4.2.7). Substituting in the integral with respect to x the potential $V(\rho, x) = \lambda/(\rho^2 + x^2)^{n/2}$, we obtain

$$\frac{1}{v}\int_{-\infty}^{\infty} V(\rho, x)\,dx = \alpha_n \frac{\lambda}{v}\rho^{-(n-1)},$$
$$\alpha_n = \sqrt{\pi}\,\Gamma\left(\frac{n-1}{2}\right)\Big/\Gamma\left(\frac{n}{2}\right) . \tag{4.2.12}$$

When

$$\rho < \rho_0 = \left(\alpha_n \frac{\lambda}{v}\right)^{1/(n-1)} , \tag{4.2.13}$$

the argument of $\sin^2[\lambda/v \int V\,dx]$ in (4.2.7) becomes larger than unity, and $\sin^2[\lambda/v \int V\,dx]$ begins to oscillate rapidly about the mean value equal to 1/2. If

$$\rho_0 > R_0 , \tag{4.2.14}$$

the replacement of the potential (4.2.3) by the potential λ/R^n does not significantly affect the magnitude of the cross section, because this replacement only leads to a change of the character of the oscillations of $\sin^2[\lambda/v \int Vdx]$ in the region where the argument of the $\sin[\lambda/v \int Vdx]$ is large. Therefore, for velocities v satisfying the relation (4.2.14) it is possible in calculating σ to use the approximation

$$W = \sin^2\left(\frac{\alpha_n \beta_n^{n/2}}{\chi^{n-1}}\right)\exp\left[-2\left(2^{2/n}\beta_n \sin^2\frac{\pi}{2n} + \chi^2\right)^{1/2}\right] . \tag{4.2.15}$$

It is interesting to compare this expression with the Born approximation for W:

$$W^{\mathrm{B}} = \left(\frac{\alpha_n \beta_n^{n/2}}{\chi^{n-1}}\right)^2 \exp(-2\chi) . \tag{4.2.16}$$

When $\beta_n \ll 1$, (4.2.15) in practice coincides with (4.2.16) for all ρ for which $W^B < 1/2$, and for smaller ρ, oscillates about a mean value close to 1/2. Therefore when $\beta_n \ll 1$, it is possible to restrict oneself to the Born approximation, supplementing it with some method of normalization, for example, assuming

$$W = \begin{cases} W^B & \text{for} \quad \rho > \rho_0 \\ 1/2 & \text{for} \quad \rho < \rho_0 \end{cases} \tag{4.2.17}$$

where $W^B(\rho_0) = 1/2$.

When $\beta_n \gg 1$, (4.2.15) differs considerably from (4.2.16) even in the region of ρ where W^B is small. The maximum value of W is restricted by the condition (4.2.10). Therefore when $\beta_n \gg 1$, it is impossible to use the Born approximation formula (4.2.16) even for rough estimates.

By substituting of (4.2.15) in (4.1.1) the following expression can be obtained for the cross section:

$$\sigma = 2\pi \left(\frac{\lambda}{v}\right)^{2/(n-1)} \exp\left(-2^{(2n+1)/n}\sqrt{\beta_n}\sin\frac{\pi}{2n}\right) I_n(\beta_n)\,, \tag{4.2.18}$$

$$I_n(\beta_n) = \int_0^\infty \sin^2\left(\alpha_n/y^{(n-1)}\right)$$

$$\times \exp\left\{-2\sqrt{\beta_n}\left[\left(2^{2/n}\sin^2\frac{\pi}{2n} + \beta_n^{1/(n-1)}y^2\right)^{1/2} - 2^{1/n}\sin\frac{\pi}{2n}\right]\right\} y\,dy\,. \tag{4.2.19}$$

When $\beta_n \to 0$,

$$I_2(\beta_2) \to \pi^2 \ln(1/\beta_2), \quad I_3(\beta_3) \to \pi/2\,,$$

$$I_{n>3}(\beta_n) \to 2^{-2(n-2)/(n-1)}\alpha^{2/(n-1)}\Gamma\left(\frac{n-3}{n-1}\right)\sin\left(\frac{\pi}{2}\frac{n-3}{n-1}\right)\,. \tag{4.2.20}$$

When $\beta_n \gg 1$,

$$I_n(\beta_n) \simeq \frac{1}{8}\left(2\sqrt{2\beta_n}\sin\frac{\pi}{2n} + 1\right)\beta^{-n/(n-1)}\,. \tag{4.2.21}$$

The values of $I_n(\beta_n)$ for $n = 2$ and 3 are given in Table 4.1.

At low velocities when β_n is large, the cross section decreases very rapidly (exponentially) with decreasing v. Such behavior of the cross section in the region of low velocities is typical for the inelastic collisions of heavy particles. An exception is the case of small values of ω, when values $\beta_n \lesssim 1$ are possible also at low velocities.

We shall consider the dependence of σ on ω at a fixed velocity. When $\omega = 0$ (zero energy defect), $\beta_n = 0$ and

$$\sigma = 2\pi\left(\frac{\lambda}{v}\right)^{2/(n-1)} I_n(0)\,. \tag{4.2.22}$$

Table 4.1. Factors I_2 and I_3; 8.8 (−3) denotes 8.8×10^{-3}

β	0.02	0.04	0.08	0.16	0.32	0.64	1.28	2.56	5.12	10.24
I_2	21.6	15.9	10.7	6.25	2.94	1.01	0.268	8.3(−2)	2.7(−2)	8.8(−3)
I_3	1.42	1.32	1.16	0.95	0.70	0.44	0.224	9.8(−2)	4.2(−2)	1.9(−2)

When $\omega \lesssim (v\lambda^{-1/n})^{n/(n-1)}$, the cross section remains close to (4.2.22), and when $\omega > (v\lambda^{-1/n})^{n/(n-1)}$, it decreases exponentially with increasing ω.

At high velocities when $\beta_n \ll 1$, formula (4.2.18) gives

$$n \neq 2 \qquad \sigma \sim (\lambda/v)^{2/(n-1)}, \tag{4.2.23}$$

$$n = 2 \qquad \sigma \sim (\lambda/v)^2 \ln\,(\text{const} \times v)\,. \tag{4.2.24}$$

This dependence is obviously valid only in the case when λ is so large and ω so small that β_n can be small at velocities satisfying the condition (4.2.14). At higher velocities when the condition (4.2.14) is violated, the approximation (4.2.15) based on the polar potential, which tends to infinity at $\rho \to 0$, becomes illegitimate. It can be shown that the potentials

$$V(R) = \lambda R^{n-1}/(R_0^2 + R^2)^{n-1/2}\,, \tag{4.2.25}$$

or

$$V(R) = \lambda/(R_0^2 + R^2)^{n/2}\,, \tag{4.2.26}$$

which do not have a pole when $R \to 0$, lead to the following dependence on v in the Born region

$$n > 2 \qquad \sigma \sim (\lambda/vR_0^{n-2})^2\,, \tag{4.2.27}$$

$$n = 2 \qquad \sigma \sim (\lambda/v)^2 \ln(\text{const}\ vR_0^{-1})\,. \tag{4.2.28}$$

It must be noted that at small λ when (4.2.14) can be infringed even at small v, the presence of the pole in the potential $V(R)$ leads to a considerable shift of the cross section in the region of low velocities. Thus the range of applicability of formula (4.2.18) for the cross section is limited by the conditions (4.2.9, 14). The range in which (4.2.17) is valid is determined by the second of these conditions, and the first one does not depend on velocity and is satisfied all the better the larger λ is and the smaller ω is.

The case of small values of ω corresponds to collisions which are accompanied by a quasi-resonant transfer of excitation energy. Assume, as a result of the collision, the first atom passes from the level E_1 to the level E_1', and the second atom from the level E_2 to the level E_2', where $E_1 > E_1'$ and $E_2 < E_2'$. In this case,

$$\hbar\omega = E_1 + E_2 - E_1' - E_2'\,. \tag{4.2.29}$$

In the case of collisions of identical atoms when $E_1 = E_2'$, $E_1' = E_2$, $\omega = 0$ (exact resonance). In collisions between different atoms, cases are possible when the resonance defect $\hbar\omega$ is small.

4.2.2 Two-Levels and Rotating-Axis Approximations

We shall now discuss the applicability of the two-level approximation. It is usually applicable for the transitions between two nondegenerate levels. When one of the levels is degenerate the problem must involve many levels. To reduce it to the two-level problems, some additional assumptions must be adopted.

The simplest and most physically clear approximation consists in the use of the body fixed coordinate system, in which the quantization axis (the z axis) is directed towards the perturbing particle. In the course of a collision the direction of the quantization axis is changed. In such coordinates the direction of the electric field does not change and therefore the electric field does not mix the states with different magnetic quantum numbers m. Then for the multipole transitions of the type $ns \to n'l'$, we obtain the two-level system of equations describing the transition $ns \to n'l'm = 0$.

Such a description is not complete. The rotation of the internuclear axis leads to the transitions $nlm \to n'l'm'$ with $\Delta m \neq 0$ in a system of coordinates fixed in space. In the rotating coordinate system these transitions can be treated by introducing the rotation operator

$$\hat{T}_\varphi = -\dot{\varphi}\hat{L}_x \, , \tag{4.2.30}$$

where $\dot{\varphi}$ is the angular velocity of rotation of the internuclear axis, and $\hat{L}_x$ is the operator of the x component of the orbital momentum where the x axis is perpendicular to the collision plane. It is not difficult to see that $T_\varphi \propto v$. Therefore when $v \to 0$, the rotation of the axis can be neglected, the problem being reduced to the two-level system with $\Delta m = 0$. When the velocity increases, the expression (4.2.30) increases and hence the neglect of $\Delta m \neq 0$ transitions is no longer possible. For example, for excitation of the optically allowed transition by the charged particle, the rotating axis approximation gives in the Born region a result $\pi^2/4$ times lower than the exact asymptotic value.

To estimate the cross sections for the transitions between two degenerate levels $aJ - a'J'$, the simple two-level approximation is however often used. In this case one has to substitute in the equations the potential averaged over M, M':

$$V_{aJ,\,a'J'} = \left(\frac{1}{g} \sum_{MM'} |V_{aJM,\,a'J'M'}|^2 \right)^{1/2} \, , \tag{4.2.31}$$

where g is the statistical weight of the initial level. The matrix elements in (4.2.31) have to be written in the coordinate system fixed in space. The cross section obtained in such a way has to be considered as the averaged one:

$$\sigma_{aJ,\,a'J'} = \frac{1}{g} \sum_{MM'} \sigma(aJM,\, a'J'M')\,. \tag{4.2.32}$$

Since the operator $\hat{T}_\varphi$ (4.2.30) is proportional to v then at $v \to 0$ the results obtained with the averaged potential (4.2.31) are close to the results of the solution of the strong coupling equations in the system of coordinates fixed in space.

We noted above, see (4.4.22–28), that the presence of the pole in the potential leads in the Born region to the difference with the accurate values of the cross section. Therefore the parameter R_0 has to be chosen so as to ensure the coincidence of the cross section with an accurate Born asymptotic value. The latter can be obtained in the framework of the usual perturbation theory.

In several cases it is of interest to obtain formulas for the averaged multipole potentials. For the excitation of an optically allowed transition in collision with a structureless charged particle with charge Ze, $V = \lambda/R^2$

$$\lambda = \frac{Ze}{\hbar}\left(\frac{S}{3g}\right)^{1/2} \tag{4.2.33}$$

where S is the transition line strength [Ref. 4.14, Sect. 9.2.2] and g is the statistical weight.

Consider the case of a collision in which one of the colliding atoms passes from the level J_1 to the level J_1' and the second atom from the level J_2 to the level J_2'. We assume that electric multipole transitions of order κ_1 and κ_2 are allowed between the levels J_1, J_1' and J_2, J_2'. The interaction constant is defined by the relationship.

$$V = \lambda/R^n, \quad n = \kappa_1 + \kappa_2 + 1,$$
$$\lambda = \frac{1}{\hbar}\left(\frac{S_{\kappa_1} S_{\kappa_2}}{g_1 g_2 (2\kappa_1 + 1)(2\kappa_2 + 1)}\right)^{1/2}, \tag{4.2.34}$$

where S_{κ_1} and S_{κ_2} are the line strengths of the electric multipole transitions $J_1 - J_1'$, $J_2 - J_2'$ and g_1, g_2 are the statistical weights of the levels J_1, J_2. The line strength is defined by the formula

$$\frac{1}{g} S_\kappa = Q_\kappa(J, J')(2l' + 1)\begin{pmatrix} l & \kappa & l' \\ 0 & 0 & 0 \end{pmatrix}^2 \left| \int_0^\infty P_l(r) P_{l'}(r) r^\kappa dr \right|^2 e^2\,. \tag{4.2.35}$$

To conclude this section, we give an estimate of the cross sections for the transfer of excitation in the case of small energy defect using the interaction potential (4.2.34). The condition for applicability of small energy defect approximation can be written in the form

$$2\sqrt{2^{2/n}\beta_n}\,\sin\frac{\pi}{2n} = 2^{(n+1)/n}\,\frac{\lambda^{1/n}\omega^{(n-1)/n}}{v}\sin\frac{\pi}{2n} \ll 1\,. \tag{4.2.36}$$

In this case

$$\exp\left(-2^{(n+1)/n}\sqrt{\beta_n}\sin\frac{\pi}{2n}\right) \simeq 1, \quad I_n(\beta_n) \simeq I_n(0),$$

and formulas (4.2.22, 34) give the following expression for the cross section:

$$\sigma \simeq 2\pi\left(\frac{e^2}{\hbar v}\right)^{2/(n-1)}\left[\frac{S_{\kappa_1}S_{\kappa_2}e^{-4}}{g_1g_2(2\kappa_1+1)(2\kappa_2+1)}\right]^{1/(n-1)} I_n(0). \qquad (4.2.37)$$

4.2.3 Treatment of the Coulomb Repulsion of Nuclei

The previous results have been obtained using the assumption of rectilinear trajectories of the perturbing particles. In the case of collisions between the two positive ions one has to take into consideration the Coulomb repulsion, i.e., to use the hyperbolic trajectories. For low collision velocities the cross section for excitation by a structureless particle with charge Z_pe ($\kappa_2 = 0$) can be evaluated using the semiclassical first-order perturbation theory described in detail by Alder et al. [4.15]. In this approximation the cross section can be written in the form[1]

$$\sigma = \frac{3(2n-1)}{4\pi}\frac{\lambda_n^2}{v^2}\frac{f_{E\,n-1}(\xi)}{a^{2n-4}}, \qquad (4.2.38)$$

$$a = Z_iZ_p/Mv^2, \qquad (4.2.39)$$

$$\xi = Z_iZ_p\omega/Mv^3. \qquad (4.2.40)$$

Here Z_i and Z_p are the target and projectile ion charges, M is the reduced mass of a colliding pair. The functions $f_{E\lambda}(\xi)$ are tabulated in [4.15] ($\lambda = n-1$). At low velocities, i.e., for large values of ξ, $f_{E\lambda}(\xi) \propto \exp(-2\pi\xi)$, and is exponentially small. In case of dipole transitions, $n = 2, \lambda = 1$, the asymptotic value is

$$f_{E1} \simeq \frac{32\pi^3}{9\sqrt{3}}e^{-2\pi\xi}, \quad \xi \gg 1. \qquad (4.2.41)$$

When $\xi > 1$ the cross sections are very small. They become comparable in magnitude with the electronic cross sections only for $\xi < 1$. In plasmas usually $v < Z_i/M^{1/2}$. Therefore, the excitation by positive ions is substantial for small energy splittings

$$\omega \ll Z_i^2/Z_pM^{1/2}.$$

When $\mathscr{E} \to \infty$ the Coulomb repulsion may be neglected, $\xi \to 0$. Moreover, at very high energies the Born approximation is valid provided the correct interaction potential is used. In this limiting case the Born approximation and the first-order impact parameter method (used with the same potential) coincide

[1] We recall that in this section the atomic units with the energy unit $me^4/\hbar^2 = 2$ Ry are used.

asymptotically. Hence, at $\mathscr{E} \to \infty$ the order of magnitude of the cross section can be evaluated using the simple analytic formulas obtained by the impact-parameter method with rectilinear paths and with the model potential $V_n \propto R^{n-1}/(R_0^2 + R^2)^{n-1/2}$ (Sect. 5.1.4). For dipole transitions (n=2)

$$\sigma = \frac{8\pi Z_p^2}{v^2} \lambda_2^2 \Phi_1(x_0) , \tag{4.2.42}$$

and for quadrupole transitions (n=3),

$$\sigma = \frac{4\pi}{27} \frac{Z_p^2}{v^2} \frac{\lambda_3^2}{R_0^2} \Phi_2(x_0) , \tag{4.2.43}$$

$$x_0 = R_0 \omega / v .$$

The functions $\Phi_1(x)$ and $\Phi_2(x)$ are given by (5.1.51 and 55). The cut-off radius R_0 can be chosen according to (5.1.53).

Since at low velocities the most important dependence on v is given by the exponential factor, the qualitative behaviour of the cross sections for excitation of positive ions by another positive projectile can be described by (4.2.42, 43) multiplied by the exponential factor $\exp(-2\pi\xi)$.

4.3 Charge Exchange

4.3.1 Special Features of Charge Exchange Processes

Charge exchange in atomic collisions, i.e., the process

$$A + B^+ \to A^+ + B \tag{4.3.1}$$

in which an electron goes over from the atom A to the ion B, is an example of a process with redistribution of particles. The electron wave functions of initial state $\Psi(\boldsymbol{r}_A)$ and final state $\Phi(\boldsymbol{r}_B)$, where $\boldsymbol{r}_A$ and $\boldsymbol{r}_B$ are respectively the distances between the electron and the nuclei A and B, are not orthogonal. They correspond to two different complete sets of wave functions. Moreover, the electron transition from the nucleus A to the nucleus B is accompanied by transfer of the momentum $m\boldsymbol{v}$, where m is the electron mass and $\boldsymbol{v}$ is the relative velocity of the nuclei A and B.

As a consequence of all these particular features, the charge exchange differs very much from other collisional processes considered above [4.5, 6, 16, 17].

When the relative velocity of the nuclei is small $v \ll v_0$, where $v_0 = e^2/\hbar$ is the atomic unit of velocity, it is useful in describing the charge exchange to expand the wave function of the system in terms of the wave functions of the quasimolecule $(AB)^+$. Such an approach proves to be convenient and gives good results in describing the resonance and quasi-resonance charge exchange at low velocities [4.18].

In the limiting case of high velocities, perturbation theory can be used. In the framework of perturbation theory, the velocity dependence of the charge exchange cross section is given by the Brinkman-Kramers formula [4.19].

$$\sigma(v) \propto v^{-12}, \quad (v \gg v_0). \tag{4.3.2}$$

4.3.2 Resonance Charge Exchange

We shall consider the collision of the atom A in the state γ with the ion of the same atom A^+

$$A(\gamma) + A^+ \to A^+ + A(\gamma)\,. \tag{4.3.3}$$

We shall assume that the state γ is the state ns and that the ion A^+ contains only completely filled electronic shells. At low relative velocities, the system $A(ns) + A^+$ can be considered as molecular ion A_2^+. As $R \to \infty$, the ground term of the system is doubly degenerate because the electron can be located near either of the two ions A^+. We shall denote the corresponding wave functions by Ψ_1 and Ψ_2. When R decreases, the ground term splits into two terms, an even term $V(^2\Sigma_g) = V_g$, and an odd term $V(^2\Sigma_u) = V_u$. In the adiabatic approximation the wave functions of these terms are expressed in terms of wave functions Ψ_1 and Ψ_2 in the following way

$$\begin{aligned} \Psi_g(t) &= \frac{1}{\sqrt{2}}(\Psi_1 + \Psi_2)\exp\left[-\frac{i}{\hbar}\int_{-\infty}^{t} V_g(t')\,dt'\right], \\ \Psi_u(t) &= \frac{1}{\sqrt{2}}(\Psi_1 - \Psi_2)\exp\left[-\frac{i}{\hbar}\int_{-\infty}^{t} V_u(t')\,dt'\right], \end{aligned} \tag{4.3.4}$$

The wave function satisfying the initial condition $\Psi(t \to -\infty) = \Psi_1$ has the form

$$\Psi(t) = \frac{1}{\sqrt{2}}[\Psi_g(t) + \Psi_u(t)]\,. \tag{4.3.5}$$

By substituting (4.3.4) into (4.3.5) we have in the limit $t \to \infty$

$$\Psi(t) \underset{t\to\infty}{\longrightarrow} \exp\left[-\frac{i}{2\hbar}\int_{-\infty}^{\infty}[V_g + V_u)\,dt\right](\Psi_1\cos\eta - i\Psi_2\sin\eta)\,, \tag{4.3.6}$$

$$\eta = \frac{1}{2\hbar}\int_{-\infty}^{\infty}[V_u(t) - V_g(t)]\,dt\,. \tag{4.3.7}$$

The probability of the transition $1 \to 2$ is

$$W = \sin^2\eta\,. \tag{4.3.8}$$

The corresponding cross section can be obtained by integrating W over the impact parameter,

$$\sigma = 2\pi \int_0^\infty \rho d\rho \sin^2\eta(\rho) . \tag{4.3.9}$$

If $\eta \gg 1$, the function $\sin^2\eta$ can be replaced by its mean value 1/2. Therefore it is enough to know the function at comparatively large ρ when $\eta < 1$.

In the case of the molecular ion H_2^+, the exchange splitting at large distances has the form [4.8, 20]

$$\frac{1}{2}[V_u(R) - V_g(R)] = 2R \exp(-1 - R) . \tag{4.3.10}$$

Here atomic units are used.

In the rectilinear-trajectory approximation, $R = \sqrt{\rho^2 + v^2 t^2}$,

$$\eta = 2\sqrt{2\pi} \frac{v_0}{v} x^{3/2} \exp(-1 - x), \; x = \rho/a_0 . \tag{4.3.11}$$

Using (4.3.9) we have

$$\sigma = \pi a_0^2 \frac{x_0^2}{2} , \tag{4.3.12}$$

where x_0 is determined using the equation $\sin\eta(x_0) \simeq \eta(x_0) = \exp(-C)/2 = 0.28$($C = 0.5772$ is Euler's constant):

$$x_0^{3/2} \exp(-x_0) \simeq \frac{0.15v}{v_0} . \tag{4.3.13}$$

A detailed discussion of the resonance and quasi-resonance charge exchange is given in [4.6, 20, 21]. The expression (4.3.12) for the charge exchange cross section is valid if $v < v_0 = e^2/\hbar$. In the limiting case $v/v_0 \gg 1$, see [4.6, 21],

$$\sigma = \pi a_0^2 \frac{2^8}{15} \left(\frac{v_0}{v}\right)^2 \left[\frac{1}{4}\left(\frac{v}{v_0}\right)^2 + 1\right]^{-5} \to \pi a_0^2 \frac{2^{18}}{15} \left(\frac{v_0}{v}\right)^{12} . \tag{4.3.14}$$

4.3.3 Contribution of Inner Shells

If the atom A in (4.3.1) is a multielectron atom there exists a range of velocities v for which the condition $v > v_0\sqrt{I}$ for the valence electrons and the condition $v < v_0\sqrt{I}$ for the electrons of the inner shell are valid. Here, I is the ionization potential expressed in atomic units, i.e., in units of 2 Ry.

In this range of velocities, inner electronic shells give the main contribution to the total charge exchange cross section.

Now we shall consider the cross section of the charge exchange process

$$A(n_a l_a^N) + H^+ \to A^+(n_a l_a^{N-1}) + H(n) , \tag{4.3.15}$$

in which an electron from the shell $n_a l_a^N$ is captured by a proton, and as a result the hydrogen atom in the state with principal quantum number n is formed. If the relative velocity of the colliding particles is not too low, the cross section can be described by a simple generalization of the Brinkman–Kramers formula [4.22, 23]

$$\sigma(n_a l_a^N - n) = \pi a_0^2 \frac{2^8 \eta}{5} \frac{N[2I(n_a l_a)]^{5/2}}{n^3 f(v)} \left(\frac{v_0}{v}\right)^2 , \tag{4.3.16}$$

where

$$f(v) = \left\{ \left[\left(\frac{v_0}{v}\right) \Delta I + \frac{1}{2} \left(\frac{v}{v_0}\right) \right]^2 + 2I_n \right\}^5 , \tag{4.3.17}$$

$$I_n = 1/2n^2, \ \Delta I = I - I_n .$$

The ionization potential I of the shell $n_a l_a^N$ is expressed in atomic units. In the original Brinkman–Kramers formula the factor η is equal to unity. The accurate first-order perturbation theory [4.24] gives in the limit of high velocity, $\eta \simeq 1/3$. In the estimates given below, we shall assume $\eta = 1/3$.

At low velocities, the cross section decreases more rapidly than follows from (4.3.16); in fact exponential decrease occurs:

$$\sigma(n_a l_a^N - n) = \pi a_0^2 \frac{\pi}{8\gamma^2} \exp\left(- \frac{\pi \Delta I}{2\gamma} \frac{v_0}{v} \right) . \tag{4.3.18}$$

This result can be obtained by the method described in Sect. 4.1.2.

We shall denote the velocity for which the curves $\sigma(v)$ defined by (4.3.16) and (4.3.18) intersect by v_c. When $v > v_c$, the cross section can be estimated by means of (4.3.16), and when $v < v_c$, one has to use (4.3.18). The cross section for capture from the inner shells is usually given by (4.3.16). In this case, $I \simeq \Delta I \gg 1/2n^2$, and the maximum value of the cross section of capture from the shell $n_a l_a^N$ is achieved at the velocity $v_m = v_0\sqrt{4I/3}$:

$$\sigma(v_\mathrm{m}) \simeq \pi a_0^2 \frac{25N}{n^3} (2I)^{-7/2} . \tag{4.3.19}$$

When $v > v_\mathrm{m}$, the cross section decreases rapidly ($\propto v^{-12}$). In the range of velocities $v'_\mathrm{m} < v < v''_\mathrm{m}$, where v'_m corresponds to the outer shell and v''_m corresponds to the inner shell $n_a l_a$, the dependence of the total cross section on energy $\mathscr{E}$ (summed over all shells $n_a l_a$) has the form

$$\sigma(\mathscr{E}) \propto \mathscr{E}^{-p} (2.5) \le p \le 6) . \tag{4.3.20}$$

The limiting value of $p = 2.5$ corresponds to the heaviest atoms.

The charge-exchange cross sections for the Ne and Ar atoms are shown in Figs. 4.1, 2. At energies $\mathscr{E} \lesssim 100$ keV, the outer shell gives the main contribution. At higher energies, the capture of electrons from inner shells becomes significant.

In the case of alkali elements, inner shells give the main contribution to the cross section of charge exchange beginning from energies $\mathscr{E} \gtrsim 20\,\mathrm{keV}$ [4.27]. Experimental data on the electron capture cross sections in collisions of protons with neutral targets were given in [4.28, 29].

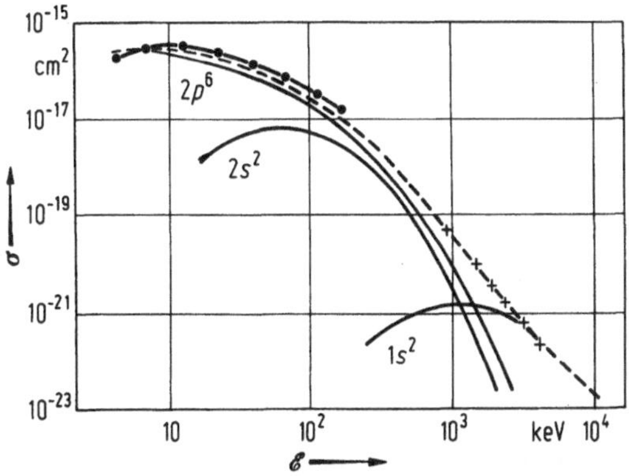

Fig. 4.1. Effective cross section of the charge transfer $H^+ + Ne \rightarrow H + Ne^+$. Solid curves show the capture from different shells. Dotted line is the total theoretical cross section. Points correspond to experiment [4.25]. Crosses correspond to experiment [4.26]

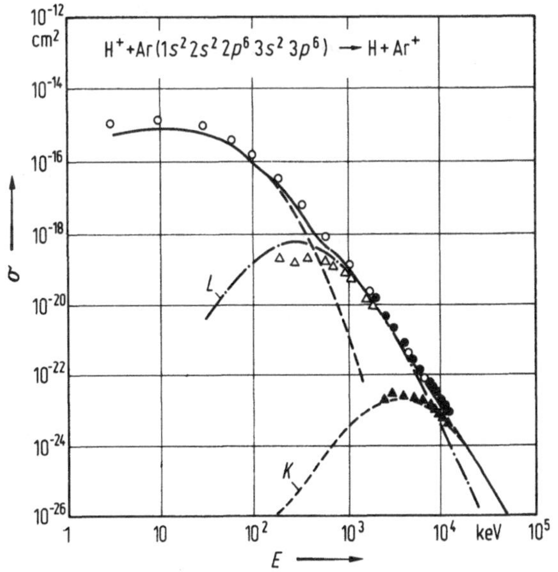

Fig. 4.2. Electron capture cross section in collision $H^+ + Ar$: broken curves show the contribution from K- and L- shells, the solid curve shows the total capture cross section calculated in a modified Brinkman–Kramers approximation. Circles and crosses are various experimental data (for details see [4.23]).

4.3.4 Charge Exchange in the Case of Multicharged Ions

The Brinkman–Kramers formula can be easily generalized to the case of charge transfer from an arbitrary ion to a neutral atom [4.23]:

$$A + B_{z+1} \to A^+ + B_z ,$$

$$\sigma(n_a l_a^N - n) = \pi a_0^2 \frac{2^8 \eta}{5} N \frac{(2I)^{5/2}(2I_n)^{5/2} n^2}{f(v)} \left(\frac{v_0}{v}\right)^2 . \qquad (4.3.21)$$

The function $f(v)$ is given by (4.3.17) when $\Delta I = I - I_n$, $I_n = z^2/2n^2$. Here the levels n of an ion B_z, are assumed to be hydrogenlike. The cross section of quasi-resonance charge exchange to the level of an ion B_z with the principal quantum number

$$n_0 = z\sqrt{2I_A} \qquad (4.3.22)$$

has evidently the maximum magnitude. However, applicability of (4.3.21) for the resonance charge exchange is limited to the range of high velocities $v \gg v_m$. The Coulomb field induces a strong perturbation of the terms in the final channel. Therefore, the exponential decrease of the cross section with decrease of velocity occurs at velocities considerably larger than v_c defined by (4.3.16, 18).

The Coulomb perturbation of terms is responsible for another mechanism of charge exchange in multicharged ions. If the ionization potential I_n of the ion B_z is larger than the ionization potential of the atom A, then electronic terms of the system $A^+ + B_z(n)$ intersect the ground term of the system $A + B_{z+1}$ at the points X_n determined by the condition

$$\frac{z^2}{2n^2} - I_A - \frac{z-1}{X_n} = 0 . \qquad (4.3.23)$$

In the vicinity of the point of intersection of the terms, nonadiabatic Landau–Zener transitions can take place. At small velocities such transitions give the principal contribution to the transition probability. At higher velocities, transitions at other values of R also must be taken into consideration.

The calculations [4.30, 31] show that the most effective transitions at high velocities are those in the states of ion B_z with principal quantum number n_0 for which the quasi-resonance condition holds at high X [4.5, 25]. When v decreases, capture to other levels with $n < n_0$ becomes efficient. For $n < n_0$ the cross section can have two maxima corresponding to the two above-mentioned mechanisms.

The total cross section of electron transfer to all excited levels of an ion B_z has relatively weak dependence on velocity up to those velocities at which an ion B_z in the state n_0 is mainly formed. The magnitude of the total cross section σ_{tot} increases with z. Although the different models give somewhat different dependences on z, the order of magnitude of σ_{tot} can be estimated by means of formula

$$\sigma_{tot} \cong z \cdot 10^{-15}\ \mathrm{cm}^2$$

A more detailed discussion of charge exchange in the case of multiply charged ions is given in [4.5, 32–35]. The available experimental data on electron capture by multicharged ions from simple H- and He- targets are given in [4.7, 36, 37]. An extensive bibliography on electron transfer processes was recently published by Tawara [4.38].

5 Some Problems of Excitation Kinetics

This chapter may be considered as an introduction to some problems of plasma spectroscopy. In the beginning we give a summary of the analytic formulas which can approximate the dependence of the rate coefficients $\langle v\sigma \rangle$ on temperature and atomic characteristics. These formulas contain two fitting parameters which should be determined from the results of numerical calculations. Tables of these parameters will be described in Chap. 6. Some useful semiempirical formulas are also given below.

Special attention is paid to the process of dielectronic recombination and formation of the dielectronic satellites. These processes are very important for the spectroscopy of high-temperature laboratory and astrophysical plasmas.

The last section of the chapter is devoted to the simplest approaches to the kinetics of level populations in a plasma.

5.1 Rate Coefficients for Elementary Processes in a Plasma. Approximation of Cross Sections and Rate Coefficients by Analytic Formulas

5.1.1 Excitation of Atoms and Ions

In this section we summarize the formulas expressing cross sections σ and rate coefficients $\langle v\sigma \rangle$ in terms of radial integrals and angular factors Q. It is convenient for applications to plasma kinetics problems to use the approximate analytic formulas. In this section, simple fitting formulas are proposed which describe the results of numerical calculations by means of two or three adjustable parameters and provide the correct asymptotic behavior of the quantities σ and $\langle v\sigma \rangle$. Some semiempirical formulas, which are often used, are also presented. When constructing the approximate formulas we are mainly concerned with the energy region $\mathscr{E} \lesssim E_z$ (E_z is the ionization energy of the ion X_z) which is of the most importance for plasma diagonostics. In particular, the asymptotic behavior of the cross sections for optically allowed transition is taken in the form $\mathscr{E}^{-1}$, instead of the form $\mathscr{E}^{-1} \ln \mathscr{E}$ (Sect. 3.1.3). Therefore one has to be cautious when applying the formulas given here to the problems of excitation and ionization in beam experiments. In these cases, the factor $\ln \mathscr{E}$ can be substantial.

In this section the initial state is denoted by index 0, and the final state by index 1. The final results are given in CGS units.

We shall assume everywhere that the electron energy distribution function $\mathcal{F}(\mathcal{E})$ is the Maxwellian, see (1.2.1).

For excitation $0 \to 1$ by electron impact the rate coefficient $\langle v\sigma_{01}\rangle$ averaged over the electron distribution function is

$$\langle v\sigma_{01}\rangle = \int_{\Delta E}^{\infty} v\sigma_{01}\mathcal{F}(\mathcal{E})\,d\mathcal{E}\,, \tag{5.1.1}$$

where $\mathcal{E}$ is the energy of the incident electron, and ΔE is the threshold excitation energy. On substituting the Maxwellian distribution function (1.2.1) we obtain

$$\begin{aligned}\langle v\sigma_{01}\rangle &= K\int_{\Delta E}^{\infty} \frac{\sigma}{\pi a_0^2}\frac{\exp(-\mathcal{E}/T)\,\mathcal{E}\,d\mathcal{E}}{(\mathrm{Ry})^{1/2}T^{3/2}}\,,\\ K &= \frac{2\sqrt{\pi}\,\hbar a_0}{m} = 2.17\times 10^{-8}\mathrm{cm}^3\,\mathrm{s}^{-1}\,.\end{aligned} \tag{5.1.2}$$

It is convenient to express the energy and temperature in the scaled units (Sect. 3.1.3). On setting

$$u = (\mathcal{E} - \Delta E)/DE, \quad \beta = DE/T\,. \tag{5.1.3}$$

Eq. (5.1.2) can be written as

$$\langle v\sigma_{01}\rangle = K\left(\frac{DE}{\mathrm{Ry}}\right)^{1/2}\beta^{3/2}e - \beta p\int_0^{\infty}\frac{\sigma(u)}{\pi a_0^2}(u+p)\,e - \beta u du \tag{5.1.4}$$

with $p = \Delta E/DE$.

The rate of the inverse process (deexcitation) can be expressed through the excitation rate by (1.2.7). As was shown in Sects. 2.3 and 3.1, 2 the cross section for the transition $a_0 - a_1$ can be written, in the general case, in the form

$$\sigma_{a_0a_1} = \sum_{\kappa}[Q'_\kappa(a_0,a_1)\,\sigma'_\kappa(l_0,l_1) + Q''_\kappa(a_0,a_1)\sigma''_\kappa(l_0,l_1)]\,, \tag{5.1.5}$$

where the quantities σ'_κ and σ''_κ depend only on the quantum numbers n_0l_0 and n_1l_1 of the optical electron. The dependence on the total angular momenta of an atom $(J,S,L,\ldots)$ is described by the factors Q'_κ and Q''_κ. For the cross section summed with respect to $J,S,L,\ldots)$ in the case of configurations $n_0l_0^m$–$n_0l_0^{m-1}n_1l_1$, we have

$$\sigma(n_0l_0^m, n_0l_0^{m-1}n_1l_1) = m\sum_{\kappa}[\sigma'_\kappa(l_0,l_1) + \sigma''_\kappa(l_o,l_1)]\,. \tag{5.1.6}$$

Thus the quantities σ'_κ and σ''_κ correspond to the single-electron transition cross sections.

The quantity σ'_κ includes the direct and interference terms

$$\sigma'_\kappa = \sigma_\kappa + \sigma_\kappa^{\mathrm{int}}, \quad Q'_\kappa(a_0,a_1) \propto \delta_{S_0S_1}\,; \tag{5.1.7}$$

the quantity σ''_κ representing purely the exchange term. The expressions of σ'_κ and σ''_κ in terms of radial integrals are given in Sect. 2.3. These quantities, evidently, can be obtained only by means of numerical calculations.

The index κ varies in the interval of $\kappa_{\min}$ to $\kappa_{\max}$, where

$$\kappa_{\min} = |l_0 - l_1|, \quad \kappa_{\max} = l_0 + l_1 . \tag{5.1.8}$$

In accordance with (2.3.9, 10), $\sigma'_\kappa \neq 0$ only for κ with the same parity as $\kappa_{\min}$.

We shall use the analytical fitting formulas to represent the results of the numerical calculations. In computer codes one can explore formulas with many adjusted parameters providing very accurate fitting to numerically calculated or measured cross sections. However, good accuracy is based on mutual compensations of essential contributions from nearby terms of opposite signs. Therefore, a small change of cross section results in a much larger change of parameters. In particular, the extrapolation of parameters along an isoelectronic sequence becomes very difficult if possible at all.

We use comparatively simple formulas with true asymptotic and 2 or 3 adjusted parameters. The accuracy of these formulas is usually a few percents, but an interpolation and extrapolation of parameters for similar transitions, in particular in isoelectronic sequence, is very simple.

Generally we use (5.1.5) and the fitting formulas for $\sigma'_\kappa, \sigma''_\kappa$. For cross sections, summed over J, the factors Q' and Q'' differ only in the spin parts:

$$\begin{aligned} &Q'_\kappa = Q_\kappa(L_0, L_1) A_0, \quad Q''_\kappa = Q_\kappa(L_0, L_1) A_2 , \\ &A_0 = \delta(S_0, S_1), \qquad A_2 = (2S_1 + 1)/2(2S_p + 1) . \end{aligned} \tag{5.1.9}$$

In this case the fitting formula may be applied to the total one-electron cross section

$$\sigma^t_\kappa = A_0 \sigma'_\kappa(l_0, l_1) + A_2 \sigma''_\kappa(l_0, l_1) . \tag{5.1.10}$$

This provides often the better accuracy of fitting since for a direct transition σ^t_κ is energy function smoother than σ'_κ, which may be negative at small and medium energies. It should be noted, however, that for transitions between fine-structure components the $J_0 - J_1$ orbital parts of Q' and Q'' are different and the form of (5.1.10) is not valid.

Taking into account properties described in Sects. 3.1–2 the one-electron excitation cross sections are written in the form

$$\begin{aligned} &\sigma'_\kappa(l_0, l_1) = \frac{\pi a_0^2}{2l_0 + 1} \left(\frac{\mathrm{Ry}}{DE}\right)^2 \left(\frac{E_1}{E_0}\right)^{3/2} \cdot \frac{C\Phi'(u)}{u + \varphi} \\ &\qquad\qquad\qquad\qquad\qquad\qquad\qquad\qquad u = \frac{\mathscr{E} - \Delta E}{DE} \\ &\sigma''_\kappa(l_0, l_1) = \frac{\pi a_0^2}{2l_0 + 1} \left(\frac{\mathrm{Ry}}{DE}\right)^2 \left(\frac{E_1}{E_0}\right)^{3/2} \cdot \frac{C\Phi''(u)}{u + \varphi} \end{aligned} \tag{5.1.11}$$

where E_0, E_1 are atomic level energies (from the ionization limit). For σ^t_κ the function $\Phi^t = \Phi'$ is used. C, φ are the fitting parameters, DE is a scaling factor.

The functions Φ depend on the type of transition, and generally can include a third fitting parameter D. If one prefers to use a two-parameter formula $D = 0$ should be assumed. For the excitation cross sections, we have

$$\begin{aligned}\Phi'(u) &= \left[1 - D/(u+1)^2\right] \eta_1(u)\\ \Phi''(u) &= [u + 0.4 + D]^{-2}\, \eta_1(u)\end{aligned} \tag{5.1.12}$$

where

$$\eta_1(u) = \begin{cases} \left(\dfrac{u}{u+p}\right)^{1/2} & , z = 1 \text{ (neutrals)} \\ 1 & , z > 1 \text{ (ions)} \end{cases} \qquad p = \Delta E/DE \tag{5.1.13}$$

The excitation rate coefficient can be written in a similar way:

$$\langle v\sigma(a_0, a_1)\rangle = \sum_\kappa \left[Q'_\kappa(a_0, a_1)\langle v\sigma'_\kappa(l_0, l_1)\rangle + Q''_\kappa(a_0, a_1)\langle v\sigma''_\kappa(l_0, l_1)\rangle\right] , \tag{5.1.14}$$

$$\begin{aligned}\langle v\sigma'_\kappa(l_0, l_1)\rangle &= \frac{10^{-8}}{2l_0+1}\left(\frac{\text{Ry}}{DE}\cdot\frac{E_1}{E_0}\right)^{3/2} \cdot \frac{AG'(\beta)}{\beta+\chi}\, e - \beta p \;\left[\text{cm}^3\ \text{s}^{-1}\right] ,\\ \langle v\sigma''_\kappa(l_0, l_1)\rangle &= \frac{10^{-8}}{2l_0+1}\left(\frac{\text{Ry}}{DE}\cdot\frac{E_1}{E_0}\right)^{3/2} \cdot \frac{AG''(\beta)}{\beta+\chi}\, e - \beta p \;\left[\text{cm}^3\ \text{s}^{-1}\right] ,\end{aligned} \tag{5.1.15}$$

$$\beta = DE/T .$$

Here A and χ are adjusted parameters, the functions $G(\beta)$ can include the third adjusted parameter

$$\begin{aligned}G'(\beta) &= \beta^{1/2}(\beta + 1 + D)\eta_1(\beta^{-1}) ,\\ G''(\beta) &= \beta^{1/2}(\beta + D)\eta_1(\beta^{-1}) .\end{aligned} \tag{5.1.16}$$

For the total one-electron rate coefficient

$$\langle v\sigma^t_\kappa\rangle = A_0\langle v\sigma'_\kappa\rangle + A_2\langle v\sigma''_\kappa\rangle \tag{5.1.17}$$

the function $G^t = G'$ is used.

Eqs. (5.1.11–16) can be employed in any of the scales DE given by (3.1.30a, b, c). But in the case of closely spaced levels only $DE = |E_0|$ or z^2 Ry are appropriate.

The functions Φ and G are symmetric with respect to the initial and final states. Therefore, in the case of deexcitation collisions $1 \to 0$, we have

$$\langle v\sigma(a_1, a_0)\rangle = \sum_\kappa \left[Q'_\kappa(a_1, a_0)\langle v\sigma'_\kappa(l_1, l_0)\rangle + Q''_\kappa(a_1, a_0)\langle v\sigma''_\kappa(l_1, l_0)\rangle\right] , \tag{5.1.18}$$

$$\begin{aligned}\langle v\sigma'_\kappa(l_1, l_0)\rangle &= \frac{10^{-8}}{2l_1+1}\left(\frac{\text{Ry}}{DE}\cdot\frac{E_1}{E_0}\right)^{3/2} \cdot \frac{A\,G'(\beta)}{\beta+\chi} \left[\text{cm}^3\ \text{s}^{-1}\right] ,\\ \langle v\sigma''_\kappa(l_1, l_0)\rangle &= \frac{10^{-8}}{2l_0+1}\left(\frac{\text{Ry}}{DE}\cdot\frac{E_1}{E_0}\right)^{3/2} \cdot \frac{A\,G''(\beta)}{\beta+\chi} \left[\text{cm}^3\ \text{s}^{-1}\right] .\end{aligned} \tag{5.1.19}$$

5.1.2 Ionization

Cross sections and rate coefficient of ionization can be described in the way similar to that outlined in Sect. 5.1.1. In this book we do not consider exchange effects in the case of ionization. The ionization cross section summed with respect to the quantum numbers J_1, L_1, S_1 is usually of interest. In this case the angular factor Q_i does not depend on κ. For the transition

$$n_0 l_0^m L_0 S_0 \rightarrow n_0 l_0^{m-1} L_i S_i + \mathrm{e} \tag{5.1.20}$$

We have

$$Q_\mathrm{i} = m \left(G_{L_i S_i}^{L_0 S_0} \right)^2 , \quad \sigma_\mathrm{i}(a_0, a_1) = Q_\mathrm{i} \sigma(l_0) . \tag{5.1.21}$$

Thus we can write fitting formula for the total one-electron ionization cross section summed over κ and l_1 :

$$\sigma_\mathrm{i}(l_0) = \frac{\pi a_0^2}{2l_0 + 1} \left(\frac{\mathrm{Ry}}{DE} \right)^2 \frac{C\Phi(u)}{u + \varphi} \qquad u = \frac{\mathscr{E} - E_z}{DE} \tag{5.1.22}$$

$$\langle v\sigma_i(l_0) \rangle = \frac{10^{-8}}{2l_0 + 1} \left(\frac{\mathrm{Ry}}{DE} \right)^{3/2} \cdot \frac{AG(\beta)}{\beta + \chi} e - \beta p \left[\mathrm{cm}^3 \ \mathrm{s}^{-1} \right] , \tag{5.1.23}$$

$$\beta = \frac{DE}{T} ,$$

where E_z is the ionization energy of the level a_0, the scale DE is E_z or z^2Ry. The functions Φ and G are

$$\Phi(u) = \frac{u}{u + 1 + D} \eta_1(u) , \tag{5.1.24}$$

$$G(\beta) = \beta^{1/2} \frac{\beta + 1 + D}{\beta + 1} \eta_1(\beta^{-1}) . \tag{5.1.25}$$

It should be noted that in the case of neutral atoms ($z = 1$), the formulas for Φ and G are based on the properties of the Born cross sections. The real behavior of the ionization cross section of neutral atoms when $u \rightarrow 0$ is just the same as for ions (see Sect. 3.1.4). Parameters C, φ, and A, χ obtained from the results of numerical calculations of cross sections are given in Sect. 6.1.6. The rate coefficient of three-body recombination to the level 0 is determined by (1.2.7)

$$\kappa_r = \frac{g_z}{2g_{z+1}} \left(\frac{2\pi\hbar^2}{mT} \right)^{3/2} \exp{(\beta p)} \langle v\sigma_i \rangle . \tag{5.1.26}$$

5.1.3 Recombination

In high-density plasmas the principal recombination process is three-body recombination. The number of these recombination events per second is $N_{z+1}N_e^2\kappa_r$, and the rate coefficient of recombination is related to the ionization rate coefficient by (5.1.26). In plasmas with moderate and low densities, two-body recombination prevails. The two-body processes are the radiative recombination and dielectronic recombination, see (1.1.4, 5). The number of two-body recombination events per second is $N_{z+1}N_e\kappa$, where $\kappa = \kappa_v + \kappa_d$. In two-body recombination a photon is involved, and therefore it is a weaker process than three-body recombination. When the density is not high, the probability of triple collisions is substantially smaller than the probability of binary collisions. This compensates the weakness of the interaction.

Dielectronic recombination will be considered in detail in Sect. 5.2. For the rate coefficient of dielectronic recombination

$$X_{z+1}(\alpha_0) + e \to X_z(\alpha_1 nl) \to X_z(\alpha_0 nl) + \hbar\omega\,, \qquad (5.1.27)$$

the fitting formula has the form (Sect. 5.2.5)

$$\kappa_d = 10^{-13} Q_d A_d \beta^{3/2} \exp(-\beta\chi_d)\ [\mathrm{cm^3\ s^{-1}}]\,,$$
$$\beta = (z+1)^2 \mathrm{Ry}/T\,, \qquad (5.1.28)$$

where A_d and χ_d are the adjusted parameters. The parameters A_d, χ_d for several types of transitions $\alpha_0 - \alpha_1$, are given in Sect. 6.1.7. The factor Q_d depends on the type of the transition $\alpha_0 - \alpha_1$. For transitions $\alpha_0 = l_0^m \to \alpha_1 = l_0^{m-1} l_1$,

$$Q_d = m/(2l_0+1)\,, \qquad (5.1.29a)$$

and for transitions $l_0^m l_1^k \to l_0^{m-1} l_1^{k+1}$,

$$Q_d = \frac{m}{2l_0+1}\left(1 - \frac{k}{2(2l_1+1)}\right)\,. \qquad (5.1.29b)$$

In most cases, recombination occurs onto the highly excited levels nl.

We shall consider now the radiative recombination process,

$$X_{z+1}(a_i) + \mathrm{e} \to X_z(a) + \hbar\omega \qquad (5.1.30)$$

[Ref. 1.1, Sect. 9.5]. Taking into account the known properties of the radiative recombination cross section [1.1], we adopt the following fitting formula for the rate coefficient of recombination to the level a:

$$\kappa_v(a) = 10^{-13} Q_v \left|\frac{E_a}{\mathrm{Ry}}\right|^{1/2} \frac{A}{\beta+\chi} \beta^{1/2}(\beta + D)\,[\mathrm{cm^3\ s^{-1}}]\,, \quad \beta = E_a/T \qquad (5.1.31)$$

where A, χ and D are the adjusted parameters [5.1.], and E_a is the ionization energy of the level a, which corresponds to the term $L_i S_i$ of an ion X_{z+1}. (We note that the numerical factor adopted here is different from the factor used in [Ref. 5.1, Sect. 9.7.4]. The factor Q_v depends on the angular momenta of the level a.

When $a = a_i\, nl\, LS$, $a_i = n_0 l_0^{m-1} L_i S_i$,

$$Q_v = \frac{(2S+1)(2L+1)}{2(2l+1)(2S_i+1)(2L_i+1)} . \tag{5.1.32}$$

For the cross section summed with respect to L, S,

$$Q_v(n_0 l_0^{m-1}[L_i S_i]nl) = Q_v(n_0 l_0^{m-1} nl) = 1 . \tag{5.1.33}$$

When $a = n_0 l_0^m LS$ and $a_i = n_0 l_0^{m-1} L_i S_i$,

$$Q_v = \frac{(2S+1)(2L+1)}{2(2l+1)(2S_i+1)(2L_i+1)} m \left(G\begin{matrix} LS \\ L_i S_i \end{matrix}\right)^2 . \tag{5.1.34}$$

For the cross section summed with respect to $L_i S_i$ and LS

$$Q_v(n_0 l_0^m) = 1 - \frac{m-1}{2(2l_0+1)} . \tag{5.1.35}$$

For the total rate coefficient of radiative recombination $X_{z+1} \to X_z$ one can adopt a formula similar to that of (5.1.31):

$$\kappa_v = \sum_a \kappa_v(a) = 10^{-13} \left(\frac{E_z}{\mathrm{Ry}}\right)^{1/2} A_v \frac{\beta^{1/2}(\beta + D)}{\beta + \chi} \,[\mathrm{cm^3\, s^{-1}}] ,$$
$$\beta = E_z/T , \tag{5.1.36}$$

where E_z is the ionization potential of an ion X_z. The factor Q_v should be taken into account in the summation with respect to a.

It can be shown (see below) that radiative recombinations occurs mainly to those levels a for which $|E_a| > T$. Here and anywhere in this book the energy of a level $E_a < 0$ is measured from the ionization limit. For these levels $\kappa_v(a)$ decreases slowly with increasing n, being proportional to n^{-1}. The higher levels with $|E_a| < T$ produce only small contributions to κ_v, since for these levels $\kappa_v(a) \propto n^{-3}$. In the case of high temperature $T > |E_z|$, radiative recombination to the states with minimal value of n provides the major contribution to the total recombination rate.

We shall now give for reference the formulas which allow the calculation of the cross section and the rate coefficient of radiative recombination (for details see [5.1]). The rate coefficient of radiative recombination is given by expression

$$\kappa_v(a) = K \left|\frac{E_a}{\mathrm{Ry}}\right|^{1/2} \beta^{3/2} \int_0^\infty \frac{\sigma_{rv}(u)}{\pi a_0^2} u \exp(-\beta u)\, du, \quad u = \mathscr{E}/|E_a| , \tag{5.1.37}$$

where $\beta = |E_a|/T$, and σ_{rv} is the cross section of recombination to the level a:

$$\sigma_{rv} = \pi a_0^2 \frac{2}{3(137)^3} Q_v(a) \frac{(|E_a| + \mathcal{E})^3}{\mathcal{E}} \sum_{\lambda=l\pm 1} l_m \rho^2(k\lambda, nl) ,$$
$$\hbar\omega = |E_a| + \mathcal{E}, \; l_m = \max(l, \lambda) . \qquad (5.1.38)$$

The radial integral $\rho(k\lambda, nl)$ is equal to

$$\rho(k\lambda, nl) = \int_0^\infty P_{nl}(r) P_{k\lambda}(r) r \, dr , \qquad (5.1.39)$$

where P_{nl} and $P_{k\lambda}$ are the radial wave functions of the optical electron in the discrete and continuous spectra of an atom X_z. The function P_{nl} is normalized, as usually, to unity, and the function $P_{k\lambda}$ is normalized in accordance with (3.2.2):

$$P_{k\lambda} \underset{r\to\infty}{\simeq} \frac{1}{\sqrt{k}} \sin\left(kr - \frac{\lambda\pi}{2} + \frac{z}{k} \ln kr + \eta \right) . \qquad (5.1.40)$$

In (5.1.38–40) atomic units with the Ry unit for the energy are used.

Recombination to highly excited levels $n \gg 1$ is well described by the quasi-classical formula [5.2]

$$\sigma_{\mathrm{rv}}(nl) = \pi a_0^2 \frac{16\pi z^2}{3(137)^3} \cdot \frac{\Delta n \, \mathrm{Ry}}{n\mathcal{E}} \sqrt{1 + \varepsilon_l^2} \, \Theta(\Delta n, \varepsilon_l) , \qquad (5.1.41)$$

$$\Theta(\Delta n, \varepsilon_l) = \left(\frac{d}{dx} J_{\Delta n}(x) \right)^2 - \frac{(\Delta n)^2 - x^2}{x^2} [J_{\Delta n}(x)]^2, \; x = \varepsilon_l \Delta n ,$$
$$\Delta n = \frac{\hbar\omega n^2}{2z^2 \mathrm{Ry}}, \; \varepsilon_l = \sqrt{1 - \frac{(l + 1/2)^2}{n^2}} . \qquad (5.1.42)$$

On summing (5.1.41) with respect to l we get the known Kramers formula for the radiative recombination to the hydrogenic levels:

$$\sigma_{\mathrm{rv}}^{\mathrm{Kr.}}(n) = \pi a_0^2 \frac{16 z^2}{3\sqrt{3}(137)^3} \frac{\mathrm{Ry}}{\Delta n \, \mathcal{E}} . \qquad (5.1.43)$$

On substituting (5.1.43) into (5.1.37) and using (5.1.2) we find

$$\kappa_v(n) = \frac{64\sqrt{\pi} \, a_0 \hbar}{3\sqrt{3} \, m \, 137^3} z \, \beta_n^{3/2} [-\exp(\beta_n) \, \mathrm{Ei}(-\beta_n)] ,$$
$$\beta_n = z^2 \mathrm{Ry}/n^2 T , \qquad (5.1.44)$$

where $\mathrm{Ei}(-\beta)$ is the exponential integral. With the use of (5.1.44) it is not difficult to show that

$$\kappa_v(n) \propto n^{-1} \quad \text{when} \quad \beta_n \gg 1 ,$$
$$\kappa_v(n) \propto n^{-3} \quad \text{when} \quad \beta_n \ll 1 . \qquad (5.1.45)$$

Using (5.1.44) for the levels with $n \geq n_1$ and the more accurate method for $n < n_1$, one can write the radiative recombination rate coefficient summed over all levels in the form

$$\kappa_v = \sum_{a(n<n_1)} \kappa_v(a) + \kappa_v^{\text{Kr.}}(n \geq n_1)\,, \tag{5.1.46}$$

$$\kappa_v^{\text{Kr.}}(n \geq n_1) = \frac{32\sqrt{\pi}\,a_0\hbar}{3\sqrt{3}\cdot m\cdot 137^3} z\, n_1 \beta_1^{1/2} \times\left[\ln 1.78\beta_1 - \exp(\beta_1)\,\mathrm{Ei}(-\beta_1)\left(1+\frac{\beta_1}{n_1}\right)\right],\quad \beta_1 = z^2\mathrm{Ry}/n_1^2 T\,. \tag{5.1.47}$$

5.1.4 Semiempirical Formulas for the Rates of Excitation, Ionization and Dielectronic Recombination

a) *Excitation*

In the range of high energies of the electron, the excitation cross section for optically allowed transition ($\Delta l = \pm 1, \Delta S = 0$) is expressed in terms of the oscillator strength by the formula (3.1.24). There are a number of semiempirical formulas for estimating cross sections which are based on the Bethe formula. Here we shall give the *Van Regemorter* formula [5.3] which is currently the most frequently used:

$$\sigma_{a_0a_1} = \pi a_0^2 \frac{8\pi}{\sqrt{3}} f_{a_0a_1} \left(\frac{\mathrm{Ry}}{\Delta E}\right)^2 \frac{\gamma(u)}{u+1}\,, \tag{5.1.48}$$

where $f_{a_0a_1}$ is the oscillator strength for the transition $a_0 \rightarrow a_1$. The factor $\gamma(u)$ is determined from the experimental data and the results of numerical calculations of excitation cross sections. Its values for neutral atoms and ions are given in Table 5.1. When $u \rightarrow \infty$,

$$\gamma(u) \simeq \frac{\sqrt{3}}{2\pi} \ln(1+u)\,,$$

and (5.1.48) becomes the Bethe formula.

Table 5.1. Factor $\gamma(u)$ for atoms ($z = 1$) and ions ($z > 1$)

	$\sqrt{u}$	0.0	0.2	0.4	0.6	0.8	1.0
γ,	$z = 1$	0.000	0.015	0.034	0.057	0.084	0.124
γ,	$z > 1$	0.200	0.200	0.200	0.200	0.200	0.200

	$\sqrt{u}$	2.0	3.0	4.0	5.0	6.0
γ,	$z = 1$	0.328	0.561	0.775	0.922	1.040
γ,	$z > 1$	0.328	0.561	0.775	0.922	1.040

On using (5.1.48) for the cross section the excitation rate coefficient can be written in the form

$$\langle v\sigma_{a_0a_1}\rangle = 10^{-8} \times 32 f_{a_0a_1} \left(\frac{\mathrm{Ry}}{\Delta E}\right)^{3/2} \beta^{1/2} \exp(-\beta) \cdot p(\beta)\ [\mathrm{cm}^3\ \mathrm{s}^{-1}]. \quad (5.1.49)$$

When $\beta \ll 1$,

$$p(\beta) \simeq -\frac{\sqrt{3}}{2\pi}\,\mathrm{Ei}(-\beta)\,.$$

The values of the factor $p(\beta)$ are given in Table 5.2. In various applications also the formulas by Drawin [5.4], Mewe [5.5], and Gryzinsky [5.6, 7] are often used. In fact, all these formulas practically do not differ.

The analytic formulas for the excitation cross sections can be obtained in the Born approximation using the model interaction potential $V_\kappa(R) \propto R^\kappa/(R_0^2+R^2)^{\kappa+1/2}$ [5.8]. For transitions between the levels with small energy spacing (with the same principal quantum number) such a potential is often fairly close to the real potential.

In case of optically allowed transitions ($\Delta l = 1$, $\kappa = 1$)

$$\sigma = 8\pi a_0^2 Z_p^2 \frac{M}{m}\frac{\mathrm{Ry}}{\mathscr{E}}\frac{\mathrm{Ry}}{\Delta E} f_{a_0a_1}\,[\Phi_1(x_{\min}) - \Phi_1(x_{\max})]\,, \quad (5.1.50)$$

where Z_p is the projectile charge, M is the reduced mass of the colliding pair, $f_{a_0a_1}$ is the oscillator strength for transition $a_0 \to a_1$

$$\Phi_1(x) = \frac{x^2}{2}[\mathrm{K}_2(x)\mathrm{K}_0(x) - \mathrm{K}_1^2(x)]\,, \quad (5.1.51)$$

$\mathrm{K}_n(x)$ are the modified Bessel functions,

$$x_{\substack{\min\\ \max}} = \frac{R_0}{a_0}\sqrt{\frac{M}{m}}\left(\sqrt{\frac{\mathscr{E}}{\mathrm{Ry}}} \mp \sqrt{\frac{\mathscr{E}-\Delta E}{\mathrm{Ry}}}\right)\,, \quad (5.1.52)$$

$$R_0 = a_0 z \mathrm{Ry}/(E_0E_1)^{1/2} \quad (5.1.53)$$

Table 5.2. Factor $p(\beta)$ for atoms ($z = 1$) and ions ($z > 1$)

	β	0.01	0.02	0.04	0.1	0.2	0.4
p,	$z = 1$	1.160	0.956	0.758	0.493	0.331	0.209
p,	$z > 1$	1.160	0.977	0.788	0.554	0.403	0.290

	β	1	2	4	10	>10
p,	$z = 1$	0.100	0.063	0.040	0.023	$0.066\beta^{-1/2}$
p,	$z > 1$	0.214	0.201	0.200	0.200	0.200

In the case of quadrupole transitions ($\kappa = 2$)

$$\sigma_{a_0a_1} = \frac{4}{135}\pi a_0^2 Z_p^2 \frac{M}{m}\frac{\mathrm{Ry}}{\mathcal{E}} Q_2'(a_0, a_1) \times (2l_1+1)\begin{pmatrix} l_0 & 2 & l_1 \\ 0 & 0 & 0 \end{pmatrix}^2 \frac{\langle l_0|r^2|l_1\rangle^2}{R_0^2 a_0^2} \times [\Phi_2(x_{\min}) - \Phi_2(x_{\max})] , \tag{5.1.54}$$

$$\Phi_2(x) = x^4[\mathrm{K}_2^2(x) - \mathrm{K}_1^2(x)] , \tag{5.1.55}$$

$$\langle l_0|r^2|l_1\rangle = \int_0^\infty \mathrm{P}_{n_0l_0}(r)\,\mathrm{P}_{n_1l_1}(r)\,r^2\,dr \tag{5.1.56}$$

When both $\mathcal{E} \gg \Delta E$ and $x_{\max} > 1$, $\Phi_1(x_{\max})$ and $\Phi_2(x_{\max})$ can be neglected, and

$$x_{\min} \cong \frac{1}{2}\frac{R_0}{a_0}\left(\frac{M}{m}\right)^{1/2}\frac{\Delta E}{\mathrm{Ry}}\left(\frac{\mathrm{Ry}}{\mathcal{E}}\right)^{1/2} . \tag{5.1.57}$$

This approximation corresponds to the first-order impact parameter approximation with rectilinear trajectories. In this case the functions $\Phi_1(x_{\min})$ and $\Phi_2(x_{\max})$ can be approximated by

$$\Phi_1(x) \cong \exp(-2x) \cdot \ln(2.25 + 0.681/x) , \tag{5.1.58}$$

$$\Phi_2(x) \cong \exp(-2x) \cdot (2 + x\sqrt{3\pi/2})^2 . \tag{5.1.59}$$

The rate coefficients, averaged over the Maxwellian distribution, can be written in the form

$$\langle v\sigma_\kappa\rangle = \frac{16\pi^{1/2}}{2\kappa+1}\frac{e^2}{\hbar}a_0^2\frac{Q_\kappa'(a_0a_1)}{[(2\kappa-1)!!]^2} \times(2l_1+1)\begin{pmatrix} l_0 & \kappa & l_1 \\ 0 & 0 & 0 \end{pmatrix}^2 \frac{\langle l_0|r^\kappa|l_1\rangle^2}{R_0^{2\kappa-2}a_0^2} \times\exp(-\Delta E/T)\cdot(\mathrm{Ry}/T)^{1/2} I_\kappa(\Delta E/T, y_0) , \tag{5.1.60}$$

$$I_\kappa(x,y) = \exp(x/2)\int_0^\infty dt\, t^{2\kappa-1}\mathrm{K}_1^2(t)\exp\left[-\frac{x}{4}\left(\frac{y^2}{t^2}+\frac{t^2}{y^2}\right)\right] , \tag{5.1.61}$$

$$y_0 = \frac{R_0}{a_0}\left(\frac{\Delta E}{\mathrm{RY}}\right)^{1/2} .$$

For the dipole transitions ($\Delta l = 1$) $\langle v\sigma\rangle$ can be written in the following form

$$< v\sigma >= 1.74\,10^{-7} f_{a_0a_1}(\mathrm{Ry}/\Delta E)(\mathrm{Ry}/T)^{1/2} \times\exp(-\Delta E/T)\cdot I_1(\Delta E/T, y_0)\ [\mathrm{cm}^3\ \mathrm{s}^{-3}] , \tag{5.1.62}$$

When the argument of K_0 does not exceed 3.0 the integral $I_1(x, y)$ is approximated by the following asymptotic formula

$$I_1(x,y) \cong \exp(x/2)\ K_0(\sqrt{x(y^2+x/4)})\,. \tag{5.1.63}$$

For transitions with small energy spacing in plasmas usually $\mathcal{E} \gg \Delta E$. Therefore, the Coulomb attraction of the electrons is not important. However, in cases of neutral atoms and ions with low z the effect of normalization can be very substantial. In such cases the formulas given above can be used only for $\mathcal{E} \gg E_0$, E_1, i.e., when the Born approximation is valid. For multiply-charged ions ($z > 3$) and $\mathcal{E} > \Delta E$ they give fairly reasonable values of cross sections for the dipole transitions, and allow to estimate the order of magnitude of the cross section for quadrupole transitions.

b) *Ionization*

The well-known classical Thomson formula for the cross section of ionization from the shell $n_0 l_0^m$ corresponds to (5.1.22 and 24) when $C = 4m(2l_0 + 1)$, $\varphi = 1, DE = E_z$ and $D = 0$:

$$\sigma_{\mathrm{i}} = \pi a_0^2 \cdot 4m \left(\frac{\mathrm{Ry}}{E_z}\right)^2 \frac{u}{(u+1)^2}\,. \tag{5.1.64}$$

To estimate the rate coefficient of ionization for atoms and ions from the ground state, the *Seaton* formula [5.9] is often used:

$$\langle v\sigma_i \rangle = 10^{-8} \times 4.3\mathrm{m} \left(\frac{\mathrm{Ry}}{E_z}\right)^{3/2} \beta^{-1/2} \exp(-\beta)\ [\mathrm{cm}^3\ \mathrm{s}^{-1}],$$
$$\beta = E_z/T\,, \quad \beta \gtrsim 1\,, \tag{5.1.65}$$

where E_z is the ionization potential of an ion X_z. This formula corresponds to (5.1.23, 24) when $z > 1, \chi = 0$ and $A = 4.3 \cdot (2l_0 + 1)$.

Expression (5.1.65) is valid only for $\beta \geq 1$. Sufficiently universal semi-empirical formula was suggested by Lotz [5.10]:

$$\langle v\sigma_i \rangle = 10^{-8} \cdot 6m \left|\frac{\mathrm{Ry}}{E_z}\right|^{3/2} \beta^{-1/2} \exp(-\beta) f(\beta)\quad [\mathrm{cm}^3\ \mathrm{s}^{-1}],$$
$$f(\beta) = -\beta \exp(\beta)\, \mathrm{Ei}(-\beta)\,. \tag{5.1.66}$$

This formula corresponds approximately to (5.1.23, 24) when $z > 1, \chi = 0.4$, and $A = 6(2l_0 + 1)$. The values of the factor $f(\beta)$ are given in Table 5.3.

Table 5.3. Factor $f(\beta)$

β	$= 1/4$	1	4	8
$f(\beta)$	$= 0.34$	0.59	0.83	0.90

The compact semiempirical formula which is also often used is given in [5.4]. Classical formulas for the ionization cross sections are given in [5.7, 11].

c) *Dielectronic Recombination*
A detailed treatment of dielectronic recombination is given in Sect. 5.2. To exhaust the list of analytic formulas for bound-bound and free-bound electronic collisional processes we present here a semiempirical formula for the rate coefficient of dielectronic recombination proposed by Burgess [5.12]. This formula can be written in a form similar to (5.1.28)

$$\kappa_d(\alpha) = 10^{-13} B_d \beta^{3/2} \exp(-\beta \chi_d) \quad [\mathrm{cm}^3\,\mathrm{s}^{-1}]\,,$$
$$\beta = (z+1)^2 \mathrm{Ry}/T\,, \tag{5.1.67}$$

with

$$\mathrm{B_d} = 480 f_{\alpha 0 \alpha} \left(\frac{z\chi}{z^2+13.4}\right)^{1/2} [1 + 0.105(z+1)\chi + 0.015(z+1)^2\chi^2]^{-1}\,,$$
$$\chi_d = \chi \left(1 + 0.015\frac{z^3}{(z+1)^2}\right)^{-1}\,, \quad \chi = \frac{E_{\alpha\alpha_0}}{(z+1)^2 \mathrm{Ry}}\,. \tag{5.1.68}$$

5.2 Dielectronic Recombination

In this section and the following one we discuss some problems related to dielectronic recombination[1] and formation of dielectronic satellites. We have used cgs units here. We recall also that for ions which are members of the isoelectronic sequence of an atom A the designation $[A]$ is used. For example, the designation $[H]$ is used for a set He^+, Li^{2+} and so on.

5.2.1 Electron Capture and Underthreshold Resonances (Simplified Model)

As noted above, the excitation cross section for positive ions has a nonzero value at threshold due to the long-range Coulomb attraction. This attraction also allows the excitation of an ion X_{z+1} at an energy below threshold, the electron being captured on some level nl of the ion X_z. For example, at an energy lower than the excitation threshold for the resonance level of the He-like ion O^{6+}, the following process is possible:

$$O^{6+}(1s^2) + e \rightarrow O^{5+}(1s2p\, nl)\,.$$

The doubly excited state which is the result of electron capture is unstable, and may decay either through autoionization or spontaneous emission of the resonance

[1] Dielectronic recombination is widely discussed in the literature, see e.g. the review articles [5.13–15]. In [5.15] one can find an excellent historical review and numerous references to original articles.

photon $2p - 1s$. In the latter case, the atom X_z is transferred to the stationary state, i.e., recombination occurs. This process is called dielectronic recombination (abbreviated to DR below).

Generally, the process of dielectronic recombination of an ion X_{z+1} via the intermediate doubly excited state of an ion X_z is written in the form

$$X_{z+1}(\alpha_0) + e \rightarrow X_z^{**}(\gamma) \begin{array}{l} \nearrow X_z^*(\gamma') + \hbar\omega \\ \searrow X_{z+1}(\alpha') + e \end{array} , \tag{5.2.1}$$

$$\gamma = \alpha\, nl\, LSJ, \quad \gamma' = \alpha' nlL'S'J' .$$

Below, the LS-coupling scheme is adopted. Besides it is assumed that photon emission occurs due to transition of the "inner" electron $\alpha - \alpha'$, and the state nl of the "outer" electron does not change. Radiative transitions of the "outer" electron cause the complementary satellites (Sect. 5.3), but they do not play an essential role in the total balance of dielectronic recombination.

The three most important effects of a DR process are:

A. Dielectronic recombination for all ions other than bare nuclei is an additional recombination process. In many actual cases, as shown by Burgess and Seaton [5.16] the rate of DR can considerably exceed the rate of radiative recombination. Therefore, in low density plasmas dielectronic recombination should be necessarily taken into account.
B. Satellites to the resonance and other lines of an ion X_{z+1} originating from radiative transitions in reaction (5.2.1).
C. Complementary excitation of levels α' when the autoionization occurs in reaction (5.2.1) with $\alpha' \neq \alpha_0$.

The latter has been considered in detail in Sect. 3.4, the satellites will be discussed in Sect. 5.3. Here we shall consider the intrinsic dielectronic recombination. For simplicity we discuss here only the process (5.2.1) neglecting the secondary ionization of the excited ion $X_z(\gamma')$. Both collisional and radiative secondary ionization in a plasma are considered in [5.17]. In some cases the secondary autoionization of ion $X_z(\gamma')$ is also possible. (For a discussion of the secondary processes see also [5.15].) In Sect. 5.2.1 we confine ourselves to the description of the simplified model making the following assumptions:

a) the state of ion X_z is described by quantum numbers αnl without specifying the terms LS;
b) the value of n is large enough, so that the influence of electron nl on the state α of the core can be neglected; the levels nl can be considered as hydrogenic, and capture cross section can be expressed in terms of excitation cross section for transition $\alpha_0 - \alpha$ using the correspondence principle; and
c) in the process of photon emission, the electron returns to the initial state α_0. In this case the formulas for calculation of dielectronic recombination cross

sections prove to be sufficiently simple. Discussion of these assumptions will be given in Sect. 5.2.3. The general case will be discussed in Sect. 5.2.2.

Within the frame of our simplified model, DR process is written in the form

$$X_{z+1}(\alpha_0) + e \to X_z^{**}(\alpha\, nl) \begin{cases} \nearrow X_z^{*}(\alpha_0 nl) + \hbar\omega \\ \searrow X_{z+1}(\alpha_0) + e\,, \end{cases} \tag{5.2.2}$$

the lower branch of the reaction (autoionization) being the competing process. Therefore the cross section for dielectronic recombination via the state αnl is

$$\sigma_{\mathrm{d}}(\alpha\, nl) = \sigma_{\mathrm{d}}'(\alpha_0,\, \alpha\, nl)\frac{A(\alpha, \alpha_0)}{A(\alpha) + W_{\mathrm{a}}(\alpha\, nl)}\,, \tag{5.2.3}$$

where $A(\alpha\,, \alpha_0)$ is the probability of a radiative transition $\alpha - \alpha_0$ in an ion X_{z+1}, W_{a} is the autoionization probability for the level αnl of an atom X_z, $A(\alpha) = \sum_{\alpha_1} A(\alpha, \alpha_1)$ is the total probability of radiative decay of the level $\alpha, \sigma_{\mathrm{d}}'(\alpha_0, \alpha nl)$ is the cross section for electron capture to the level nl when the transition $\alpha_0 - \alpha$ is excited. This cross section is represented by a set of resonances at the energies

$$\mathscr{E} \simeq \Delta E - \frac{z^2\mathrm{Ry}}{n^2} < \Delta E, \quad \Delta E = E_{\alpha\alpha_0} = E_\alpha - E_{\alpha_0}\,. \tag{5.2.4}$$

It is convenient in this section to use again CGS units. The resonance width equals $\Gamma = \hbar W_{\mathrm{a}}$. The cross section averaged over the resonances can be obtained with the aid of the correspondence principle by extrapolating below the threshold the partial cross section for the excitation $\alpha_0 - \alpha$:

$$\overline{\sigma_d'(\alpha_0,\, \alpha nl)\Gamma} = \sigma(\alpha_0,\, \alpha l)\frac{2z^2\mathrm{Ry}}{n^3}\,. \tag{5.2.5}$$

Here $\sigma(\alpha_0,\, \alpha l) = \sum_{\lambda_0} \sigma(\alpha_0\lambda_0,\, \alpha l)$, where $\sigma(\alpha_0\lambda_0,\, \alpha l)$ is the partial cross section for the transition $\alpha_0 - \alpha$ in the threshold $\mathscr{E} = \Delta E$; λ_0, l are the orbital momenta of the outer electron. In accordance with condition (b) of the model we should sum the cross section over total angular momenta $L_T S_T$. The corresponding formulas are given in Sects. 2.3 and 3.2, in which the sum over λ is to be replaced by one definite value of $\lambda = l$.

The values of W_{a} and σ_{d}'(or σ) are related to each other as characteristics of direct and reverse processes. To derive this relation it should be noted that at $A = 0$, the ratio of the populations of $X_{z+1}(\alpha_0)$ and $X_z(\alpha nl)$ is given by the Saha formula. Using this formula, we obtain

$$(2l+1)g_\alpha W_{\mathrm{a}}(\alpha nl) = \frac{z^2\mathscr{E}}{\pi\hbar n^3}\frac{g_0\sigma(\alpha_0, \alpha l)}{\pi a_0^2}\,, \tag{5.2.6}$$

where g_α and g_0 are the statistical weights of the states α and α_0.

The rate coefficient of dielectronic recombination is

$$\kappa_{\mathrm{d}} = \sum_\alpha \kappa_{\mathrm{d}}(\alpha), \quad \kappa_{\mathrm{d}}\,(\alpha) = \sum_{nl} v\sigma_{\mathrm{d}}(\alpha_0,\, \alpha\, nl)\Gamma\mathscr{F}(\mathscr{E})\,, \tag{5.2.7}$$

where Γ is the resonance width, and $\mathcal{F}(\mathcal{E})$ is the Maxwellian distribution for the energies of the electrons. The value of $\mathcal{E}$ is given by (5.2.4). Substituting (5.2.3) and (5.2.5) in (5.2.7) we obtain

$$\kappa_d(\alpha) = \sum_{n>n_1} \sum_{l<n} \frac{2z^2\mathrm{Ry}}{n^3} v\sigma(\alpha_0, \alpha l)\mathcal{F}(\mathcal{E}) \left[1 + \left(\frac{n_s}{n}\right)^3\right]^{-1}, \tag{5.2.8}$$

where n_s and n_1 are determined by the relations

$$\frac{z^2\mathrm{Ry}}{n_1^2} = \Delta E, \quad \left(\frac{n_s}{n}\right)^3 = \frac{W_a(\alpha nl, \alpha_0)}{A(\alpha, \alpha_0)}. \tag{5.2.9}$$

One can see that n_1 is in fact the minimum value of n, i.e., it determines the lowest level at which the capture of an electron is possible in accordance with (5.2.4).

We now transfer (5.2.8) into a form more convenient for applications. We substitute in (5.2.8) the explicit expression for $\mathcal{F}(\mathcal{E})$ and use the relation (5.2.6), and the relation between A and the oscillator strength f:

$$g_\alpha A(\alpha, \alpha_0) = \frac{1}{137^3}\left(\frac{\mathrm{Ry}}{\hbar}\right)\left(\frac{\Delta E}{\mathrm{Ry}}\right)^2 g_0 f_{\alpha_0\alpha}. \tag{5.2.10}$$

We write the result in the form

$$\kappa_d(\alpha) = 10^{-13} B_d(\alpha)\beta^{3/2}\exp(-\beta\chi) \quad [\mathrm{cm^3\, s^{-1}}],$$
$$\beta = \frac{(z+1)^2\mathrm{Ry}}{T}, \quad \chi = \frac{\Delta E}{(z+1)^2\mathrm{Ry}}, \tag{5.2.11}$$

where $\beta^{3/2}\exp(-\beta\chi)$ provides the main temperature dependence of κ_d, and $B_d(\alpha)$ only slightly depends on temperature and is equal to

$$B_d(\alpha) = C' \cdot \frac{z}{n_1^4} \cdot f_{\alpha_0\alpha} \sum_{n>n_1} \exp(\delta\beta) \sum_{l<n} \frac{2l+1}{1+(n/n_s)^3} \equiv \sum_{n>n_1} B_n \exp(\delta\beta),$$
$$n_s = n_s(l) = 137\left(\frac{n_1^2\sigma(\alpha_0, \alpha l)}{\pi^2 a_0^2(2l+1) f_{\alpha_0\alpha}}\right)^{1/3}, \quad \delta\beta = \frac{z^2\mathrm{Ry}}{n^2 T}, \tag{5.2.12}$$
$$C' = 10^{13} \times \frac{4\pi^{3/2} a_0 \hbar}{137^3 m}\left(\frac{z}{z+1}\right)^3 = 0.53\left(\frac{z}{z+1}\right)^3 [\mathrm{cm^3\, s^{-1}}].$$

The value of n_s separates all the levels into two parts. For $n > n_s(l), W_a < A$ and after the capture of an electron the ion X_z transfers to a stable state, i.e., recombination occurs. The contribution of these levels to κ_d is

$$\kappa_d'(\alpha) \simeq \int_{\Delta E - |E_s|}^{\Delta E} \sum_l v\sigma(\alpha_0, \alpha l)\mathcal{F}(\mathcal{E})\, d\mathcal{E}, \quad E_s = -\frac{z^2\mathrm{Ry}}{n_s^2}. \tag{5.2.13}$$

For $n < n_s(l), W_a > A$ and most of captures are followed by autoionization; only a small part $\sim (n/n_s)^3$ provides the DR. Nevertheless these levels $(n < n_s)$

contribute mainly to the total rate κ_d. If we temporarily omit the factor $\exp(\delta\beta)$ in (5.2.12), then all levels with $n < n_s(l)$ will almost equally contribute to κ_d. If, as usually is the case, $n_s \gg 1$ the ratio of contributions of the levels with $n < n_s$ and $n > n_s$ is

$$\frac{\kappa_d''}{\kappa_d'} \simeq 2\frac{n_s - n_l}{n_s} \simeq 2 .$$

If now we take into account the factor $\exp(\delta\beta)$, this ratio will be even larger since $\delta\beta \sim n^{-2}$.

The levels $n < n_s(l)$ of the ion X_z are evidently in thermodynamical equilibrium with the ion X_{z+1} and hence the rate of DR is proportional to the probability of radiative transition, $A(\alpha, \alpha_0) \propto \Delta E^2 f_{\alpha_0\alpha} \propto n_l^{-4} f_{\alpha_0\alpha}$, in accordance with (5.2.12).

As seen above DR proceeds mainly in the levels $n_l - n_s$, and since in most cases $n_s \gg 1$ the condition (b) of our model ($n \gg 1$) is justified. The exception is the case of the $1s - 2p$ transition at $z \gtrsim 20$ when $n_s < n_l$ and $n > n_s$ for all possible values of n.

We consider now as an illustrative example the recombination of a [Li] ion with excitation of $2s - 2p$ transition and recombination of [H] ion with excitation of $1s - 2p$ transition. In the first case, the energy level distance ΔE is small, so that $\chi = \Delta E/(z+1)\text{Ry} \sim 1/z \ll 1$, and $\sigma(\alpha_0, \alpha l)$ is large. That means the factor $\exp(-\beta\chi) \simeq 1$ in (5.2.11), $n_s \gg n_l \gg 1$, and a great number of levels concentrated in very narrow energy band ($\backsim \Delta E$) contribute to κ_d. Because of the small value of $A \backsim \chi^2 f$, the value of κ_d is comparatively small in spite of a great number of levels.

In the case of recombination of [H] ion or [He] ion $\chi \simeq 3/4$, and due to the factor $\exp(-\beta\chi)$, the rate of DR at small temperatures is negligible. The value of n_s isn't large and for $z \gtrsim 20, n_s$ is even smaller than $n_l \simeq 2$. For this reason, a comparatively small number of levels contribute to B_d in (5.2.12), but the contribution of each one is great because of the high value of A. According to numerical calculations the total values of B_d for the $1s - 2p$ transition usually exceed those for the $2s - 2p$ transition. The value of κ_d, however, is greatly dependent on temperature in the case of the $1s - 2p$ transition.

5.2.2 General Case

The formulas (5.2.11, 12) obtained above provide a useful method of calculation of the DR rate coefficient within the simplified model. In this section we derive a general expression for the DR rate coefficient without the assumptions of the simplified model. We shall consider the process (5.2.1) again using the detailed balance principle (Sect. 1.2) to derive the general formula for recombination rate coefficient. The total DR rate is

$$\kappa_d = \sum_\gamma \kappa_d(\gamma) = \sum_\gamma \tilde{\kappa}(\gamma) \frac{A(\gamma)}{A(\gamma) + W_a(\gamma)} , \tag{5.2.14}$$

where $\tilde{\kappa}(\gamma)$ is the probability of electron capture into the state γ of X_z; and $A(\gamma) = \sum_{\gamma'} A(\gamma, \gamma')$ and $W_a(\gamma) = \sum_{\gamma'} W_a(\gamma, \gamma')$ are the probabilities of radiative and autoionization decays. Since the latter decay is associated with internal electrostatic interaction, it cannot change the total momenta LSJ.

If we suppose that $A = 0$, and hence the system is in thermodynamical equilibrium, then according to detailed balance principle, we can write

$$N_e N_{z+1} \tilde{\kappa}(\gamma) = N_z(\gamma) W_a(\gamma, \alpha_0) , \quad W_a(\gamma, \alpha_0) = \sum_{\lambda_0} W_a(\gamma, \gamma_0) ,$$

where the ratio N_{z+1}/N_z is determined by Saha equation (Sect. 1.2). The value of $\tilde{\kappa}$, of course, is independent of any assumption concerning the radiative decay associated with an electromagnetic interaction. Therefore we can use the last equation to determine the value of $\tilde{\kappa}$ in the general case, and substituting it in (5.2.14), we obtain

$$\begin{aligned} \kappa_d(\gamma) &= \frac{1}{2g_0} \cdot \frac{8\pi^{3/2} a_0^3}{(z+1)^3} \beta^{3/2} \exp(-\beta\chi + \delta\beta) \\ &\times \frac{g_\gamma A(\gamma)\, W_a(\gamma, \alpha_0)}{A(\gamma) + W_a(\gamma)} ; \quad \delta\beta = \frac{\Delta E - E_{\gamma\alpha_0}}{T} , \end{aligned} \tag{5.2.15}$$

the values of β and χ being determined by (5.2.11); g_0 and g_γ are the statistical weights of the states $\alpha_0(X_{z+1})$ and $\gamma(X_z)$, and $\Delta E = E_{\alpha\alpha_0}$ and $E_{\gamma\alpha_0}$ are the excitation energies of the states α and γ. The difference of these energies δE is equal to the bound energy of the captured electron,

$$\delta E = E_{\alpha\alpha_0} - E_{\gamma\alpha_0} \simeq \frac{z^2 \mathrm{Ry}}{n^2} . \tag{5.2.16}$$

The radiative decay probability $A(\gamma)$ summed over all final states γ' does not depend on LSJ and is denoted below by $A(\alpha)$. The values of β and $\delta\beta$ are in fact also independent of LSJ. The autoionization probability in the LS coupling scheme does not depend on J, but essentially depends on LS. Therefore we shall write the rate coefficient of DR in the form (5.2.11), the factor B_d being equal to

$$\begin{aligned} & B_d(\alpha) = C \sum_{nlLS} \exp(\delta\beta)\, q(\gamma), \quad \gamma = \alpha n l L S, \\ & q(\gamma) = \frac{g_\gamma W_a(\gamma, \alpha_0) A(\alpha)}{A(\alpha) + W_a(\gamma)}, \quad C = 10^{13} \frac{4\pi^{3/2} a_0^3}{g_0 (z+1)^3}\ [\mathrm{cm^3\ s^{-1}}] . \end{aligned} \tag{5.2.17}$$

Due to the nonlinear dependence of B_d on $W_a(\gamma)$ we cannot explicitly sum over LS in (5.2.17). For this reason, the use of this equation requires a great deal of computation. In most applications an approximate formula is used, in which an averaged value of $W_a(\gamma)$,

$$W_a(\alpha n l) = \sum_{LS} \frac{(2L+1)(2S+1)}{2(2l+1)(2L_\alpha+1)(2S_\alpha+1)} W_a(\gamma) , \tag{5.2.18}$$

is substituted in the denominator of (5.2.17). After this substitution we can write (5.2.17) in the form

$$B_d(\alpha) = C \sum_{nl} \exp(\delta\beta)\, q(\alpha nl)\,,$$

$$q(\alpha nl) = \frac{2(2l+1)g_\alpha A(\alpha)\, W_a(\alpha nl, \alpha_0)}{A(\alpha) + W_a(\alpha nl)}\,. \tag{5.2.19}$$

Calculations with these formulas are much simpler than (5.2.17). Summing of linear expressions of the type (5.2.18) can be accomplished analytically, and only the sum over nl has to be done numerically. Besides, the expression for $W_a(\alpha nl)$ is much simpler than that for $W_a(\gamma)$ (Sect. 5.2.3). Approximation (5.2.19) corresponds to the assumption (a) of the simplified model.

If $n \gg 1$ [assumption (b)], we can use the relation (5.2.6) and substitute the threshold excitation cross section in place of W_a in (5.2.19). Thus we obtain

$$B_d(\alpha) = C' \cdot \frac{z}{n_1^4} \cdot f_{\alpha_0\alpha} \sum_{n>n_1} \exp(\delta\beta) \sum_{l<n} \frac{(2l+1)B'}{B + B'(n/n_s)^3}\,, \tag{5.2.20}$$

where

$$B' = \sum_{\alpha'} \frac{g_{\alpha'} f_{\alpha'\alpha}}{g_0 f_{\alpha_0\alpha}} \left(\frac{E_{\alpha\alpha'}}{E_{\alpha\alpha_0}}\right)^2\,,$$

$$B = \frac{W_a(\alpha nl)}{W_a(\alpha nl, \alpha_0)}\,, \tag{5.2.21}$$

C', n_1, n_s, being defined by (5.2.9 and 12).

The simplified model (5.2.11, 12) is readily obtained from (5.2.20, 21) if we assume that $B = B' = 1$. This corresponds to the assumption (c) with the single final state α_0 for all decays.

5.2.3 Formulas for Autoionization Probability

In this subsection we shall give the formulas for W_a without using the approximation (5.2.6). First of all we consider the probability $W_a(\gamma)$ for decay of the state $\gamma = \alpha nlLSJ$.

The equation for probability of autoionization was derived in Sect. 3.4 on the base of the general theory of the excitation of highly charged atoms. We can obtain the same result more directly as the transition probability $\gamma - \gamma_0$ in first-order perturbation theory. The latter way is similar to deriving the formula for the partial $\gamma - \gamma_0$ excitation cross section of the ion X_{z+1}. In this case, however, we deal with transitions at given values of LSJ which correspond to total angular momenta $L_T S_T J_T$ in the problem of excitation of the ion X_{z+1}. Therefore, we cannot use the simple formulas from Sect. 2.3.2 where the cross section was summed over $L_T S_T J_T$. In the case of LS coupling, $W_a(\gamma)$ does not depend on J, but depends on LS. Using (2.3.27, 28) we obtain

$$W_a(\gamma, \alpha_0) = \frac{2\mathrm{Ry}}{\hbar} \sum_{\lambda_0} | \sum_{\kappa} A_\kappa (\delta_{S_{\alpha_0} S_\alpha} R^d_\kappa - B_{S_{\alpha_0} S_\alpha} \sum_{\kappa''} R^e_{\kappa'' \kappa})|^2 , \tag{5.2.22}$$

$$\begin{aligned} A_\kappa &= (-1)^{L+L_{\alpha_0}+L_\alpha+L_p} \Pi(\kappa L_{\alpha_0} L_\alpha) \\ &\times \begin{Bmatrix} \kappa & L_{\alpha_0} & L_\alpha \\ L_p & l_\alpha & l_{\alpha_0} \end{Bmatrix} \begin{Bmatrix} \kappa & L_{\alpha_0} & L_\alpha \\ L & l & \lambda_0 \end{Bmatrix} G^{L_{\alpha_0} S_{\alpha_0}}_{L_p S_p} \sqrt{m} , \end{aligned} \tag{5.2.23}$$

$$B_{S_{\alpha_0} S_\alpha} = (-1)^{1-S_{\alpha_0} S_\alpha} \Pi(S_{\alpha_0} S_\alpha) \begin{Bmatrix} \frac{1}{2} S_{\alpha_0} S \\ \frac{1}{2} S_\alpha S_p \end{Bmatrix} .$$

In the radial integrals R^d and R^e, the functions $F_\lambda(r)$ for the electron in continuum are replaced by the functions $P_l(r)$ for a discrete spectrum, P_l being normalized to unity similarily to P_{l_0} and P_{l_α}. The functions $\tilde{P}_l$ and $\tilde{F}_{\lambda_0}$ in the exchange integral R^e are orthogonalized functions related to P_l and F_{λ_0} according to (3.2.5).

For the averaged quantity $W_a(\alpha nl)$, we can use simpler formulas, similar to those given in Sect. 2.3.2 for the cross section summed over $L_T S_T$. Thus we obtain [cf. (2.3.8, 9)]

$$\begin{aligned} W_a(\alpha\, nl, \alpha_0) = \frac{2\mathrm{Ry}}{\hbar} \sum_{\kappa} [&Q'_\kappa(\alpha, \alpha_0)\, W'_\kappa(l_\alpha\, nl,\, l_0) \\ &+ Q''_\kappa(\alpha, \alpha_0)\, W''_\kappa(l_\alpha\, nl,\, l_0)] , \end{aligned} \tag{5.2.24}$$

$$\begin{aligned} W'_\kappa(l_\alpha\, nl,\, l_0) &= \frac{1}{(2l_\alpha+1)(2l+1)} \sum_{\lambda_0} R^d_\kappa (R^d_\kappa - \sum_{\kappa''} R^e_{\kappa''\kappa}) , \\ W''_\kappa(l_\alpha\, nl,\, l_0) &= \frac{1}{(2l_\alpha+1)(2l+1)} \sum_{\lambda_0} (\sum_{\kappa''} R^e_{\kappa''\kappa})^2 . \end{aligned} \tag{5.2.25}$$

Formulas for Q factors are given in Sect. 2.3.2, but the direction of transition has to be reversed to $\alpha \to \alpha_0$ instead of $\alpha_0 \to \alpha$.

In the limit $n \gg 1$, the expression (5.2.24) is proportional to the partial excitation cross section for the transition α_0–α in the ion X_{z+1}. To derive an explicit relation we should use the following relation of the discrete and continuous spectra functions:

$$P_{nl}(r) \underset{n\to\infty}{\to} \left(\frac{2z^2}{\pi n^3}\right)^{1/2} F_{kl}\,(k \to 0) . \tag{5.2.26}$$

Thus we obtain (5.2.8).

5.2.4 Some Inaccuracies of the Simplified Model

The error introduced in the results of calculation of κ_d due to substitution of the average value $W_a(\alpha nl)$ in place of $W_a(\gamma)$ depends on the relation between the probabilities W_a and A. In the case $W_a \ll A$ (large z or n), we have

$$q(\alpha nl) = 2(2l+1)g_\alpha W_a(\alpha nl, \alpha_0)\,,$$
$$\sum_{LS} q(\gamma) = \sum_{LS} g_\gamma W_a(\gamma, \alpha_0) = q(\alpha nl)\,, \tag{5.2.27}$$

the last equality resulting from (5.2.18). Thus the replacement of $W_a(\gamma)$ by its average value does not introduce any additional errors in this case.

The alternative case, $W_a \gg A$, is more complicated. If $W_a \gg A$ for every possible values of L, S, then $q(\gamma) = g_\gamma A(\alpha)$ and $\sum q(\gamma) = q(\alpha nl)$, and using the average value $W_a(\alpha nl)$ does not introduce any additional error. It is possible, however, that for some values of L, S $W_a(\gamma) = 0$ in accordance with selection rules, or $W_a(\Gamma)$ is very small, and the terms with these $W_a(\gamma)$ do not contribute to $\sum_\gamma q(\gamma)$.

As an illustration we consider an $s-p$ transition $\alpha_0 - \alpha$ in an ion X_{z+1} with one electron outside the closed shells. At first we consider the orbital momentum only. If $\gamma_0 = s\lambda_0 LS$, $\gamma = plLS$, then from orbital momentum and parity conservation it follows that $L = \lambda_0 = l \pm 1$ and $W_a(\gamma) = 0$ for $L = l$. Therefore we have

$$\sum_{L} q(\gamma) = \sum_{L \neq 1} g_\gamma A(\alpha) = (2S+1)(4l+2)A(\alpha)\,,$$
$$q(\alpha nl) = g_s \cdot 3(2l+1)A(\alpha)\,, \tag{5.2.28}$$

where g_s is the spin part of the statistical weight $g_{\alpha nl}$. We see that $q(\alpha nl)$ overestimates the result by a factor of 1.5. We emphasize that to produce an accurate result isn't a simple problem since LS coupling is replaced by jl coupling with increasing values of n. In jl coupling, the selection rule $L \neq l$ fails. The quantitative treatment of the coupling transfer is difficult. We note however that this inaccuracy is essential only for $s-p$ transitions. In other cases, the inaccuracy introduced by substitution of $W_a(\alpha nl)$ in place of $W_a(\gamma)$ is small with the possible exception of the lowest values of n.

If we neglect the above-mentioned change of a coupling scheme, we can derive a sufficiently simple and accurate expression for the DR rate coefficient in an $s-p$ transition. In this case $L = \lambda_0$, and therefore it is sufficient to average $W_a(\gamma)$ over S and to substitute in the denominator of (5.2.17) the quantity

$$W_a(\alpha nl) = \sum_{s} \frac{2S+1}{2(2S_\alpha+1)} W_a(\gamma) = \frac{3(2l+1)}{2\lambda_0+1} W_a(\alpha nl)\,.$$

In the way, which is similar to that used for the derivation of (5.2.20), we obtain for both cases $\alpha_0 = ns$ and $\alpha_0 = ns^2$:

$$B_d(\alpha) = C' \frac{z}{n_1^4} f_{\alpha_0\alpha} \sum_{n>n_1} \exp(\delta\beta) \sum_{l<n,\lambda_0} \frac{2\lambda_0+1}{3[1+(n/n_s')^3]}\,. \tag{5.2.29}$$

This formula can be used in place of (5.2.12) n_s' differing from n_s by the substitution

$$\sigma(\alpha_0, \alpha l) = \sum_{\lambda_0} \sigma(\alpha_0\lambda_0, \alpha l) \to \frac{3(2l+1)}{2\lambda_0+1} \sigma(\alpha_0\lambda_0, \alpha l)\,. \tag{5.2.30}$$

We consider now the dependence on spin momentum S. For a single electron outside the closed shells, in accordance with (2.3.27, 28),

$$W_a(\gamma) \propto \sigma(\gamma_0, \gamma)\frac{4}{2S+1} \propto [f + (-1)^s g]^2 ,$$

where f and g are the direct and exchange scattering amplitudes. For a $1s - 2p$ transition, $f \simeq g$ (cf. Sect. 3.2.3) and $W_a(^3L)$ is very small. If $W_a(^1L) \gg A$, we have

$$\sum_S q(\gamma) = (2L+1)A(\alpha), \quad q(\alpha nl) = 4\, g_L\, A(\alpha) ,$$

where g_L is the orbital part of the statistical weight $g_{\alpha nl}$. Therefore using $q(\alpha nl)$ in this case, one can overestimate the result up to a factor of 4, if we assume LS coupling and $f \simeq g$.

The second assumption of the simplified model is the use of an extrapolation formula (5.2.16). This formula is applicable for $n \geq 4$, but for $n = 2$ or 3 it may not be sufficiently accurate. In most cases terms with large n make the main contribution to the DR rate coefficient. However at $z \gtrsim 15$ for recombination of [H] and [He] ions, the levels with $n = 2$ and 3 are the most important. Numerical calculations show that in this case the extrapolation formula underestimates the result.

The third assumption of the simplified model implies that only transitions to the ground state α_0 take place. For the resonance level α any other transitions are, evidently, impossible. For higher levels the additional radiative transitions $\alpha - \alpha'$ lead to some increase of κ_d through the factor B' in (5.2.20). As a rule, this effect is small because $A \propto E^2_{\alpha\alpha'}$, and $E_{\alpha\alpha'} \ll E_{\alpha\alpha_0}$. On the other hand, the additional channel of autoionization $\gamma - \alpha'$ can affect the magnitude of the factor B' in (5.2.20) substantially, because $W_a(\alpha nl, \alpha')$ increases rapidly with decrease of $E_{\alpha\alpha'}$. As a result, the contribution of the levels nl satisfying the condition

$$z^2\mathrm{Ry}/n^2 < E_{\alpha\alpha'}$$

decreases considerably. Accurate treatment of this effect makes the computations considerably more cumbersome (see [5.10]). It can be estimated approximately assuming $W_a \gg A$ in the cases when autoionization to the such level α' is possible. In other words, summation over n in (5.2.12) can be restricted to

$$n < \left(z^2\mathrm{Ry}/E_{\alpha\alpha'}\right)^{1/2}, \quad E_{\alpha\alpha'} f \ll E_{\alpha\alpha_0}$$

An example of such a process is the dielectronic recombination

$$X_{z+1}(2p^k) + \mathrm{e} \to \mathrm{X}_z(2p^{k-1}3dnl) \to \begin{cases} \mathrm{X}_z(2p^k nl) + \hbar\omega \\ \mathrm{X}_{z+1}(2p^k) + \mathrm{e} \\ \mathrm{X}_{z+1}(2p^{k-1}3p) + \mathrm{e} \\ \mathrm{X}_{z+1}(2p^{k-1}3s) + \mathrm{e}. \end{cases}$$

5.2.5 Numerical Calculations and Analytical Approximation Formulas

Equation (5.2.11) describes a general temperature dependence of the DR rate coefficient. The temperature dependence of the factors $\exp(\delta\beta)$ in B_d (5.2.12), although small, can nevertheless be important in some cases. To include this effect in a simple analytical formula, we replace the quantity $\chi = \Delta E/(z+1)^2\mathrm{Ry}$ by an adjustable parameter χ_d, and rewrite (5.2.11, 12) in the form

$$\kappa_d(\alpha) = 10^{-13} B_d \beta^{3/2} \exp(-\beta\chi_d) \quad [\mathrm{cm^3\ s^{-1}}],$$
$$\beta = \frac{(z+1)^2\mathrm{Ry}}{T}, \quad B_d = \sum_{n \le n_1} B_n, \quad \chi_d = \chi - \Delta\,. \tag{5.2.31}$$

According to (5.2.31) the value of B_d is equal to its limit at high $T(\delta\beta = 0)$. The value of χ_d can be determined by the condition that κ_d is equal to its true value for some value of $T = T_1$.

The value of B_d can be calculated numerically using the simplified model, i.e., via threshold values of the partial cross sections for the transition $\alpha_0 - \alpha$.

Burgess [5.7] proposed a simple empirical formula for the DR rate coefficient which is given in (5.1.55, 56). In spite of the absence of any adjustable parameter the *Burgess* formula in most cases gives results in a good agreement with the results of numerical calculations. This is due to the general structure of (5.2.12). If we replace $\exp(\delta\beta)$ by its average value, the DR rate coefficient is proportional to

$$\sum_{\substack{nl \\ n_s \gg n_1}} \frac{2l+1}{1+(n/n_s)^3} \simeq \sum_l (2l+1)\,\Delta n(l), \quad \Delta n(l) = n_s - \max\begin{pmatrix} l+1 \\ n_1 \end{pmatrix}. \tag{5.2.32}$$

We can see from (5.2.12) that $n_s \propto (\sigma/f)^{1/3}$ and is almost independent of the type of ion and transition. Nevertheless in some cases the error of the *Burgess* formula isn't negligible.

5.3 Satellites of Resonance Lines in Spectra of Highly Charged Atoms

5.3.1 Excitation by Means of DR

The photon $\gamma - \gamma_1(\alpha n l L S J - \alpha_0 n l_1 L_1 S_1 J_1)$ emitted during DR has an energy slightly different from that of the transition $\alpha - \alpha_0$ in the ion X_{z+1}. This corresponds to additional spectral lines in the vicinity of the line $\alpha - \alpha_0$. These lines are called satellites or dielectronic satellites [5.18]. In this section we consider the simplest and most important case of the satellites of [He]- and [H]-ion[1] resonance lines.

[1] We recall that the designation [H] is used for the hydrogenlike ion, [He] for the heliumlike ion and so on.

Satellites of the resonance line $2p-1s$ in [H] ions X_{z+1} correspond to transitions

$$2pnl - 1snl_1 \tag{5.3.1}$$

in [He] ions X_z.

Satellites of the line $1s2p\,^1P-1s^2\,^1S$ in [He] ions X'_{z+1} correspond to transitions

$$1s2pnl - 1s^2\,nl_1 \tag{5.3.2}$$

in [Li] ions, and also to numerous transitions in ions of lower charge:

$$\begin{aligned} &1s2s2pnl - 1s^2\,2snl_1 \quad [\mathrm{Be}] \\ &1s2s^22p^knl - 1s^22s^22p^{k-1}nl_1 \quad (k=1\dots6) \quad [\mathrm{B}]\dots[\mathrm{Ne}]\,. \end{aligned} \tag{5.3.3}$$

The satellites (5.3.1–3) were observed in the spectra of ions with $z \gtrsim 10$ in laboratory and astrophysical plasmas. They proved to be an effective method of plasma diagonostics, since the satellite-to-resonance-line intensity ratio is essentially dependent on temperature and in some cases on electron density. At the same time the satellites are situated very close to the corresponding resonance line, and are excited by means of DR from the same initial state as the resonance line, see (5.2.1). These are important features for plasma diagnostics.

In a plasma in the state of ionization equilibrium, the satellites (5.3.3) are usually very weak. Their intensity relative to the resonance line becomes considerable at very low temperature when the absolute intensities are negligible. In transient plasmas with small characteristic time the satellites due to ions in the isoelectronic sequences [Be], [B], ... , [Ne] can be sufficiently strong. They provide a convenient method of the estimation of the ionization state in a plasma.

The energy distance between satellites and the resonance line decreases rapidly (as n^{-3}) with increasing n. In fact only satellites with $n = 2$ or 3 have been resolved from the resonance line up to now. Satellites with $n \geq 4$ can however change the shape of the line.

Apart from the transition $2p - 1s$ of an internal electron, the transitions of the outer one ($nl - 1s$) are possible, although less probable. The satellites of the members of principal spectral series $np - 1s$, $1snp - 1s^2$ are due to such transitions.

Below we confine ourselves to the most important cases of $n = 2$ and 3 satellites of the types (5.3.1,2).

It was mentioned above that the principal way of satellite excitation is DR of the ion X_{z+1}. Direct excitation of an inner-shell electron $1s$ of the ion X_z can also provide satellites of the type (5.3.2).

We consider now the relative intensity of the satellite

$$i(\gamma,\gamma_1) = \frac{I_s(\gamma,\gamma_1)}{I_{\mathrm{res}}(\alpha,\alpha_0)} = i'(\gamma,\gamma_1) + i''(\gamma,\gamma_1)\,, \tag{5.3.4}$$

where I_s and I_{res} are the absolute intensities of the satellite due to the transition $\gamma - \gamma_1$, and of the resonance line $\alpha - \alpha_0$. We suppose that $\gamma = \alpha nlLSJ$ and

$\gamma_1 = \alpha_0 n l_1 L_1 S_1 J_1$, where l_1 can be unequal to l due to configuration mixing (corresponding satellites usually being weak). The terms i' and i'' in (5.3.4) correspond to two mechanisms of excitation, namely DR and direct excitation of the $1s$ electron; for satellites of [H]-ion resonance line, $i'' = 0$. We begin with the DR excitation mechanism. In this case,

$$i'(\gamma, \gamma_1) = \frac{\kappa_\mathrm{d}(\gamma, \gamma_1)}{\langle v\sigma_{\alpha_0\alpha}\rangle} \tag{5.3.5}$$

where $\kappa_\mathrm{d}(\gamma, \gamma_1)$ is defined by (5.2.15) with substitution of the probability of radiative transition $\gamma - \gamma_1$ in place of $A(\gamma)$ in the numerator. (In this section we shall use the notation W for the radiative transition probabilities instead of A in order to avoid mixing the radiative probabilities and parameters of approximation of collisional rate coefficients.) In our case, $\alpha_0 = 1s$ or $1s^2$, $\alpha = 2p$ or $1s2p$, and only one autoionization channel is possible; hence $W_\mathrm{a}(\gamma, \alpha_0) = W_\mathrm{a}(\gamma)$. Taking this into account and using the approximation from Sect. 5.1 for $\langle v\sigma_{\alpha_0\alpha}\rangle$, we obtain

$$\begin{aligned} & i'(\gamma, \gamma_1) = a'\delta\beta \exp(\delta\beta)\, q(\gamma, \gamma_1)\,, \\ & q(\gamma, \gamma_1) = 10^{-13} g_\gamma \frac{W(\gamma, \gamma_1) W_\mathrm{a}(\gamma)}{W(\gamma) + W_\mathrm{a}(\gamma)}\,, \\ & \delta\beta = \frac{E_{\alpha\alpha_0} - E_{\gamma\alpha_0}}{T} \simeq \frac{\beta}{n^2}\,, \quad \beta = \frac{(z+1)^2 \mathrm{Ry}}{T}\,. \end{aligned} \tag{5.3.6}$$

Here the factor 10^{-13} is introduced for convenience: for $z \backsim 20$, q is of the order of unity. The factor a' is

$$a' = 2.4 \times 10^{22} \pi^{3/2} a_0^3 \xi \frac{\beta' + \chi'}{\beta' + 1} \cdot \frac{n^2}{g_0 A' Q'}; \; \xi = \frac{4}{3} \frac{E_{\alpha\alpha_0}}{n^2 \delta E} \cdot \left(\frac{E_0}{4E_\alpha}\right)^{3/2}, \tag{5.3.7}$$

where A', χ', Q' are approximation parameters (Sect. 5.1) for $\langle v\sigma_{\alpha_0\alpha}\rangle$; at large z, $\xi \to 1$. In the most interesting case $\beta' > 1$, the dependence of a' on temperature is very slight. Using the results of numerical calculation in the Born–Coulomb approximation with exchange, we obtain for $\beta' = 3$, $\xi = 1$,

$$\begin{aligned} & a' = 0.75 \times 10^{-3} \text{ for satellites of [He] ions}\,, \\ & a' = 0.47 \times 10^{-3} \text{ for satellites of [H] ions}\,. \end{aligned}$$

For applications it is very important that for all satellites with given n the dependence of i' on T is practically the same.

The value of $i'(\gamma, \gamma_1)$ is determined by a factor of $q(\gamma, \gamma_1)$. For small z the value of W_a exceeds the transition probability W by a factor of the order of $(\hbar c/e^2)^3 \sim 10^6$:

$$W \ll W_\mathrm{a}\,, \quad q(\gamma, \gamma_1) \simeq 10^{-13} g_\gamma W(\gamma, \gamma_1) \propto z^4\,. \tag{5.3.8}$$

We see that for small z, $i' \ll 1$ but rapidly increases with z. At large z,

$$W \gg W_a\,, \quad q(\gamma, \gamma_1) \simeq 10^{-13} g_\gamma W_a(\gamma) \frac{W(\gamma, \gamma_1)}{W(\gamma)} \tag{5.3.9}$$

and i' is almost independent of z. In this case i' becomes $\gtrsim 1$ for $\delta\beta \gtrsim 1$. The transfer from (5.3.8) to (5.3.9) occurs at $z \sim 20$.

As an illustration, in Table 5.4 the calculated and experimental results for the satellite structure of the [He]-ion Fe XXV are shown. The value of $\delta\beta$ was determined according to the observed value of i for $\gamma = 1s\,2p^2\,{}^2D_{5/2}$ (the brightest and best separated satellite). The calculations of wavelengths and spontaneous transition probabilities were made using the perturbation theory expansion in the parameter $1/\mathscr{Z}$. We note that for $\mathscr{Z} = 26$ one should take into account all relativistic interactions, even the Lamb shift of the electron 1s, which is equal to 0.08 Å.

Table 5.4. Wavelengths λ and relative intensities for satellites of Fe XXV resonance line in X-ray spectra of the solar corona. The temperature of the plasma was determined from the intensity of $\lambda = 1.8662$ Å satellite

Experiment	[5.19]	Theory			
λ[Å]	i[%]	λ[Å]	Transition	i'[%]	i''[%]
1.8510	100	1.8504	$1s\,2p\,{}^1P_1 \to 1s^2\,{}^1S_0$	—	100
1.8564	21	1.8555	$1s\,2p\,{}^3P_2 \to 1s^2\,{}^1S_0$	—	29
1.8571	15	1.8566	$a\,{}^2S_{1/2} \to m\,{}^2P_{3/2}$	4.7	—
—	—	1.8571	$b_1\,{}^2P_{1/2} \to n\,{}^2S_{1/2}$	10.2	2.6
1.8579	23	1.8578	$a\,{}^2P_{3/2} \to m\,{}^2P_{1/2}$	0.18	—
1.8594		1.8595	$1s\,2p\,{}^3P_1 \to 1s^2\,{}^1S_0$	—	17
1.8608	12	1.8610	$b_2\,{}^2P_{3/2} \to n\,{}^2S_{1/2}$	0.18	19.7
1.8618	22	1.8622	$a\,{}^2P_{3/2} \to m\,{}^2P_{3/2}$	13	—
1.8634	47	1.8630	$a\,{}^2D_{3/2} \to m\,{}^2P_{1/2}$	32	—
—	—	1.8635	$b_2\,{}^2P_{1/2} \to n\,{}^2S_{1/2}$	4.9	6
1.8660	47	1.8659	$a\,{}^2D_{5/2} \to m\,{}^2P_{3/2}$	47	—
1.8674	38	1.8674	$a\,{}^2D_{3/2} \to m\,{}^2P_{3/2}$	3.8	—
1.8679		1.8685	$1s\,2s\,{}^3S_1 \to 1s^2\,{}^1S_0$	—	14
1.8700	21	1.8699	$1s\,2p^3\,{}^3P_{1,2} \to 1s\,{}^22p^2\,{}^3P_1$	—	—
1.8729		1.8727	$a\,{}^4P_{5/2} \to m\,{}^2P_{3/2}$	7.7	—
			$\sum_\lambda$	127	

$a = 1s\,2p^2$, $m = 1s^2 2p$, $b_1 = 1s\,(2s2pP^1)$, $n = 1s^2 2s$. $b_2 = 1s(2s2pP^3)$,

5.3.2 Direct Inner-Shell Excitation

We consider now a direct inner-shell excitation of satellites of the [He]-ion resonance line

$$X_z(1s^2 2l_0) + e \to X_z(1s\,nl\,2l_0) + e\,. \tag{5.3.10}$$

In the case of DR, both the resonance line and satellites are excited from the same state, namely the ground state α_0 of the ion X_{z+1}. In contrast, the initial state for direct excitation is the ground state of the ion X_z; hence the relative intensity i'' is proportional to the ratio N_z/N_{z+1} of the populations of [Li] to [He] ions.

In the case of low-density (e.g., astrophysical) plasma, almost all the X_z ions are in the state $1s^2 2s$. In high-temperature laboratory plasma, the electron density is as high as $N_e \sim 10^{20}$ cm^{-3}, and the levels $2s$ and $2p$ are populated in accordance with the Boltzmann formula. Since $\Delta E \gg T$, $N_{2p}/N_{2s} = g_{2p}/g_{2s} = 3$.

It is evident that the direct excitation of satellite levels isn't effective in the case of [H]-ion resonance line satellites since the population of the corresponding initial state $1s\,2l$ of [He] is too small.

The relative intensity i'' of the [He] resonance line satellite due to the direct inner-shell excitation is

$$i'' = \sum_{\gamma_0} \frac{N_z(\gamma_0)}{N_{z+1}} \cdot \frac{\langle v\sigma_{\gamma_0\gamma}\rangle}{\langle v\sigma_{\alpha_0\alpha}\rangle} \cdot \frac{W(\gamma,\gamma_1)}{W(\gamma) + W_a(\gamma)}\,. \tag{5.3.11}$$

For a low-density plasma,

$$\frac{N_z}{N_{z+1}} = \frac{\kappa_v}{\langle v\sigma_i(2s)\rangle}[1 + D(T)], \quad D(T) = \frac{\kappa_d}{\kappa_v}\,, \tag{5.3.12}$$

where κ_v and κ_d are the total rate coefficients for radiative and dielectronic recombination, and $\langle v\sigma_i\rangle$ is the ionization rate coefficient for the ion X_z. Using approximations from Sect. 5.1 for these quantities and the approximate equality, see (5.3.6),

$$E_z = E(1s^2 2l - 1s^2) \simeq E(1s2l2l_0 - 1s^2) - E(1s2p - 1s^2) = \delta E\,, \tag{5.3.13}$$

we obtain

$$\frac{N_z}{N_{z+1}} = (\delta E)^2 \frac{A_v}{A_i}[1 + D(T)]\delta\beta \exp(\delta\beta)\frac{\delta\beta + \chi_i}{\delta\beta + \chi_v}\,. \tag{5.3.14}$$

The relative intensity i'' will be written then in the form

$$i''(\gamma,\gamma_1) = a''[1 + D(T)]\ \delta\beta \exp(\delta\beta) q''(\gamma,\gamma_1)\,,$$

$$q''(\gamma,\gamma_1) = \frac{A(\gamma_0,\gamma) \cdot W(\gamma,\gamma_1)}{W(\gamma) + W_a(\gamma)}\,,$$

$$a'' = (\delta E)^2 \frac{A_v}{A_i(2s)} \frac{\delta\beta + \chi_i(2s)}{\delta\beta + \chi_v} \frac{\beta' + \chi'}{\beta' + \chi(\gamma_0, \gamma)} \frac{\xi''}{2A'} ,$$

$$\xi'' = \left(\frac{E_\alpha E_{\gamma_0}}{E_{\alpha_0} \; E_\gamma} \right)^{3/2} , \quad \beta' = \frac{E_{\alpha\alpha_0}}{\delta E} \delta\beta . \tag{5.3.15}$$

Here A_v, χ_v, $A_i(\gamma_0)$, $\chi_i(\gamma_0)$, A', χ', $A(\gamma_0, \gamma)$, and $\chi(\gamma_0, \gamma)$ are parameters of the approximations (Sect. 5.1) for κ_v, $\langle v\sigma_i(\gamma_0)\rangle$, $\langle v\sigma_{\alpha_0\alpha}\rangle$, and $\langle v\sigma_{\gamma_0\gamma}\rangle$; the value of $\xi'' \to 1$ at $z \gg 1$.

It is clear that in the low-density case, only the satellites with $\gamma = 1s\,2s\,2l$ are excited, of which only satellites beginning from the levels of electronic configuration $1s\,2s\,2p$ can be strong enough.

In the case of a high-density plasma both the $1s^2\,2s$ and $1s^2\,2p$ levels are populated. Therefore

$$i'' = \frac{N_z}{N_{z+1}} \sum_{\gamma_0} \tilde{g}_{\gamma_0} \frac{\langle v\sigma_{\gamma_0\gamma}\rangle}{\langle v\sigma_{\alpha_0\alpha}\rangle} \frac{W(\gamma, \gamma_1)}{W(\gamma) + W_a(\gamma)} ,$$

$$\frac{N_z}{N_{z+1}} = \frac{\kappa_v}{\sum \tilde{g}_{\gamma_0} \langle v\sigma_i(\gamma_0)\rangle} , \; \tilde{g}_{\gamma_0} = \frac{g_{\gamma_0}}{\sum\limits_{\gamma_0} g_{\gamma_0}} , \tag{5.3.16}$$

and all satellites can be excited. We obtain an equation similar to (5.3.15) with the change

$$A(\gamma_0, \gamma) \to \sum_{\gamma_0} \tilde{g}_{\gamma_0} A(\gamma_0, \gamma) ,$$

$$\frac{A_i(2s)}{\delta\beta + \chi_i(2s)} \to \sum_{l_0} \frac{A_i(2l_0)}{4[\delta\beta + \chi_i(2l_0)]} . \tag{5.3.17}$$

Substitution of numerical values in the expression for a'' in (5.3.15) shows that the values of a'' in both cases of low and high densities are practically the same, and for $\delta\beta = 1$, $\beta' = 3$, $\xi'' = 1$ are equal to

$$a'' = 1.4 \times 10^{-3} \left(\frac{z - 0.6}{10n} \right)^4 . \tag{5.3.18}$$

By comparing (5.3.15) and (5.3.6), one can see that except for the value of $D(T)$ the temperature dependences $i'(T)$ and $i''(T)$ are the same. The factor $D(T)$ corresponds to the influence of DR on the ionization equilibrium. For temperature $T \sim T_m$, at which I_{res} has the maximum value, $D(T)$ significantly exceeds unity, and hence influences $i''(T)$. However, at lower temperatures, $D(T)$ rapidly decreases as $\exp(-3\delta\beta)$ and the dependences $i'(T)$ and $i''(T)$ become similar.

If the plasma is in the state of ionization equilibrium, satellite excitation due to DR prevails, especially at $z \lesssim 15$ according to (5.3.18). In the case of transient (ionizing) plasma, the role of direct excitation can be very important at $z > 15$. It should be noted that the characteristic ionization time increases with increasing z.

In Table 5.4, values of i'' for the ion Fe XXV in ionization equilibrium are given.

5.4 Populations of Excited Levels in a Plasma

By the intensity of a spectral line is usually understood the energy emitted per second by a unit volume of a plasma as a result of spontaneous transitions. For the transition $k \rightarrow i$ of an atom, this quantity is

$$I_{ki} = \hbar\omega_{ki} A_{ki} N(k) \; [\text{erg}\,\text{cm}^{-3}\text{s}^{-1}] , \tag{5.4.1}$$

where A_{ki} is the spontaneous emission probability, and $N(k)$ is the population density of atoms excited to the level k. We shall consider the case of an optically thin plasma in which the line radiation may freely escape and does not affect the level populations.

The transition probabilities A_{ki} are the atomic characteristics. The excited level populations are, in general case, dependent upon the variety of collisional and radiative processes. In this section we shall consider some approaches to the calculation of the level populations and also the ionization and recombination coefficients S and α. These coefficients determine the rates of ion production and loss and are expressed in units [$\text{cm}^3\,\text{s}^{-1}$]. Using the example of the hydrogen atom (and hydrogenlike ions) we shall describe below the approach to the numerical solution of this problem. A survey of the analytic methods of calculating the populations and recombination coefficients is given in [5.20]. For a detailed discussion of the problem of line intensities and populations, see [5.21–24]. Article [5.24] contains an extensive list of references.

Three types of models of a plasma are usually considered: the coronal model in the limit of low density, the model of local thermodynamic equilibrium (LTE) at high density, and the models of a plasma in the intermediate region which are often referred to as the collisional-radiative models. (For the conditions of applicability of the coronal and LTE approximations see Sect. 1.2.)

Simple relations between the atomic level populations exist in the high-density limit when relaxation for any atomic level is determined by collisional processes. In LTE plasmas populations are completely determined by the temperature, total density of atoms and chemical composition of a plasma, and are independent of the cross sections of elementary processes, see (1.2.2, 3). A comprehensive discussion of the validity conditions for LTE can be found in [5.25].

When LTE conditions are violated, the populations of atomic levels are determined by all collisional and radiative processes. The processes to be included in calculations are the following: collisional excitation and deexcitation, ionization and the inverse process of three-body recombination, radiative and dielectronic recombination, spontaneous radiative decay, and collisional and radiative cascades. When the degree of ionization is not too low, only the electronic collisions are of importance.

5.4.1 Populations of the Hydrogen Levels at Low Plasma Density[1]

In the low-density limit (coronal limit, see Sect. 1.2), decay of the excited levels is provided exclusively by spontaneous radiative transitions[2]. Hence it follows that the population densities for all the levels except the ground state are extremely low. An excited level n is populated by direct processes which are the excitation from the ground state by electron impact and radiative recombination, and by cascade from the higher levels. For nonhydrogenic ions, dielectronic recombination has to be taken into account as a direct process.

The steady-state rate equations may be written in the form

$$N(n)\,A_n = \sum_{n'>n} N(n')A_{n'n} + q_n \;. \tag{5.4.2}$$

In this equation, $N(n)$ is the density of atoms in the level with the principal quantum number n, $A_n = \sum_{n''<n} A_{nn''}$ is the total probability of radiative decay of a level n,

$$q_n = N(1)N_e\langle v\sigma_{1n}\rangle + N(\mathrm{H}^+)N_e\kappa_\nu(n) \;, \tag{5.4.3}$$

where $N(1)$ is the population density of the ground level, $\langle v\sigma_{1n}\rangle$ is the rate coefficient [$\mathrm{cm^3\,s^{-1}}$] of excitation by electron impact averaged over the Maxwellian distribution, $N(\mathrm{H}^+)$ is the density of bare nuclei, and $\kappa_\nu(n)$ is the rate coefficient of radiative recombination to form an atom in level n. It is assumed that all levels with the different orbital quantum numbers are populated proportionally to their statistical weights:

$$N(nl) = \frac{2l+1}{n^2}N(n) \;. \tag{5.4.4}$$

The solution of the equilibrium equations (5.4.2) may be written in the form

$$N(n) = \frac{1}{A_n}\sum_{n'\geq n} C(n',n)q_{n'} \;, \tag{5.4.5}$$

where $C(n,n) = 1$. The element of the cascade matrix $C(n',n)$ is the probability that excitation of n' is followed by all possible cascade radiative transitions to level n. Let $P(n',n)$ be the probability that the direct radiative transition from n' to n takes place:

$$P(n',n) = A_{n'n}/A_{n'} \;.$$

---

[1] The content of this section is based on [5.26].

[2] The relaxation of metastable levels or of levels with large principal quantum numbers $n \gg 1$ is determined by collisional processes even at very low density.

For $C(n', n)$, we then have

$$\begin{aligned}
&C(n+1, n) = P(n+1, n) , \\
&C(n+2, n) = P(n+2, n+1)\, C(n+1, n) + P(n+2, n) , \qquad (5.4.6) \\
&\cdots\cdots\cdots\cdots \\
&\cdots\cdots\cdots\cdots \\
&\cdots\cdots\cdots\cdots \\
&C(n+m, n) = P(n+m, n+m-1)\, C(n+m-1, n) \\
&\quad + P(n+m, n+m-2) C(n+m-2, n) + \cdots + P(n+m, n) .
\end{aligned}$$

With $C(n, n) = 1$ we obtain generally

$$C(n+m, n) = \sum_{n'=n}^{n+m-1} P(n+m, n')\, C(n', n) . \qquad (5.4.7)$$

Using (5.4.7) one may calculate similarly the radiative cascade matrix for any other atom once the matrix elements $P(n', n)$ (or radiative transition probabilities) are known. The number n in this case may be considered as an index labelling the atomic levels in the order of increase of the energy.

Table 5.5 gives the cascade matrix elements $C(n+m, n)$ for $n \leq 10$, $m \leq 10$ calculated by Seaton [5.26]. For $m \geq 5$, the calculated values can be fitted to

$$C(n+m, n) = C(\infty, n) + \Delta_n/m .$$

The quantities $C(\infty, n)$ are also given in Table 5.5. For application to the excited levels with $n > 10$, one may use an analytical formula obtained in [5.27] using the Kramers approximation:

$$C(n+m, n) = \frac{2}{n^3} \ln\left(\frac{n^3}{2}\right) \times \left(\frac{1}{n^2} - \frac{1}{(n+m)^2}\right)^{-1} \left[0.5772 - \ln\left(\frac{1}{n^2} - \frac{1}{(n+m)^2}\right)\right]^{-2} \qquad (5.4.8)$$

Table 5.5. Cascade matrix $C(n+m, n)$ for hydrogen in the case of optically thin plasma

m \ n	2	3	4	5	6	7	8	9	10
1	0.4418	0.2978	0.2336	0.1975	0.1744	0.1581	0.1464	0.1369	0.1295
2	0.4105	0.2601	0.1947	0.1587	0.1361	0.1203	0.1093	0.1006	0.0939
3	0.3991	0.2454	0.1791	0.1430	0.1206	0.1052	0.0942	0.0858	0.0791
4	0.3934	0.2380	0.1710	0.1347	0.1120	0.0968	0.0858	0.0775	0.0711
5	0.3903	0.2336	0.1661	0.1294	0.1068	0.0916	0.0806	0.0723	0.0660
6	0.3883	0.2308	0.1627	0.1260	0.1033	0.0881	0.0771	0.0688	0.0624
7	0.3870	0.2289	0.1605	0.1236	0.1008	0.0854	0.0745	0.0662	0.0598
8	0.3860	0.2274	0.1589	0.1217	0.0990	0.0835	0.0725	0.0642	0.0578
9	0.3853	0.2264	0.1575	0.1203	0.0975	0.0821	0.0710	0.0627	0.0563
10	0.3848	0.2255	0.1564	0.1191	0.0964	0.0809	0.0698	0.0615	0.0551
$C(\infty,n)$	0.3796	0.2174	0.1468	0.1088	0.0860	0.0703	0.0590	0.0505	0.0441

The error of formula (5.4.8) does not exceed 15 percent. When m is large, the error decreases. This is in accordance with the accuracy of the Kramers approximation.

The formula (5.4.5) for the population density includes the total spontaneous decay probability A_n. Let A_n be written in the form

$$\begin{aligned} A_n &= A_0 \frac{z^4}{n^5} \tau(n) , \\ A_0 &= 8\alpha^4 c / 3\pi\sqrt{3}\, a_0 = 0.789 \times 10^{10} \mathrm{s}^{-1} , \end{aligned} \tag{5.4.9}$$

where $\alpha = e^2/\hbar c = 1/137$ is the fine-structure constant, c is the velocity of light, and $a_0 = \hbar^2/me^2 = 0.529 \times 10^{-8}$ cm is the Bohr radius. The values of $\tau(n)$ are given in Table 5.6. For $n \geq 20$, $\tau(n)$ may be fitted to

$$\tau(n) \simeq 3 \ln n - 0.247 .$$

For applications, the analytical formulas of the Kramers approximation are often useful:

$$\begin{aligned} A_{nn'} &= \frac{2A_0 z^4}{n^3 n'(n^2 - n'^2)} , \\ A_n &= \frac{A_0 z^4}{n^5} \ln\left(\frac{n^3 - n}{2}\right) . \end{aligned}$$

According to (5.4.3, 5) in the limit of $N_e \to 0$ the population density for any excited level is negligibly small compared to the electron density, densities of the ground level atoms and bare nuclei. The total number density of the excited levels is also much smaller than these quantities. Therefore at low density, the ionization rate is provided by the direct collisional ionization from the ground state. The only process of recombination for hydrogen and hydrogenlike ions is radiative recombination. The rate coefficient of recombination is determined by the sum over principal quantum numbers,

$$\alpha = \kappa_\nu = \sum_{n=1}^{\infty} \kappa_\nu(n) .$$

Table 5.6. Factor $\tau(n)$

n	2	3	4	5	6	7	8	9
$\tau(n)$	1.911	3.084	3.929	4.589	5.132	5.590	5.989	6.389
n	10	11	12	13	14	15	20	25
$\tau(n)$	6.657	6.943	7.205	7.442	7.668	7.876	8.740	9.410

5.4.2 Intermediate Density. Collisional-Radiative Model of a Plasma

At intermediate densities one has to take into account both radiative and collisional cascade transitions. Here we shall also consider hydrogenlike ions. It is assumed that the relationship (5.4.4) is satisfied, the free electrons have a Maxwellian distribution, and that the plasma is optically thin for the line radiation. The rate equations which determine the population densities $N^{(z)}(n)$ of the bound levels of an ion X_z may be written in the form

$$\frac{d}{dt}N^{(z)}(n) = -N^{(z)}(n)\Gamma_n + \sum_{m\neq n} N^{(z)}(m)\Gamma_{mn} + N^{(z+1)}R_n \ , \tag{5.4.10}$$

where Γ_n is the total decay probability of the level $n[\mathrm{s}^{-1}]$ including the radiative decay probability $A_n = \sum_{n'<n} A_{nn'}$, the frequency of transitions to the other bound states due to electronic collisions $W_n = \sum_{m\neq n} N_e \langle v\sigma_{nm}\rangle$, the frequency of ionization by electron impact $W_n^i = N_e\langle v\sigma_i(n)\rangle$, and, generally speaking, the frequency of recombination to form an ion X_{z-1} with lower charge. Recombination is substantial for the ground state only. The electronic transitions due to the heavy particle collisions may be neglected. Γ_{mn} is given by

$$\Gamma_{mn} = \begin{cases} W_{mn} + A_{mn} & \text{when} \quad m < n \ , \\ W_{mn} & \text{when} \quad m < n \ . \end{cases}$$

The recombination frequency R_n includes the radiative recombination term $N_e \kappa_v(n)$ and three-body recombination term $\kappa_r(n) N_e^2$. $N^{(z+1)}$ is the bare-nuclei number density.

The calculation of the instantaneous population densities requires the solution of a set of equations (5.4.10). Due to the electric microfield of a plasma, the higher levels merge into a continuum and the number of bound levels is finite (this problem is beyond the scope of this text, see, e.g. [5.21, 28–30]). Nevertheless this number is large enough, especially at low densities. In actual numerical calculations one can confine oneself to considerably smaller principal quantum number n_0. Here we give a simple explicit estimate of n_0 (for a detailed analysis see [5.31]). The values of n_0 can be chosen taking into account the fact that for n large enough, the collisional processes are much more important than the radiative processes. The radiative decay probability decreases rapidly with n : $A_n \propto n^{-5}$. The collisional frequency for the transitions $n \to n \pm 1$ increase $\propto n^4$, and the collisional ionization frequency $W_n^i \propto n^2$. Therefore the population and decay of highly excited levels are determined by exclusively electronic collisions. When $n \gg 1$, $W_{n,n\pm 1} \gg W_n^i$, and the electron stepwise motion between the highly excited states is like a slow diffusive motion [5.32–34]. Hence for n greater than some value n_0, the populations are determined by the balance of three-body recombination and ionization by electron impact, and therefore satisfy the Saha-Boltzmann equations, see (1.2.3, 4),

$$N^{(z)}(n) = N_{\mathrm{E}}(n) = N^{(z+1)} \cdot N_{\mathrm{e}} B_n(T) \,,$$
$$B_n(T) = \frac{2n^2}{SN_{\mathrm{e}}} \exp\left(\frac{z^2 \mathrm{Ry}}{n^2 T}\right) \,, \quad S = 2\left(\frac{mT}{2\hbar^2 \pi}\right)^{3/2} \frac{1}{N_{\mathrm{e}}} \,. \tag{5.4.11}$$

The value of n_0 may be roughly estimated from the condition

$$W^{\mathrm{i}}_{n_0} \gg A_{n_0} \,. \tag{5.4.12}$$

The frequency of ionization $W^{\mathrm{i}}_n = N_{\mathrm{e}} \langle v\sigma_{\mathrm{i}}(n)\rangle$ may be estimated using the classical Thomson formula for the cross section (Sect. 5.1). In this case

$$\langle v\sigma_i(n)\rangle = 8\sqrt{\pi}\frac{e^2}{\hbar} a_0^2 \left|\frac{\mathrm{Ry}}{E_n}\right|^{3/2} \beta^{3/2} \exp(-\beta)$$
$$\times [1 + \beta \exp(\beta) \mathrm{Ei}(-\beta)] \,, \quad \beta = |E_n|/T = z^2 \mathrm{Ry}/n^2 T \,. \tag{5.4.13}$$

For highly excited levels $\beta \ll 1$, and it follows from (5.4.13)

$$W^{\mathrm{i}}_n \simeq \frac{8.7 \times 10^{-8}}{z^3} n^2 \left(\frac{z^2 \mathrm{Ry}}{T}\right)^{1/2} N_{\mathrm{e}} \,. \tag{5.4.14}$$

Using (5.4.9) for estimation of the radiative decay probability A_n [assuming for simplicity $\tau(n) = 10$] we obtain from the condition (5.4.12)

$$n_0 \simeq z \left(\frac{10^{18}}{N_{\mathrm{e}}}\right)^{1/7} \left(\frac{T}{z^2 \mathrm{Ry}}\right)^{1/14} \,. \tag{5.4.15}$$

For the range of electron temperature of 0.01–100 eV and an electron density of 10^8–$10^{15}\,\mathrm{cm}^{-3}$, it follows from (5.4.15) that

$$2 \le n_0/z \le 30 \,.$$

To find the level populations and the ionization state of a plasma one has to solve the several sets of equations (5.4.10) for ions of the different stages of ionization z under the conditions that the density of the nuclei of each element is conserved and the plasma is electrically neutral. If temperature and density change not too rapidly, the computational problem can be simplified.

5.4.3 Quasi-Stationary Approach for Hydrogen

The small perturbations from the steady state relax in a time of order $1/\Gamma_n$. The relaxation time of any of the excited levels is always much shorter than that of the ground level[1]. Hence the time in which the population densities of the excited

[1] The estimate for the relaxation time given above is rather rough. In actual cases the relaxation time for one level depends on that for all other levels and on initial conditions, and may be an order of magnitude larger than $1/\Gamma_n$ [5.35]. For metastable levels of atoms and ions other than hydrogen and hydrogenlike ions the relaxation time can be great.

levels come into correspondence with the instantaneous population density of the ground level, free electrons and bare nuclei H^+, is very short. Thus the quasi-steady approximation is obtained by setting $dN^{(z)}(n)/dt = 0$ for all the levels except the ground level. The population density of the ground level should change comparatively slow, on a time scale much longer than the longest relaxation time for the excited levels, but in general will not be necessarily in equilibrium with the electron density and the number density of bare nuclei. The quasi-stationary approach given in the works of Bates et al. [5.36, 37] enables population densities and coefficients of recombination and ionization to be tabulated in a wide range of electron temperature and density. These quantities are tabulated for hydrogen and hydrogenlike ions (see, e.g. [5.37–41]. More recent values are given in [5.40]. New results for helium and additional references may be found in [5.42].

It is convenient to express the populations in terms of Saha-Boltzmann equilibrium values $N_E(n)$, see (5.4.11),

$$b(n) = N^{(z)}(n)/N_E(n) \ .$$

For hydrogen $N^{(z+1)}$ in (5.4.11) is equal to $N(H^+)$. Using this notation the rate equations for the levels with $2 \leq n \leq n_0$ can be rewritten

$$b(n)\,\Gamma_n - \sum_{m\neq n} b(m)\frac{N_E(m)}{N_E(n)}\Gamma_{mn} - \frac{N(H^+)}{N_E(n)}R_n = 0 \ . \tag{5.4.16}$$

Taking into account (5.4.11) and the relationship

$$N_E(1)W_{1n} = N_E(n)W_{n1} \ ,$$

(5.4.16) may be reduced to

$$b(n)\,\Gamma_n - \sum_{\substack{m\neq n\\ m\neq 1}} b(m)\frac{B_m(T)}{B_n(T)}\cdot\Gamma_{mn} = \frac{R_n}{N_e B_n(T)} + \frac{N(1)}{N_E(1)}W_{n1} \ . \tag{5.4.17}$$

The first term in the right-hand side of (5.4.17) corresponds to the direct recombination to the level n and depends on N_e and T_e only. The second term corresponds to the excitation from the ground state and is proportional to $N(1)$. The solution of this set of $n_0 - 1$ equations may be expressed by the sum of two terms

$$b(n) = \frac{N(n)}{N_E(n)} = r_0(n) + r_1(n)\frac{N(1)}{N_E(1)} \ . \tag{5.4.18}$$

The first term determines both direct and cascade populating from the continuum and the second term corresponds to the direct and stepwise excitation from the ground state.

With the use of (5.4.18) the differential equation for $N(1)$ may be written in the form

$$dN(1)/dt = N(H^+)N_e\alpha - N(1)N_eS \ . \tag{5.4.19}$$

The quantity

$$\alpha = \frac{1}{N_e}\left[R_1 + \sum_{n>1}\frac{N_E(n)}{N(H^+)}r_0(n)\Gamma_{n1}\right] ,$$

following the works by Bates et al. [5.36, 37] is referred to as the collisional-radiative recombination coefficient, and the quantity

$$S = \frac{1}{N_e}\left[\Gamma_1 - \sum_{n>1}\frac{N_E(n)}{N_E(1)}r_1(n)\Gamma_{n1}\right]$$

is referred to as collisional-radiative ionization coefficient.

The steady-state value of $N(1)$ is given by the relation

$$N(1) = N(H^+)\frac{\alpha}{S} . \quad (5.4.20)$$

As was noted in Sect. 1.2, the total population of all the excited levels is usually considerably smaller than the ground level population density[1]

$$\sum_{n>1} N(n) \ll N(1) , N_e , N(H^+) . \quad (5.4.21)$$

Therefore the coefficients α and S may be regarded as the total rate coefficients of recombination and ionization[2]. When one of the inequalities (5.4.21) is not satisfied the steady state solution may not be valid.

In the low-density limit the net ionization rate is equal to the rate coefficient of ionization from the ground state $\langle v\sigma_i(1)\rangle$, and α is equal to the radiative recombination rate. At low densities,

$$r_1(n) = \frac{W_{n1}}{A_{n1}} \propto N_e ,$$

and $r_0(n)$ depends on T only.

In the high-density limit, collisional cascades are of major importance and the quantities r_0 and r_1 become independent of N_e.

Some numerical data for r_0, r_1, α, and S from the work by Johnson and Hinnov [5.39] are given in Table 5.7 (for more recent results see [5.40]). The quasi-steady level population densities at a given temperature T are determined by (5.4.18) on substituting the values of electron density N_e, the density of protons $N(H^+)$ and the population density of the ground level $N(1)$. In a plasma composed of hydrogen only $N_e = N(H^+)$. The steady-state population densities are to be found with the condition (5.4.20) taken into account. The steady-state populations for the ground level are given in [5.43].

---

[1] The problem of an upper limit of n for the sum over excited levels in (5.4.21) is outside the scope of this text.

[2] S and α do not correspond to coefficients giving the rate at which electrons leave the ground level and come to it. They are smaller than such coefficients would be.

Table 5.7. Parameters $r_0(n)$, $r_1(n)$ and coefficients S and α for optically thin hydrogen plasma [5.39]. T=10^3 K. 4.9 −6 denotes 4.9×10^{-6}.

N_e	0	10^9	10^{10}	10^{11}	10^{12}	10^{13}	10^{14}	10^{15}	10^{16}	10^{17}	∞
					$T = 4\times10^3$K.						
$r_0(2)$	4.9–6	6.2–6	7.6–6	1.1–5	1.9–5	4.9–5	2.4–4	2.2–3	1.8–2	6.2–2	8.5–2
$r_1(2)$	2.5–17N_e	2.5–8	2.5–7	2.5–6	2.5–5	2.5–4	2.5–3	2.4–2	2.0–1	6.7–1	9.2–1
$r_0(3)$	1.4–3	1.8–3	2.2–3	3.1–3	6.0–3	2.2–2	1.3–1	3.5–1	4.2–1	4.5–1	4.7–1
$r_1(3)$	1.0–17N_e	1.0–8	1.0–7	1.0–6	1.0–5	1.0–4	1.1–3	1.3–2	1.1–1	3.9–1	5.3–1
$r_0(4)$	1.1–2	1.5–2	1.8–2	2.8–2	7.3–2	3.1–1	6.0–1	7.4–1	7.7–1	7.9–1	7.9–1
$r_1(4)$	7.2–18N_e	7.2–9	7.2–8	7.1–7	6.9–6	5.7–5	4.8–4	5.3–3	4.5–2	1.5–1	2.1–1
$r_0(5)$	3.2–2	4.2–2	5.5–2	1.0–1	3.3–1	6.8–1	8.5–1	9.0–1	9.2–1	9.2–1	9.2–1
$r_1(5)$	6.1–18N_e	6.0–9	6.0–8	5.7–7	4.4–6	2.5–5	1.8–4	2.0–3	1.6–2	5.6–2	7.7–2
$r_0(6)$	5.8–2	8.1–2	1.1–1	2.7–1	6.4–1	8.6–1	9.4–1	9.6–1	9.7–1	9.7–1	9.7–1
$r_1(6)$	5.5–18N_e	5.4–9	5.2–8	4.3–7	2.3–6	1.0–5	7.2–5	7.7–4	6.5–3	2.2–2	3.0–2
S	1.2–26	1.6–26	2.1–26	3.2–26	6.5–26	2.1–25	1.3–24	1.4–23	1.2–22	4.0–22	5.4–22
α	7.9–13	1.0–12	1.2–12	1.7–12	2.9–12	7.1–12	2.7–11	1.6–10	1.4–9	1.3–8	1.3–25N_e

N_e	0	10^{10}	10^{11}	10^{12}	10^{13}	10^{14}	10^{15}	10^{16}	10^{17}	10^{18}	∞
					$T = 8\times10^3$K.						
$r_0(2)$	1.2–3	1.5–3	1.8–3	2.5–3	4.5–3	1.3–2	7.1–2	3.7–1	7.0–1	7.7–1	7.1–1
$r_1(2)$	1.9–17N_e	1.9–7	1.9–6	1.9–5	1.9–4	1.9–3	1.8–2	1.0–1	2.0–1	2.2–1	2.2–1
$r_0(3)$	2.0–2	2.6–2	3.3–2	5.0–2	1.2–1	4.3–1	7.2–1	8.5–1	9.3–1	9.5–1	9.5–1
$r_1(3)$	8.2–18N_e	8.2–8	8.1–7	8.0–6	7.7–5	6.1–4	4.5–3	2.4–2	4.7–2	5.1–2	5.2–2
$r_0(4)$	6.1–2	8.2–2	1.1–1	2.2–1	5.6–1	8.3–1	9.3–1	9.6–1	9.8–1	9.9–1	9.9–1
$r_1(4)$	6.0–18N_e	5.9–8	5.7–7	5.1–6	3.1–5	1.7–4	1.1–3	5.9–3	1.1–2	1.2–2	1.3–2
$r_0(5)$	1.1–1	1.5–1	2.4–1	5.5–1	8.4–1	9.5–1	9.8–1	9.9–1	1.0	1.0	1.0
$r_1(5)$	5.0–18N_e	4.8–8	4.4–7	2.7–6	1.1–5	5.0–5	3.2–4	1.7–3	3.3–3	3.6–3	3.6–3
$r_0(6)$	1.5–1	2.4–1	4.5–1	7.9–1	9.4–1	9.8–1	9.9–1	1.0	1.0	1.0	1.0
$r_1(6)$	4.5–18N_e	4.1–8	3.0–7	1.2–6	4.0–6	1.7–5	1.1–4	5.9–4	1.1–3	1.2–3	1.3–3
S	7.5–18	1.0–17	1.3–17	2.0–17	4.3–17	1.5–16	9.4–16	5.0–15	9.5–15	1.0–14	1.1–14
α	4.9–13	6.1–13	7.3–13	1.0–12	1.7–12	3.9–12	1.4–11	7.1–11	3.2–10	2.4–9	2.3–27N_e

Table 5.7. (*continued*)

N_e	0	10^{10}	10^{11}	10^{12}	10^{13}	10^{14}	10^{15}	10^{16}	10^{17}	10^{18}	∞
					$T = 1.6 \times 10^4$K.						
$r_0(2)$	2.3–2	2.6–2	2.9–2	3.5–2	4.9–2	9.6–2	3.2–1	7.8–1	9.4–1	9.6–1	9.6–1
$r_1(2)$	1.6–17N_e	1.6–7	1.6–6	1.6–5	1.6–4	1.5–3	1.1–2	3.2–2	3.9–2	4.0–2	4.0–2
$r_0(3)$	9.8–2	1.1–1	1.3–1	1.6–1	3.0–1	6.8–1	8.9–1	9.7–1	9.9–1	1.0	1.0
$r_1(3)$	7.2–18N_e	7.1–8	7.0–7	6.8–6	5.9–5	3.3–4	1.6–3	4.1–3	4.9–3	5.1–3	5.1–3
$r_0(4)$	1.8–1	2.0–1	2.4–1	3.9–1	7.4–1	9.2–1	9.8–1	9.9–1	1.0	1.0	1.0
$r_1(4)$	5.2–18N_e	5.1–8	4.9–7	4.0–6	1.9–5	7.1–5	3.0–4	7.8–4	9.4–4	9.6–4	9.6–4
$r_0(5)$	2.4–1	2.9–1	4.0–1	7.0–1	9.1–1	9.8–1	9.9–1	1.0	1.0	1.0	1.0
$r_1(5)$	4.4–18N_e	4.1–8	3.5–7	1.8–6	5.9–6	1.9–5	8.1–5	2.0–4	2.5–4	2.5–4	2.5–4
$r_0(6)$	2.9–1	3.8–1	6.0–1	8.7–1	9.7–1	9.9–1	1.0	1.0	1.0	1.0	1.0
$r_1(6)$	3.8–18N_e	3.4–8	2.2–7	7.7–7	2.1–6	6.6–6	2.7–5	6.8–5	8.3–5	8.4–5	8.5–5
S	2.6–13	3.0–13	3.4–13	4.4–13	7.1–13	1.7–12	6.1–12	1.5–11	1.8–11	1.9–11	1.9–11
α	3.0–13	3.3–13	3.6–13	4.3–13	5.7–13	9.2–13	2.0–12	4.8–12	1.2–11	8.0–11	7.5–29N_e
					$T = 3.2 \times 10^4$K.						
$r_0(2)$	1.3–1	1.3–1	1.4–1	1.5–1	1.9–1	2.8–1	6.1–1	9.2–1	9.8–1	9.8–1	9.9–1
$r_1(2)$	1.5–17N_e	1.5–7	1.5–6	1.5–5	1.5–4	1.3–3	7.2–3	1.3–2	1.4–2	1.4–2	1.4–2
$r_0(3)$	2.5–1	2.7–1	2.9–1	3.4–1	5.0–1	8.2–1	9.5–1	9.9–1	1.0	1.0	1.0
$r_1(3)$	6.9–18N_e	6.8–8	6.7–7	6.3–6	4.9–5	2.1–4	7.3–4	1.2–3	1.3–3	1.3–3	1.3–3
$r_0(4)$	3.4–1	3.7–1	4.1–1	5.7–1	8.4–1	9.6–1	9.9–1	1.0	1.0	1.0	1.0
$r_1(4)$	5.0–18N_e	4.8–8	4.5–7	3.4–6	1.4–5	4.2–5	1.3–4	2.1–4	2.3–4	2.3–4	2.3–4
$r_0(5)$	4.0–1	4.5–1	5.5–1	8.0–1	9.5–1	9.9–1	1.0	1.0	1.0	1.0	1.0
$r_1(5)$	4.1–18N_e	3.8–8	3.2–7	1.5–6	4.2–6	1.1–5	3.4–5	5.5–5	5.9–6	6.0–5	5.0–5
$r_0(6)$	4.5–1	5.3–1	7.2–1	9.1–1	9.8–1	1.0	1.0	1.0	1.0	1.0	1.0
$r_1(6)$	3.6–18N_e	3.1–8	1.9–7	6.0–7	1.5–6	3.8–6	1.1–5	1.8–5	2.0–5	2.0–5	2.0–5
S	6.2–11	6.7–11	7.3–11	8.6–11	1.1–10	2.0–10	4.9–10	7.6–10	8.2–10	8.2–10	8.2–10
α	1.8–13	1.8–13	1.9–13	2.1–13	2.4–13	3.1–13	4.8–13	7.0–13	1.5–12	9.0–12	8.3–30N_e

Table 5.7. (*continued*)

N_e	0	10^{10}	10^{11}	10^{12}	10^{13}	10^{14}	10^{15}	10^{16}	10^{17}	10^{18}	∞
					$T = 1.28 \times 10^5$K.						
$r_0(2)$	6.8–1	6.9–1	6.9–1	7.0–1	7.3–1	7.9–1	9.2–1	9.8–1	9.9–1	9.9–1	9.9–1
$r_1(2)$	1.8–17N_e	1.8–7	1.8–6	1.8–5	1.7–4	1.4–3	5.1–3	7.0–3	7.3–3	7.3–3	7.3–3
$r_0(3)$	7.2–1	7.3–1	7.4–1	7.7–1	8.5–1	9.5–1	9.9–1	1.0	1.0	1.0	1.0
$r_1(3)$	8.2–18N_e	8.1–8	7.8–7	7.2–6	5.1–5	1.9–4	4.5–4	5.8–4	6.0–4	6.0–4	6.0–4
$r_0(4)$	7.6–1	7.7–1	7.9–1	8.5–1	9.5–1	9.9–1	1.0	1.0	1.0	1.0	1.0
$r_1(4)$	5.9–18N_e	5.6–8	5.2–7	3.7–6	1.4–5	3.6–5	8.1–5	1.0–4	1.1–4	1.1–4	1.1–4
$r_0(5)$	7.8–1	8.0–1	8.4–1	9.3–1	9.8–1	1.0	1.0	1.0	1.0	1.0	1.0
$r_1(5)$	4.8–18N_e	4.4–8	3.6–7	1.6–6	4.3–6	1.0–5	2.2–5	2.8–5	2.8–5	2.9–5	2.9–5
$r_0(6)$	7.9–1	8.2–1	9.0–1	9.7–1	9.9–1	1.0	1.0	1.0	1.0	1.0	1.0
$r_1(6)$	4.2–18N_e	3.6–8	2.1–7	6.7–7	1.6–6	3.5–6	7.5–6	9.5–6	9.8–6	9.8–6	9.8–6
S	6.7–9	6.9–9	7.2–9	7.7–9	8.9–9	1.2–8	1.9–8	2.2–8	2.2–8	2.2–8	2.2–8
α	5.6–14	5.6–14	5.7–14	5.7–14	5.9–14	6.1–41	6.5–14	7.2–14	1.4–13	7.7–13	7.0–31N_e
					$T = 5.12 \times 10^5$K.						
$r_0(2)$	1.5	1.5	1.5	1.5	1.5	1.4	1.1	1.0	1.0	1.0	1.0
$r_1(2)$	2.3–17N_e	2.3–7	2.3–6	2.3–5	2.2–4	1.7–3	6.3–3	8.7–3	9.0–3	9.0–3	9.0–3
$r_0(3)$	1.3	1.3	1.3	1.2	1.1	1.0	1.0	1.0	1.0	1.0	1.0
$r_1(3)$	1.0–17N_e	9.7–8	9.5–7	8.8–6	6.5–5	2.5–4	6.0–4	7.6–4	7.9–4	7.9–4	7.9–4
$r_0(4)$	1.2	1.2	1.2	1.1	1.1	1.0	1.0	1.0	1.0	1.0	1.0
$r_1(4)$	7.0–18N_e	6.7–8	6.3–7	4.6–6	1.9–5	5.2–5	1.1–4	1.4–4	1.5–4	1.5–4	1.5–4
$r_0(5)$	1.2	1.2	1.1	1.1	1.0	1.0	1.0	1.0	1.0	1.0	1.0
$r_1(5)$	5.7–18N_e	5.3–8	4.5–7	2.2–6	6.2–6	1.5–5	3.1–5	3.9–5	4.0–5	4.0–5	4.0–5
$r_0(6)$	1.1	1.1	1.1	1.0	1.0	1.0	1.0	1.0	1.0	1.0	1.0
$r_1(6)$	5.0–18N_e	4.4–8	2.9–7	9.8–7	2.4–6	5.3–6	1.1–5	1.4–5	1.4–5	1.4–5	1.4–5
S	2.8–8	2.8–8	2.9–8	3.0–8	3.3–8	4.1–8	5.8–8	6.7–8	6.8–8	6.8–8	6.8–8
α	1.5–14	1.5–14	1.5–14	1.5–14	1.5–14	1.5–14	1.5–14	1.5–14	2.4–14	1.2–13	1.0–31N_e

Table 5.7. (*continued*)

N_e	0	10^{10}	10^{11}	10^{12}	10^{13}	10^{14}	10^{15}	10^{16}	10^{17}	10^{18}	∞
					$T = 2.048 \times 10^6$K.						
$r_0(2)$	2.5	2.4	2.4	2.3	2.1	1.4	1.4	1.1	1.0	9.9–1	9.9–1
$r_1(2)$	2.2–17N_e	2.2–7	2.2–6	2.1–5	2.1–4	1.7–3	7.4–3	1.1–2	1.2–2	1.2–2	1.2–2
$r_0(3)$	1.9	1.9	1.9	1.8	1.6	1.2	1.1	1.0	1.0	1.0	1.0
$r_1(3)$	0.9–18N_e	8.9–8	8.7–7	8.2–6	6.5–5	2.8–4	7.6–4	1.0–3	1.1–3	1.1–3	1.1–3
$r_0(4)$	1.7	1.7	1.6	1.5	1.2	1.1	1.0	1.0	1.0	1.0	1.0
$r_1(4)$	6.3–18N_e	6.1–8	5.8–7	4.7–6	2.2–5	6.3–5	1.5–4	2.0–4	2.1–4	2.1–4	2.1–4
$r_0(5)$	1.6	1.5	1.5	1.3	1.1	1.0	1.0	1.0	1.0	1.0	1.0
$r_1(5)$	5.1–18N_e	4.8–8	4.3–7	2.4–6	7.6–6	1.9–5	4.3–5	5.7–5	5.9–5	5.9–6	5.9–5
$r_0(6)$	1.5	1.5	1.3	1.1	1.0	1.0	1.0	1.0	1.0	1.0	1.0
$r_1(6)$	4.4–18N_e	4.0–8	3.0–7	1.2–6	3.0–6	7.0–6	1.5–5	2.1–5	2.1–5	2.1–5	2.1–5
S	3.3–8	3.4–8	3.4–8	3.6–8	3.9–8	4.7–8	6.5–8	7.7–8	7.8–8	7.9–8	7.9–8
α	3.4–15	3.4–15	3.4–15	3.3–15	3.3–15	3.2–15	3.0–15	3.0–15	4.1–15	1.5–14	1.2–32N_e
					$T = 8.192 \times 10^6$K.						
$r_0(2)$	3.4	3.4	3.4	3.3	3.2	2.9	2.0	1.2	1.0	9.9–1	9.9–1
$r_1(2)$	1.6–17N_e	1.6–7	1.6–6	1.6–5	1.6–4	1.4–3	7.2–3	1.3–2	1.4–2	1.4–2	1.4–2
$r_0(3)$	2.5	2.5	2.4	2.4	2.1	1.5	1.1	1.0	1.0	1.0	1.0
$r_1(3)$	6.6–18N_e	6.5–8	6.4–7	6.1–6	5.2–5	2.7–4	8.0–4	1.2–3	1.3–3	1.3–3	1.3–3
$r_0(4)$	2.2	2.1	2.1	1.9	1.5	1.1	1.0	1.0	1.0	1.0	1.0
$r_1(4)$	4.5–18N_e	4.4–8	4.2–7	3.7–6	2.0–5	6.4–5	1.6–4	2.5–4	2.6–4	2.7–4	2.7–4
$r_0(5)$	2.0	1.9	1.9	1.6	1.2	1.0	1.0	1.0	1.0	1.0	1.0
$r_1(5)$	3.7–18N_e	3.5–8	3.2–7	2.1–6	7.5–6	2.0–5	4.8–5	7.2–5	7.6–5	7.7–5	7.7–5
$r_0(6)$	1.9	1.8	1.7	1.3	1.1	1.0	1.0	1.0	1.0	1.0	1.0
$r_1(6)$	3.2–18N_e	3.0–8	2.4–7	1.1–6	3.1–6	7.5–6	1.8–5	2.6–5	2.8–5	2.8–5	2.8–5
S	2.5–8	2.5–8	2.6–8	2.6–8	2.8–8	3.3–8	4.6–8	5.8–8	6.0–8	6.0–8	6.6–8
α	6.5–16	6.5–16	6.5–16	6.4–16	6.4–16	6.2–16	5.8–16	5.7–16	6.6–16	1.6–15	1.1–33N_e

5.4.4 Hydrogenlike Ions

Both in the coronal limit and at high density, the temperature at which the ions X_z exist is proportional to the ionization potential E_z. For hydrogenlike ions, $E_z = z^2\mathrm{Ry}$. When $T \propto z^2$, the quantities $\langle v\sigma \rangle \propto z^{-3}$. The spontaneous radiative transition probabilities $A_{nn'} \propto z^4$. Therefore the reduced density and temperature

$$\eta_e = N_e/z^7 , \quad \tilde{T} = T/z^2$$

are convenient when considering the ions. At given η_e and $\tilde{T}$, the quantities r_0, r_1, α/z, and z^3S do not depend on z.

The Saha-Boltzman equation (5.4.11) may be rewritten in the form

$$N_E(n) = \frac{N_e^2}{Q} n^2 8\pi^{3/2} \frac{a_0^3}{z^3} \left(\frac{z^2\mathrm{Ry}}{T}\right)^{3/2} \exp\left(\frac{E_n}{T}\right) , \tag{5.4.22}$$

where $Q = N_e/N^{(z+1)}$. From (5.4.18), it follows that the reduced population density

$$\eta(n) = QN^{(z)}(n)/z^{11} \tag{5.4.23}$$

will not depend on z. Table 5.8 gives some values of r_0, r_1, α/z, and z^3S from data of the work by McWhirter and Hearn [5.38]. The population densities are obtained by substituting $N_E(n)$ from (5.4.22) into (5.4.18) with $N^{(z+1)}$ instead of $N(H^+)$. The more recent quantitative results can be found in [5.40, 43].

The difference between two sets of coefficients r_0, r_1 given by Tables 5.7 and 5.8 is great at low temperature $T \ll z^2\mathrm{Ry}$. It can be explained by different threshold behaviour of the cross sections for neutral atom and for ion (the excitation cross

Table 5.8. Parameters $r_0(n), r_1(n)$ and coefficients S and α for hydrogenlike ions in the optically thin plasma [5.38]. 4.7–6 denotes 4.7×10^{-6}

η_e	0	10^8	10^{10}	10^{12}	10^{13}	10^{14}	10^{15}	10^{16}	10^{18}	∞
					$T = 4 \times 10^3 z^2$K.					
$r_0(2)$	4.7–6	5.9–6	8.8–6	3.6–5	1.6–4	1.4–3	1.3–2	6.1–2	1.1–1	1.1–1
$r_1(2)$		1.2–8	1.2–6	1.2–4	1.2–3	1.2–2	1.1–1	5.1–1	8.9–1	8.9–1
$r_0(3)$	1.3–3	1.6–3	2.5–3	1.5–2	9.6–2	3.2–1	4.3–1	4.7–1	5.0–1	5.0–1
$r_1(3)$		5.5–9	5.5–7	5.5–5	5.7–4	6.4–3	6.0–2	2.9–1	5.1–1	5.1–1
$r_0(4)$	1.0–2	1.4–2	2.2–2	2.1–1	5.2–1	6.9–1	7.5–1	7.7–1	7.8–1	7.8–1
$r_1(4)$		5.1–9	5.0–7	4.2–5	2.9–4	2.9–3	2.7–2	1.3–1	2.2–1	2.2–1
$r_0(5)$	2.9–2	3.9–2	7.0–2	5.5–1	7.8–1	8.7–1	8.9–1	9.0–1	9.0–1	9.0–1
$r_1(5)$		4.3–9	4.1–7	2.3–5	1.3–4	1.3–3	1.2–2	5.8–2	1.0–1	1.0–1
$r_0(7)$	7.9–2	1.1–1	3.2–1	8.8–1	9.5–1	9.7–1	9.8–1	9.8–1	9.8–1	9.8–1
$r_1(7)$		3.4–9	2.7–7	6.3–6	3.0–5	3.0–4	2.8–3	1.4–2	2.4–2	2.4–2
$S \cdot z^3$	9.1–26	1.1–25	1.9–25	1.0–24	6.1–24	4.6–23	3.7–22	2.0–21	3.5–21	3.8–21
α/z	7.9–13	9.2–13	1.4–12	5.2–12	2.0–11	1.1–10	8.6–10	8.0–9	7.6–7	7.6–25η_e

Table 5.8. (*continued*)

η_e	0	10^8	10^{10}	10^{12}	10^{13}	10^{14}	10^{15}	10^{16}	10^{18}	∞
					$T = 8 \times 10^3 z^2$K.					
$r_0(2)$	1.1–3	1.3–3	1.5–3	3.1–3	7.4–3	3.4–2	2.1–1	6.0–1	7.6–1	7.6–1
$r_1(2)$		8.9–9	8.9–7	8.8–5	8.8–4	8.5–3	6.4–2	1.9–1	2.4–1	2.4–1
$r_0(3)$	1.9–2	2.2–2	2.7–2	6.9–2	2.5–1	6.1–1	7.7–1	8.9–1	9.4–1	9.4–1
$r_1(3)$		4.2–9	4.1–7	4.0–5	3.6–4	2.6–3	1.8–2	5.1–2	6.5–2	6.5–2
$r_0(4)$	5.9–2	6.7–2	7.5–2	3.3–1	6.9–1	8.7–1	9.3–1	9.7–1	9.8–1	9.8–1
$r_1(4)$		3.5–9	3.4–7	2.6–5	1.4–4	8.4–4	5.5–3	1.6–2	2.0–2	2.0–2
$r_0(5)$	1.0–1	1.2–1	1.6–1	6.5–1	8.8–1	9.5–1	9.8–1	9.9–1	9.9–1	9.9–1
$r_1(5)$		3.2–9	3.0–7	1.3–5	5.4–5	3.1–4	2.0–3	5.7–3	7.3–3	7.3–3
$r_0(7)$	1.8–1	2.2–1	4.0–1	9.2–1	9.8–1	9.9–1	1.0	1.0	1.0	1.0
$r_1(7)$		2.5–9	2.0–7	3.3–6	9.9–6	5.9–5	3.8–4	1.1–3	1.4–3	1.4–3
$S \cdot z^3$	4.9–17	5.3–17	6.8–17	1.7–16	4.1–16	2.2–15	1.6–14	3.8–14	4.5–14	4.5–14
α/z	4.8–13	5.1–13	6.1–13	1.2–12	2.5–12	7.6–12	3.7–11	1.7–10	1.0–8	1.0–26η_e
					$T = 1.6 \times 10^4 z^2$K.					
$r_0(2)$	2.2–2	2.4–2	2.6–2	3.6–2	5.7–2	1.5–1	5.2–1	8.6–1	9.4–1	9.4–1
$r_1(2)$		6.8–9	6.8–7	6.7–5	6.6–4	6.0–3	3.2–2	5.7–2	6.3–2	6.3–2
$r_0(3)$	9.3–2	1.0–1	1.1–1	1.8–1	3.8–1	7.4–1	9.7–1	9.8–1	9.9–1	9.9–1
$r_1(3)$		3.2–9	3.2–7	3.0–5	2.4–4	1.3–3	5.8–3	9.9–3	1.1–2	1.1–2
$r_0(4)$	1.7–1	1.8–1	2.0–1	4.3–1	7.6–1	9.3–1	9.8–1	9.9–1	1.0	1.0
$r_1(4)$		2.6–9	2.5–7	1.8–5	8.6–5	3.6–4	1.5–3	2.5–3	2.8–3	2.8–3
$r_0(5)$	2.3–1	2.5–1	2.9–1	7.1–1	9.1–1	9.8–1	9.9–1	1.0	1.0	1.0
$r_1(5)$		2.3–9	2.2–7	9.3–6	3.2–5	1.2–4	4.9–4	8.4–4	9.1–4	9.1–4
$r_0(7)$	3.1–1	3.5–1	4.8–1	9.3–1	9.8–1	1.0	1.0	1.0	1.0	1.0
$r_1(7)$		1.8–9	1.5–7	2.1–6	5.7–6	2.1–5	8.5–5	1.5–4	1.6–4	1.6–4
$S \cdot z^3$	1.3–12	1.4–12	1.5–12	2.4–12	4.2–12	1.2–11	4.3–11	7.2–11	7.8–11	7.8–11
α/z	2.9–13	3.0–13	3.2–13	4.3–13	6.3–13	1.2–12	3.0–12	7.9–12	3.1–10	3.1–28η_e
					$T = 3.2 \times 10^4 z^2$K.					
$r_0(2)$	1.2–1	1.3–1	1.3–1	1.5–1	1.9–1	3.1–1	6.8–1	9.3–1	9.7–1	9.7–1
$r_1(2)$		5.4–9	5.4–7	5.4–5	5.2–4	4.5–3	2.0–2	3.0–2	3.1–2	3.1–2
$r_0(3)$	2.5–1	2.6–1	2.7–1	3.3–1	5.1–1	8.1–1	9.5–1	9.9–1	1.0	1.0
$r_1(3)$		2.6–9	2.5–7	2.4–5	1.8–4	8.8–4	2.9–3	4.2–3	4.4–3	4.4–3
$r_0(4)$	3.3–1	3.5–1	3.7–1	5.4–1	8.1–1	9.5–1	9.9–1	1.0	1.0	1.0
$r_1(4)$		1.9–9	1.9–7	1.4–5	6.4–5	2.3–4	6.8–4	9.8–4	1.0–3	1.0–3
$r_0(5)$	3.9–1	4.1–1	4.4–1	7.5–1	9.3–1	9.8–1	1.0	1.0	1.0	1.0
$r_1(5)$		1.7–9	1.6–7	7.6–6	2.3–5	7.4–5	2.2–4	3.1–4	3.3–4	3.3–4
$r_0(7)$	4.6–1	5.0–1	5.9–1	9.4–1	9.9–1	1.0	1.0	1.0	1.0	1.0
$r_1(7)$		1.3–9	1.1–7	2.0–6	4.1–6	1.2–5	3.6–5	5.1–5	5.4–5	5.4–5
$S \cdot z^3$	2.4–10	2.4–10	2.5–10	3.2–10	4.4–10	8.5–10	2.0–9	2.8–9	2.9–9	2.9–9
α/z	1.7–13	1.8–13	1.8–13	1.8–13	2.0–13	2.4–13	3.3–13	5.0–13	2.9–11	2.9–29η_e

Table 5.8. (*continued*)

η_e	0	10^8	10^{10}	10^{12}	10^{13}	10^{14}	10^{15}	10^{16}	10^{18}	∞
					$T = 6.4 \times 10^4 z^2$K.					
$r_0(2)$	3.1–1	3.5–1	3.6–1	3.8–1	4.2–1	5.2–1	7.8–1	9.5–1	9.8–1	9.8–1
$r_1(2)$		4.5–9	4.5–7	4.4–5	4.3–4	3.6–3	1.5–2	2.3–2	2.4–2	2.4–2
$r_0(3)$	4.6–1	4.8–1	4.9–1	5.3–1	6.5–1	8.6–1	9.7–1	9.9–1	1.0	1.0
$r_1(3)$		2.1–9	2.1–7	1.9–5	1.5–4	7.1–4	2.1–3	2.9–3	3.1–3	3.1–3
$r_0(4)$	5.3–1	5.5–1	5.7–1	6.7–1	8.6–1	9.6–1	9.9–1	1.0	1.0	1.0
$r_1(4)$		1.5–9	1.5–7	1.2–5	5.5–5	1.8–4	4.8–4	6.7–4	7.0–4	7.0–4
$r_0(5)$	5.7–1	6.0–1	6.2–1	8.1–1	9.4–1	9.9–1	1.0	1.0	1.0	1.0
$r_1(5)$		1.2–9	1.1–7	5.9–6	2.0–5	5.8–5	1.5–4	2.1–4	2.2–4	2.2–4
$r_0(7)$	6.2–1	6.6–1	7.1–1	9.6–1	9.9–1	1.0	1.0	1.0	1.0	1.0
$r_1(7)$		9.2–10	7.9–8	1.3–6	3.3–6	9.7–6	2.4–5	3.4–5	3.5–5	3.5–5
$S \cdot z^3$	3.4–9	3.4–9	3.5–9	4.0–9	4.8–9	7.3–9	1.4–8	1.8–8	1.8–8	1.8–8
α/z	1.0–13	1.0–13	1.0–13	1.0–13	1.1–13	1.2–13	1.6–13	2.1–13	5.6–12	5.6–30η_e
					$T = 2.56 \times 10^5 z^2$K.					
$r_0(2)$	1.1	1.1	1.1	1.1	1.1	1.1	1.0	9.8–1	9.7–1	9.7–1
$r_1(2)$		3.5–9	3.5–7	3.5–5	3.4–4	2.9–8	1.4–2	2.4–2	2.6–2	2.6–2
$r_0(3)$	1.0	1.0	1.0	1.0	1.0	1.0	1.0	1.0	1.0	1.0
$r_1(3)$		1.6–9	1.6–7	1.5–5	1.3–4	6.7–4	2.0–3	3.0–3	3.2–3	3.2–3
$r_0(4)$	1.0	1.0	1.0	1.0	1.0	1.0	1.0	1.0	1.0	1.0
$r_1(4)$		1.1–9	1.1–7	9.2–6	5.2–5	1.8–4	4.5–4	6.8–4	7.2–4	7.2–4
$r_0(5)$	9.9–1	1.0	1.0	1.0	1.0	1.0	1.0	1.0	1.0	1.0
$r_1(5)$		6.0–10	5.8–8	3.9–6	1.8–5	5.4–5	1.4–4	2.1–4	2.2–4	2.2–4
$r_0(7)$	9.7–1	1.0	1.0	1.0	1.0	1.0	1.0	1.0	1.0	1.0
$r_1(7)$		4.6–10	4.1–8	9.0–7	3.0–6	8.5–6	2.2–5	3.2–5	3.2–5	3.2–5
	—	—	—	—	—	—	—	—	—	—
	—	—	—	—	—	—	—	—	—	—

section for a neutral atom at the threshold is equal to zero, but, for ions, the threshold value is not zero: see Sect. 3.2). In the range $T \sim z^2$Ry, the values of r_0, r_1 from the Tables 5.7 and 5.8 are found to be in reasonable agreement.

5.4.5 Population Densities of Highly Excited Levels at High Density; Steady-Flow Regime

The highly excited bound levels with $n \gg n_0$ are populated and evacuated exclusively by collisions. After an electron has been transferred to a level through three-body recombination it may be either reionized through electron impact, or transferred to another bound level through inelastic or super-elastic (quenching) collision. If $n > n_0$, the collisional transition frequency between the bound levels is n^2 times larger than the frequency of reionization. The most probable are collisional transitions between adjacent levels, followed by transfer of a small energy amount. The electron wandering between the highly excited levels can be treated as diffusion in the space of quantum numbers which can be described by Focker-Planck equation [5.32–34, 44].

Neglecting the radiative processes the electron flux $j(n)$ [$\mathrm{cm^{-3}s^{-1}}$] in the space of quantum numbers can be determined by

$$j(n) = \sum_{k' \geq 0}^{\infty} \sum_{k \geq 0} [N(n + k' + 1)W_{n+k'+1,\, n-k} - N(n - k)W_{n-k,\, n+k'+1}] \; , \quad (5.4.24)$$

the first term in the square brackets representing the electron flux via n directed to the ground state, the second term representing the flux directed to the continuum. Using the ratio $N(n)/N_{\mathrm{E}}(n) = b(n)$ which shows the departure of the level population from the Saha-Boltzmann population density, (5.4.24) may be rewritten in the form

$$\begin{aligned} j(n) = N^{(z+1)} \frac{2}{S} \sum_{k,k'} [&(n + k' + 1)^2 \exp(E_{n+k'+1}/T) b(n + k' + 1) W_{n+k'+1,\, n-k} \\ &-(n - k)^2 \exp(E_{n-k}/T)\, b(n - k) W_{n-k,\, n+k'+1}] \; , \end{aligned} \quad (5.4.25)$$

where $S = z^3 \Theta^{3/2}/4\pi^{3/2} a_0^3 N_{\mathrm{e}}$, $\Theta = T/z^2\mathrm{Ry}$.

Assuming $b(n + k) = b(n) + k\partial b/\partial n$, retaining the terms of the order $\sim k^2$, and neglecting the weak functions of n, one can obtain from (5.4.25)

$$j(n) = N^{(z+1)} N_{\mathrm{e}} \frac{2}{S} \exp\left(\frac{E_n}{T}\right) \frac{\partial b}{\partial n} n^2 \sum_{|k| \geq 1} k^2 \langle v\sigma_{n,\, n+k} \rangle \; . \quad (5.4.26)$$

In $n \gg 1$ the sum with respect to k in (5.4.26) may be extended to infinity. Let the basic dependence of $\langle v\sigma_{n,\, n+k} \rangle$ on temperature and the numbers n and k, which determine the rate coefficient order of magnitude, be written in explicit form (Sect. 3.5)

$$\langle v\sigma_{n,\, n+k} \rangle = \frac{2}{\sqrt{\pi}} \frac{e^2}{\hbar} \frac{\pi a_0^2}{z^3} \Theta^{-1/2} \frac{n^4}{k^3} \varphi(n, k, \Theta) \; , \quad (5.4.27)$$

where $\varphi(n, k, \Theta)$ is a weak function of its arguments. It is convenient to introduce the continuous variable $\varepsilon = 1/n^2$. Using (5.4.27) we can write the formula for the flux as

$$j(\varepsilon) = -N^{(z+1)} N_{\mathrm{e}}^2 \frac{32\pi^2}{z^6} \frac{e^2}{\hbar} a_0^5 \Theta^{-2} \varepsilon^{-3/2} \exp\left(\frac{\varepsilon}{\Theta}\right) L \frac{\partial b}{\partial \varepsilon} \; . \quad (5.4.28)$$

The factor $L = \sum_k k^{-1} \varphi(n = \varepsilon^{-1/2}, k, \Theta)$ is a weak function of ε and Θ. (In reference [5.20] it is estimated to be $\simeq 0.2$).

Using the variable ε the diffusion equation can be written

$$2\varepsilon^{3/2} \frac{dj(\varepsilon)}{d\varepsilon} = q(\varepsilon) - N(\varepsilon) \tilde{\Gamma}_{\varepsilon} \; . \quad (5.4.29)$$

Here $q(\varepsilon)$ [$\mathrm{cm^{-3}\ s^{-1}}$] is the rate for the direct population of the level ε, and

$\tilde{\Gamma}_\varepsilon[s^{-1}]$ is the total frequency of electron transfer to the continuum and to the lower levels n' with $n' < n_0$. The collisional transfer to the highly excited levels is already taken into account in the expressions (5.4.26, 28) for the electron flux $j(\varepsilon)$ and should be excluded from $\tilde{\Gamma}_\varepsilon$.

In order to obtain $b(\varepsilon)$ and $j(\varepsilon)$, the equation (5.4.29) should be supplemented with appropriate boundary conditions. The boundary condition at $\varepsilon = 0$ is found easily. In the limit of $\varepsilon \to 0\,(n \to \infty)$, the function $b(\varepsilon)$ should correspond to the continuum distribution. When Maxwellian velocity distribution is valid,

$$\lim_{\varepsilon \to 0} b(\varepsilon) = 1 \; . \tag{5.4.30}$$

The second boundary condition in general must be chosen by fitting the solution of (5.4.29) to the solution of the set of rate equations (5.4.10) for lower levels where the discrete structure and radiative transitions are of importance. In some cases it is possible to obtain the explicit form of this boundary condition. Somewhat later we shall consider low-temperature recombining plasmas when the condition of total absorption at some value ε_1 can be treated as the second boundary condition.

Without any particular pumping or evacuation of the highly excited levels, $q(\varepsilon)$ is equal to the rate of three-body recombination to the level ε, and $\tilde{\Gamma}_\varepsilon$ is the probability of the inverse process, ionization by electron impact.

Assuming the departure of $b(\varepsilon)$ from the Saha-Boltzmann equilibrium value (equal to unity) to be small, the right-hand part of (5.4.29) may be put equal to zero, because $q(\varepsilon) \simeq N(\varepsilon)\tilde{\Gamma}_\varepsilon$ due to detailed balance. Thus we obtain the constant-flow approximation

$$j(\varepsilon) = \text{const.} = j$$

Then solving the equation (5.4.28) together with the boundary condition (5.4.30) yields the relationship

$$b(\varepsilon) = 1 - j\frac{z^6 L \Theta^2}{N^{(z+1)} N_e^2 C} \int_0^\varepsilon (\varepsilon')^{3/2} \exp\left(-\frac{\varepsilon'}{\Theta}\right) d\varepsilon' \tag{5.4.31}$$

$$C = 32\pi^2 a_0^5 e^2/\hbar \; .$$

If $\varepsilon \ll \Theta$

$$b(\varepsilon) \simeq 1 - j\frac{z^6 L \Theta^2}{N^{(z+1)} N_e^2 C} \varepsilon^{5/2} \; . \tag{5.4.32}$$

The flux value should be determined using the second boundary condition.

Now we shall consider low-temperature recombining plasmas. The flux $j(\varepsilon)$ is positive definite, and $b(\varepsilon)$ decreases with ε. As the second boundary condition for equation (5.4.29), one can use the condition of total absorption of the flux at some value ε_1,

$$b(\varepsilon_1) = 0 \; . \tag{5.4.33}$$

If we deal only with low temperature $\Theta \ll \varepsilon_1$, the value of j determined from (5.4.31) and (5.4.33) does not depend on ε_1, and the integration with respect to ε' may be extended to infinity. Thus

$$j = N^{(z+1)} N_e \alpha \, , \tag{5.4.34}$$

where

$$\begin{aligned} \alpha &= \frac{N_e}{z^6} \frac{2^7 \pi^{3/2}}{3} \frac{e^2}{\hbar} a_0^5 L \Theta^{-9/2} \\ &= N_e z^3 \frac{4\sqrt{2}\pi^{3/2}}{3} \frac{e^{10}}{m^{1/2} T^{9/2}} L \end{aligned} \tag{5.4.35}$$

is the recombination coefficient. The order of magnitude and the dependence of the coefficient α on temperature appear to be in agreement with the results of numerical calculations by Johnson and Hinnov, and Bates et al. given in Tables 5.7 and 5.8.

6 Tables and Formulas for the Estimation of Effective Cross Sections

In this chapter, tables of cross sections and rate coefficients of excitation and ionization by electron impact and rate coefficients of dielectronic recombination are given. The cross sections and rate coefficients are presented in the form of products of angular and radial factors, the latter being expressed in analytical form containing two or three adjusted parameters. The tabulated parameters are obtained from the results of numerical calculations. The first section contains a description of the contents of the tables and relevant fitting formulas. In the second and third sectins of the chapter the formulas for angular factors are given which are necessary for applying the tables.

6.1 Tables of Numerical Results

6.1.1 Methods of Calculations and Survey of the Tables

In this section the results of numerical calculations for the cross sections σ, collision rate coefficients $\langle v\sigma\rangle$, and dielectronic recombination rate coefficient κ_d are given. The calculations are made using the Born method (see Sects. 2.3 and 3.1) and its modifications which are described in Sect. 3.2. The atomic wave functions are assumed to be constructed from single electron wave functions in accordance with a specific scheme of angular momenta coupling. A single-configuration approximation is used. The radial wave functions for all levels of a given electronic configuration are assumed to be the same. Under these assumptions the cross sections may be expressed in the form (2.3.4, 8). In this chapter, however, we use the formulas somewhat different from (2.3.8).

For the excitation cross sections of the transitions $a_0 \to a_1$ we write in the general case

$$\begin{aligned}\sigma(a_0, a_1) = \sigma'(a_0, a_1) + \sigma''(a_0, a_1) = \sum_\kappa Q'_\kappa(a_0, a_1)\sigma'_\kappa(l_0, l_1)\\ + \sum_\kappa Q''_\kappa(a_0, a_1)\sigma''_\kappa(l_0, l_1)\end{aligned} \tag{6.1.1}$$

where σ' consists of the direct plus interference terms, and σ'' is the purely exchange contribution to the cross section (Sects. 2.3, 3.1, 2). A summary of formulas for the Q-factors is given in Sect. 6.2. In order to simplify the use of the tables given below the subsequent subsections comprise the specific formulas for analytical approximation and Q-factors which can be used for specific tables.

The exchange cross sections are given either summed over κ or for those cases in which index κ has a single value. Therefore, in most cases the exchange cross sections can be determined only for transitions between the atomic terms as a whole, so

$$\sigma''(a_0, a_1) = Q''(a_0, a_1)\sigma''(l_0, l_1) \qquad (6.1.2)$$

where index κ is not included, and $\sigma''(l_0, l_1) = \sum_\kappa \sigma''_\kappa(l_0, l_1)$.

The calculated quantities σ'_κ, σ''_κ have been approximated by means of simple analytic formulas which contain the two or three fitting parameters: C, φ, D for σ and A, χ, D for $\langle v\sigma\rangle$. The fitting parameters have been found from the results of numerical calculations by the method of least squares. The errors of analytic approximation R are also given in the tables.

We pass on to a brief description of the tables and of the approximate methods which have been used for calculations.

The Born approximation with normalization has been used for calculation of the cross sections for hydrogen atom summed over l_0 and l_1 (Table 6.1). Tables 6.2 and 6.3 contain the fitting parameters for cross sections and rate coefficients calculated in the Born approximation with the Bates–Damgaard approximation for the atomic wave functions.

The Bates–Damgaard approximation used for calculations of Tables 6.2 and 6.3 is most valid in cases when the maxima of both radial functions lie outside the atomic core. This condition is usually formulated explicitly as

$$n_0^* = \sqrt{z^2\mathrm{Ry}/|E_0|} > l_0 + 1/2\,, \quad n_1^* = \sqrt{z^2\mathrm{Ry}/|E_1|} > l_1 + 1/2\,, \qquad (6.1.3)$$

$$n_0^* > n_c\,, \quad n_1^* > n_c\,, \qquad (6.1.4)$$

where n_c is the largest of the principal quantum numbers of the electrons of the atomic core. The condition (6.1.4) is sometimes stricter than (6.1.3). However, in many cases when conditions (6.1.3) are fulfilled, but (6.1.4) are not fulfilled, the error does not exceed the factor of 2. Such errors are inherent in the Born method itself.

The excitation cross sections for specific atoms and ions (Tables 6.1, 4–10), have been normalized with the use of the K matrix. For calculation of radial

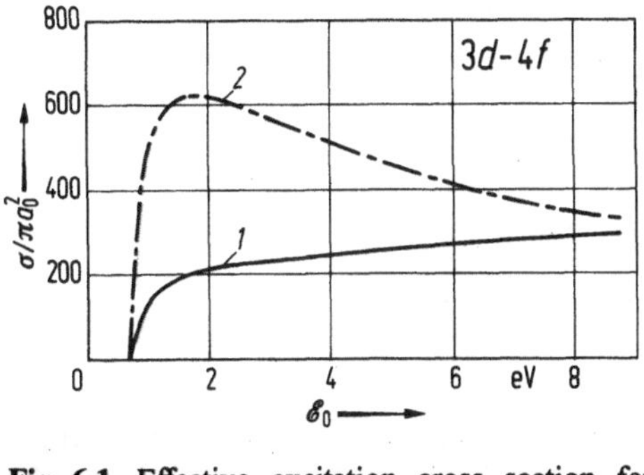

Fig. 6.1. Effective excitation cross section for the transition $3d$–$4f$ of the H atom: (1) Normalized cross section; (2) Born approximation

integrals, the semiempirical numerical wave functions [6.1] have been used. These functions provide considerably better accuracy in all cases, especially for transitions from the shells of equivalent electrons, than the Bates–Damgaard radial functions. In the case of neutral atoms, the normalized Born cross sections are tabulated for transitions without spin change. For intercombination transitions the method of orthogonalized functions (Sect. 3.2) with normalization has been used.

In case of ions for transitions with no change of spin the normalized Coulomb–Born cross sections have been calculated for Table 6.4. The Coulomb–Born-exchange method (Sect. 3.2) has been used for the cross sections tabulated in Tables 6.5–10. The exchange interaction has been taken into account within the framework of the method of orthogonalized functions.

The ionization cross sections have been calculated in the Coulomb–Born approximation.

We shall give a brief description of corrections to the Born approximation which are induced by treating the Coulomb attraction and the exchange interaction, and by normalization of the cross sections.

The difference between the Born and Coulomb–Born cross sections is great only near the threshold $\mathcal{E} \lesssim \Delta E$, $T \lesssim \Delta E$; see Figs. 3.5, 6.2, 6.3. The treatment of exchange affects the value of the cross section for transitions with no change of spin also not far from the threshold (see Fig. 3.6). Therefore the treatment of exchange and Coulomb attraction are necessary for the cross sections of excitation from the ground state of ions. In the *LS*-coupling approximation, the intercombination-transition excitation is solely due to exchange. The effect of normalization upon the cross sections of excitation from the ground state is not substantial.

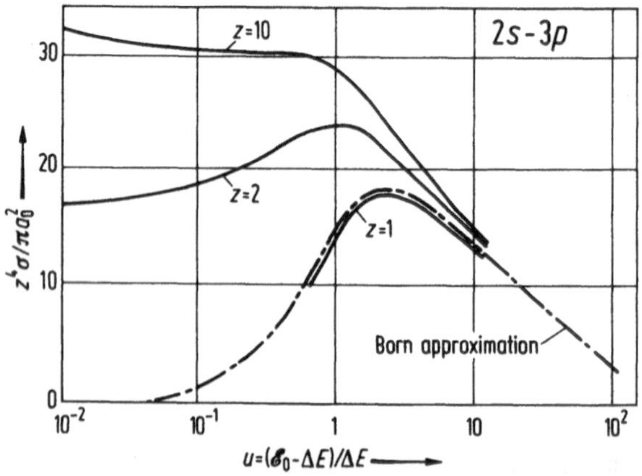

Fig. 6.2. Effective cross section for the transition $2s$–$3p$ of the hydrogen and hydrogenlike ions. (– • – • –) Born approximation. The effect of normalization is very small even for neutral hydrogen

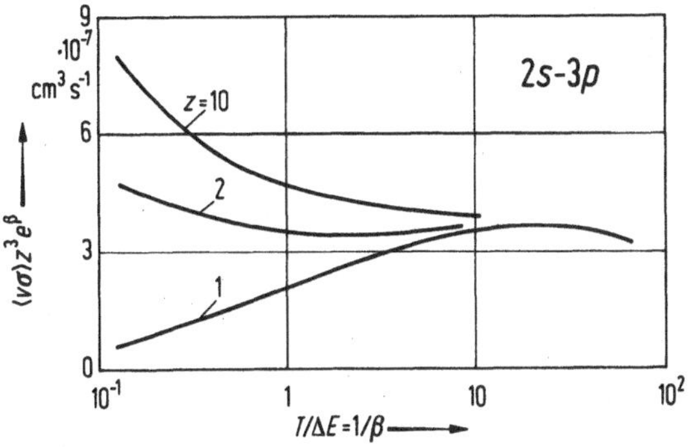

Fig. 6.3. Excitation rate coefficient $\langle v\sigma \rangle$ for the transition $2s$–$3p$ of the hydrogen and hydrogenlike ions

For transitions between the excited levels of neutral atoms, first order perturbation theory can give considerably overestimated results at low energies $\mathscr{E} < \mathrm{Ry}$ (Figs. 3.7, 6.1). Normalization removes this deficiency. In the case of multiply charged ions, the matrix element of interaction of the outer electron with the bound one is z times smaller than the interaction of the bound electron with the atomic core. In Coulomb units, this matrix element is proportional to the parameter $1/z$. Therefore at $z \gg 1$ (actually at $z > 3$) the normalization effect is negligible even for transitions between closely spaced levels. The Coulomb attraction for such transitions is also not important. Hence when $\mathscr{E} > \Delta E$, $T > \Delta E$, the Born cross sections may be used for multiply charged ions.

In cases not covered by the tables which are given in this section, one can use the semiempirical formulas given in Sect. 5.1.

6.1.2 Excitation Cross Sections for Neutral Hydrogen. Transitions $n_0 \to n_1$ (Table 6.1)

The tabulated cross sections are the sums of normalized Born cross sections for $n_0 l_0 \to n_1 l_1$ transitions. They are fitted by means of the following formula

$$\sigma = \pi a_0^2 \left(\frac{n_0}{n_1}\right)^3 \frac{C}{n_0^2} \frac{\Phi'(u)}{u+\varphi}, \qquad (6.1.5)$$

$$u = (\varepsilon - \Delta E)/\mathrm{Ry}.$$

The rate coefficients are given by

$$\langle v\sigma \rangle = 10^{-8} \left(\frac{n_0}{n_1}\right)^3 \frac{A}{n_0^2} \frac{G'(\beta)}{\beta+\chi} \exp(-p\beta)\,[\mathrm{cm}^3/\mathrm{s}],$$

$$\beta = \mathrm{Ry}/T, \quad p = \Delta E/\mathrm{Ry}. \qquad (6.1.6)$$

Table 6.1. Normalized Born cross sections for neutral hydrogen

Transition n_0–n_1	C	φ	D	R	A	χ	D	R
1–2	68.28	3.26	0.00	0.07	32.17	0.26	0.00	0.01
1–3	39.61	2.70	0.00	0.08	25.66	0.38	0.00	0.01
1–4	33.62	2.58	0.00	0.08	23.82	0.42	0.00	0.02
1–5	31.27	2.53	0.00	0.08	23.03	0.44	0.00	0.02
1–6	30.10	2.50	0.00	0.08	22.61	0.45	0.00	0.02
2–3	1275.0	1.23	0.00	0.10	332.0	0.70	3.80	0.02
2–4	454.0	0.62	0.00	0.16	316.0	0.59	0.80	0.02
2–5	318.0	0.53	0.00	0.16	281.0	0.52	0.30	0.02
2–6	268.0	0.50	0.00	0.15	264.0	0.43	0.00	0.02
3–4	8928.0	0.88	0.00	0.08	1437.0	0.96	9.90	0.02
3–5	2298.0	0.08	0.70	0.17	1337.0	1.19	3.30	0.02
3–6	1435.0	0.09	0.60	0.14	1202.0	1.02	1.60	0.01
4–5	35494.0	0.19	0.70	0.13	7505.0	1.14	9.90	0.04
4–6	8492.0	0.06	0.70	0.12	4201.0	1.59	5.50	0.02
5–6	118366.0	0.42	0.00	0.19	30951.0	1.49	9.90	0.04

The formulas for $\Phi'(u)$ and $G'(\beta)$ are expressed by (5.1.12, 13 and 16). The set of parameters C, φ, D is adjusted for the range $0.02 < u < 16$, and the set A, χ, D for $0.25 < \beta < 8$. For the transitions $n_0 \to n_0 + 1$ with $n_0 > 5$ the quasiclassical cross sections given in Tables 3.3 and 3.4 may be used.

6.1.3 Born Cross Sections Calculated in the Bates–Damgaard Approximation for Atomic Wave Functions (Tables 6.2, 6.3)

The cross sections for transitions with no change of spin of the atom, $\Delta S = 0$, are tabulated in a form analogous to the tables of oscillator strengths in Bates–Damgaard approximation given in [6.2]. The quantities σ and $\langle v\sigma\rangle$ are expressed in the form

$$\sigma(a_0, a_1) = \pi a_0^2 \left(\frac{\mathrm{Ry}}{\Delta E}\right)^2 \left(\frac{E_1}{E_0}\right)^{3/2} \frac{1}{2l_0+1} \sum_\kappa Q'_\kappa(a_0, a_1)\ \Phi_\kappa(u),\ u = \frac{\mathcal{E} - \Delta E}{\Delta E}, \tag{6.1.7}$$

$$\langle v\sigma(a_0, a_1)\rangle = 10^{-8} \left(\frac{\mathrm{Ry}}{\Delta E}\right)^{3/2} \left(\frac{E_1}{E_0}\right)^{3/2} \frac{\exp(-\beta)}{2l_0+1} \sum_\kappa Q'_\kappa(a_0, a_1) \times G_\kappa(\beta)\ [\mathrm{cm^3 s^{-1}}]\,, \beta = \Delta E/T$$

$$\kappa = |l_0 - l_1|,\quad |l_0 - l_1| + 2, \ldots |l_0 + l_1|\,. \tag{6.1.8}$$

Table 6.2. Transitions with no change of spin $\Delta S = 0$; Born cross sections in the Bates–Damgaard approximation for atomic wave functions. Asterisks indicate cases where the errors exceed 10%. Parameters C and φ.

	The effective principal quantum number n_0^*									
Δn	0.5	1.0	1.5	2.0	2.5	3.0	3.5	4.0	4.5	5.0
	Transition $s - s$, parameter C.[a]									
0.6	67 + 1	55 + 1	57 + 1	61 + 1	66 + 1	71 + 1	77 + 1	83 + 1	89 + 1	95 + 1
0.7	72 + 1	55 + 1	55 + 1	57 + 1	61 + 1	65 + 1	69 + 1	74 + 1	79 + 1	84 + 1
0.8	69 + 1	49 + 1	48 + 1	49 + 1	51 + 1	54 + 1	57 + 1	60 + 1	64 + 1	67 + 1
0.9	58 + 1	39 + 1	37 + 1	37 + 1	38 + 1	40 + 1	42 + 1	44 + 1	46 + 1	49 + 1
1.0	42 + 1	26 + 1	25 + 1	24 + 1	25 + 1	26 + 1	27 + 1	28 + 1	30 + 1	31 + 1
1.1	16 + 1	15 + 1	14 + 1	14 + 1	15 + 1	15 + 1	16 + 1	16 + 1*	17 + 1*	18 + 1*
1.2	67 + 0	81 + 0	84 + 0	85 + 0	87 + 0	90 + 0	94 + 0	97 + 0	10 + 1*	10 + 1
1.3	60 + 0*	65 + 0*	68 + 0	69 + 0	70 + 0	72 + 0	75 + 0	77 + 0	80 + 0	83 + 0
1.4	13 + 1*	10 + 1	97 + 0	94 + 0	93 + 0	94 + 0	96 + 0	98 + 0	10 + 1	10 + 1
1.5	27 + 1	17 + 1	15 + 1	14 + 1	13 + 1	13 + 1	13 + 1	14 + 1	14 + 1	14 + 1
1.6	41 + 1	25 + 1	21 + 1	19 + 1	18 + 1	18 + 1	18 + 1	18 + 1	18 + 1	18 + 1
1.7	52 + 1	30 + 1	25 + 1	22 + 1	21 + 1	20 + 1	20 + 1	20 + 1	20 + 1	21 + 1
1.8	57 + 1	31 + 1	26 + 1	22 + 1	21 + 1	20 + 1	20 + 1	20 + 1	20 + 1	20 + 1
1.9	54 + 1	28 + 1	23 + 1	20 + 1	18 + 1	18 + 1	17 + 1	17 + 1	17 + 1	18 + 1
2.0	43 + 1	21 + 1	18 + 1	15 + 1	14 + 1	13 + 1	13 + 1	13 + 1	13 + 1	13 + 1
2.1	15 + 1	13 + 1	11 + 1	10 + 1	99 + 0	95 + 0	93 + 0	93 + 0*	93 + 0*	94 + 0*
2.2	71 + 0	78 + 0	73 + 0	68 + 0	64 + 0	63 + 0	62 + 0	61 + 0	61 + 0*	62 + 0*
2.3	57 + 0*	60 + 0*	58 + 0	54 + 0	52 + 0	50 + 0	49 − 0	49 − 0	49 − 0	49 − 0
2.4	12 + 1*	84 + 0	75 + 0	68 + 0	63 + 0	60 + 0	58 + 0	58 + 0	57 + 0	57 + 0
2.5	24 + 1	14 + 1	11 + 1	10 + 1	91 + 0	86 + 0	82 + 0	80 + 0	79 + 0	79 + 0
2.6	38 + 1	20 + 1	16 + 1	13 + 1	12 + 1	11 + 1	11 + 1	10 + 1	10 + 1	10 + 1
2.7	50 + 1	26 + 1	20 + 1	16 + 1	14 + 1	13 + 1	13 + 1	12 + 1	12 + 1	12 + 1
2.8	55 + 1	27 + 1	21 + 1	17 + 1	15 + 1	14 + 1	13 + 1	13 + 1	12 + 1	12 + 1
2.9	54 + 1	25 + 1	20 + 1	16 + 1	14 + 1	13 + 1	12 + 1	12 + 1	11 + 1	11 + 1
3.0	45 + 1	19 + 1	16 + 1	12 + 1	11 + 1	10 + 1	10 + 1	96 + 0	94 + 0	92 + 0*
3.1	15 + 1	12 + 1	10 + 1	91 + 0	82 + 0	76 + 0	72 + 0	69 + 0*	67 + 0*	66 + 0*
3.2	72 + 0	76 + 0	69 + 0	61 + 0	55 + 0	52 + 0	49 − 0	48 − 0	47 − 0*	46 + 0*
3.3	56 + 0*	58 + 0*	54 + 0	49 − 0	45 − 0	42 − 0	40 − 0	39 − 0	38 − 0	37 − 0
3.4	11 + 1*	79 + 0	68 + 0	59 + 0	53 + 0	49 − 0	46 − 0	44 − 0	43 − 0	42 − 0
3.5	23 + 1	13 + 1	10 + 1	87 + 0	76 + 0	69 + 0	64 + 0	61 + 0	59 + 0	57 + 0
3.6	37 + 1	19 + 1	14 + 1	12 + 1	10 + 1	93 + 0	86 + 0	81 + 0	78 + 0	76 + 0
3.7	49 + 1	24 + 1	18 + 1	14 + 1	12 + 1	11 + 1	10 + 1	98 + 0	93 + 0	90 + 0
3.8	55 + 1	26 + 1	19 + 1	15 + 1	13 + 1	11 + 1	11 + 1	10 + 1	98 + 0	95 + 0
3.9	54 + 1	24 + 1	18 + 1	14 + 1	12 + 1	11 + 1	10 + 1	96 + 0	92 + 0	88 + 0
4.0	46 + 1	19 + 1	15 + 1	11 + 1	10 + 1	91 + 0	84 + 0	79 + 0	75 + 0	72 + 0*
4.1	15 + 1	12 + 1	10 + 1	84 + 0	74 + 0	67 + 0	62 + 0	58 + 0*	55 + 0*	54 + 0*
4.2	72 + 0	76 + 0	66 + 0	57 + 0	51 + 0	47 − 0	43 − 0	41 − 0	39 − 0*	38 − 0*
4.3	56 + 0*	57 + 0*	52 + 0	46 − 0	42 − 0	38 − 0	35 − 0	33 − 0	32 − 0	31 − 0*
4.4	11 + 1*	76 + 0	65 + 0	55 + 0	48 − 0	44 − 0	40 − 0	38 − 0	36 − 0	35 − 0

[a] This table can be applied to transitions between the levels of two different electronic configurations *ns–n′s* with no change of any of the angular quantum numbers

Table 6.2. (*continued*)

	The effective principal quantum number n_0^*									
Δn	0.5	1.0	1.5	2.0	2.5	3.0	3.5	4.0	4.5	5.0
	Transition $s-s$, parameter φ									
0.6	92 + 0	80 + 0	76 + 0	78 + 0	81 + 0	86 + 0	91 + 0	96 + 0	10 + 1	10 + 1
0.7	88 + 0	78 + 0	78 + 0	82 + 0	88 + 0	94 + 0	10 + 1	10 + 1	11 + 1	12 + 1
0.8	85 + 0	78 + 0	81 + 0	89 + 0	97 + 0	10 + 1	11 + 1	12 + 1	13 + 1	14 + 1
0.9	84 + 0	81 + 0	89 + 0	10 + 1	11 + 1	12 + 1	13 + 1	15 + 1	16 + 1	17 + 1
1.0	84 + 0	89 + 0	10 + 1	11 + 1	13 + 1	15 + 1	17 + 1	18 + 1	20 + 1	22 + 1
1.1	93 + 0	10 + 1	12 + 1	14 + 1	16 + 1	19 + 1	21 + 1	24 + 1*	26 + 1*	29 + 1*
1.2	15 + 1	15 + 1	15 + 1	16 + 1	18 + 1	20 + 1	23 + 1	25 + 1	27 + 1*	30 + 1*
1.3	20 + 1*	15 + 1*	12 + 1	11 + 1	12 + 1	12 + 1	13 + 1	13 + 1	14 + 1	14 + 1
1.4	13 + 1*	10 + 1	77 + 0	69 + 0	67 + 0	66 + 0	64 + 0	64 + 0	63 + 0	62 + 0
1.5	10 + 1	78 + 0	62 + 0	56 + 0	55 + 0	53 + 0	52 + 0	52 + 0	52 + 0	52 + 0
1.6	91 + 0	70 + 0	60 + 0	57 + 0	57 + 0	57 + 0	58 + 0	59 + 0	60 + 0	62 + 0
1.7	85 + 0	68 + 0	62 + 0	63 + 0	65 + 0	67 + 0	69 + 0	72 + 0	75 + 0	79 + 0
1.8	82 + 0	69 + 0	68 + 0	71 + 0	75 + 0	80 + 0	85 + 0	90 + 0	96 + 0	10 + 1
1.9	81 + 0	72 + 0	76 + 0	83 + 0	90 + 0	98 + 0	10 + 1	11 + 1	12 + 1	13 + 1
2.0	81 + 0	81 + 0	89 + 0	10 + 1	11 + 1	12 + 1	13 + 1	15 + 1	16 + 1	17 + 1
2.1	89 + 0	99 + 0	11 + 1	12 + 1	14 + 1	15 + 1	17 + 1	19 + 1*	21 + 1*	23 + 1*
2.2	13 + 1	13 + 1	13 + 1	14 + 1	16 + 1	18 + 1	20 + 1	21 + 1	23 + 1*	25 + 1*
2.3	20 + 1*	15 + 1*	12 + 1	12 + 1	12 + 1	13 + 1	14 + 1	15 + 1	15 + 1	16 + 1
2.4	13 + 1*	10 + 1	81 + 0	74 + 0	74 + 0	74 + 0	73 + 0	73 + 0	73 + 0	73 + 0
2.5	10 + 1	78 + 0	61 + 0	56 + 0	55 + 0	53 + 0	52 + 0	51 + 0	51 + 0	51 + 0
2.6	91 + 0	68 + 0	57 + 0	55 + 0	54 + 0	53 + 0	53 + 0	53 + 0	54 + 0	55 + 0
2.7	85 + 0	66 + 0	59 + 0	59 + 0	60 + 0	61 + 0	63 + 0	64 + 0	67 + 0	69 + 0
2.8	82 + 0	67 + 0	65 + 0	67 + 0	70 + 0	73 + 0	77 + 0	81 + 0	86 + 0	90 + 0
2.9	81 + 0	70 + 0	73 + 0	78 + 0	84 + 0	90 + 0	97 + 0	10 + 1	11 + 1	12 + 1
3.0	81 + 0	78 + 0	85 + 0	95 + 0	10 + 1	11 + 1	12 + 1	13 + 1	14 + 1	16 + 1*
3.1	88 + 0	96 + 0	10 + 1	11 + 1	13 + 1	14 + 1	16 + 1	17 + 1*	19 + 1*	21 + 1*
3.2	13 + 1	13 + 1	13 + 1	14 + 1	15 + 1	17 + 1	18 + 1	20 + 1	22 + 1*	24 + 1*
3.3	20 + 1*	15 + 1*	12 + 1	12 + 1	12 + 1	13 + 1	14 + 1	15 + 1	15 + 1	16 + 1
3.4	13 + 1*	10 + 1	82 + 0	76 + 0	77 + 0	77 + 0	77 + 0	78 + 0	78 + 0	79 + 0
3.5	10 + 1	78 + 0	61 + 0	57 + 0	55 + 0	54 + 0	53 + 0	52 + 0	52 + 0	52 + 0
3.6	91 + 0	68 + 0	56 + 0	54 + 0	53 + 0	52 + 0	52 + 0	52 + 0	52 + 0	53 + 0
3.7	85 + 0	65 + 0	58 + 0	58 + 0	58 + 0	59 + 0	60 + 0	61 + 0	63 + 0	65 + 0
3.8	82 + 0	66 + 0	63 + 0	65 + 0	68 + 0	70 + 0	73 + 0	77 + 0	81 + 0	85 + 0
3.9	81 + 0	69 + 0	71 + 0	76 + 0	81 + 0	87 + 0	93 + 0	99 + 0	10 + 1	11 + 1
4.0	81 + 0	78 + 0	84 + 0	92 + 0	10 + 1	11 + 1	11 + 1	13 + 1	14 + 1	15 + 1*
4.1	88 + 0	95 + 0	10 + 1	11 + 1	12 + 1	14 + 1	15 + 1	16 + 1*	18 + 1*	19 + 1*
4.2	13 + 1	13 + 1	12 + 1	13 + 1	15 + 1	16 + 1	18 + 1	19 + 1	21 + 1*	22 + 1*
4.3	20 + 1*	15 + 1*	12 + 1	12 + 1	12 + 1	13 + 1	14 + 1	15 + 1	15 + 1	16 + 1
4.4	14 + 1*	10 + 1	82 + 0	77 + 0	79 + 0	79 + 0	79 + 0	80 + 0	81 + 0	82 + 0

Table 6.2. (*continued*)

Δn	The effective principal quantum number n_0^* 0.5	1.0	1.5	2.0	2.5	3.0	3.5	4.0	4.5	5.0
	Transition $s-p$, parameter C									
0.1			11 + 1	16 + 1	21 + 1	25 + 1	29 + 1	32 + 1	36 + 1	39 + 1
0.2			27 + 1	37 + 1	44 + 1	50 + 1	56 + 1	61 + 1	67 + 1	72 + 1
0.3			49 + 1	59 + 1	67 + 1	73 + 1	80 + 1	86 + 1	92 + 1	98 + 1
0.4			73 + 1	80 + 1	86 + 1	92 + 1	98 + 1	10 + 2	10 + 2	11 + 2
0.5			96 + 1	97 + 1	10 + 2	10 + 2	10 + 2	11 + 2	11 + 2	12 + 2
0.6		11 + 2	11 + 2	10 + 2	10 + 2	10 + 2	10 + 2	10 + 2	11 + 2	11 + 2
0.7		14 + 2	11 + 2	10 + 2	98 + 1	96 + 1	95 + 1	96 + 1*	97 + 1*	99 + 1*
0.8		15 + 2	11 + 2	94 + 1	84 + 1	80 + 1*	77 + 1*	77 + 1*	77 + 1*	77 + 1*
0.9		15 + 2	93 + 1	75 + 1	63 + 1*	59 + 1*	55 + 1*	54 + 1*	53 + 1*	53 + 1*
1.0		13 + 2	69 + 1	52 + 1*	41 + 1*	37 + 1*	34 + 1*	32 + 1*	31 + 1*	31 + 1*
1.1	30 + 2	10 + 2	46 + 1	30 + 1*	22 + 1*	19 + 1	16 + 1	15 + 1*	14 + 1*	13 + 1*
1.2	25 + 2	61 + 1	23 + 1	12 + 1	86 + 0*	68 + 0*	58 + 0*	53 + 0*	50 + 0*	48 − 0*
1.3	18 + 2	27 + 1	74 + 0*	42 − 0	37 − 0	37 − 0	38 − 0*	40 − 0*	43 − 0*	45 − 0*
1.4	10 + 2	71 + 0	43 − 0	55 + 0*	63 + 0*	70 + 0*	75 + 0*	80 + 0*	84 + 0*	88 + 0*
1.5	51 + 1	53 + 0*	11 + 1	13 + 1*	13 + 1*	14 + 1*	14 + 1*	14 + 1*	14 + 1*	15 + 1*
1.6	14 + 1	19 + 1*	24 + 1	24 + 1	22 + 1	21 + 1	21 + 1	21 + 1*	20 + 1*	21 + 1*
1.7	10 − 0	42 + 1	38 + 1	34 + 1	29 + 1	27 + 1	25 + 1	25 + 1*	24 + 1*	24 + 1*
1.8	76 + 0	68 + 1	47 + 1	39 + 1	32 + 1	29 + 1	26 + 1*	25 + 1*	24 + 1*	23 + 1*
1.9	26 + 1	87 + 1	49 + 1	38 + 1	29 + 1	26 + 1*	23 + 1*	21 + 1*	20 + 1*	19 + 1*
2.0	48 + 1	95 + 1	44 + 1	31 + 1	23 + 1*	20 + 1*	17 + 1*	15 + 1*	14 + 1*	13 + 1*
2.1	19 + 2	80 + 1	36 + 1	21 + 1*	15 + 1*	12 + 1	10 + 1	91 + 0	82 + 0	75 + 0
2.2	18 + 2	55 + 1	21 + 1	11 + 1	74 + 0	55 + 0*	45 − 0*	39 − 0*	35 − 0*	32 − 0*
2.3	14 + 2	29 + 1	87 + 0	42 − 0	30 − 0*	26 − 0	24 − 0	24 − 0	24 − 0	24 − 0
2.4	10 + 2	10 + 1	34 − 0	34 − 0*	35 − 0*	37 − 0*	38 − 0*	39 − 0	40 − 0	41 − 0
2.5	54 + 1	37 − 0	72 + 0	77 + 0	77 + 0*	76 + 0*	75 + 0*	74 + 0*	74 + 0*	74 + 0*
2.6	20 + 1	11 + 1	16 + 1*	15 + 1	13 + 1	12 + 1	12 + 1*	11 + 1*	11 + 1*	11 + 1*
2.7	31 − 0	30 + 1	27 + 1	23 + 1	19 + 1	17 + 1	15 + 1	14 + 1*	14 + 1*	13 + 1*
2.8	28 − 0	52 + 1	36 + 1	29 + 1	22 + 1	19 + 1	17 + 1*	16 + 1*	14 + 1*	14 + 1*
2.9	15 + 1	73 + 1	40 + 1	30 + 1	22 + 1	18 + 1*	16 + 1*	14 + 1*	13 + 1*	12 + 1*
3.0	33 + 1	84 + 1	37 + 1	26 + 1	18 + 1*	15 + 1*	12 + 1*	11 + 1*	10 + 1*	93 + 0*
3.1	16 + 2	73 + 1	32 + 1	18 + 1*	12 + 1*	10 + 1	81 + 0*	70 + 0*	61 + 0	55 + 0
3.2	16 + 2	53 + 1	20 + 1	10 + 1	67 + 0	49 − 0*	39 − 0*	32 − 0*	28 − 0*	25 − 0*
3.3	13 + 2	30 + 1	90 + 0	41 − 0*	28 − 0	23 − 0	20 − 0	19 − 0	18 − 0	17 − 0
3.4	96 + 1	11 + 1	34 − 0	28 − 0*	28 − 0*	28 − 0*	28 − 0*	27 − 0*	27 − 0*	28 − 0*
3.5	55 + 1	35 − 0	59 + 0	62 + 0	59 + 0*	57 + 0*	55 + 0*	53 + 0*	51 + 0*	50 + 0*
3.6	32 + 1	93 + 0	13 + 1*	12 + 1*	10 + 1*	99 + 0*	90 + 0*	85 + 0*	80 + 0*	77 + 0*
3.7	46 − 0	25 + 1	23 + 1	20 + 1	15 + 1	13 + 1	12 + 1	11 + 1	10 + 1	98 + 0
3.8	16 − 0*	47 + 1	32 + 1	25 + 1	18 + 1	16 + 1	13 + 1	12 + 1*	11 + 1*	10 + 1*
3.9	11 + 1	67 + 1	36 + 1	26 + 1	18 + 1*	15 + 1*	13 + 1*	11 + 1*	10 + 1*	97 + 0*
4.0	27 + 1	80 + 1	34 + 1	23 + 1	15 + 1*	13 + 1*	10 + 1*	93 + 0*	81 + 0*	74 + 0*
4.1	15 + 2	71 + 1	31 + 1	17 + 1	11 + 1*	89 + 0	71 + 0	59 + 0	51 + 0	46 − 0
4.2	15 + 2	52 + 1	20 + 1	99 + 0	62 + 0	45 − 0*	35 − 0*	29 − 0*	25 − 0*	22 − 0*
4.3	13 + 2	30 + 1	92 + 0	41 − 0*	27 − 0	21 − 0	18 − 0	16 − 0	15 − 0	14 − 0
4.4	94 + 1	11 + 1	34 − 0	26 − 0*	25 − 0	24 − 0*	23 − 0*	22 − 0*	22 − 0*	22 − 0*

Table 6.2. (*continued*)

Δn	The effective principal quantum number n_0^*									
	0.5	1.0	1.5	2.0	2.5	3.0	3.5	4.0	4.5	5.0
	Transition $s-p$, parameter φ									
0.1			10 + 1	11 + 1	12 + 1	13 + 1	15 + 1	16 + 1	17 + 1	18 + 1
0.2			14 + 1	16 + 1	18 + 1	20 + 1	22 + 1	24 + 1	26 + 1	28 + 1
0.3			17 + 1	21 + 1	24 + 1	27 + 1	30 + 1	33 + 1	36 + 1	39 + 1
0.4			21 + 1	25 + 1	29 + 1	34 + 1	38 + 1	42 + 1	46 + 1	50 + 1
0.5			25 + 1	30 + 1	36 + 1	41 + 1	46 + 1	52 + 1	57 + 1	62 + 1
0.6		22 + 1	29 + 1	36 + 1	43 + 1	49 + 1	56 + 1	63 + 1	70 + 1	76 + 1
0.7		25 + 1	34 + 1	42 + 1	51 + 1	60 + 1	68 + 1	77 + 1*	85 + 1*	94 + 1*
0.8		28 + 1	40 + 1	50 + 1	62 + 1	73 + 1*	84 + 1*	94 + 1*	10 + 2*	11 + 2*
0.9		32 + 1	47 + 1	61 + 1	76 + 1*	89 + 1*	10 + 2*	11 + 2*	12 + 2*	14 + 2*
1.0		37 + 1	58 + 1	75 + 1*	93 + 1*	11 + 2*	12 + 2*	14 + 2*	15 + 2*	17 + 2*
1.1	26 + 1	44 + 1	70 + 1	93 + 1*	11 + 2*	12 + 2	12 + 2	13 + 2*	12 + 2*	11 + 2*
1.2	29 + 1	56 + 1	85 + 1	84 + 1	61 + 1*	40 + 1*	27 + 1*	19 + 1*	14 + 1*	10 + 1*
1.3	33 + 1	73 + 1	38 + 1*	49 − 0	10 − 0	10 − 0	10 − 0*	10 − 0*	10 − 0*	10 − 0*
1.4	38 + 1	37 + 1	10 − 0	10 − 0*	10 − 0*	10 − 0*	10 − 0*	10 − 0*	10 − 0*	10 − 0*
1.5	47 + 1	10 − 0*	10 − 0	13 − 0*	37 − 0*	59 + 0*	79 + 0*	98 + 0*	11 + 1*	13 + 1*
1.6	70 + 1	15 − 0*	74 + 0	11 + 1	14 + 1	18 + 1	21 + 1	24 + 1*	27 + 1*	30 + 1*
1.7	57 + 0	10 + 1	16 + 1	21 + 1	26 + 1	31 + 1	40 + 1*	40 + 1*	45 + 1*	50 + 1*
1.8	45 − 0	17 + 1	25 + 1	31 + 1	38 + 1	45 + 1	52 + 1*	58 + 1*	65 + 1*	71 + 1*
1.9	16 + 1	23 + 1	34 + 1	43 + 1	53 + 1	62 + 1*	71 + 1*	79 + 1*	88 + 1*	96 + 1*
2.0	23 + 1	29 + 1	45 + 1	57 + 1	69 + 1*	81 + 1*	92 + 1*	10 + 2*	11 + 2*	12 + 2*
2.1	24 + 1	37 + 1	56 + 1	73 + 1*	86 + 1*	96 + 1	10 + 2	10 + 2	10 + 2	10 + 2
2.2	28 + 1	48 + 1	71 + 1	78 + 1	68 + 1*	53 + 1*	39 + 1*	30 + 1*	23 + 1*	19 + 1*
2.3	32 + 1	64 + 1	56 + 1	17 + 1	44 − 0*	10 − 0	10 − 0	10 − 0	10 − 0	10 − 0
2.4	57 + 1	59 + 1	10 − 0	10 − 0*	10 − 0	10 − 0*	10 − 0*	10 − 0	10 − 0	10 − 0
2.5	46 + 1	10 − 0	10 − 0	10 − 0	10 − 0*	10 − 0*	23 − 0*	37 − 0*	50 − 0*	63 − 0*
2.6	63 + 1	10 − 0	33 − 0*	67 + 0	91 + 0	11 + 1	14 + 1*	16 + 1*	18 + 1*	21 + 1*
2.7	73 + 1	69 + 0	12 + 1	16 + 1	20 + 1	23 + 1	27 + 1	31 + 1*	35 + 1*	38 + 1*
2.8	10 − 0	14 + 1	21 + 1	26 + 1	32 + 1	37 + 1	43 + 1*	48 + 1*	53 + 1*	59 + 1*
2.9	11 + 1	21 + 1	30 + 1	38 + 1	46 + 1	53 + 1*	61 + 1*	68 + 1*	75 + 1*	82 + 1*
3.0	20 + 1	27 + 1	41 + 1	51 + 1	62 + 1*	71 + 1*	80 + 1*	89 + 1*	97 + 1*	10 + 2*
3.1	23 + 1	35 + 1	52 + 1	67 + 1*	78 + 1*	86 + 1*	92 + 1*	95 + 1	96 + 1	95 + 1
3.2	27 + 1	46 + 1	67 + 1	74 + 1	67 + 1	56 + 1*	44 + 1*	34 + 1*	72 + 1*	22 + 1*
3.3	32 + 1	61 + 1	59 + 1	23 + 1*	80 + 0	21 − 0	10 − 0	10 − 0	10 − 0	10 − 0
3.4	37 + 1	63 + 1	10 − 0	10 − 0*	10 − 0*	10 − 0*	10 − 0*	10 − 0*	10 − 0*	10 − 0*
3.5	45 + 1	10 − 0	10 − 0	10 − 0	10 − 0*	10 − 0*	10 − 0*	14 − 0*	24 − 0*	35 − 0*
3.6	61 + 1	10 − 0	18 − 0*	49 − 0*	69 + 0*	90 + 0*	11 + 1*	13 + 1*	15 + 1*	17 + 1*
3.7	88 + 1	54 + 0	10 + 1	14 + 1	17 + 1	20 + 1	24 + 1	27 + 1	30 + 1	33 + 1
3.8	10 − 0*	13 + 1	19 + 1	24 + 1	29 + 1	34 + 1	39 + 1	43 + 1*	48 + 1*	52 + 1*
3.9	91 + 0*	27 + 1	28 + 1	35 + 1	42 + 1*	49 + 1*	56 + 1*	62 + 1*	68 + 1*	74 + 1*
4.0	19 + 1	27 + 1	38 + 1	48 + 1	58 + 1*	66 + 1*	74 + 1*	82 + 1*	89 + 1*	96 + 1*
4.1	23 + 1	34 + 1	57 + 1	63 + 1	73 + 1*	81 + 1	87 + 1	90 + 1	91 + 1*	91 + 1
4.2	27 + 1	45 + 1	65 + 1	71 + 1	66 + 1	56 + 1*	45 + 1*	36 + 1*	29 + 1*	24 + 1*
4.3	31 + 1	60 + 1	61 + 1	25 + 1*	10 + 1	70 − 0	10 − 0	10 − 0	10 − 0	10 − 0
4.4	37 + 1	64 + 1	20 − 0	10 − 0*	10 − 0	10 − 0*	10 − 0*	10 − 0*	10 − 0*	10 − 0*

Table 6.2. (*continued*)

Δn	The effective principal quantum number n_0^*									
	0.5	1.0	1.5	2.0	2.5	3.0	3.5	4.0	4.5	5.0
	Transition $s-d$, parameter C									
0.1					28 − 0	47 − 0	63 + 0	79 + 0	93 + 0	10 + 1
0.2					79 + 0	11 + 1	15 + 1	18 + 1	21 + 1	23 + 1
0.3					15 + 1	21 + 1	26 + 1	30 + 1	33 + 1	37 + 1
0.4					25 + 1	32 + 1	37 + 1	42 + 1	46 + 1	50 + 1
0.5					37 + 1	43 + 1	48 + 1	52 + 1	56 + 1	60 + 1
0.6				39 + 1	49 + 1	53 + 1	57 + 1	60 + 1	62 + 1	65 + 1
0.7				53 + 1	59 + 1	60 + 1	62 + 1	63 + 1	64 + 1	66 + 1
0.8				66 + 1	66 + 1	63 + 1	62 + 1	61 + 1	61 + 1	61 + 1
0.9				75 + 1	67 + 1	61 + 1	57 + 1	55 + 1	53 + 1	53 + 1*
1.0				78 + 1	65 + 1	54 + 1	51 + 1	45 + 1	48 + 1	41 + 1*
1.1			99 + 1	74 + 1	54 + 1	43 + 1	37 + 1	34 + 1*	31 + 1*	30 + 1*
1.2			10 + 2	63 + 1	42 + 1	31 + 1	26 + 1*	23 + 1	21 + 1*	20 + 1*
1.3			95 + 1	49 + 1	29 + 1	21 + 1*	17 + 1	15 + 1	13 + 1	13 + 1
1.4			80 + 1	33 + 1	18 + 1	13 + 1	11 + 1	10 + 1	99 + 0*	98 + 0*
1.5			60 + 1	19 + 1	11 + 1	93 + 0	90 + 0	92 + 0	95 + 0	99 + 0
1.6		14 + 2	34 + 1	11 + 1	90 + 0	96 + 0	10 + 1	11 + 1	11 + 1	12 + 1
1.7		11 + 2	16 + 1	89 + 0	11 + 1	13 + 1	14 + 1	14 + 1	15 + 1	15 + 1
1.8		75 + 1	91 + 0	13 + 1	17 + 1	18 + 1	18 + 1	18 + 1	18 + 1	18 + 1
1.9		40 + 1	12 + 1	21 + 1	24 + 1	22 + 1	21 + 1	20 + 1	20 + 1	19 + 1
2.0		16 + 1	23 + 1	31 + 1	29 + 1	25 + 1	23 + 1	20 + 1	20 + 1	18 + 1*
2.1	26 + 2	40 − 0	33 + 1	38 + 1	30 + 1	24 + 1	21 + 1	18 + 1	17 + 1*	16 + 1*
2.2	14 + 2	14 + 1	46 + 1	40 + 1	28 + 1	21 + 1	17 + 1*	15 + 1*	13 + 1*	12 + 1*
2.3	60 + 1	38 + 1	55 + 1	37 + 1	23 + 1	16 + 1	13 + 1	11 + 1*	10 + 1*	94 + 0*
2.4	15 + 1	62 + 1	57 + 1	29 + 1	16 + 1	11 + 1	95 + 0	83 + 0	75 + 0	71 + 0
2.5	83 − 1	79 + 1	51 + 1	20 + 1	11 + 1	83 + 0	72 + 0	67 + 0	65 + 0	64 + 0
2.6	29 − 0	83 + 1	35 + 1	12 + 1	80 + 0	71 + 0	70 + 0	71 + 0	72 + 0	73 + 0
2.7	11 + 1	75 + 1	20 + 1	84 + 0	83 + 0	85 + 0	88 + 0	90 + 0	91 + 0	92 + 0
2.8	19 + 1	58 + 1	10 + 1	90 + 0	11 + 1	11 + 1	11 + 1	11 + 1	11 + 1	11 + 1
2.9	23 + 1	37 + 1	90 + 0	14 + 1	16 + 1	15 + 1	14 + 1	13 + 1	13 + 1	12 + 1
3.0	23 + 1	19 + 1	15 + 1	22 + 1	21 + 1	18 + 1	16 + 1	14 + 1	13 + 1	12 + 1*
3.1	12 + 2	50 + 0	22 + 1	29 + 1	23 + 1	18 + 1	16 + 1	14 + 1	12 + 1*	11 + 1*
3.2	78 + 1	85 + 0	34 + 1	33 + 1	23 + 1	17 + 1	14 + 1	12 + 1*	10 + 1*	98 + 0*
3.3	37 + 1	24 + 1	44 + 1	32 + 1	20 + 1	14 + 1	11 + 1	96 + 0*	84 + 0*	76 + 0*
3.4	11 + 1	43 + 1	49 + 1	27 + 1	15 + 1	10 + 1	85 + 0	72 + 0	64 + 0	59 + 0
3.5	11 − 0	59 + 1	47 + 1	20 + 1	10 + 1	77 + 0	64 + 0	57 + 0	64 + 0	51 + 0
3.6	13 − 0	66 + 1	34 + 1	13 + 1	77 + 0	63 + 0	59 + 0	57 + 0	56 + 0	55 + 0
3.7	66 + 0	63 + 1	21 + 1	85 + 0	73 + 0	70 + 0	70 + 0	69 + 0	69 + 0	68 + 0
3.8	12 + 1	52 + 1	11 + 1	79 + 0	96 + 0	94 + 0	93 + 0	89 + 0	87 + 0	84 + 0
3.9	16 + 1	36 + 1	83 + 0	11 + 1	13 + 1	12 + 1	11 + 1	10 + 1	10 + 1	97 + 0
4.0	17 + 1	20 + 1	12 + 1	19 + 1	18 + 1	15 + 1	13 + 1	12 + 1	11 + 1	10 + 1
4.1	97 + 1	58 + 0	17 + 1	25 + 1	20 + 1	16 + 1	13 + 1	11 + 1	10 + 1	96 + 0*
4.2	61 + 1	66 + 0	28 + 1	29 + 1	21 + 1	15 + 1	12 + 1	10 + 1*	92 + 0*	83 + 0*
4.3	30 + 1	18 + 1	39 + 1	30 + 1	19 + 1	13 + 1	10 + 1	86 + 0	74 + 0*	66 + 0*
4.4	10 + 1	35 + 1	45 + 1	26 + 1	15 + 1	10 + 1	79 + 0	66 + 0	58 + 0	52 + 0

Table 6.2. (*continued*)

	The effective principal quantum number n_0^*									
Δn	0.5	1.0	1.5	2.0	2.5	3.0	3.5	4.0	4.5	5.0
	Transition $s-d$, parameter φ									
0.1					81 + 0	70 + 0	64 + 0	62 + 0	60 + 0	60 + 0
0.2					62 + 0	61 + 0	62 + 0	64 + 0	66 + 0	69 + 0
0.3					63 + 0	66 + 0	70 + 0	74 + 0	79 + 0	83 + 0
0.4					68 + 0	74 + 0	80 + 0	86 + 0	93 + 0	99 + 0
0.5					74 + 0	83 + 0	91 + 0	99 + 0	10 + 1	11 + 1
0.6				72 + 0	82 + 0	92 + 0	10 + 1	11 + 1	12 + 1	13 + 1
0.7				78 + 0	90 + 0	10 + 1	11 + 1	13 + 1	14 + 1	16 + 1
0.8				84 + 0	10 + 1	11 + 1	13 + 1	15 + 1	17 + 1	19 + 1
0.9				91 + 0	11 + 1	13 + 1	15 + 1	17 + 1	20 + 1	22 + 1*
1.0				10 + 1	14 + 1	15 + 1	22 + 1	21 + 1	34 + 1	27 + 1*
1.1			85 + 0	11 + 1	14 + 1	17 + 1	21 + 1	25 + 1*	28 + 1*	32 + 1*
1.2			91 + 0	12 + 1	16 + 1	21 + 1	25 + 1*	29 + 1	33 + 1*	36 + 1*
1.3			99 + 0	14 + 1	19 + 1	24 + 1*	28 + 1	30 + 1	31 + 1	31 + 1
1.4			10 + 1	17 + 1	22 + 1	24 + 1	23 + 1	20 + 1	17 + 1*	15 + 1*
1.5			12 + 1	20 + 1	20 + 1	15 + 1	11 + 1	87 + 0	74 + 0	66 + 0
1.6		88 + 0	15 + 1	19 + 1	98 + 0	64 + 0	53 + 0	49 − 0	48 − 0	49 − 0
1.7		94 + 0	18 + 1	89 + 0	49 − 0	45 − 0	48 − 0	52 + 0	57 + 0	63 + 0
1.8		10 + 1	13 + 1	46 − 0	47 − 0	54 + 0	62 + 0	70 + 0	80 + 0	90 + 0
1.9		11 + 1	56 + 0	48 − 0	60 + 0	71 + 0	83 + 0	96 + 0	11 + 1	12 + 1
2.0		14 + 1	50 + 0	61 + 0	77 + 0	93 + 0	11 + 1	12 + 1	15 + 1	16 + 1*
2.1	84 + 0	14 + 1	57 + 0	75 + 0	96 + 0	11 + 1	14 + 1	16 + 1	19 + 1*	21 + 1*
2.2	86 + 0	59 + 0	68 + 0	92 + 0	11 + 1	14 + 1	17 + 1*	20 + 1*	23 + 1*	26 + 1*
2.3	90 + 0	65 + 0	80 + 0	11 + 1	14 + 1	18 + 1	21 + 1	24 + 1*	26 + 1*	27 + 1*
2.4	98 + 0	71 + 0	92 + 0	13 + 1	17 + 1	21 + 1	22 + 1	22 + 1	21 + 1	20 + 1
2.5	16 + 1	77 + 0	10 + 1	16 + 1	19 + 1	18 + 1	15 + 1	13 + 1	11 + 1	97 + 0
2.6	66 + 0	84 + 0	12 + 1	18 + 1	14 + 1	10 + 1	76 + 0	63 + 0	56 + 0	53 + 0
2.7	78 + 0	92 + 0	16 + 1	14 + 1	67 + 0	52 + 0	47 − 0	46 − 0	48 − 0	51 + 0
2.8	85 + 0	10 + 1	16 + 1	62 + 0	45 − 0	46 − 0	50 + 0	55 + 0	62 + 0	68 + 0
2.9	90 + 0	11 + 1	83 + 0	44 − 0	50 + 0	58 + 0	66 + 0	76 + 0	86 + 0	97 + 0
3.0	96 + 0	13 + 1	48 − 0	52 + 0	65 + 0	76 + 0	92 + 0	10 + 1	12 + 1	13 + 1*
3.1	86 + 0	18 + 1	49 − 0	65 + 0	82 + 0	99 + 0	11 + 1	13 + 1	15 + 1*	17 + 1*
3.2	89 + 0	62 + 0	61 + 0	81 + 0	10 + 1	12 + 1	15 + 1	17 + 1*	20 + 1*	22 + 1*
3.3	94 + 0	59 + 0	73 + 0	99 + 0	13 + 1	16 + 1	18 + 1	21 + 1*	23 + 1*	24 + 1*
3.4	10 + 1	67 + 0	86 + 0	12 + 1	16 + 1	19 + 1	20 + 1	21 + 1	21 + 1	20 + 1
3.5	16 + 1	74 + 0	10 + 1	14 + 1	18 + 1	18 + 1	17 + 1	14 + 1	13 + 1	11 + 1
3.6	57 + 0	82 + 0	12 + 1	17 + 1	15 + 1	11 + 1	91 + 0	75 + 0	65 + 0	59 + 0
3.7	75 + 0	90 + 0	15 + 1	15 + 1	80 + 0	59 + 0	51 + 0	48 − 0	47 − 0	48 − 0
3.8	83 + 0	10 + 1	17 + 1	76 + 0	47 − 0	45 − 0	47 − 0	50 + 0	55 + 0	60 + 0
3.9	90 + 0	11 + 1	10 + 1	45 − 0	47 − 0	53 + 0	59 + 0	67 + 0	76 + 0	85 + 0
4.0	96 + 0	13 + 1	49 − 0	49 − 0	60 + 0	69 + 0	82 + 0	92 + 0	10 + 1	11 + 1
4.1	87 + 0	19 + 1	47 − 0	61 + 0	76 + 0	90 + 0	10 + 1	12 + 1	14 + 1	15 + 1*
4.2	90 + 0	68 + 0	57 + 0	76 + 0	97 + 0	11 + 1	13 + 1	16 + 1*	18 + 1*	20 + 1*
4.3	96 + 0	57 + 0	70 + 0	94 + 0	12 + 1	14 + 1	17 + 1	19 + 1	21 + 1*	23 + 1*
4.4	10 + 1	64 + 0	83 + 0	11 + 1	15 + 1	17 + 1	19 + 1	20 + 1	21 + 1	20 + 1

Table 6.2. (*continued*)

	The effective principal quantum number n_0^*							
Δn	1.5	2.0	2.5	3.0	3.5	4.0	4.5	5.0
	Transition $p-s$, parameter C							
0.1	83 + 0	13 + 1	18 + 1	22 + 1	26 + 1	29 + 1	33 + 1	36 + 1
0.2	15 + 1	25 + 1	32 + 1	39 + 1	45 + 1	51 + 1	57 + 1	62 + 1
0.3	21 + 1	33 + 1	43 + 1	51 + 1	59 + 1	66 + 1	73 + 1	79 + 1
0.4	24 + 1	37 + 1	48 + 1	57 + 1	65 + 1	72 + 1	79 + 1	86 + 1
0.5	24 + 1	37 + 1	47 + 1	56 + 1	63 + 1	70 + 1	77 + 1	84 + 1
0.6	21 + 1	33 + 1	42 + 1	49 + 1	56 + 1	62 + 1*	67 + 1*	73 + 1*
0.7	16 + 1	26 + 1	33 + 1	39 + 1*	44 + 1*	48 + 1*	53 + 1*	57 + 1*
0.8	11 + 1	18 + 1	22 + 1*	27 + 1*	30 + 1*	33 + 1*	36 + 1*	39 + 1*
0.9	58 + 0	10 + 1	13 + 1*	15 + 1	17 + 1	19 + 1	21 + 1*	23 + 1*
1.0	24 − 0	47 − 0*	58 + 0*	71 + 0*	80 + 0*	89 + 0*	97 + 0*	10 + 1*
1.1	15 − 0	21 − 0	26 − 0	30 − 0	33 − 0	36 − 0	39 − 0	42 − 0
1.2	21 − 0*	25 − 0*	27 − 0*	29 − 0*	31 − 0*	32 − 0*	33 − 0*	35 − 0*
1.3	41 − 0*	47 − 0*	51 + 0*	53 + 0*	55 + 0*	56 + 0*	58 + 0*	60 + 0*
1.4	64 + 0*	77 + 0*	83 + 0*	87 + 0*	90 + 0*	93 + 0*	96 + 0*	99 + 0*
1.5	81 + 0	10 + 1*	11 + 1	11 + 1*	12 + 1*	12 + 1*	13 + 1*	13 + 1*
1.6	85 + 0	11 + 1	12 + 1	13 + 1	13 + 1*	14 + 1*	14 + 1*	15 + 1*
1.7	75 + 0	10 + 1	11 + 1	12 + 1*	13 + 1*	13 + 1*	14 + 1*	14 + 1*
1.8	53 + 0	83 + 0	91 + 0	10 + 1*	10 + 1*	11 + 1*	11 + 1*	12 + 1*
1.9	30 − 0	55 + 0	60 + 0	68 + 0	71 + 0	76 + 0	79 + 0	83 + 0
2.0	17 − 0	29 − 0*	31 − 0*	36 − 0*	38 − 0*	41 − 0*	42 − 0*	45 − 0*
2.1	12 − 0*	16 − 0	18 − 0	19 − 0	20 − 0	21 − 0	22 − 0	22 − 0
2.2	17 − 0*	18 − 0*	19 − 0*	19 − 0*	19 − 0*	19 − 0*	19 − 0*	19 − 0*
2.3	32 − 0*	33 − 0*	33 − 0*	32 − 0*	32 − 0*	31 − 0*	31 − 0*	31 − 0*
2.4	49 − 0*	53 − 0*	54 + 0*	52 + 0*	52 + 0*	51 + 0*	51 + 0*	51 + 0*
2.5	63 + 0	71 + 0*	73 + 0*	72 + 0*	71 + 0*	71 + 0*	70 + 0*	70 + 0*
2.6	66 + 0	81 + 0	83 + 0	83 + 0	83 + 0	83 + 0*	83 + 0*	83 + 0*
2.7	58 + 0	78 + 0	80 + 0	82 + 0*	82 + 0*	83 + 0*	83 + 0*	83 + 0*
2.8	41 − 0	63 + 0	65 + 0	69 + 0*	69 + 0*	70 + 0*	70 + 0*	72 + 0*
2.9	23 − 0	43 − 0	43 − 0	48 − 0	48 − 0	50 + 0*	50 + 0	51 + 0
3.0	15 − 0	24 − 0*	24 − 0*	27 − 0*	27 − 0*	28 − 0*	28 − 0*	29 − 0*
3.1	12 − 0*	14 − 0	15 − 0	15 − 0	15 − 0	16 − 0	16 − 0	16 − 0
3.2	16 − 0*	16 − 0*	16 − 0*	15 − 0*	15 − 0*	15 − 0*	14 − 0*	14 − 0*
3.3	29 − 0*	28 − 0*	27 − 0*	25 − 0*	24 − 0*	23 − 0*	23 − 0*	22 − 0*
3.4	44 − 0*	45 − 0*	44 − 0*	41 − 0*	40 − 0*	38 − 0*	37 − 0*	36 − 0*
3.5	56 + 0	61 + 0	60 + 0*	57 + 0*	55 + 0*	53 + 0*	51 + 0*	50 + 0*
3.6	59 + 0	70 + 0	68 + 0	67 + 0	64 + 0	63 + 0	61 + 0	60 + 0
3.7	51 + 0	68 + 0	66 + 0	66 + 0	64 + 0	63 + 0*	62 + 0*	61 + 0*
3.8	36 − 0	56 + 0	54 + 0	56 + 0	55 + 0*	55 + 0*	53 + 0*	53 + 0*
3.9	20 − 0	38 − 0	37 − 0	40 − 0	39 − 0	39 − 0	39 − 0	39 − 0
4.0	14 − 0	21 − 0*	21 − 0	23 − 0*	22 − 0*	23 − 0*	23 − 0*	23 − 0*
4.1	11 − 0*	13 − 0	14 − 0	14 − 0	13 − 0	13 − 0	13 − 0	13 − 0
4.2	15 − 0*	15 − 0*	15 − 0*	14 − 0*	13 − 0*	12 − 0*	12 − 0*	12 − 0*
4.3	27 − 0*	26 − 0*	25 − 0*	22 − 0*	21 − 0*	20 − 0*	19 − 0*	18 − 0*
4.4	42 − 0*	42 − 0*	40 − 0*	36 − 0*	34 − 0*	32 − 0*	30 − 0*	29 − 0*

Table 6.2. (*continued*)

Δn	The effective principal quantum number n_0^*							
	1.5	2.0	2.5	3.0	3.5	4.0	4.5	5.0
	Transition $p-s$, parameter φ							
0.1	10 + 1	11 + 1	12 + 1	14 + 1	15 + 1	16 + 1	17 + 1	18 + 1
0.2	14 + 1	17 + 1	19 + 1	21 + 1	23 + 1	25 + 1	27 + 1	29 + 1
0.3	19 + 1	22 + 1	25 + 1	29 + 1	32 + 1	34 + 1	37 + 1	40 + 1
0.4	24 + 1	28 + 1	33 + 1	37 + 1	41 + 1	45 + 1	49 + 1	53 + 1
0.5	30 + 1	35 + 1	41 + 1	46 + 1	51 + 1	57 + 1	62 + 1	67 + 1
0.6	37 + 1	44 + 1	51 + 1	58 + 1	64 + 1	71 + 1*	78 + 1*	84 + 1*
0.7	46 + 1	54 + 1	63 + 1	72 + 1*	80 + 1*	89 + 1*	97 + 1*	10 + 2*
0.8	55 + 1	66 + 1	78 + 1*	88 + 1*	99 + 1*	11 + 2*	12 + 2*	13 + 2*
0.9	52 + 1	74 + 1	87 + 1*	10 + 2	11 + 2	12 + 2	14 + 2*	15 + 2*
1.0	11 + 1	45 + 1*	49 + 1*	66 + 1*	72 + 1*	84 + 1*	90 + 1*	98 + 1*
1.1	10 − 0	10 − 0	29 − 0	52 + 0	70 + 0	86 + 0	99 + 0	11 + 1
1.2	10 − 0*	10 − 0*	10 − 0*	10 − 0*	10 − 0*	10 − 0*	10 − 0*	10 − 0*
1.3	10 − 0*	10 − 0*	10 − 0*	10 − 0*	10 − 0*	10 − 0*	10 − 0*	10 − 0*
1.4	67 + 0*	66 + 0*	69 + 0*	72 + 0*	77 + 0*	82 + 0*	88 + 0*	94 + 0*
1.5	14 + 1	15 + 1*	17 + 1	18 + 1*	20 + 1*	21 + 1*	23 + 1*	25 + 1*
1.6	23 + 1	25 + 1	28 + 1	31 + 1	34 + 1*	37 + 1*	40 + 1*	43 + 1*
1.7	32 + 1	36 + 1	41 + 1	46 + 1*	50 + 1*	55 + 1*	59 + 1*	64 + 1*
1.8	38 + 1	47 + 1	54 + 1	61 + 1*	67 + 1*	74 + 1*	80 + 1*	87 + 1*
1.9	27 + 1	51 + 1	57 + 1	68 + 1	75 + 1	85 + 1	92 + 1	10 + 2
2.0	10 − 0	27 + 1*	27 + 1*	40 + 1*	42 + 1*	51 + 1*	54 + 1*	60 + 1*
2.1	10 − 0*	10 − 0	10 − 0	28 − 0	42 − 0	57 + 0	68 + 0	79 + 0
2.2	10 − 0*	10 − 0*	10 − 0*	10 − 0*	10 − 0*	10 − 0*	10 − 0*	10 − 0*
2.3	10 − 0*	10 − 0*	10 − 0*	10 − 0*	10 − 0*	10 − 0*	10 − 0*	10 − 0*
2.4	49 − 0*	43 − 0*	41 − 0*	41 − 0*	42 − 0*	44 − 0*	46 − 0*	49 − 0*
2.5	12 + 1	12 + 1*	13 + 1*	14 + 1*	15 + 1*	16 + 1*	17 + 1*	18 + 1*
2.6	20 + 1	22 + 2	24 + 1	26 + 1	28 + 1	20 + 1*	32 + 1*	34 + 1*
2.7	28 + 1	32 + 1	35 + 1	39 + 1*	43 + 1*	46 + 1*	50 + 1*	53 + 1*
2.8	33 + 1	42 + 1	46 + 1	52 + 1*	57 + 1*	63 + 1*	68 + 1*	73 + 1*
2.9	19 + 1	44 + 1	47 + 1	57 + 1	62 + 1	70 + 1	76 + 1	83 + 1
3.0	10 − 0	22 + 1*	19 + 1*	33 + 1*	33 + 1*	41 + 1*	42 + 1*	48 + 1*
3.1	10 − 0*	10 − 0	10 − 0	17 − 0	30 − 0	43 − 0	53 + 0	63 + 0
3.2	10 − 0*	10 − 0*	10 − 0*	10 − 0*	10 − 0*	10 − 0*	10 − 0*	10 − 0*
3.3	10 − 0*	10 − 0*	10 − 0*	10 − 0*	10 − 0*	10 − 0*	10 − 0*	10 − 0*
3.4	42 − 0*	34 − 0*	32 − 0*	30 − 0*	30 − 0*	31 − 0*	32 − 0*	33 − 0*
3.5	11 + 1	11 + 1	12 + 1*	12 + 1*	13 + 1*	14 + 1*	14 + 1*	15 + 1*
3.6	19 + 1	20 + 1	22 + 1	23 + 1	25 + 1	27 + 1	29 + 1	30 + 1
3.7	27 + 1	30 + 1	33 + 1	36 + 1	39 + 1	42 + 1*	45 + 1*	48 + 1*
3.8	30 + 1	39 + 1	43 + 1	48 + 1	53 + 1	57 + 1*	62 + 1*	66 + 1*
3.9	14 + 1	41 + 1	42 + 1	52 + 1	56 + 1	64 + 1	68 + 1	74 + 1
4.0	10 − 0	20 + 1*	16 + 1	29 + 1*	28 + 1*	36 + 1*	37 + 1*	43 + 1*
4.1	10 − 0*	10 − 0	10 − 0	12 − 0	23 − 0	35 − 0	45 − 0	54 + 0
4.2	10 − 0*	10 − 0*	10 − 0*	10 − 0*	10 − 0*	10 − 0*	10 − 0*	10 − 0*
4.3	10 − 0*	10 − 0*	10 − 0*	10 − 0*	10 − 0*	10 − 0*	10 − 0*	10 − 0*
4.4	39 − 0*	30 − 0*	27 − 0*	25 − 0*	24 − 0*	24 − 0*	24 − 0*	25 − 0*

Table 6.2. (*continued*)

	The effective principal quantum number n_0^*							
Δn	1.5	2.0	2.5	3.0	3.5	4.0	4.5	5.0
	Transition $p-p$, parameter C $\kappa = 0$							
0.6	10 + 2	14 + 2	16 + 2	19 + 2	21 + 2	23 + 2	25 + 2	27 + 2
0.7	11 + 2	13 + 2	15 + 2	17 + 2	19 + 2	21 + 2	22 + 2	24 + 2
0.8	10 + 2	12 + 2	13 + 2	14 + 2	16 + 2	17 + 2	18 + 2	19 + 2
0.9	87 + 1	96 + 1	10 + 2	11 + 2	12 + 2	12 + 2	13 + 2	14 + 2
1.0	65 + 1	67 + 1	73 + 1	77 + 1	81 + 1	86 + 1	91 + 1	96 + 1
1.1	39 + 1	42 + 1	44 + 1	46 + 1	49 + 1	51 + 1	54 + 1	57 + 1
1.2	22 + 1	25 + 1	26 + 1	27 + 1	28 + 1	30 + 1	31 + 1	32 + 1
1.3	14 + 1	18 + 1	19 + 1	20 + 1	21 + 1	22 + 1	23 + 1	24 + 1
1.4	16 + 1	21 + 1	22 + 1	24 + 1	25 + 1	27 + 1	28 + 1	29 + 1
1.5	26 + 1	31 + 1	33 + 1	35 + 1	36 + 1	38 + 1	40 + 1	41 + 1
1.6	39 + 1	43 + 1	45 + 1	47 + 1	48 + 1	50 + 1	52 + 1	54 + 1
1.7	51 + 1	52 + 1	54 + 1	55 + 1	56 + 1	58 + 1	59 + 1	61 + 1
1.8	58 + 1	55 + 1	56 + 1	56 + 1	57 + 1	58 + 1	59 + 1	61 + 1
1.9	57 + 1	51 + 1	51 + 1	51 + 1	51 + 1	51 + 1	52 + 1	53 + 1
2.0	49 + 1	41 + 1	41 + 1	40 + 1	40 + 1	40 + 1	41 + 1	41 + 1
2.1	31 + 1	30 + 1	29 + 1	28 + 1	28 + 1	28 + 1	28 + 1	29 + 1
2.2	20 + 1	20 + 1	19 + 1	19 + 1	19 + 1	19 + 1	19 + 1	19 + 1
2.3	13 + 1	15 + 1	14 + 1	14 + 1	14 + 1	14 + 1	14 + 1	14 + 1
2.4	13 + 1	16 + 1	15 + 1	16 + 1	15 + 1	16 + 1	16 + 1	16 + 1
2.5	20 + 1	22 + 1	22 + 1	22 + 1	22 + 1	22 + 1	22 + 1	22 + 1
2.6	31 + 1	31 + 1	30 + 1	30 + 1	29 + 1	29 + 1	29 + 1	29 + 1
2.7	42 + 1	39 + 1	38 + 1	36 + 1	35 + 1	35 + 1	35 + 1	35 + 1
2.8	49 + 1	43 + 1	41 + 1	39 + 1	38 + 1	37 + 1	37 + 1	36 + 1
2.9	50 + 1	41 + 1	39 + 1	36 + 1	35 + 1	34 + 1	34 + 1	34 + 1
3.0	45 + 1	34 + 1	33 + 1	30 + 1	29 + 1	28 + 1	28 + 1	27 + 1
3.1	29 + 1	26 + 1	24 + 1	22 + 1	21 + 1	21 + 1	20 + 1	20 + 1
3.2	19 + 1	18 + 1	16 + 1	16 + 1	15 + 1	14 + 1	14 + 1	14 + 1
3.3	13 + 1	14 + 1	12 + 1	12 + 1	11 + 1	11 + 1	11 + 1	11 + 1
3.4	12 + 1	14 + 1	13 + 1	13 + 1	12 + 1	12 + 1	12 + 1	11 + 1
3.5	18 + 1	19 + 1	18 + 1	17 + 1	17 + 1	16 + 1	16 + 1	16 + 1
3.6	28 + 1	27 + 1	25 + 1	24 + 1	23 + 1	22 + 1	21 + 1	21 + 1
3.7	39 + 1	34 + 1	32 + 1	29 + 1	28 + 1	27 + 1	26 + 1	25 + 1
3.8	46 + 1	38 + 1	35 + 1	32 + 1	30 + 1	29 + 1	28 + 1	27 + 1
3.9	48 + 1	37 + 1	34 + 1	31 + 1	29 + 1	27 + 1	27 + 1	26 + 1
4.0	44 + 1	31 + 1	29 + 1	26 + 1	25 + 1	23 + 1	22 + 1	22 + 1
4.1	28 + 1	24 + 1	21 + 1	20 + 1	18 + 1	17 + 1	17 + 1	16 + 1
4.2	19 + 1	17 + 1	15 + 1	14 + 1	13 + 1	12 + 1	12 + 1	12 + 1
4.3	13 + 1	13 + 1	11 + 1	11 + 1	10 + 1	10 + 1	96 + 0	93 + 0
4.4	12 + 1	13 + 1	12 + 1	11 + 1	11 + 1	10 + 1	10 + 1	99 + 0

Table 6.2. (*continued*)

	The effective principal quantum number n_0^*							
Δn	1.5	2.0	2.5	3.0	3.5	4.0	4.5	5.0
	Transition $p-p$, parameter φ $\kappa=0$							
0.6	90 + 0	77 + 0	78 + 0	83 + 0	88 + 0	94 + 0	99 + 0	10 + 1
0.7	84 + 0	77 + 0	82 + 0	90 + 0	97 + 0	10 + 1	11 + 1	12 + 1
0.8	81 + 0	80 + 0	90 + 0	10 + 1	11 + 1	12 + 1	13 + 1	14 + 1
0.9	80 + 0	87 + 0	10 + 1	11 + 1	13 + 1	14 + 1	16 + 1	18 + 1
1.0	82 + 0	10 + 1	12 + 1	14 + 1	16 + 1	19 + 1	22 + 1	24 + 1
1.1	94 + 0	12 + 1	15 + 1	18 + 1	22 + 1	26 + 1	29 + 1	34 + 1
1.2	13 + 1	16 + 1	18 + 1	21 + 1	25 + 1	28 + 1	31 + 1	34 + 1
1.3	20 + 1	15 + 1	14 + 1	14 + 1	14 + 1	14 + 1	13 + 1	13 + 1
1.4	17 + 1	92 + 0	74 + 0	70 + 0	67 + 0	64 + 0	61 + 0	59 + 0
1.5	11 + 1	64 + 0	56 + 0	54 + 0	53 + 0	52 + 0	51 + 0	52 + 0
1.6	85 + 0	57 + 0	55 + 0	56 + 0	57 + 0	58 + 0	60 + 0	62 + 0
1.7	74 + 0	58 + 0	60 + 0	63 + 0	67 + 0	71 + 0	75 + 0	80 + 0
1.8	70 + 0	62 + 0	69 + 0	75 + 0	81 + 0	88 + 0	96 + 0	10 + 1
1.9	69 + 0	71 + 0	81 + 0	91 + 0	10 + 1	11 + 1	12 + 1	13 + 1
2.0	72 + 0	84 + 0	99 + 0	11 + 1	13 + 1	15 + 1	17 + 1	19 + 1
2.1	82 + 0	10 + 1	12 + 1	15 + 1	17 + 1	20 + 1	23 + 1	26 + 1
2.2	11 + 1	13 + 1	15 + 1	18 + 1	21 + 1	24 + 1	27 + 1	30 + 1
2.3	17 + 1	14 + 1	14 + 1	15 + 1	15 + 1	16 + 1	16 + 1	16 + 1
2.4	17 + 1	96 + 0	83 + 0	81 + 0	77 + 0	74 + 0	72 + 0	70 + 0
2.5	11 + 1	64 + 0	57 + 0	56 + 0	53 + 0	51 + 0	50 + 0	50 + 0
2.6	86 + 0	55 + 0	53 + 0	52 + 0	52 + 0	52 + 0	53 + 0	55 + 0
2.7	73 + 0	54 + 0	56 + 0	58 + 0	60 + 0	63 + 0	66 + 0	70 + 0
2.8	68 + 0	58 + 0	64 + 0	68 + 0	73 + 0	79 + 0	85 + 0	92 + 0
2.9	67 + 0	66 + 0	75 + 0	83 + 0	92 + 0	10 + 1	11 + 1	12 + 1
3.0	69 + 0	79 + 0	92 + 0	10 + 1	12 + 1	13 + 1	15 + 1	17 + 1
3.1	78 + 0	10 + 1	11 + 1	13 + 1	16 + 1	18 + 1	21 + 1	23 + 1
3.2	10 + 1	13 + 1	14 + 1	17 + 1	19 + 1	22 + 1	25 + 1	27 + 1
3.3	16 + 1	13 + 1	13 + 1	15 + 1	15 + 1	16 + 1	17 + 1	17 + 1
3.4	17 + 1	97 + 0	87 + 0	86 + 0	83 + 0	80 + 0	78 + 0	76 + 0
3.5	12 + 1	65 + 0	59 + 0	57 + 0	54 + 0	52 + 0	51 + 0	50 + 0
3.6	87 + 0	54 + 0	52 + 0	51 + 0	51 + 0	51 + 0	51 + 0	52 + 0
3.7	73 + 0	53 + 0	55 + 0	56 + 0	57 + 0	59 + 0	62 + 0	65 + 0
3.8	67 + 0	57 + 0	62 + 0	65 + 0	69 + 0	74 + 0	80 + 0	86 + 0
3.9	66 + 0	64 + 0	73 + 0	80 + 0	87 + 0	96 + 0	10 + 1	11 + 1
4.0	68 + 0	76 + 0	89 + 0	10 + 1	11 + 1	12 + 1	14 + 1	16 + 1
4.1	76 + 0	97 + 0	11 + 1	13 + 1	15 + 1	17 + 1	19 + 1	22 + 1
4.2	10 + 1	12 + 1	14 + 1	16 + 1	19 + 1	21 + 1	23 + 1	26 + 1
4.3	12 + 1	13 + 1	13 + 1	15 + 1	15 + 1	16 + 1	17 + 1	17 + 1
4.4	17 + 1	97 + 0	88 + 0	89 + 0	85 + 0	83 + 0	81 + 0	80 + 0

Table 6.2. (*continued*)

Δn	The effective principal quantum number n_0^*							
	1.5	2.0	2.5	3.0	3.5	4.0	4.5	5.0
	Transition $p-p$, parameter C $\kappa=2$							
0.1	19 − 0	42 − 0	59 + 0	76 + 0	91 + 0	10 + 1	12 + 1	13 + 1
0.2	47 − 0	93 + 0	12 + 1	15 + 1	18 + 1	21 + 1	24 + 1	26 + 1
0.3	83 + 0	14 + 1	19 + 1	23 + 1	27 + 1	31 + 1	35 + 1	38 + 1
0.4	12 + 1	19 + 1	25 + 1	30 + 1	34 + 1	38 + 1	43 + 1	47 + 1
0.5	15 + 1	23 + 1	29 + 1	34 + 1	38 + 1	43 + 1	47 + 1	51 + 1
0.6	18 + 1	25 + 1	30 + 1	35 + 1	39 + 1	43 + 1	47 + 1	51 + 1
0.7	19 + 1	25 + 1	29 + 1	33 + 1	37 + 1	40 + 1	43 + 1	47 + 1
0.8	18 + 1	22 + 1	26 + 1	29 + 1	31 + 1	34 + 1	37 + 1	39 + 1
0.9	15 + 1	19 + 1	21 + 1	23 + 1	25 + 1	27 + 1	29 + 1	31 + 1
1.0	12 + 1	14 + 1	16 + 1	17 + 1	18 + 1	19 + 1	21 + 1	22 + 1
1.1	84 + 0	10 + 1	11 + 1	12 + 1	12 + 1	13 + 1	14 + 1	15 + 1
1.2	57 + 0	73 + 0	80 + 0	86 + 0	91 + 0	95 + 0	10 + 1	10 + 1
1.3	43 − 0	60 + 0	67 + 0	72 + 0	76 + 0	81 + 0	84 + 0	88 + 0
1.4	44 − 0	64 + 0	72 + 0	78 + 0	83 + 0	87 + 0	92 + 0	96 + 0
1.5	59 + 0	79 + 0	88 + 0	94 + 0	10 + 1	10 + 1	11 + 1	11 + 1
1.6	79 + 0	99 + 0	10 + 1	11 + 1	11 + 1	12 + 1	12 + 1	13 + 1
1.7	99 + 0	11 + 1	12 + 1	12 + 1	13 + 1	13 + 1	14 + 1	14 + 1
1.8	11 + 1	12 + 1	12 + 1	12 + 1	13 + 1	13 + 1	14 + 1	14 + 1
1.9	10 + 1	11 + 1	11 + 1	11 + 1	12 + 1	12 + 1	12 + 1	13 + 1
2.0	96 + 0	97 + 0	10 + 1	10 + 1	10 + 1	10 + 1	10 + 1	11 + 1
2.1	70 + 0	78 + 0	80 + 0	82 + 0	83 + 0	84 + 0	86 + 0	87 + 0
2.2	52 + 0	62 + 0	63 + 0	65 + 0	65 + 0	66 + 0	67 + 0	68 + 0
2.3	40 − 0	52 + 0	55 + 0	56 + 0	57 + 0	57 + 0	58 + 0	59 + 0
2.4	39 − 0	53 + 0	56 + 0	58 + 0	59 + 0	59 + 0	60 + 0	61 + 0
2.5	50 + 0	64 + 0	66 + 0	67 + 0	68 + 0	69 + 0	70 + 0	71 + 0
2.6	67 + 0	78 + 0	80 + 0	80 + 0	81 + 0	81 + 0	82 + 0	83 + 0
2.7	84 + 0	91 + 0	91 + 0	91 + 0	91 + 0	90 + 0	91 + 0	91 + 0
2.8	96 + 0	98 + 0	97 + 0	95 + 0	94 + 0	94 + 0	94 + 0	94 + 0
2.9	98 + 0	95 + 0	94 + 0	91 + 0	90 + 0	89 + 0	89 + 0	89 + 0
3.0	90 + 0	84 + 0	83 + 0	81 + 0	80 + 0	79 + 0	78 + 0	78 + 0
3.1	66 + 0	70 + 0	69 + 0	68 + 0	67 + 0	66 + 0	65 + 0	65 + 0
3.2	58 + 0	57 + 0	56 + 0	56 + 0	55 + 0	54 + 0	53 + 0	53 + 0
3.3	39 − 0	49 − 0	49 − 0	49 − 0	48 − 0	48 − 0	47 − 0	47 − 0
3.4	38 − 0	50 + 0	50 + 0	50 + 0	49 − 0	49 − 0	48 − 0	48 − 0
3.5	47 − 0	58 + 0	58 + 0	57 + 0	56 + 0	56 + 0	55 + 0	55 + 0
3.6	63 + 0	71 + 0	69 + 0	68 + 0	66 + 0	65 + 0	64 + 0	63 + 0
3.7	79 + 0	83 + 0	80 + 0	77 + 0	75 + 0	73 + 0	72 + 0	71 + 0
3.8	92 + 0	89 + 0	85 + 0	81 + 0	79 + 0	76 + 0	75 + 0	74 + 0
3.9	95 + 0	87 + 0	84 + 0	79 + 0	77 + 0	74 + 0	72 + 0	71 + 0
4.0	87 + 0	78 + 0	76 + 0	72 + 0	69 + 0	67 + 0	65 + 0	64 + 0
4.1	64 + 0	66 + 0	63 + 0	61 + 0	59 + 0	57 + 0	55 + 0	54 + 0
4.2	49 − 0	54 + 0	53 + 0	51 + 0	49 − 0	48 − 0	46 − 0	45 − 0
4.3	39 − 0	47 − 0	47 − 0	45 − 0	44 − 0	42 − 0	41 − 0	40 − 0
4.4	37 − 0	48 − 0	47 − 0	46 − 0	45 − 0	43 − 0	42 − 0	41 − 0

Table 6.2. (*continued*)

Δn	The effective principal quantum number n_0^*							
	1.5	2.0	2.5	3.0	3.5	4.0	4.5	5.0
	Transition $p-p$, parameter φ $\kappa=2$							
0.1	18 + 1	13 + 1	98 + 0	81 + 0	71 + 0	66 + 0	62 + 0	60 + 0
0.2	10 + 1	75 + 0	64 + 0	61 + 0	61 + 0	63 + 0	65 + 0	68 + 0
0.3	79 + 0	64 + 0	63 + 0	66 + 0	70 + 0	75 + 0	79 + 0	84 + 0
0.4	70 + 0	65 + 0	70 + 0	76 + 0	83 + 0	89 + 0	97 + 0	10 + 1
0.5	68 + 0	71 + 0	79 + 0	88 + 0	98 + 0	10 + 1	11 + 1	12 + 1
0.6	70 + 0	79 + 0	91 + 0	10 + 1	11 + 1	12 + 1	14 + 1	15 + 1
0.7	75 + 0	91 + 0	10 + 1	12 + 1	13 + 1	15 + 1	17 + 1	19 + 1
0.8	84 + 0	10 + 1	12 + 1	14 + 1	16 + 1	19 + 1	22 + 1	24 + 1
0.9	96 + 0	12 + 1	14 + 1	17 + 1	20 + 1	23 + 1	27 + 1	30 + 1
1.0	11 + 1	14 + 1	17 + 1	20 + 1	24 + 1	27 + 1	32 + 1	35 + 1
1.1	14 + 1	16 + 1	18 + 1	21 + 1	24 + 1	26 + 1	29 + 1	32 + 1
1.2	16 + 1	14 + 1	14 + 1	15 + 1	16 + 1	17 + 1	17 + 1	17 + 1
1.3	12 + 1	87 + 0	84 + 0	83 + 0	82 + 0	80 + 0	79 + 0	78 + 0
1.4	72 + 0	51 + 0	49 − 0	48 − 0	47 − 0	47 − 0	47 − 0	48 − 0
1.5	49 − 0	42 − 0	43 − 0	43 − 0	45 − 0	47 − 0	49 − 0	52 + 0
1.6	46 − 0	47 − 0	49 − 0	53 + 0	57 + 0	61 + 0	66 + 0	72 + 0
1.7	51 + 0	58 + 0	63 + 0	69 + 0	77 + 0	85 + 0	94 + 0	10 + 1
1.8	61 + 0	72 + 0	80 + 0	91 + 0	10 + 1	11 + 1	13 + 1	14 + 1
1.9	73 + 0	90 + 0	10 + 1	11 + 1	13 + 1	15 + 1	17 + 1	19 + 1
2.0	90 + 0	10 + 1	12 + 1	14 + 1	16 + 1	18 + 1	21 + 1	23 + 1
2.1	11 + 1	12 + 1	14 + 1	16 + 1	18 + 1	20 + 1	22 + 1	24 + 1
2.2	13 + 1	12 + 1	12 + 1	13 + 1	14 + 1	15 + 1	16 + 1	16 + 1
2.3	12 + 1	88 + 0	88 + 0	89 + 0	90 + 0	89 + 0	89 + 0	88 + 0
2.4	78 + 0	56 + 0	55 + 0	54 + 0	53 + 0	52 + 0	52 + 0	51 + 0
2.5	58 + 0	43 − 0	43 − 0	43 − 0	43 − 0	44 − 0	45 − 0	46 − 0
2.6	44 − 0	44 − 0	45 − 0	47 − 0	49 − 0	52 + 0	56 + 0	59 + 0
2.7	48 − 0	52 + 0	55 + 0	60 + 0	65 + 0	71 + 0	77 + 0	84 + 0
2.8	56 + 0	65 + 0	70 + 0	78 + 0	87 + 0	97 + 0	10 + 1	11 + 1
2.9	68 + 0	81 + 0	89 + 0	10 + 1	11 + 1	12 + 1	14 + 1	16 + 1
3.0	83 + 0	98 + 0	11 + 1	12 + 1	14 + 1	16 + 1	18 + 1	20 + 1
3.1	10 + 1	11 + 1	12 + 1	14 + 1	15 + 1	17 + 1	19 + 1	21 + 1
3.2	12 + 1	11 + 1	12 + 1	13 + 1	13 + 1	14 + 1	15 + 1	16 + 1
3.3	12 + 1	87 + 0	88 + 0	90 + 0	91 + 0	91 + 0	92 + 0	91 + 0
3.4	79 + 0	58 + 0	57 + 0	56 + 0	55 + 0	55 + 0	54 + 0	54 + 0
3.5	51 + 0	44 − 0	43 − 0	43 − 0	43 − 0	44 − 0	44 − 0	45 − 0
3.6	44 − 0	43 − 0	44 − 0	45 − 0	47 − 0	49 − 0	52 + 0	55 + 0
3.7	47 − 0	50 + 0	52 + 0	56 + 0	60 + 0	65 + 0	70 + 0	76 + 0
3.8	54 + 0	62 + 0	66 + 0	73 + 0	81 + 0	89 + 0	98 + 0	10 + 1
3.9	66 + 0	77 + 0	84 + 0	95 + 0	10 + 1	11 + 1	13 + 1	14 + 1
4.0	80 + 0	94 + 0	10 + 1	11 + 1	13 + 1	14 + 1	16 + 1	18 + 1
4.1	10 + 1	10 + 1	11 + 1	13 + 1	14 + 1	16 + 1	17 + 1	19 + 1
4.2	12 + 1	10 + 1	11 + 1	12 + 1	13 + 1	14 + 1	14 + 1	15 + 1
4.3	11 + 1	86 + 0	87 + 0	90 + 0	92 + 0	92 + 0	92 + 0	93 + 0
4.4	80 + 0	58 + 0	58 + 0	58 + 0	57 + 0	56 + 0	55 + 0	55 + 0

Table 6.2. (*continued*)

Δn	1.5	2.0	2.5	3.0	3.5	4.0	4.5	5.0
	The effective principal quantum number n_0^*							
	Transition $p-d$, parameter C $\kappa = 1$							
0.1			24 + 1	37 + 1	47 + 1	57 + 1	65 + 1	73 + 1
0.2			58 + 1	80 + 1	98 + 1	11 + 2	12 + 2	13 + 2
0.3			10 + 2	13 + 2	15 + 2	16 + 2	18 + 2	19 + 2
0.4			15 + 2	17 + 2	19 + 2	21 + 2	23 + 2	24 + 2
0.5			20 + 2	22 + 2	23 + 2	24 + 2	26 + 2	27 + 2
0.6		22 + 2	23 + 2	24 + 2	25 + 2	26 + 2	26 + 2	27 + 2*
0.7		27 + 2	25 + 2	25 + 2	25 + 2	25 + 2*	25 + 2*	25 + 2*
0.8		30 + 2	25 + 2	24 + 2	22 + 2*	21 + 2*	21 + 2*	20 + 2*
0.9		30 + 2	23 + 2	20 + 2*	18 + 2*	17 + 2*	15 + 2*	14 + 2*
1.0		28 + 2	19 + 2	15 + 2*	13 + 2*	11 + 2*	99 + 1*	89 + 1*
1.1	34 + 2	23 + 2	14 + 2*	10 + 2*	79 + 1*	67 + 1*	61 + 1*	56 + 1*
1.2	31 + 2	16 + 2	86 + 1*	55 + 1*	39 + 1	26 + 1*	18 + 1*	15 + 1*
1.3	26 + 2	94 + 1*	40 + 1	17 + 1*	10 + 1*	91 + 0*	87 + 0	89 + 0
1.4	19 + 2	40 + 1	10 + 1*	83 + 0*	96 + 0*	11 + 1*	13 + 1*	14 + 1*
1.5	12 + 2	10 + 1	10 + 1*	16 + 1*	20 + 1*	23 + 1*	25 + 1*	27 + 1*
1.6	50 + 1*	14 + 1*	25 + 1*	34 + 1*	37 + 1	40 + 1	41 + 1	43 + 1*
1.7	99 + 0	38 + 1*	49 + 1	55 + 1	55 + 1	56 + 1	55 + 1*	55 + 1*
1.8	76 + 0*	73 + 1	73 + 1	74 + 1	68 + 1	66 + 1*	62 + 1*	61 + 1*
1.9	31 + 1	11 + 2	89 + 1	82 + 1	71 + 1*	66 + 1*	59 + 1*	55 + 1*
2.0	68 + 1	13 + 2	93 + 1	78 + 1	63 + 1*	55 + 1*	47 + 1*	42 + 1*
2.1	13 + 2	13 + 2	86 + 1*	63 + 1	48 + 1*	39 + 1*	33 + 1*	30 + 1*
2.2	16 + 2	11 + 2	64 + 1*	41 + 1	29 + 1*	22 + 1	16 + 1*	12 + 1*
2.3	16 + 2	80 + 1*	37 + 1*	19 + 1	10 + 1*	76 + 0*	62 + 0*	56 + 0*
2.4	14 + 2	44 + 1*	13 + 1*	67 + 0*	57 + 0	58 + 0*	62 + 0*	66 + 0*
2.5	10 + 2	14 + 1	65 + 0*	79 + 0*	94 + 0*	10 + 1*	11 + 1*	12 + 1*
2.6	58 + 1	79 + 0*	12 + 1*	17 + 1*	18 + 1*	20 + 1*	20 + 1*	21 + 1*
2.7	20 + 1*	19 + 1*	27 + 1	31 + 1	30 + 1	31 + 1	30 + 1*	30 + 1*
2.8	41 − 0	43 + 1	45 + 1	46 + 1	41 + 1	39 + 1*	37 + 1*	35 + 1*
2.9	13 + 1*	74 + 1	61 + 1	56 + 1	47 + 1*	43 + 1*	38 + 1*	35 + 1*
3.0	39 + 1	10 + 2	69 + 1	57 + 1*	45 + 1*	39 + 1*	33 + 1*	29 + 1*
3.1	94 + 1	10 + 2	69 + 1	49 + 1*	37 + 1*	30 + 1*	25 + 1*	22 + 1*
3.2	12 + 2	98 + 1	55 + 1*	35 + 1*	24 + 1*	19 + 1	14 + 1	10 + 1*
3.3	13 + 2	74 + 1	35 + 1*	18 + 1	10 + 1*	71 + 0*	54 + 0*	46 − 0*
3.4	12 + 2	45 + 1	15 + 1	65 + 0*	48 − 0	45 − 0*	44 − 0*	45 − 0*
3.5	10 + 2	18 + 1	57 + 0	59 + 0*	66 + 0*	73 + 0*	78 + 0*	82 + 0*
3.6	61 + 1	70 + 0	92 + 0*	12 + 1*	13 + 1*	14 + 1*	14 + 1*	14 + 1*
3.7	24 + 1*	13 + 1*	20 + 1*	24 + 1	22 + 1	22 + 1	21 + 1*	21 + 1*
3.8	47 − 0	33 + 1	36 + 1	36 + 1	31 + 1	30 + 1	27 + 1*	26 + 1*
3.9	88 + 0*	61 + 1	50 + 1	46 + 1	37 + 1*	34 + 1*	29 + 1*	27 + 1*
4.0	28 + 1	89 + 1	59 + 1	48 + 1	37 + 1*	32 + 1*	27 + 1*	23 + 1*
4.1	77 + 1	97 + 1	62 + 1	43 + 1*	32 + 1*	25 + 1*	21 + 1*	18 + 1*
4.2	10 + 2	91 + 1	51 + 1*	31 + 1*	22 + 1*	17 + 1*	12 + 1	99 + 0*
4.3	12 + 2	71 + 1	34 + 1*	18 + 1	10 + 1	68 + 0*	50 + 0*	41 − 0*
4.4	11 + 2	45 + 1	15 + 1	64 + 0*	45 − 0	39 − 0	37 − 0*	37 − 0*

Table 6.2. (*continued*)

Δn	The effective principal quantum number n_0^*							
	1.5	2.0	2.5	3.0	3.5	4.0	4.5	5.0
	Transition $p-d$, parameter φ $\kappa=1$							
0.1			13 + 1	14 + 1	15 + 1	16 + 1	17 + 1	18 + 1
0.2			16 + 1	19 + 1	21 + 1	23 + 1	25 + 1	27 + 1
0.3			20 + 1	24 + 1	27 + 1	30 + 1	34 + 1	37 + 1
0.4			24 + 1	29 + 1	34 + 1	38 + 1	43 + 1	48 + 1
0.5			28 + 1	35 + 1	41 + 1	48 + 1	55 + 1	62 + 1
0.6		24 + 1	33 + 1	41 + 1	50 + 1	59 + 1	68 + 1	78 + 1*
0.7		28 + 1	39 + 1	50 + 1	62 + 1	73 + 1*	86 + 1*	98 + 1*
0.8		32 + 1	46 + 1	60 + 1	76 + 1*	92 + 1*	10 + 2*	12 + 2*
0.9		36 + 1	56 + 1	74 + 1*	96 + 1*	11 + 2*	12 + 2*	13 + 2*
1.0		42 + 1	69 + 1	94 + 1*	11 + 2*	13 + 2*	15 + 2*	16 + 2*
1.1	28 + 1	52 + 1	87 + 1*	11 + 2*	14 + 2*	17 + 2*	20 + 2*	22 + 2*
1.2	32 + 1	66 + 1	11 + 2*	14 + 2*	14 + 2	93 + 1*	54 + 1*	35 + 1*
1.3	37 + 1	88 + 1*	12 + 2	49 + 1*	16 + 1*	70 + 0*	35 − 0*	22 − 0
1.4	43 + 1	98 + 1	15 + 1*	12 − 0*	10 − 0*	10 − 0*	22 − 0*	36 − 0*
1.5	54 + 1	10 + 1	10 − 0*	24 − 0*	52 + 0*	81 + 0*	11 + 1*	14 + 1*
1.6	83 + 1*	10 − 0*	56 + 0*	10 + 1*	15 + 1	19 + 1	24 + 1	28 + 1*
1.7	57 + 1	67 + 0*	14 + 1	20 + 1	27 + 1	33 + 1	40 + 1*	47 + 1*
1.8	10 − 0*	14 + 1	23 + 1	31 + 1	41 + 1	51 + 1*	61 + 1*	70 + 1*
1.9	85 + 0	21 + 1	33 + 1	45 + 1	59 + 1*	72 + 1*	85 + 1*	96 + 1*
2.0	15 + 1	29 + 1	46 + 1	62 + 1	81 + 1*	97 + 1*	11 + 2*	12 + 2*
2.1	21 + 1	38 + 1	61 + 1*	85 + 1*	10 + 2*	12 + 2*	14 + 2*	16 + 2*
2.2	26 + 1	50 + 1	83 + 1*	11 + 2	13 + 2*	12 + 2*	92 + 1*	63 + 1*
2.3	32 + 1	68 + 1*	10 + 2*	91 + 1	42 + 1*	20 + 1*	11 + 1*	67 + 0*
2.4	39 + 1	88 + 1*	47 + 1*	90 + 0*	21 − 0	10 − 0*	10 − 0*	10 − 0*
2.5	49 + 1	40 + 1	11 − 0*	10 − 0*	14 − 0*	33 − 0*	53 + 0*	73 + 0*
2.6	69 + 1	10 − 0*	17 − 0*	55 + 0*	87 + 0*	12 + 1*	15 + 1*	19 + 1*
2.7	96 + 1*	25 − 0*	90 + 0	14 + 1	19 + 1	24 + 1	29 + 1*	35 + 1*
2.8	28 − 0	98 + 0	17 + 1	24 + 1	31 + 1	39 + 1*	47 + 1*	56 + 1*
2.9	34 − 0*	17 + 1	27 + 1	37 + 1	48 + 1*	59 + 1*	70 + 1*	80 + 1*
3.0	11 + 1	25 + 1	39 + 1	52 + 1*	68 + 1*	82 + 1*	95 + 1*	10 + 2*
3.1	18 + 1	34 + 1	53 + 1	73 + 1*	92 + 1*	10 + 2*	12 + 2*	14 + 2*
3.2	24 + 1	45 + 1	73 + 1*	98 + 1*	11 + 2*	12 + 2	10 + 2	79 + 1*
3.3	30 + 1	61 + 1	95 + 1*	98 + 1	57 + 1*	30 + 1*	16 + 1*	10 + 1*
3.4	37 + 1	81 + 1	64 + 1	15 + 1*	48 − 0	15 − 0*	10 − 0*	10 − 0*
3.5	46 + 1	58 + 1	35 − 0	10 − 0*	10 − 0*	17 − 0*	32 − 0*	49 − 0*
3.6	65 + 1	25 − 0	10 − 0*	36 − 0*	63 + 0*	92 + 0*	12 + 1*	15 + 1*
3.7	93 + 1*	10 − 0*	69 + 0*	11 + 1	15 + 1	20 + 1	24 + 1*	29 + 1*
3.8	12 + 1	78 + 0	15 + 1	21 + 1	27 + 1	34 + 1	41 + 1*	48 + 1*
3.9	13 − 0*	15 + 1	24 + 1	33 + 1	42 + 1*	52 + 1*	62 + 1*	71 + 1*
4.0	94 + 0	23 + 1	36 + 1	48 + 1	62 + 1*	75 + 1*	87 + 1*	96 + 1*
4.1	17 + 1	32 + 1	49 + 1	67 + 1*	85 + 1*	10 + 2*	11 + 2*	12 + 2*
4.2	23 + 1	43 + 1	68 + 1*	91 + 1*	10 + 2*	11 + 2*	10 + 2	87 + 1*
4.3	29 + 1	58 + 1	90 + 1*	98 + 1	66 + 1	36 + 1*	21 + 1*	13 + 1*
4.4	36 + 1	77 + 1	71 + 1	20 + 1*	68 + 0	25 − 0	10 − 0*	10 − 0*

Table 6.2. (*continued*)

Δn	1.5	2.0	2.5	3.0	3.5	4.0	4.5	5.0
	The effective principal quantum number n_0^*							
	Transition $p-d$, parameter C $\kappa = 3$							
0.1			10 − 0	16 − 0	22 − 0	27 − 0	31 − 0	35 − 0
0.2			26 − 0	38 − 0	48 − 0	57 + 0	66 + 0	74 + 0
0.3			47 − 0	64 + 0	78 + 0	91 + 0	10 + 1	11 + 1
0.4			73 + 0	93 + 0	10 + 1	12 + 1	31 + 1	14 + 1
0.5			10 + 1	11 + 1	13 + 1	15 + 1	16 + 1	17 + 1
0.6		95 + 0	12 + 1	14 + 1	15 + 1	16 + 1	17 + 1	18 + 1
0.7		12 + 1	14 + 1	15 + 1	16 + 1	17 + 1	18 + 1	18 + 1
0.8		14 + 1	15 + 1	15 + 1	15 + 1	16 + 1	16 + 1	17 + 1
0.9		15 + 1	14 + 1	14 + 1	14 + 1	14 + 1	14 + 1	15 + 1
1.0		15 + 1	13 + 1	12 + 1	12 + 1	12 + 1	12 + 1	12 + 1
1.1	13 + 1	13 + 1	11 + 1	10 + 1	98 + 0	97 + 0	97 + 0	98 + 0
1.2	12 + 1	10 + 1	88 + 0	79 + 0	76 + 0	75 + 0	75 + 0	76 + 0
1.3	11 + 1	82 + 0	66 + 0	60 + 0	59 + 0	59 + 0	60 + 0	62 + 0
1.4	90 + 0	58 + 0	50 + 0	49 − 0	50 + 0	51 + 0	53 + 0	55 + 0
1.5	63 + 0	42 − 0	44 − 0	46 − 0	49 − 0	51 + 0	54 + 0	57 + 0
1.6	37 − 0	36 − 0	46 − 0	50 + 0	54 + 0	57 + 0	59 + 0	62 + 0
1.7	20 − 0	41 − 0	54 + 0	59 + 0	61 + 0	64 + 0	66 + 0	68 + 0
1.8	15 − 0	54 + 0	65 + 0	68 + 0	68 + 0	70 + 0	70 + 0	72 + 0
1.9	21 − 0	70 + 0	74 + 0	73 + 0	72 + 0	71 + 0	71 + 0	72 + 0
2.0	34 − 0	84 + 0	79 + 0	73 + 0	70 + 0	69 + 0	68 + 0	67 + 0
2.1	59 + 0	88 + 0	77 + 0	68 + 0	64 + 0	62 + 0	61 + 0	60 + 0
2.2	71 + 0	83 + 0	68 + 0	60 + 0	56 + 0	54 + 0	52 + 0	52 + 0
2.3	76 + 0	71 + 0	57 + 0	50 + 0	47 − 0	46 − 0	45 − 0	45 − 0
2.4	71 + 0	56 + 0	46 − 0	42 − 0	41 − 0	41 − 0	40 − 0	41 − 0
2.5	60 + 0	42 − 0	40 − 0	39 − 0	39 − 0	40 − 0	39 − 0	41 − 0
2.6	40 − 0	35 − 0	39 − 0	40 − 0	41 − 0	42 − 0	42 − 0	44 − 0
2.7	24 − 0	35 − 0	43 − 0	45 − 0	45 − 0	46 − 0	46 − 0	47 − 0
2.8	16 − 0	43 − 0	51 + 0	52 + 0	51 + 0	51 + 0	49 − 0	50 + 0
2.9	16 − 0	56 + 0	59 + 0	57 + 0	54 + 0	53 + 0	51 + 0	52 + 0
3.0	24 − 0	70 + 0	65 + 0	59 + 0	55 + 0	53 + 0	51 + 0	50 + 0
3.1	44 − 0	75 + 0	66 + 0	57 + 0	53 + 0	50 + 0	48 − 0	46 − 0
3.2	57 + 0	74 + 0	61 + 0	52 + 0	47 − 0	45 − 0	42 − 0	42 − 0
3.3	64 + 0	66 + 0	52 + 0	45 − 0	41 − 0	40 − 0	38 − 0	38 − 0
3.4	65 + 0	54 + 0	44 − 0	39 − 0	37 − 0	36 − 0	34 − 0	33 − 0
3.5	57 + 0	42 − 0	38 − 0	35 − 0	34 − 0	35 − 0	33 − 0	34 − 0
3.6	41 − 0	34 − 0	36 − 0	36 − 0	35 − 0	36 − 0	33 − 0	37 − 0
3.7	26 − 0	33 − 0	39 − 0	40 − 0	39 − 0	39 − 0	38 − 0	40 − 0
3.8	17 − 0	38 − 0	46 − 0	45 − 0	43 − 0	44 − 0	41 − 0	40 − 0
3.9	15 − 0	50 + 0	53 + 0	50 + 0	47 − 0	46 − 0	41 − 0	43 − 0
4.0	20 − 0	63 + 0	59 + 0	52 + 0	48 − 0	46 − 0	43 − 0	43 − 0
4.1	38 − 0	70 + 0	61 + 0	51 + 0	47 − 0	44 − 0	41 − 0	39 − 0
4.2	50 + 0	70 + 0	57 + 0	47 − 0	43 − 0	41 − 0	37 − 0	37 − 0
4.3	59 + 0	64 + 0	50 + 0	42 − 0	38 − 0	36 − 0	33 − 0	33 − 0
4.4	61 + 0	54 + 0	42 − 0	37 − 0	34 − 0	34 − 0	30 − 0	28 − 0

Table 6.2. (*continued*)

	The effective principal quantum number n_0^*							
Δn	1.5	2.0	2.5	3.0	3.5	4.0	4.5	5.0
	Transition $p-d$, parameter φ $\kappa = 3$							
0.1			35 + 1	23 + 1	17 + 1	13 + 1	11 + 1	92 + 0
0.2			13 + 1	90 + 0	68 + 0	57 + 0	51 + 0	47 − 0
0.3			81 + 0	58 + 0	49 − 0	45 − 0	44 − 0	44 − 0
0.4			60 + 0	48 − 0	45 − 0	45 − 0	45 − 0	46 − 0
0.5			51 + 0	45 − 0	45 − 0	46 − 0	48 − 0	49 − 0
0.6		64 + 0	47 − 0	45 − 0	47 − 0	49 − 0	51 + 0	53 + 0
0.7		56 + 0	46 − 0	47 − 0	49 − 0	52 + 0	55 + 0	58 + 0
0.8		51 + 0	46 − 0	49 − 0	52 + 0	56 + 0	60 + 0	64 + 0
0.9		49 − 0	48 − 0	52 + 0	56 + 0	61 + 0	67 + 0	73 + 0
1.0		48 − 0	51 + 0	56 + 0	62 + 0	68 + 0	76 + 0	84 + 0
1.1	62 + 0	48 − 0	55 + 0	61 + 0	68 + 0	77 + 0	86 + 0	95 + 0
1.2	56 + 0	51 + 0	60 + 0	67 + 0	75 + 0	84 + 0	94 + 0	10 + 1
1.3	53 + 0	57 + 0	66 + 0	71 + 0	79 + 0	86 + 0	92 + 0	98 + 0
1.4	51 + 0	67 + 0	69 + 0	70 + 0	73 + 0	75 + 0	77 + 0	78 + 0
1.5	51 + 0	78 + 0	63 + 0	60 + 0	60 + 0	60 + 0	59 + 0	57 + 0
1.6	56 + 0	78 + 0	53 + 0	50 + 0	49 − 0	48 − 0	47 − 0	46 − 0
1.7	76 + 0	63 + 0	45 − 0	44 − 0	43 − 0	43 − 0	42 − 0	42 − 0
1.8	10 + 1	51 + 0	42 − 0	42 − 0	42 − 0	42 − 0	43 − 0	44 − 0
1.9	89 + 0	45 − 0	42 − 0	42 − 0	43 − 0	45 − 0	46 − 0	49 − 0
2.0	67 + 0	43 − 0	43 − 0	45 − 0	47 − 0	49 − 0	53 + 0	56 + 0
2.1	60 + 0	43 − 0	46 − 0	48 − 0	52 + 0	56 + 0	61 + 0	66 + 0
2.2	53 + 0	45 − 0	49 − 0	53 + 0	58 + 0	64 + 0	70 + 0	77 + 0
2.3	49 − 0	49 − 0	54 + 0	58 + 0	65 + 0	70 + 0	77 + 0	85 + 0
2.4	48 − 0	56 + 0	58 + 0	62 + 0	67 + 0	71 + 0	75 + 0	78 + 0
2.5	48 − 0	65 + 0	59 + 0	60 + 0	62 + 0	63 + 0	64 + 0	65 + 0
2.6	51 + 0	70 + 0	54 + 0	54 + 0	54 + 0	53 + 0	54 + 0	54 + 0
2.7	63 + 0	64 + 0	48 − 0	47 − 0	47 − 0	46 − 0	45 − 0	43 − 0
2.8	86 + 0	53 + 0	43 − 0	43 − 0	43 − 0	42 − 0	42 − 0	42 − 0
2.9	90 + 0	45 − 0	42 − 0	42 − 0	42 − 0	42 − 0	43 − 0	44 − 0
3.0	78 + 0	42 − 0	42 − 0	43 − 0	44 − 0	45 − 0	48 − 0	49 − 0
3.1	60 + 0	42 − 0	44 − 0	45 − 0	47 − 0	50 + 0	53 + 0	59 + 0
3.2	52 + 0	44 − 0	47 − 0	49 − 0	53 + 0	57 + 0	62 + 0	68 + 0
3.3	48 − 0	47 − 0	50 + 0	54 + 0	59 + 0	64 + 0	69 + 0	86 + 0
3.4	46 − 0	53 + 0	55 + 0	58 + 0	63 + 0	67 + 0	69 + 0	77 + 0
3.5	47 − 0	60 + 0	56 + 0	59 + 0	62 + 0	62 + 0	59 + 0	77 + 0
3.6	49 − 0	66 + 0	54 + 0	55 + 0	55 + 0	52 + 0	48 − 0	73 + 0
3.7	59 + 0	63 + 0	49 − 0	48 − 0	48 − 0	46 − 0	48 − 0	56 + 0
3.8	78 + 0	53 + 0	44 − 0	44 − 0	44 − 0	43 − 0	44 − 0	39 − 0
3.9	88 + 0	46 − 0	42 − 0	42 − 0	42 − 0	43 − 0	40 − 0	37 − 0
4.0	71 + 0	42 − 0	42 − 0	42 − 0	43 − 0	45 − 0	44 − 0	46 − 0
4.1	60 + 0	42 − 0	43 − 0	44 − 0	46 − 0	48 − 0	50 + 0	47 − 0
4.2	52 + 0	43 − 0	45 − 0	48 − 0	51 + 0	55 + 0	59 + 0	60 + 0
4.3	48 − 0	46 − 0	49 − 0	52 + 0	57 + 0	61 + 0	67 + 0	63 + 0
4.4	46 − 0	51 + 0	53 + 0	56 + 0	61 + 0	64 + 0	71 + 0	68 + 0

Table 6.2. (*continued*)

	The effective principal quantum number n_0^*					
Δn	2.5	3.0	3.5	4.0	4.5	5.0
	Transition $d-s$, parameter C					
0.1	17 − 0	33 − 0	48 − 0	62 + 0	76 + 0	89 + 0
0.2	31 − 0	60 + 0	88 + 0	11 + 1	14 + 1	16 + 1
30.3	41 − 0	79 + 0	11 + 1	15 + 1	18 + 1	21 + 1
0.4	45 − 0	88 + 0	12 + 1	16 + 1	20 + 1	24 + 1
0.5	45 − 0	87 + 0	12 + 1	16 + 1	20 + 1	24 + 1
0.6	41 − 0	79 + 0	11 + 1	15 + 1	18 + 1	22 + 1
0.7	34 − 0	66 + 0	97 + 0	12 + 1	15 + 1	18 + 1
0.8	27 − 0	52 + 0	76 + 0	99 + 0	12 + 1	14 + 1
0.9	21 − 0	40 − 0	57 + 0	74 + 0	90 + 0	10 + 1
1.0	30 − 0*	31 − 0	13 + 1*	55 + 0	16 + 1*	76 + 0
1.1	16 − 0	26 − 0	36 − 0	44 − 0	51 + 0	58 + 0
1.2	16 − 0	26 − 0	34 − 0	41 − 0	47 − 0	52 + 0
1.3	17 − 0	28 − 0	36 − 0	43 − 0	49 − 0	54 + 0
1.4	19 − 0	31 − 0	40 − 0	48 − 0	54 + 0	60 + 0
1.5	19 − 0	33 − 0	43 − 0	51 + 0	59 + 0	66 + 0
1.6	19 − 0	33 − 0	43 − 0	52 + 0	60 + 0	67 + 0
1.7	18 − 0	31 − 0	41 − 0	50 + 0	57 + 0	65 + 0
1.8	16 − 0	28 − 0	37 − 0	44 − 0	51 + 0	58 + 0
1.9	14 − 0	24 − 0	32 − 0	38 − 0	44 − 0	49 − 0
2.0	25 − 0*	21 − 0	33 − 0*	33 − 0	44 − 0*	41 − 0
2.1	13 − 0	20 − 0	25 − 0	29 − 0	33 − 0	36 − 0
2.2	13 − 0	20 − 0	25 − 0	29 − 0	31 − 0	34 − 0
2.3	14 − 0	22 − 0	26 − 0	30 − 0	33 − 0	35 − 0
2.4	15 − 0	23 − 0	29 − 0	33 − 0	36 − 0	38 − 0
2.5	16 − 0	24 − 0	31 − 0	35 − 0	39 − 0	42 − 0
2.6	16 − 0	25 − 0	31 − 0	36 − 0	40 − 0	43 − 0
2.7	15 − 0	24 − 0	30 − 0	35 − 0	39 − 0	42 − 0
2.8	14 − 0	22 − 0	28 − 0	32 − 0	36 − 0	39 − 0
2.9	13 − 0	20 − 0	25 − 0	29 − 0	32 − 0	35 − 0
3.0	14 − 0*	18 − 0	23 − 0	26 − 0	29 − 0	30 − 0
3.1	12 − 0	18 − 0	21 − 0	24 − 0	26 − 0	27 − 0
3.2	12 − 0	18 − 0	21 − 0	24 − 0	25 − 0	27 − 0
3.3	13 − 0	19 − 0	23 − 0	25 − 0	27 − 0	28 − 0
3.4	14 − 0	20 − 0	24 − 0	27 − 0	29 − 0	30 − 0
3.5	14 − 0	21 − 0	26 − 0	29 − 0	31 − 0	32 − 0
3.6	14 − 0	21 − 0	26 − 0	29 − 0	32 − 0	34 − 0
3.7	14 − 0	21 − 0	26 − 0	29 − 0	31 − 0	33 − 0
3.8	13 − 0	20 − 0	24 − 0	27 − 0	29 − 0	31 − 0
3.9	12 − 0	18 − 0	22 − 0	24 − 0	26 − 0	28 − 0
4.0	13 − 0*	17 − 0	21 − 0	22 − 0	25 − 0	25 − 0
4.1	11 − 0	16 − 0	19 − 0	21 − 0	22 − 0	23 − 0
4.2	12 − 0	17 − 0	19 − 0	21 − 0	22 − 0	23 − 0
4.3	12 − 0	18 − 0	20 − 0	22 − 0	23 − 0	24 − 0
4.4	13 − 0	19 − 0	22 − 0	24 − 0	25 − 0	26 − 0

Table 6.2. (*continued*)

Δn	The effective principal quantum number n_0^*					
	2.5	3.0	3.5	4.0	4.5	5.0
	Transition $d-s$, parameter φ					
0.1	76 + 0	68 + 0	63 + 0	61 + 0	59 + 0	59 + 0
0.2	63 + 0	62 + 0	63 + 0	66 + 0	68 + 0	71 + 0
0.3	70 + 0	72 + 0	75 + 0	79 + 0	84 + 0	88 + 0
0.4	80 + 0	84 + 0	89 + 0	95 + 0	10 + 1	10 + 1
0.5	92 + 0	98 + 0	10 + 1	11 + 1	12 + 1	13 + 1
0.6	10 + 1	11 + 1	12 + 1	13 + 1	14 + 1	16 + 1
0.7	11 + 1	13 + 1	14 + 1	16 + 1	17 + 1	19 + 1
0.8	12 + 1	14 + 1	16 + 1	18 + 1	20 + 1	23 + 1
0.9	13 + 1	14 + 1	17 + 1	19 + 1	22 + 1	25 + 1
1.0	10 − 0*	12 + 1	93 + 1*	17 + 1	87 + 1*	22 + 1
1.1	81 + 0	90 + 0	10 + 1	11 + 1	13 + 1	15 + 1
1.2	61 + 0	59 + 0	63 + 0	69 + 0	75 + 0	81 + 0
1.3	52 + 0	47 − 0	47 − 0	48 − 0	49 − 0	51 + 0
1.4	53 + 0	47 − 0	47 − 0	48 − 0	48 − 0	49 − 0
1.5	59 + 0	56 + 0	57 + 0	59 + 0	61 + 0	63 + 0
1.6	68 + 0	69 + 0	73 + 0	77 + 0	82 + 0	86 + 0
1.7	79 + 0	84 + 0	91 + 0	99 + 0	10 + 1	11 + 1
1.8	92 + 0	96 + 0	10 + 1	12 + 1	13 + 1	14 + 1
1.9	10 + 1	10 + 1	11 + 1	13 + 1	14 + 1	16 + 1
2.0	58 + 1*	92 + 0	22 + 1*	12 + 1	24 + 1*	15 + 1
2.1	71 + 0	73 + 0	82 + 0	93 + 0	10 + 1	11 + 1
2.2	59 + 0	55 + 0	59 + 0	64 + 0	68 + 0	73 + 0
2.3	52 + 0	46 − 0	47 − 0	48 − 0	49 − 0	51 + 0
2.4	51 + 0	46 − 0	46 − 0	46 − 0	47 − 0	47 − 0
2.5	54 + 0	52 + 0	53 + 0	54 + 0	55 + 0	57 + 0
2.6	62 + 0	62 + 0	65 − 0	68 + 0	71 + 0	75 + 0
2.7	72 + 0	74 + 0	80 + 0	86 + 0	92 + 0	98 + 0
2.8	85 + 0	85 + 0	95 + 0	10 + 1	11 + 1	12 + 1
2.9	97 + 0	89 + 0	10 + 1	11 + 1	12 + 1	13 + 1
3.0	21 + 1*	82 + 0	12 + 1	10 + 1	14 + 1	13 + 1
3.1	68 + 0	68 + 0	75 + 0	84 + 0	92 + 0	10 + 1
3.2	58 + 0	53 + 0	57 + 0	61 + 0	65 + 0	69 + 0
3.3	51 + 0	46 − 0	46 − 0	48 − 0	49 − 0	50 + 0
3.4	50 + 0	45 − 0	45 − 0	46 − 0	46 − 0	47 − 0
3.5	53 + 0	50 + 0	51 + 0	52 + 0	53 + 0	54 + 0
3.6	60 + 0	60 + 0	62 + 0	65 + 0	67 + 0	70 + 0
3.7	70 + 0	71 + 0	76 + 0	81 + 0	86 + 0	91 + 0
3.8	82 + 0	80 + 0	89 + 0	96 + 0	10 + 1	11 + 1
3.9	96 + 0	84 + 0	96 + 0	10 + 1	11 + 1	12 + 1
4.0	25 + 1*	78 + 0	11 + 1	98 + 0	13 + 1	11 + 1
4.1	66 + 0	65 + 0	72 + 0	79 + 0	87 + 0	95 + 0
4.2	57 + 0	52 + 0	56 + 0	59 + 0	63 + 0	67 + 0
4.3	51 + 0′	45 − 0	46 − 0	47 − 0	49 − 0	50 + 0
4.4	49 − 0	45 − 0	45 − 0	45 − 0	46 − 0	46 − 0

Table 6.2. (*continued*)

Δn	The effective principal quantum number n_0^*					
	2.5	3.0	3.5	4.0	4.5	5.0
	Transition $d-p$, parameter C $\kappa = 1$					
0.1	16 + 1	27 + 1	37 + 1	46 + 1	55 + 1	62 + 1
0.2	25 + 1	44 + 1	62 + 1	76 + 1	90 + 1	10 + 2
0.3	38 + 1	54 + 1	75 + 1	94 + 1	11 + 2	12 + 2
0.4	31 + 1	57 + 1	79 + 1	99 + 1	11 + 2	13 + 2
0.5	27 + 1	53 + 1	74 + 1	93 + 1	11 + 2	12 + 2*
0.6	22 + 1	44 + 1	62 + 1	78 + 1*	93 + 1*	10 + 2*
0.7	15 + 1	33 + 1	46 + 1*	59 + 1*	69 + 1*	79 + 1*
0.8	87 + 0	21 + 1	29 + 1*	38 + 1*	46 + 1*	53 + 1*
0.9	44 − 0	10 + 1	14 + 1	19 + 1	22 + 1	26 + 1*
1.0	30 − 0	45 − 0*	59 + 0	74 + 0	85 + 0*	97 + 0*
1.1	27 − 0*	34 − 0*	43 − 0	49 − 0	55 + 0	60 + 0
1.2	37 − 0*	47 − 0*	60 + 0	66 + 0*	73 + 0*	78 − 0*
1.3	50 + 0*	73 + 0*	92 + 0	10 + 1*	11 + 1*	12 + 1*
1.4	59 + 0	98 + 0	12 + 1	14 + 1	16 + 1	17 + 1*
1.5	61 + 0	11 + 1	14 + 1	17 + 1*	19 + 1*	21 + 1*
1.6	56 + 0	11 + 1	14 + 1	17 + 1*	20 + 1*	22 + 1*
1.7	43 − 0	99 + 0	12 + 1	15 + 1*	18 + 1*	20 + 1*
1.8	31 − 0	72 + 0	92 + 0	11 + 1	13 + 1	15 + 1*
1.9	24 − 0	43 − 0	53 + 0	68 + 0	77 + 0	88 + 0*
2.0	24 − 0	26 − 0	32 − 0	37 − 0*	41 − 0*	45 − 0*
2.1	21 − 0*	23 − 0*	28 − 0*	30 − 0*	32 − 0*	34 − 0*
2.2	27 − 0*	31 − 0*	37 − 0*	39 − 0*	42 − 0*	43 − 0*
2.3	34 − 0*	45 − 0*	54 + 0*	59 + 0*	63 + 0*	65 + 0*
2.4	39 − 0*	60 + 0	73 + 0	81 + 0	87 + 0	92 + 0*
2.5	39 − 0	70 + 0	86 + 0	98 + 0	10 + 1	11 + 1*
2.6	35 − 0	72 + 0	88 + 0	10 + 1	11 + 1*	12 + 1*
2.7	28 − 0	63 + 0	77 + 0	94 + 0	10 + 1*	11 + 1*
2.8	22 − 0	48 − 0	57 + 0	73 + 0	79 + 0	89 + 0*
2.9	20 − 0	31 − 0	36 − 0	45 − 0	48 − 0	54 + 0*
3.0	23 − 0	20 − 0	24 − 0	27 − 0	29 − 0	31 − 0*
3.1	19 − 0	19 − 0*	22 − 0*	23 − 0	24 − 0*	25 − 0*
3.2	24 − 0	25 − 0*	29 − 0*	30 − 0*	31 − 0*	31 − 0*
3.3	29 − 0	36 − 0*	42 − 0*	44 − 0*	46 − 0*	47 − 0*
3.4	32 − 0	47 − 0	56 + 0*	60 + 0*	63 + 0	65 + 0*
3.5	31 − 0	55 + 0	65 + 0	73 + 0*	77 + 0	81 + 0*
3.6	28 − 0	57 + 0	67 + 0	78 + 0	82 + 0	87 + 0*
3.7	23 − 0	50 + 0	59 + 0	71 + 0	76 + 0	82 + 0*
3.8	19 − 0*	38 − 0	44 − 0	55 + 0	59 + 0	65 + 0*
3.9	19 − 0*	26 − 0	29 − 0	35 − 0	37 − 0	40 − 0*
4.0	24 − 0*	18 − 0	21 − 0	22 − 0*	23 − 0	24 − 0*
4.1	18 − 0*	17 − 0*	20 − 0*	20 − 0*	21 − 0	21 − 0*
4.2	23 − 0*	22 − 0*	26 − 0*	26 − 0*	26 − 0*	26 − 0*
4.3	27 − 0	31 − 0*	36 − 0*	37 − 0*	38 − 0*	38 − 0*
4.4	29 − 0	41 − 0	48 − 0	51 + 0*	52 + 0	53 + 0*

Table 6.2. (*continued*)

Δn	The effective principal quantum number n_0^*					
	2.5	3.0	3.5	4.0	4.5	5.0
	Transition $d-p$, parameter φ $\kappa = 1$					
0.1	13 + 1	14 + 1	15 + 1	16 + 1	17 + 1	19 + 1
0.2	18 + 1	20 + 1	23 + 1	25 + 1	27 + 1	29 + 1
0.3	24 + 1	27 + 1	31 + 1	34 + 1	37 + 1	41 + 1
0.4	31 + 1	36 + 1	41 + 1	46 + 1	51 + 1	56 + 1
0.5	40 + 1	46 + 1	53 + 1	61 + 1	68 + 1	75 + 1*
0.6	50 + 1	60 + 1	70 + 1	80 + 1*	90 + 1*	10 + 2*
0.7	56 + 1	76 + 1	90 + 1*	10 + 2*	11 + 2*	12 + 2*
0.8	43 + 1	85 + 1	10 + 2*	12 + 2*	14 + 2*	15 + 2*
0.9	14 + 1	55 + 1	64 + 1	88 + 1	10 + 2	11 + 2*
1.0	13 − 0	12 + 1*	12 + 1	18 + 1	19 + 1*	23 + 1*
1.1	10 − 0*	10 − 0*	10 − 0	11 − 0	15 − 0	21 − 0
1.2	15 − 0*	10 − 0*	11 − 0	12 − 0*	15 − 0*	17 − 0*
1.3	61 + 0*	61 + 0*	68 + 0	74 + 0*	80 + 0*	87 + 0*
1.4	11 + 1	13 + 1	15 + 1	16 + 1	18 + 1	20 + 1*
1.5	18 + 1	22 + 1	26 + 1	29 + 1	32 + 1*	36 + 1*
1.6	22 + 1	33 + 1	39 + 1	45 + 1*	51 + 1*	56 + 1*
1.7	20 + 1	44 + 1	52 + 1	62 + 1*	70 + 1*	79 + 1*
1.8	11 + 1	45 + 1	53 + 1	70 + 1	80 + 1	94 + 1*
1.9	25 − 0	26 + 1	27 + 1	42 + 1	46 + 1	56 + 1*
2.0	10 − 0	67 + 0	61 + 0	10 + 1*	11 + 1*	14 + 1*
2.1	10 − 0*	10 − 0*	10 − 0*	10 − 0*	10 − 0*	13 − 0*
2.2	13 − 0*	10 − 0*	10 − 0*	10 − 0*	10 − 0*	10 − 0*
2.3	48 − 0*	46 − 0*	51 + 0*	54 + 0*	59 + 0*	64 + 0*
2.4	92 + 0*	10 + 1	12 + 1	13 + 1	14 + 1	15 + 1*
2.5	13 + 1	18 + 1	21 + 1	24 + 1	26 + 1	29 + 1*
2.6	15 + 1	27 + 1	32 + 1	37 + 1	41 + 1*	46 + 1*
2.7	12 + 1	35 + 1	41 + 1	50 + 1	57 + 1*	65 + 1*
2.8	52 + 0	34 + 1	38 + 1	54 + 1	61 + 1	72 + 1*
2.9	10 − 0	19 + 1	19 + 1	31 + 1	33 + 1	41 + 1*
3.0	10 − 0	51 + 0	41 − 0	83 + 0	83 + 0	11 + 1*
3.1	10 − 0	10 − 0*	10 − 0*	10 − 0	10 − 0*	10 − 0*
3.2	14 − 0	10 − 0*	10 − 0*	10 − 0*	10 − 0*	10 − 0*
3.3	45 − 0	40 − 0*	45 − 0*	48 − 0*	52 + 0*	56 + 0*
3.4	83 + 0	96 + 0	10 + 1*	11 + 1*	13 + 1*	14 + 1*
3.5	11 + 1	16 + 1	19 + 1	21 + 1*	23 + 1	26 + 1*
3.6	12 + 1	24 + 1	28 + 1	33 + 1	37 + 1	41 + 1*
3.7	87 + 0	31 + 1	36 + 1	45 + 1	50 + 1	58 + 1*
3.8	30 − 0*	29 + 1	32 + 1	46 + 1	51 + 1	62 + 1*
3.9	10 − 0*	16 + 1	15 + 1	26 + 1	27 + 1	35 + 1*
4.0	10 − 0*	43 − 0	31 − 0	72 + 0	69 + 0	96 + 0*
4.1	10 − 0*	10 − 0*	10 − 0*	10 − 0*	10 − 0	10 − 0*
4.2	14 − 0*	10 − 0*	10 − 0*	10 − 0*	10 − 0*	10 − 0*
4.3	44 − 0	37 − 0	42 − 0*	44 − 0*	48 − 0*	51 + 0*
4.4	78 + 0	90 + 0	10 + 1	11 + 1*	12 + 1	13 + 1*

Table 6.2. (*continued*)

Δn	The effective principal quantum number n_0^*					
	2.5	3.0	3.5	4.0	4.5	5.0
	Transition $d-p$, parameter C $\kappa = 3$					
0.1	74 − 1	13 − 0	18 − 0	23 − 0	27 − 0	32 − 0
0.2	13 − 0	24 − 0	34 − 0	43 − 0	51 + 0	59 + 0
0.3	18 − 0	33 − 0	46 − 0	59 + 0	70 + 0	82 + 0
0.4	22 − 0	39 − 0	55 + 0	70 + 0	83 + 0	96 + 0
0.5	23 − 0	42 − 0	59 + 0	74 + 0	88 + 0	10 + 1
0.6	23 − 0	42 − 0	58 + 0	72 + 0	86 + 0	99 + 0
0.7	22 − 0	39 − 0	53 + 0	66 + 0	78 + 0	90 + 0
0.8	20 − 0	34 − 0	46 − 0	57 + 0	68 + 0	77 + 0
0.9	17 − 0	29 − 0	39 − 0	48 − 0	56 + 0	64 + 0
1.0	15 − 0	25 − 0	33 − 0	40 − 0	46 − 0	52 + 0
1.1	13 − 0	22 − 0	28 − 0	34 − 0	39 − 0	44 − 0
1.2	13 − 0	20 − 0	26 − 0	31 − 0	35 − 0	39 − 0
1.3	13 − 0	20 − 0	26 − 0	30 − 0	34 − 0	38 − 0
1.4	13 − 0	21 − 0	27 − 0	31 − 0	35 − 0	39 − 0
1.5	14 − 0	22 − 0	28 − 0	33 − 0	37 − 0	40 − 0
1.6	15 − 0	23 − 0	29 − 0	34 − 0	37 − 0	41 − 0
1.7	15 − 0	23 − 0	29 − 0	33 − 0	37 − 0	41 − 0
1.8	14 − 0	22 − 0	27 − 0	32 − 0	36 − 0	39 − 0
1.9	14 − 0	20 − 0	25 − 0	29 − 0	33 − 0	36 − 0
2.0	13 − 0	19 − 0	23 − 0	27 − 0	30 − 0	32 − 0
2.1	12 − 0	18 − 0	22 − 0	25 − 0	27 − 0	29 − 0
2.2	12 − 0	17 − 0	21 − 0	24 − 0	26 − 0	28 − 0
2.3	12 − 0	17 − 0	21 − 0	24 − 0	26 − 0	28 − 0
2.4	12 − 0	18 − 0	21 − 0	24 − 0	26 − 0	28 − 0
2.5	12 − 0	19 − 0	22 − 0	25 − 0	27 − 0	29 − 0
2.6	13 − 0	19 − 0	23 − 0	26 − 0	28 − 0	29 − 0
2.7	13 − 0	19 − 0	23 − 0	26 − 0	28 − 0	29 − 0
2.8	13 − 0	19 − 0	22 − 0	25 − 0	27 − 0	29 − 0
2.9	13 − 0	18 − 0	21 − 0	24 − 0	26 − 0	27 − 0
3.0	12 − 0	17 − 0	20 − 0	23 − 0	24 − 0	24 − 0
3.1	11 − 0	16 − 0	19 − 0	21 − 0	22 − 0	23 − 0
3.2	11 − 0	16 − 0	19 − 0	21 − 0	21 − 0	23 − 0
3.3	11 − 0	16 − 0	18 − 0	21 − 0	21 − 0	23 − 0
3.4	12 − 0	16 − 0	19 − 0	22 − 0	22 − 0	23 − 0
3.5	12 − 0	17 − 0	20 − 0	22 − 0	23 − 0	24 − 0
3.6	12 − 0	17 − 0	20 − 0	23 − 0	23 − 0	24 − 0
3.7	12 − 0	17 − 0	20 − 0	23 − 0	24 − 0	25 − 0
3.8	12 − 0	17 − 0	20 − 0	23 − 0	24 − 0	23 − 0
3.9	12 − 0	16 − 0	19 − 0	21 − 0	22 − 0	24 − 0
4.0	12 − 0	16 − 0	18 − 0	20 − 0	22 − 0	22 − 0
4.1	11 − 0	15 − 0	18 − 0	20 − 0	21 − 0	20 − 0
4.2	11 − 0	15 − 0	17 − 0	20 − 0	20 − 0	19 − 0
4.3	11 − 0	15 − 0	17 − 0	20 − 0	19 − 0	18 − 0
4.4	11 − 0	16 − 0	17 − 0	20 − 0	20 − 0	18 − 0

Table 6.2. (*continued*)

Δn	The effective principal quantum number n_0^*					
	2.5	3.0	3.5	4.0	4.5	5.0
	Transition $d-p$, parameter φ $\kappa = 3$					
0.1	33 + 1	22 + 1	16 + 1	12 + 1	10 + 1	89 + 0
0.2	11 + 1	81 + 0	63 + 0	54 + 0	49 − 0	46 − 0
0.3	67 + 0	52 + 0	46 − 0	45 − 0	44 − 0	45 − 0
0.4	52 + 0	47 − 0	46 − 0	46 − 0	47 − 0	47 − 0
0.5	51 + 0	49 − 0	48 − 0	49 − 0	49 − 0	51 + 0
0.6	57 + 0	53 + 0	52 + 0	52 + 0	53 + 0	55 + 0
0.7	66 + 0	58 + 0	55 + 0	55 + 0	58 + 0	61 + 0
0.8	76 + 0	60 + 0	58 + 0	60 + 0	64 + 0	69 + 0
0.9	88 + 0	62 + 0	61 + 0	65 + 0	71 + 0	77 + 0
1.0	77 + 0	61 + 0	63 + 0	69 + 0	76 + 0	83 + 0
1.1	68 + 0	59 + 0	62 + 0	68 + 0	75 + 0	82 + 0
1.2	58 + 0	56 + 0	58 + 0	62 + 0	67 + 0	72 + 0
1.3	53 + 0	52 + 0	52 + 0	54 + 0	57 + 0	59 + 0
1.4	53 + 0	50 + 0	47 − 0	47 − 0	48 − 0	48 − 0
1.5	57 + 0	48 − 0	44 − 0	43 − 0	43 − 0	43 − 0
1.6	62 + 0	47 − 0	42 − 0	42 − 0	42 − 0	43 − 0
1.7	66 + 0	47 − 0	43 − 0	43 − 0	44 − 0	44 − 0
1.8	68 + 0	47 − 0	45 − 0	47 − 0	48 − 0	51 + 0
1.9	66 + 0	48 − 0	48 − 0	51 + 0	54 + 0	57 + 0
2.0	61 + 0	49 − 0	51 + 0	55 + 0	59 + 0	61 + 0
2.1	55 + 0	51 + 0	53 + 0	57 + 0	61 + 0	63 + 0
2.2	52 + 0	52 + 0	53 + 0	56 + 0	59 + 0	64 + 0
2.3	52 + 0	51 + 0	50 + 0	52 + 0	55 + 0	60 + 0
2.4	54 + 0	50 + 0	47 − 0	48 − 0	49 − 0	51 + 0
2.5	59 + 0	48 − 0	43 − 0	43 − 0	44 − 0	45 − 0
2.6	63 + 0	46 − 0	42 − 0	42 − 0	43 − 0	42 − 0
2.7	66 + 0	45 − 0	42 − 0	43 − 0	42 − 0	43 − 0
2.8	65 + 0	44 − 0	44 − 0	45 − 0	46 − 0	48 − 0
2.9	62 + 0	45 − 0	46 − 0	47 − 0	51 + 0	49 − 0
3.0	57 + 0	47 − 0	49 − 0	51 + 0	53 + 0	52 + 0
3.1	52 + 0	49 − 0	51 + 0	53 + 0	54 + 0	57 + 0
3.2	50 + 0	51 + 0	51 + 0	53 + 0	54 + 0	62 + 0
3.3	51 + 0	50 + 0	49 − 0	50 + 0	53 + 0	71 + 0
3.4	55 + 0	49 − 0	46 − 0	48 − 0	48 − 0	54 + 0
3.5	68 + 0	47 − 0	44 − 0	44 − 0	41 − 0	53 + 0
3.6	64 + 0	45 − 0	42 − 0	41 − 0	38 − 0	52 + 0
3.7	65 + 0	44 − 0	42 − 0	42 − 0	45 − 0	50 + 0
3.8	64 + 0	43 − 0	43 − 0	44 − 0	45 − 0	44 − 0
3.9	68 + 0	44 − 0	45 − 0	47 − 0	48 − 0	53 + 0
4.0	55 + 0	46 − 0	48 − 0	50 + 0	54 + 0	60 + 0
4.1	50 + 0	48 − 0	50 + 0	54 + 0	59 + 0	54 + 0
4.2	49 − 0	50 + 0	50 + 0	55 + 0	58 + 0	39 − 0
4.3	51 + 0	50 + 0	50 + 0	53 + 0	53 + 0	41 − 0
4.4	55 + 0	49 − 0	47 − 0	47 − 0	52 + 0	41 − 0

Table 6.3. Transitions with no change of spin $\Delta S = 0$; excitations rate coefficients $\langle v\sigma \rangle$ in the Born approximation. Parameters A and x

	The effective principal quantum number n_0^*									
Δn	0.5	1.0	1.5	2.0	2.5	3.0	3.5	4.0	4.5	5.0
	Transition $s - s$, parameter A[a]									
0.6	10 + 2	10 + 2	11 + 2	13 + 2	13 + 2	14 + 2	14 + 2	14 + 2	14 + 2	14 + 2
0.7	12 + 2	11 + 2	11 + 2	12 + 2	11 + 2	11 + 2	11 + 2	10 + 2	10 + 2	10 + 2
0.8	12 + 2	10 + 2	10 + 2	94 + 1	88 + 1	81 + 1	75 + 1	70 + 1	65 + 1	61 + 1
0.9	10 + 2	80 + 1	72 + 1	62 + 1	54 + 1	47 + 1	42 + 1	37 + 1	34 + 1	31 + 1
1.0	80 + 1	51 + 1	41 + 1	32 + 1	26 + 1	22 + 1	18 + 1	61 + 1	14 + 1	12 + 1
1.1	29 + 1	23 + 1	17 + 1	12 + 1	10 + 1	83 + 0	70 + 0	61 + 0	54 + 0	49 − 0
1.2	67 + 0	65 + 0	54 + 0	50 + 0	49 − 0	50 + 0	51 + 0	53 + 0	54 + 0	55 + 0
1.3	21 − 0	32 − 0	54 + 0	80 + 0	10 + 1	12 + 1	14 + 1	15 + 1	16 + 0	17 + 1
1.4	11 + 1	11 + 1	17 + 1	23 + 1	28 + 1	32 + 1	35 + 1	36 + 1	36 + 1	36 + 1
1.5	35 + 1	29 + 1	37 + 1	44 + 1	48 + 1	49 + 1	49 + 1	47 + 1	45 + 1	43 + 1
1.6	64 + 1	52 + 1	57 + 1	60 + 1	59 + 1	55 + 1	51 + 1	47 + 1	44 + 1	40 + 1
1.7	91 + 1	69 + 1	68 + 1	64 + 1	57 + 1	51 + 1	45 + 1	40 + 1	36 + 1	33 + 1
1.8	10 + 2	74 + 1	66 + 1	56 + 1	47 + 1	39 + 1	34 + 1	29 + 1	25 + 1	23 + 1
1.9	10 + 2	65 + 1	53 + 1	41 + 1	32 + 1	25 + 1	21 + 1	17 + 1	15 + 1	13 + 1
2.0	85 + 1	45 + 1	34 + 1	23 + 1	17 + 1	13 + 1	10 + 1	88 + 0	74 + 0	63 + 0
2.1	30 + 1	22 + 1	15 + 1	10 + 1	78 + 0	60 + 0	48 − 0	40 − 0	34 − 0	29 − 0
2.2	80 + 0	74 + 0	56 + 0	46 − 0	41 − 0	38 − 0	35 − 0	33 − 0	32 − 0	30 − 0
2.3	19 − 0	30 − 0	45 − 0	60 + 0	70 + 0	77 + 0	80 + 0	82 + 0	82 + 0	81 + 0
2.4	94 + 0	84 + 0	12 + 1	16 + 1	19 + 1	20 + 1	20 + 1	20 + 1	19 + 1	18 + 1
2.5	30 + 1	23 + 1	29 + 1	33 + 1	35 + 1	34 + 1	31 + 1	29 + 1	26 + 1	24 + 1
2.6	59 + 1	44 + 1	47 + 1	47 + 1	44 + 1	39 + 1	35 + 1	30 + 1	27 + 1	24 + 1
2.7	86 + 1	61 + 1	58 + 1	52 + 1	44 + 1	37 + 1	31 + 1	27 + 1	23 + 1	20 + 1
2.8	10 + 2	67 + 1	59 + 1	47 + 1	37 + 1	29 + 1	24 + 1	20 + 1	17 + 1	14 + 1
2.9	10 + 2	61 + 1	48 + 1	35 + 1	26 + 1	20 + 1	15 + 1	12 + 1	10 + 1	89 + 0
3.0	88 + 1	43 + 1	32 + 1	20 + 1	15 + 1	10 + 1	84 + 0	66 + 0	54 + 0	45 − 0
3.1	30 + 1	22 + 1	14 + 1	97 + 0	69 + 0	51 + 0	40 − 0	32 − 0	26 − 0	22 + 0
3.2	84 + 0	76 + 0	56 + 0	44 − 0	38 − 0	33 − 0	29 − 0	26 − 0	24 − 0	22 − 0
3.3	19 − 0	30 − 0	42 − 0	53 + 0	60 + 0	62 + 0	62 + 0	60 + 0	58 + 0	56 + 0
3.4	87 + 0	76 + 0	11 + 1	14 + 1	16 + 1	16 + 1	16 + 1	15 + 1	14 + 1	13 + 1
3.5	28 + 1	21 + 1	26 + 1	29 + 1	30 + 1	28 + 1	25 + 1	22 + 1	20 + 1	18 + 1
3.6	57 + 1	41 + 1	43 + 1	42 + 1	38 + 1	33 + 1	28 + 1	24 + 1	21 + 1	18 + 1
3.7	84 + 1	58 + 1	55 + 1	47 + 1	39 + 1	31 + 1	26 + 1	21 + 1	18 + 1	15 + 1
3.8	10 + 2	65 + 1	55 + 1	43 + 1	33 + 1	25 + 1	20 + 1	16 + 1	13 + 1	11 + 1
3.9	10 + 2	59 + 1	46 + 1	32 + 1	23 + 1	17 + 1	13 + 1	10 + 1	86 + 0	71 + 0
4.0	90 + 1	42 + 1	31 + 1	19 + 1	13 + 1	97 + 0	73 + 0	56 + 0	45 − 0	37 − 0
4.1	30 + 1	22 + 1	14 + 1	93 + 0	64 + 0	47 − 0	35 − 0	28 − 0	22 − 0	19 − 0
4.2	85 + 0	77 + 0	56 + 0	43 − 0	36 − 0	30 − 0	26 − 0	23 − 0	20 − 0	19 − 0
4.3	19 − 0	30 − 0	41 − 0	51 + 0	55 + 0	55 + 0	53 + 0	50 + 0	47 − 0	44 − 0
4.4	84 + 0	72 + 0	10 + 1	13 + 1	14 + 1	14 + 1	13 + 1	12 + 1	11 + 1	10 + 1

[a] This table can be applied to transitions between the levels of two different electronic configurations *ns–n′s*

Table 6.3. (*continued*)

	The effective principal quantum number n_0^*									
Δn	0.5	1.0	1.5	2.0	2.5	3.0	3.5	4.0	4.5	5.0
	Transition $s-s$, parameter χ									
0.6	13 + 1	15 + 1	17 + 1	17 + 1	17 + 1	16 + 1	15 + 1	14 + 1	13 + 1	12 + 1
0.7	14 + 1	16 + 1	17 + 1	17 + 1	16 + 1	15 + 1	13 + 1	12 + 1	11 + 1	10 + 1
0.8	15 + 1	17 + 1	17 + 1	16 + 1	14 + 1	13 + 1	11 + 1	10 + 1	88 + 0	78 + 0
0.9	15 + 1	17 + 1	16 + 1	14 + 1	12 + 1	10 + 1	87 + 0	74 + 0	63 + 0	54 + 0
1.0	16 + 1	16 + 1	14 + 1	11 + 1	90 + 0	72 + 0	58 + 0	47 − 0	39 − 0	32 − 0
1.1	15 + 1	13 + 1	99 + 0	74 + 0	57 + 0	44 − 0	35 − 0	29 − 0	24 − 0	20 − 0
1.2	88 + 0	68 + 0	53 + 0	48 − 0	47 − 0	48 − 0	49 − 0	50 + 0	50 + 0	50 + 0
1.3	26 − 0	39 − 0	64 + 0	98 + 0	13 + 1	15 + 1	17 + 1	18 + 1	19 + 1	19 + 1
1.4	72 + 0	91 + 0	14 + 1	20 + 1	25 + 1	28 + 1	30 + 1	30 + 1	29 + 1	28 + 1
1.5	10 + 1	14 + 1	20 + 1	25 + 1	28 + 1	29 + 1	28 + 1	27 + 1	25 + 1	23 + 1
1.6	13 + 1	17 + 1	22 + 1	25 + 1	26 + 1	25 + 1	23 + 1	21 + 1	19 + 1	17 + 1
1.7	14 + 1	18 + 1	22 + 1	23 + 1	22 + 1	20 + 1	18 + 1	16 + 1	14 + 1	12 + 1
1.8	15 + 1	19 + 1	21 + 1	20 + 1	18 + 1	16 + 1	13 + 1	11 + 1	10 + 1	90 + 0
1.9	16 + 1	19 + 1	19 + 1	17 + 1	14 + 1	12 + 1	10 + 1	83 + 0	70 + 0	60 + 0
2.0	16 + 1	18 + 1	16 + 1	13 + 1	10 + 1	82 + 0	66 + 0	53 + 0	43 − 0	36 − 0
2.1	16 + 1	14 + 1	11 + 1	86 + 0	66 + 0	52 + 0	42 − 0	34 − 0	28 − 0	23 − 0
2.2	99 + 0	81 + 0	63 + 0	56 + 0	54 + 0	52 + 0	50 + 0	48 − 0	46 − 0	45 − 0
2.3	26 − 0	40 − 0	62 + 0	93 + 0	11 + 1	13 + 1	14 + 1	15 + 1	15 + 1	15 + 1
2.4	65 + 0	82 + 0	13 + 1	20 + 1	25 + 1	28 + 1	30 + 1	29 + 1	28 + 1	27 + 1
2.5	10 + 1	13 + 1	20 + 1	27 + 1	31 + 1	32 + 1	31 + 1	29 + 1	27 + 1	24 + 1
2.6	13 + 1	17 + 1	23 + 1	27 + 1	28 + 1	27 + 1	25 + 1	23 + 1	20 + 1	18 + 1
2.7	14 + 1	19 + 1	23 + 1	25 + 1	24 + 1	22 + 1	19 + 1	17 + 1	15 + 1	13 + 1
2.8	15 + 1	20 + 1	22 + 1	22 + 1	19 + 1	17 + 1	14 + 1	12 + 1	10 + 1	95 + 0
2.9	16 + 1	19 + 1	20 + 1	18 + 1	15 + 1	12 + 1	10 + 1	87 + 0	74 + 0	63 + 0
3.0	16 + 1	18 + 1	16 + 1	13 + 1	10 + 1	86 + 0	69 + 0	56 + 0	46 − 0	38 − 0
3.1	16 + 1	15 + 1	11 + 1	90 + 0	70 + 0	55 + 0	44 − 0	36 − 0	30 − 0	25 − 0
3.2	10 + 1	86 + 0	68 + 0	60 + 0	57 + 0	54 + 0	51 + 0	48 − 0	46 − 0	43 − 0
3.3	26 − 0	41 − 0	63 + 0	93 + 0	11 + 1	13 + 1	14 + 1	14 + 1	14 + 1	14 + 1
3.4	63 + 0	79 + 0	13 + 1	20 + 1	25 + 1	28 + 1	29 + 1	29 + 1	27 + 1	26 + 1
3.5	10 + 1	13 + 1	20 + 1	28 + 1	32 + 1	33 + 1	32 + 1	30 + 1	27 + 1	25 + 1
3.6	12 + 1	17 + 1	23 + 1	28 + 1	30 + 1	28 + 1	26 + 1	23 + 1	21 + 1	19 + 1
3.7	14 + 1	19 + 1	24 + 1	26 + 1	25 + 1	23 + 1	20 + 1	17 + 1	15 + 1	14 + 1
3.8	15 + 1	20 + 1	23 + 1	22 + 1	20 + 1	17 + 1	15 + 1	13 + 1	11 + 1	98 + 0
3.9	16 + 1	20 + 1	20 + 1	18 + 1	15 + 1	13 + 1	10 + 1	90 + 0	76 + 0	64 + 0
4.0	16 + 1	18 + 1	17 + 1	14 + 1	11 + 1	88 + 0	71 + 0	57 + 0	47 − 0	39 − 0
4.1	16 + 1	15 + 1	12 + 1	93 + 0	72 + 0	57 + 0	46 − 0	37 − 0	31 − 0	26 − 0
4.2	10 + 1	88 + 0	70 + 0	62 + 0	58 + 0	55 + 0	52 + 0	48 − 0	46 − 0	43 − 0
4.3	26 − 0	41 − 0	63 + 0	93 + 0	11 + 1	12 + 1	13 + 1	13 + 1	13 + 1	13 + 1
4.4	62 + 0	78 + 0	13 + 1	20 + 1	25 + 1	28 + 1	28 + 1	28 + 1	27 + 1	25 + 1

Table 6.3. (*continued*)

Δn	The effective principal quantum number n_0^*									
	0.5	1.0	1.5	2.0	2.5	3.0	3.5	4.0	4.5	5.0
	Transition $s - p$, parameter A									
0.1			14 + 1	20 + 1	25 + 1	28 + 1	31 + 1	33 + 1	35 + 1	36 + 1
0.2			30 + 1	37 + 1	40 + 1	41 + 1	41 + 1	41 + 1	41 + 1	40 + 1
0.3			46 + 1	48 + 1	46 + 1	44 + 1	42 + 1	40 + 1	37 + 1	35 + 1
0.4			58 + 1	52 + 1	46 + 1	41 + 1	37 + 1	34 + 1	31 + 1	28 + 1
0.5			64 + 1	51 + 1	41 + 1	34 + 1	30 + 1	26 + 1	23 + 1	20 + 1
0.6		89 + 1	62 + 1	44 + 1	33 + 1	26 + 1	21 + 1	18 + 1	15 + 1	13 + 1*
0.7		95 + 1	53 + 1	34 + 1	23 + 1	17 + 1	14 + 1	11 + 1*	97 + 0*	84 + 0*
0.8		91 + 1	40 + 1	23 + 1	14 + 1	10 + 1*	81 + 0*	66 + 0*	55 + 0*	48 − 0*
0.9		77 + 1	25 + 1	13 + 1	79 + 0*	57 + 0*	44 − 0*	37 − 0*	33 − 0*	30 − 0*
1.0		57 + 1	13 + 1	66 + 0*	41 − 0	33 − 0	30 − 0	28 − 0	28 − 0	27 − 0*
1.1	19 + 2	32 + 1	66 + 0	34 − 0	31 − 0	33 − 0*	36 − 0*	38 − 0*	40 − 0*	41 − 0*
1.2	14 + 2	13 + 1	32 − 0	43 − 0*	64 + 0*	86 + 0	10 + 1	10 + 1	10 + 1	97 + 0
1.3	89 + 1	37 − 0*	73 + 0	21 + 1	22 + 1	18 + 1	15 + 1	13 + 1	11 + 1	94 + 0
1.4	44 + 1	52 + 0	29 + 1	28 + 1	22 + 1	17 + 1	13 + 1	11 + 1	94 + 0	80 + 0
1.5	15 + 1	29 + 1	37 + 1	27 + 1	19 + 1	14 + 1	11 + 1	92 + 0	76 + 0	64 + 0
1.6	21 − 0	45 + 1	36 + 1	24 + 1	16 + 1	12 + 1	90 + 0	71 + 0	58 + 0	49 − 0*
1.7	34 − 0	56 + 1	33 + 1	20 + 1	12 + 1	89 + 0	65 + 0	51 + 0*	41 − 0*	34 − 0*
1.8	14 + 1	61 + 1	27 + 1	15 + 1	88 + 0	60 + 0	43 − 0*	43 − 0*	27 − 0*	23 − 0*
1.9	25 + 1	58 + 1	19 + 1	99 + 0	54 + 0	37 − 0*	27 − 0*	22 − 0*	19 − 0*	17 − 0*
2.0	34 + 1	49 + 1	11 + 1	55 + 0	32 − 0	24 − 0	20 − 0	18 − 0	17 − 0	16 − 0
2.1	13 + 2	30 + 1	66 + 0	31 − 0	24 − 0	23 − 0*	23 − 0*	23 − 0*	23 − 0*	22 − 0*
2.2	10 + 2	14 + 1	32 − 0	31 − 0	39 − 0*	46 − 0*	52 + 0	53 + 0	52 + 0	48 − 0
2.3	73 + 1	46 − 0	44 − 0	11 + 1	15 + 1	13 + 1	10 + 1	84 + 0	67 + 0	55 + 0
2.4	41 + 1	32 − 0	23 + 1	22 + 1	17 + 1	12 + 1	97 + 0	75 + 0	60 + 0	49 − 0
2.5	16 + 1	22 + 1	32 + 1	23 + 1	15 + 1	11 + 1	82 + 0	63 + 0	50 + 0	41 − 0
2.6	37 − 0	39 + 1	32 + 1	20 + 1	13 + 1	91 + 0	65 + 0	50 + 0	39 − 0	32 − 0*
2.7	75 − 1	49 + 1	29 + 1	17 + 1	10 + 1	69 + 0	49 − 0	37 − 0*	29 − 0*	23 − 0*
2.8	10 + 1	54 + 1	24 + 1	13 + 1	72 + 0	48 − 0	33 − 0*	25 − 0*	20 − 0*	16 − 0*
2.9	18 + 1	54 + 1	17 + 1	88 + 0	46 − 0	31 − 0*	22 − 0*	17 − 0*	14 − 0*	12 − 0*
3.0	26 + 1	46 + 1	10 + 1	51 + 0	28 − 0	21 − 0	17 − 0	14 − 0	13 − 0	12 − 0
3.1	11 + 2	29 + 1	66 + 0	29 − 0	22 − 0	20 − 0	19 − 0*	18 − 0*	17 − 0*	16 − 0*
3.2	96 + 1	14 + 1	32 − 0	28 − 0	32 − 0*	36 − 0*	38 − 0	38 − 0	36 − 0	34 − 0
3.3	68 + 1	50 + 0	38 − 0	87 + 0	12 + 1	11 + 1	86 + 0	66 + 0	52 + 0	42 − 0
3.4	39 + 1	29 − 0	21 + 1	20 + 1	15 + 1	11 + 1	81 + 0	61 + 0	47 − 0	38 − 0
3.5	17 + 1	19 + 1	30 + 1	21 + 1	14 + 1	96 + 0	69 + 0	52 + 0	40 − 0	32 − 0
3.6	44 − 0	36 + 1	30 + 1	19 + 1	11 + 1	79 + 0	55 + 0	41 − 0	32 − 0	25 − 0*
3.7	65 − 1	46 + 1	27 + 1	16 + 1	91 + 0	61 + 0	42 − 0	31 − 0*	24 − 0*	19 − 0*
3.8	82 + 0	51 + 1	22 + 1	12 + 1	65 + 0	43 − 0	29 − 0*	22 − 0*	17 − 0*	14 − 0*
3.9	15 + 1	51 + 1	16 + 1	84 + 0	42 − 0	28 − 0*	20 − 0*	15 − 0*	12 − 0	10 − 0
4.0	23 + 1	45 + 1	10 + 1	49 − 0	26 − 0	19 − 0	15 − 0	12 − 0	11 − 0	10 − 0
4.1	10 + 2	29 + 1	66 + 0	28 − 0	20 − 0	18 − 0	16 − 0*	15 − 0*	14 − 0*	13 − 0*
4.2	91 + 1	14 + 1	33 − 0	27 − 0	29 − 0*	31 − 0	32 − 0	31 − 0	30 − 0	27 − 0
4.3	65 + 1	52 + 0	35 − 0	76 + 0	10 + 1	97 + 0	76 + 0	57 + 0	45 − 0	35 − 0
4.4	38 + 1	28 − 0	19 + 1	19 + 1	14 + 1	10 + 1	72 + 0	53 + 0	41 − 0	32 − 0

Table 6.3. (*continued*)

	The effective principal quantum number n_0^*									
Δn	0.5	1.0	1.5	2.0	2.5	3.0	3.5	4.0	4.5	5.0
	Transition $s - p$, parameter χ									
0.1			10 + 1	10 + 1	10 + 1	10 + 1	97 + 0	93 + 0	89 + 0	85 + 0
0.2			99 + 0	92 + 0	84 + 0	77 + 0	70 + 0	64 + 0	59 + 0	54 + 0
0.3			87 + 0	76 + 0	66 + 0	58 + 0	51 + 0	45 − 0	40 − 0	35 − 0
0.4			75 + 0	62 + 0	52 + 0	44 − 0	37 − 0	31 − 0	27 − 0	23 − 0
0.5			64 + 0	51 + 0	40 − 0	32 − 0	26 − 0	22 − 0	18 − 0	15 − 0
0.6		71 + 0	54 + 0	40 − 0	30 − 0	23 − 0	18 − 0	14 − 0	11 − 0	10 − 0*
0.7		63 + 0	44 − 0	31 − 0	22 − 0	16 − 0	12 − 0	10 − 0*	10 − 0*	10 − 0*
0.8		56 + 0	35 − 0	23 − 0	15 − 0	10 − 0*	10 − 0*	10 − 0*	10 − 0*	10 − 0*
0.9		48 − 0	26 − 0	16 − 0	10 − 0*	10 − 0*	10 − 0*	10 − 0*	10 − 0*	10 − 0*
1.0		41 − 0	18 − 0	10 − 0*	10 − 0	10 − 0	10 − 0	10 − 0	10 − 0	10 − 0*
1.1	61 + 0	31 − 0	12 − 0	10 − 0	15 − 0	23 − 0*	30 − 0*	37 − 0*	43 − 0*	47 − 0*
1.2	55 + 0	20 − 0	13 − 0	44 − 0*	96 + 0*	15 + 1	19 + 1	21 + 1	20 + 1	17 + 1
1.3	48 − 0	11 − 0	11 + 1*	42 + 1	39 + 1	30 + 1	23 + 1	18 + 1	14 + 1	11 + 1
1.4	40 − 0	81 + 0	38 + 1	30 + 1	21 + 1	16 + 1	12 + 1	93 + 0	73 + 0	59 + 0
1.5	29 − 0	31 + 1	21 + 1	15 + 1	10 + 1	81 + 0	62 + 0	48 − 0	38 − 0	30 − 0
1.6	12 − 0	18 + 1	12 + 1	86 + 0	62 + 0	46 − 0	35 − 0	27 − 0	21 − 0	16 − 0*
1.7	26 + 1	11 + 1	78 + 0	54 + 0	38 − 0	27 − 0	20 − 0	15 − 0*	12 − 0*	10 − 0*
1.8	14 + 1	82 + 0	53 + 0	35 − 0	23 − 0	17 − 0	12 − 0*	10 − 0*	10 − 0*	10 − 0*
1.9	87 + 0	63 + 0	36 − 0	23 − 0	15 − 0	11 − 0*	10 − 0*	10 − 0*	10 − 0*	10 − 0*
2.0	67 + 0	50 + 0	24 − 0	14 − 0	11 − 0	10 − 0	11 − 0	11 − 0	12 − 0	12 − 0
2.1	65 + 0	36 − 0	16 − 0	12 − 0	16 − 0	22 − 0*	29 − 0*	34 − 0*	38 − 0*	41 − 0*
2.2	57 + 0	24 − 0	13 − 0	34 − 0	68 + 0*	10 + 1*	13 + 1	15 + 1	15 + 1	15 + 1
2.3	49 − 0	13 − 0	61 + 1	27 + 1	41 + 1	35 + 1	27 + 1	21 + 1	16 + 1	13 + 1
2.4	40 − 0	36 − 0	45 + 1	37 + 1	27 + 1	20 + 1	15 + 1	11 + 1	90 + 0	72 + 0
2.5	30 − 0	35 + 1	28 + 1	20 + 1	14 + 1	10 + 1	78 + 0	60 + 0	47 − 0	37 − 0
2.6	16 − 0	23 + 1	15 + 1	10 + 1	77 + 0	56 + 0	42 − 0	33 − 0	26 − 0	20 − 0*
2.7	27 − 0	13 + 1	92 + 0	64 + 0	45 − 0	33 − 0	24 − 0	19 − 0*	14 − 0*	11 − 0*
2.8	23 + 1	92 + 0	60 + 0	40 − 0	27 − 0	20 − 0	15 − 0*	11 − 0*	10 − 0*	10 − 0*
2.9	10 + 1	69 + 0	40 − 0	26 − 0	17 − 0	13 − 0*	10 − 0*	10 − 0*	10 − 0*	10 − 0*
3.0	73 + 0	53 + 0	26 − 0	16 − 0	13 − 0	11 − 0	12 − 0	12 − 0	13 − 0	13 − 0
3.1	66 + 0	39 − 0	17 − 0	13 − 0	17 − 0	23 − 0	28 − 0*	33 − 0*	37 − 0*	40 − 0*
3.2	57 + 0	25 − 0	14 − 0	32 − 0	61 + 0*	92 + 0*	11 + 1	13 + 1	14 + 1	13 + 1
3.3	49 − 0	14 − 0	50 + 0	22 + 1	37 + 1	36 + 1	28 + 1	22 + 1	17 + 1	14 + 1
3.4	40 − 0	28 − 0	45 + 1	40 + 1	30 + 1	22 + 1	16 + 1	12 + 1	99 + 0	78 + 0
3.5	30 − 0	35 + 1	31 + 1	22 + 1	16 + 1	11 + 1	87 + 0	67 + 0	52 + 0	42 − 0
3.6	17 − 0	25 + 1	17 + 1	11 + 1	85 + 0	62 + 0	47 − 0	36 − 0	28 − 0	23 − 0*
3.7	13 − 0	14 + 1	10 + 1	69 + 0	49 − 0	36 − 0	27 − 0	21 − 0*	16 − 0*	13 − 0*
3.8	28 + 1	97 + 0	64 + 0	43 − 0	30 − 0	22 − 0	16 − 0*	12 − 0*	10 − 0*	10 − 0*
3.9	11 + 1	71 + 0	42 − 0	27 − 0	18 − 0	14 − 0*	11 − 0*	10 − 0*	10 − 0	10 − 0
4.0	78 + 0	54 + 0	27 − 0	17 − 0	14 − 0	12 − 0	13 − 0	13 − 0	14 − 0	14 − 0
4.1	67 + 0	40 − 0	18 − 0	14 − 0	17 − 0	23 − 0	28 − 0*	32 − 0*	36 − 0*	38 − 0*
4.2	57 + 0	26 − 0	14 − 0	31 − 0	59 + 0*	86 + 0	11 + 1	12 + 1	13 + 1	13 + 1
4.3	49 − 0	15 − 0	45 − 0	19 + 1	35 + 1	35 + 1	29 + 1	23 + 1	18 + 1	14 + 1
4.4	40 − 0	25 − 0	43 + 1	42 + 1	32 + 1	23 + 1	17 + 1	13 + 1	10 + 1	83 + 0

Table 6.3. (*continued*)

Δn	The effective principal quantum number n_0^*									
	0.5	1.0	1.5	2.0	2.5	3.0	3.5	4.0	4.5	5.0
	Transition $s-d$, parameter A									
0.1					36 − 0	73 + 0	11 + 1	15 + 1	20 + 1	24 + 1
0.2					16 + 1	28 + 1	38 + 1	47 + 1	55 + 1	61 + 1
0.3					38 + 1	55 + 1	66 + 1	74 + 1	79 + 1	81 + 1
0.4					64 + 1	79 + 1	86 + 1	88 + 1	87 + 1	84 + 1
0.5					89 + 1	96 + 1	94 + 1	89 + 1	83 + 1	77 + 1
0.6				93 + 1	10 + 2	10 + 2	91 + 1	80 + 1	71 + 1	63 + 1
0.7				12 + 2	11 + 2	96 + 1	80 + 1	66 + 1	56 + 1	47 + 1
0.8				13 + 2	11 + 2	82 + 1	63 + 1	49 + 1	39 + 1	32 + 1
0.9				14 + 2	97 + 1	63 + 1	45 + 1	33 + 1	25 + 1	20 + 1
1.0				13 + 2	74 + 1	43 + 1	28 + 1	20 + 1	15 + 1	12 + 1
1.1			19 + 2	10 + 2	50 + 1	26 + 1	16 + 1	11 + 1	94 + 0	81 + 0
1.2			18 + 2	76 + 1	29 + 1	14 + 1	96 + 0	80 + 0	75 + 0	76 + 0
1.3			16 + 2	45 + 1	14 + 1	86 + 0	80 + 0	88 + 0	10 + 1	11 + 1
1.4			11 + 2	22 + 1	84 + 0	91 + 0	12 + 1	17 + 1	20 + 1	22 + 1
1.5			74 + 1	97 + 0	10 + 1	20 + 1	30 + 1	34 + 1	33 + 1	30 + 1
1.6		27 + 2	30 + 1	84 + 0	30 + 1	46 + 1	46 + 1	40 + 1	34 + 1	28 + 1
1.7		19 + 2	89 + 0	26 + 1	60 + 1	56 + 1	46 + 1	36 + 1	29 + 1	23 + 1
1.8		11 + 2	86 + 0	61 + 1	70 + 1	53 + 1	39 + 1	29 + 1	22 + 1	17 + 1
1.9		55 + 1	36 + 1	84 + 1	68 + 1	44 + 1	31 + 1	21 + 1	16 + 1	12 + 1
2.0		15 + 1	79 + 1	92 + 1	58 + 1	34 + 1	22 + 1	15 + 1	17 + 1*	86 + 0
2.1	49 + 2	31 − 0	10 + 2	85 + 1	43 + 1	23 + 1	14 + 1	97 + 0	73 + 0	60 + 0
2.2	26 + 2	36 + 1	12 + 2	68 + 1	29 + 1	14 + 1	89 + 0	66 + 0	56 + 0	51 + 0
2.3	10 + 2	91 + 1	11 + 2	47 + 1	16 + 1	87 + 0	65 + 0	61 + 0	61 + 0	63 + 0
2.4	25 + 1	14 + 2	10 + 2	27 + 1	93 + 0	70 + 0	76 + 0	89 + 0	10 + 1	10 + 1
2.5	66 − 1	16 + 2	75 + 1	13 + 1	77 + 0	10 + 1	14 + 1	17 + 1	18 + 1	17 + 1
2.6	69 + 0	16 + 2	37 + 1	75 + 0	15 + 1	25 + 1	29 + 1	27 + 1	23 + 1	19 + 1
2.7	24 + 1	13 + 2	13 + 1	12 + 1	39 + 1	42 + 1	35 + 1	27 + 1	21 + 1	17 + 1
2.8	38 + 1	93 + 1	65 + 0	39 + 1	58 + 1	43 + 1	32 + 1	23 + 1	17 + 1	13 + 1
2.9	43 + 1	51 + 1	19 + 1	68 + 1	59 + 1	38 + 1	26 + 1	18 + 1	13 + 1	10 + 1
3.0	40 + 1	19 + 1	55 + 1	81 + 1	53 + 1	30 + 1	19 + 1	12 + 1	93 + 0	70 + 0
3.1	23 + 2	24 − 0	79 + 1	78 + 1	40 + 1	21 + 1	13 + 1	87 + 0	64 + 0	50 + 0
3.2	14 + 2	20 + 1	10 + 2	65 + 1	28 + 1	13 + 1	84 + 0	60 + 0	48 − 0	42 − 0
3.3	63 + 1	62 + 1	10 + 2	47 + 1	17 + 1	86 + 0	60 + 0	51 + 0	49 − 0	48 − 0
3.4	17 + 1	10 + 2	96 + 1	29 + 1	98 + 0	64 + 0	62 + 0	67 + 0	72 + 0	75 + 0
3.5	91 − 1	13 + 2	75 + 1	14 + 1	71 + 0	84 + 0	10 + 1	12 + 1	13 + 1	12 + 1
3.6	35 − 0	13 + 2	40 + 1	79 + 0	11 + 1	18 + 1	22 + 1	21 + 1	18 + 1	15 + 1
3.7	14 + 1	11 + 2	16 + 1	97 + 0	31 + 1	35 + 1	30 + 1	23 + 1	18 + 1	14 + 1
3.8	24 + 1	85 + 1	67 + 0	29 + 1	51 + 1	39 + 1	28 + 1	20 + 1	15 + 1	11 + 1
3.9	30 + 1	49 + 1	13 + 1	60 + 1	55 + 1	35 + 1	23 + 1	16 + 1	11 + 1	87 + 0
4.0	29 + 1	21 + 1	44 + 1	75 + 1	50 + 1	28 + 1	17 + 1	11 + 1	84 + 0	62 + 0
4.1	17 + 2	28 − 0	69 + 1	74 + 1	39 + 1	20 + 1	12 + 1	81 + 0	58 + 0	45 − 0
4.2	10 + 2	14 + 1	90 + 1	63 + 1	27 + 1	13 + 1	80 + 0	56 + 0	44 − 0	37 − 0
4.3	50 + 1	51 + 1	98 + 1	47 + 1	17 + 1	84 + 0	57 + 0	47 − 0	42 − 0	40 − 0
4.4	14 + 1	90 + 1	92 + 1	29 + 1	10 + 1	61 + 0	56 + 0	57 + 0	59 + 0	60 + 0

Table 6.3. (*continued*)

	The effective principal quantum number n_0^*									
Δn	0.5	1.0	1.5	2.0	2.5	3.0	3.5	4.0	4.5	5.0
	Transition $s-d$, parameter χ									
0.1					10 + 1	12 + 1	14 + 1	15 + 1	17 + 1	18 + 1
0.2					17 + 1	19 + 1	20 + 1	21 + 1	21 + 1	21 + 1
0.3					20 + 1	21 + 1	21 + 1	20 + 1	19 + 1	18 + 1
0.4					20 + 1	20 + 1	19 + 1	17 + 1	16 + 1	14 + 1
0.5					20 + 1	18 + 1	16 + 1	14 + 1	12 + 1	11 + 1
0.6				19 + 1	18 + 1	16 + 1	13 + 1	11 + 1	98 + 0	83 + 0
0.7				18 + 1	16 + 1	13 + 1	11 + 1	91 + 0	74 + 0	62 + 0
0.8				17 + 1	14 + 1	11 + 1	88 + 0	69 + 0	55 + 0	44 − 0
0.9				16 + 1	12 + 1	91 + 0	67 + 0	51 + 0	39 − 0	31 − 0
1.0				14 + 1	10 + 1	70 + 0	49 − 0	36 − 0	28 − 0	23 − 0
1.1			16 + 1	12 + 1	80 + 0	51 + 0	36 − 0	27 − 0	23 − 0	21 − 0
1.2			15 + 1	10 + 1	59 + 0	37 − 0	29 − 0	29 − 0	31 − 0	36 − 0
1.3			14 + 1	81 + 0	42 − 0	34 − 0	42 − 0	58 + 0	74 + 0	89 + 0
1.4			12 + 1	57 + 0	39 − 0	66 + 0	11 + 1	16 + 1	19 + 1	21 + 1
1.5			10 + 1	41 − 0	90 + 0	21 + 1	30 + 1	32 + 1	29 + 1	25 + 1
1.6		15 + 1	77 + 0	68 + 0	29 + 1	40 + 1	36 + 1	28 + 1	22 + 1	18 + 1
1.7		14 + 1	44 − 0	25 + 1	41 + 1	34 + 1	25 + 1	19 + 1	14 + 1	11 + 1
1.8		13 + 1	82 + 0	37 + 1	32 + 1	23 + 1	17 + 1	12 + 1	97 + 0	76 + 0
1.9		11 + 1	24 + 1	31 + 1	23 + 1	16 + 1	11 + 1	85 + 0	65 + 0	50 + 0
2.0		85 + 0	27 + 1	24 + 1	16 + 1	11 + 1	79 + 0	58 + 0	10 − 0*	35 − 0
2.1	15 + 1	64 + 0	24 + 1	18 + 1	12 + 1	79 + 0	54 + 0	41 − 0	33 − 0	28 − 0
2.2	15 + 1	19 + 1	21 + 1	14 + 1	86 + 0	55 + 0	40 − 0	35 − 0	34 − 0	35 − 0
2.3	15 + 1	19 + 1	18 + 1	10 + 1	60 + 0	42 − 0	41 − 0	48 − 0	57 + 0	66 + 0
2.4	13 + 1	18 + 1	15 + 1	78 + 0	46 − 0	52 + 0	76 + 0	10 + 1	13 + 1	14 + 1
2.5	70 + 0	17 + 1	12 + 1	53 + 0	61 + 0	12 + 1	19 + 1	24 + 1	25 + 1	24 + 1
2.6	19 + 1	16 + 1	93 + 0	50 + 0	17 + 1	32 + 1	36 + 1	32 + 1	26 + 1	21 + 1
2.7	17 + 1	15 + 1	58 + 0	13 + 1	40 + 1	40 + 1	31 + 1	24 + 1	18 + 1	14 + 1
2.8	16 + 1	13 + 1	51 + 0	36 + 1	39 + 1	29 + 1	21 + 1	15 + 1	12 + 1	93 + 0
2.9	15 + 1	11 + 1	17 + 1	37 + 1	28 + 1	20 + 1	14 + 1	10 + 1	79 + 0	61 + 0
3.0	14 + 1	90 + 0	29 + 1	29 + 1	20 + 1	13 + 1	96 + 0	70 + 0	53 + 0	42 − 0
3.1	15 + 1	40 − 0	28 + 1	22 + 1	14 + 1	94 + 0	65 + 0	49 − 0	39 − 0	33 − 0
3.2	15 + 1	19 + 1	23 + 1	16 + 1	10 + 1	65 + 0	47 − 0	39 − 0	36 − 0	36 − 0
3.3	14 + 1	21 + 1	19 + 1	12 + 1	70 + 0	48 − 0	43 − 0	47 + 0	53 + 0	59 + 0
3.4	13 + 1	19 + 1	16 + 1	88 + 0	51 + 0	51 + 0	68 + 0	89 + 0	10 + 1	12 + 1
3.5	71 + 0	18 + 1	13 + 1	60 + 0	57 + 0	10 + 1	15 + 1	20 + 1	22 + 1	21 + 1
3.6	21 + 1	17 + 1	10 + 1	49 − 0	14 + 1	26 + 1	32 + 1	31 + 1	26 + 1	22 + 1
3.7	18 + 1	15 + 1	64 + 0	10 + 1	36 + 1	41 + 1	34 + 1	26 + 1	20 + 1	15 + 1
3.8	16 + 1	13 + 1	48 − 0	31 + 1	42 + 1	33 + 1	24 + 1	17 + 1	13 + 1	10 + 1
3.9	15 + 1	11 + 1	14 + 1	40 + 1	31 + 1	22 + 1	16 + 1	11 + 1	88 + 0	68 + 0
4.0	14 + 1	92 + 0	29 + 1	31 + 1	22 + 1	15 + 1	10 + 1	78 + 0	59 + 0	46 − 0
4.1	15 + 1	39 − 0	30 + 1	23 + 1	15 + 1	10 + 1	72 + 0	54 + 0	42 − 0	36 − 0
4.2	15 + 1	17 + 1	25 + 1	17 + 1	10 + 1	71 + 0	51 + 0	42 − 0	38 − 0	36 − 0
4.3	14 + 1	21 + 1	20 + 1	13 + 1	76 + 0	52 + 0	45 − 0	46 − 0	51 + 0	56 + 0
4.4	12 + 1	20 + 1	17 + 1	94 + 0	54 + 0	51 + 0	64 + 0	82 + 0	99 + 0	11 + 1

Table 6.3. (*continued*)

	The effective principal quantum number n_0^*							
Δn	1.5	2.0	2.5	3.0	3.5	4.0	4.5	5.0
	Transition $p-s$, parameter A							
0.1	10 + 1	17 + 1	21 + 1	25 + 1	28 + 1	30 + 1	32 + 1	33 + 1
0.2	16 + 1	24 + 1	28 + 1	31 + 1	32 + 1	33 + 1	33 + 1	33 + 1
0.3	18 + 1	24 + 1	27 + 1	28 + 1	28 + 1	28 + 1	27 + 1	27 + 1
0.4	16 + 1	21 + 1	22 + 1	22 + 1	21 + 1	21 + 1	20 + 1	19 + 1
0.5	12 + 1	15 + 1	15 + 1	15 + 1	14 + 1	14 + 1	13 + 1	12 + 1
0.6	84 + 0	99 + 0	99 + 0	95 + 0	89 + 0	84 + 0*	79 + 0*	74 + 0*
0.7	45 − 0	55 + 0	55 + 0	53 + 0	50 + 0*	48 − 0*	45 − 0*	43 − 0*
0.8	21 − 0	29 − 0	30 − 0	31 − 0	31 − 0	30 − 0	30 − 0	29 − 0*
0.9	16 − 0	21 − 0	26 − 0	27 − 0	28 − 0*	28 − 0*	28 − 0*	28 − 0*
1.0	51 + 0	37 − 0	49 − 0	45 − 0*	48 − 0*	45 − 0*	45 − 0*	43 − 0*
1.1	92 + 0	13 + 1	15 + 1	14 + 1	13 + 1	11 + 1	10 + 1	94 + 0
1.2	12 + 1	15 + 1	15 + 1	14 + 1	12 + 1	11 + 1	98 + 0	86 + 0
1.3	12 + 1	13 + 1	12 + 1	11 + 1	99 + 0	87 + 0	77 + 0	68 + 0
1.4	10 + 1	10 + 1	96 + 0	84 + 0	74 + 0	65 + 0	58 + 0	51 + 0
1.5	75 + 0	76 + 0	69 + 0	60 + 0	53 + 0	46 − 0	41 − 0	37 − 0*
1.6	48 − 0	51 + 0	46 − 0	40 − 0	36 − 0	32 − 0	29 − 0*	26 − 0*
1.7	27 − 0	31 − 0	29 − 0	27 − 0	24 − 0	22 − 0	20 − 0	19 − 0
1.8	15 − 0	20 − 0	21 − 0	20 − 0	19 − 0	18 − 0	17 − 0	16 − 0
1.9	19 − 0	20 − 0	23 − 0	22 − 0*	21 − 0*	20 − 0*	19 − 0*	17 − 0*
2.0	71 + 0	39 − 0	51 + 0	39 − 0	39 − 0*	33 − 0*	31 − 0*	28 − 0
2.1	85 + 0	11 + 1	12 + 1	10 + 1	89 + 0	74 + 0	62 + 0	53 + 0
2.2	11 + 1	12 + 1	11 + 1	99 + 0	83 + 0	69 + 0	58 + 0	49 − 0
2.3	11 + 1	11 + 1	97 + 0	80 + 0	67 + 0	56 + 0	47 − 0	40 − 0
2.4	89 + 0	85 + 0	73 + 0	60 + 0	50 + 0	42 − 0	36 − 0	31 − 0
2.5	65 + 0	61 + 0	52 + 0	43 − 0	36 − 0	31 − 0	27 − 0	23 − 0*
2.6	41 − 0	41 − 0	35 − 0	30 − 0	26 − 0	22 − 0	19 − 0*	17 − 0*
2.7	23 − 0	26 − 0	24 − 0	21 − 0	18 − 0	16 − 0	14 − 0	13 − 0
2.8	14 − 0	18 − 0	18 − 0	17 − 0	16 − 0	14 − 0	13 − 0	12 − 0
2.9	22 − 0	19 − 0	23 − 0	19 − 0*	19 − 0*	16 − 0*	15 − 0*	14 − 0*
3.0	79 + 0	39 − 0	53 + 0	36 − 0	36 − 0	28 − 0	26 − 0	22 − 0
3.1	83 + 0	10 + 1	10 + 1	89 + 0	72 + 0	58 + 0	47 − 0	39 − 0
3.2	10 + 1	11 + 1	10 + 1	83 + 0	67 + 0	54 + 0	44 − 0	37 − 0
3.3	10 + 1	10 + 1	85 + 0	68 + 0	55 + 0	44 − 0	36 − 0	30 − 0
3.4	85 + 0	77 + 0	64 + 0	51 + 0	41 − 0	34 − 0	28 − 0	24 − 0
3.5	61 + 0	56 + 0	46 − 0	37 − 0	30 − 0	25 − 0	21 − 0	18 − 0*
3.6	38 − 0	38 − 0	31 − 0	26 − 0	21 − 0	18 − 0	15 − 0	13 − 0*
3.7	21 − 0	24 − 0	21 − 0	19 − 0	16 − 0	14 − 0	12 − 0	11 − 0
3.8	14 − 0	17 − 0	17 − 0	16 − 0	14 − 0	12 − 0	11 − 0	10 − 0
3.9	24 − 0	19 − 0	22 − 0	18 − 0*	17 − 0*	15 − 0*	13 − 0*	12 − 0*
4.0	83 + 0	39 − 0	55 + 0	35 − 0	34 − 0	25 − 0	23 − 0	19 − 0
4.1	81 + 0	10 + 1	10 + 1	82 + 0	64 + 0	50 + 0	40 − 0	33 − 0
4.2	10 + 1	10 + 1	95 + 0	75 + 0	59 + 0	47 − 0	38 − 0	30 − 0
4.3	10 + 1	96 + 0	79 + 0	62 + 0	49 − 0	38 − 0	31 − 0	25 − 0
4.4	82 + 0	73 + 0	59 + 0	46 − 0	37 − 0	30 − 0	24 − 0	20 − 0

Table 6.3. (*continued*)

	The effective principal quantum number n_0^*							
Δn	1.5	2.0	2.5	3.0	3.5	4.0	4.5	5.0
	Transition $p-s$, parameter χ							
0.1	10 + 1	10 + 1	10 + 1	10 + 1	97 + 0	93 + 0	89 + 0	85 + 0
0.2	97 + 0	89 + 0	82 + 0	75 + 0	68 + 0	62 + 0	57 + 0	52 + 0
0.3	81 + 0	71 + 0	62 + 0	54 + 0	47 − 0	41 − 0	37 − 0	33 − 0
0.4	66 + 0	54 + 0	45 − 0	38 − 0	32 − 0	27 − 0	24 − 0	20 − 0
0.5	51 + 0	40 + 0	32 − 0	26 − 0	21 − 0	18 − 0	15 − 0	12 − 0
0.6	37 − 0	28 − 0	22 − 0	17 − 0	13 − 0	11 − 0*	10 − 0*	10 − 0*
0.7	25 − 0	19 − 0	14 − 0	11 − 0	10 − 0*	10 − 0*	10 − 0*	10 − 0*
0.8	17 − 0	13 − 0	11 − 0	10 − 0	10 − 0	10 − 0	10 − 0	10 − 0*
0.9	29 − 0	21 − 0	23 − 0	20 − 0	19 − 0*	18 − 0*	17 − 0*	15 − 0*
1.0	18 + 1	93 + 0	10 + 1	84 + 0*	81 + 0*	71 + 0*	66 + 0*	60 + 0*
1.1	39 + 1	46 + 1	45 + 1	39 + 1	33 + 1	28 + 1	24 + 1	20 + 1
1.2	33 + 1	34 + 1	31 + 1	27 + 1	23 + 1	19 + 1	16 + 1	13 + 1
1.3	21 + 1	19 + 1	17 + 1	14 + 1	12 + 1	10 + 1	85 + 0	71 + 0
1.4	12 + 1	11 + 1	91 + 0	76 + 0	63 + 0	52 + 0	44 − 0	37 − 0
1.5	81 + 0	65 + 0	53 + 0	43 − 0	35 − 0	29 − 0	24 − 0	20 − 0*
1.6	51 + 0	40 − 0	32 − 0	26 − 0	21 − 0	17 − 0	14 − 0*	12 − 0*
1.7	32 − 0	26 − 0	22 − 0	18 − 0	15 − 0	13 − 0	11 − 0	10 − 0
1.8	26 − 0	22 − 0	22 − 0	19 − 0	17 − 0	15 − 0	14 − 0	12 − 0
1.9	64 + 0	39 − 0	45 − 0	38 − 0*	37 − 0*	32 − 0*	30 − 0*	27 − 0*
2.0	30 + 1	14 + 1	17 + 1	12 + 1	12 + 1*	10 + 1*	93 + 0*	80 + 0
2.1	42 + 1	50 + 1	49 + 1	42 + 1	35 + 1	28 + 1	23 + 1	19 + 1
2.2	36 + 1	37 + 1	34 + 1	29 + 1	24 + 1	20 + 1	16 + 1	14 + 1
2.3	24 + 1	22 + 1	19 + 1	16 + 1	13 + 1	11 + 1	92 + 0	77 + 0
2.4	14 + 1	12 + 1	10 + 1	86 + 0	71 + 0	59 + 0	49 − 0	41 − 0
2.5	89 + 0	72 + 0	59 + 0	49 − 0	40 − 0	33 − 0	28 − 0	23 − 0*
2.6	55 + 0	45 − 0	36 − 0	30 − 0	25 − 0	21 − 0	17 − 0*	15 − 0*
2.7	35 − 0	30 − 0	26 − 0	22 − 0	19 − 0	16 − 0	14 − 0	12 − 0
2.8	31 − 0	27 − 0	28 − 0	24 − 0	23 − 0	20 − 0	18 − 0	16 − 0
2.9	89 + 0	48 − 0	59 + 0	47 − 0*	47 − 0*	40 − 0*	37 − 0*	33 − 0*
3.0	34 + 1	16 + 1	22 + 1	14 + 1	14 + 1	11 + 1	10 + 1	90 + 0
3.1	43 + 1	52 + 1	50 + 1	43 + 1	35 + 1	29 + 1	23 + 1	19 + 1
3.2	37 + 1	39 + 1	35 + 1	29 + 1	24 + 1	20 + 1	17 + 1	14 + 1
3.3	25 + 1	23 + 1	20 + 1	16 + 1	13 + 1	11 + 1	95 + 0	80 + 0
3.4	15 + 1	13 + 1	10 + 1	90 + 0	75 + 0	62 + 0	52 + 0	44 − 0
3.5	92 + 0	76 + 0	63 + 0	52 + 0	43 − 0	36 − 0	30 − 0	25 − 0*
3.6	57 + 0	47 − 0	39 − 0	32 − 0	27 − 0	23 − 0	19 − 0	16 − 0*
3.7	36 − 0	32 − 0	28 − 0	24 − 0	21 − 0	18 − 0	16 − 0	14 − 0
3.8	34 − 0	29 − 0	32 − 0	27 − 0	26 − 0	23 − 0	21 − 0	18 − 0
3.9	10 + 1	53 + 0	68 + 0	53 + 0*	53 + 0*	44 − 0*	42 − 0*	36 − 0*
4.0	35 + 1	18 + 1	25 + 1	16 + 1	16 + 1	12 + 1	11 + 1	97 + 0
4.1	43 + 1	53 + 1	51 + 1	43 + 1	36 + 1	29 + 1	24 + 1	19 + 1
4.2	38 + 1	39 + 1	35 + 1	30 + 1	24 + 1	20 + 1	17 + 1	14 + 1
4.3	25 + 1	24 + 1	20 + 1	17 + 1	14 + 1	11 + 1	97 + 0	82 + 0
4.4	15 + 1	13 + 1	11 + 1	93 + 0	77 + 0	64 + 0	54 + 0	45 − 0

Table 6.3. (*continued*)

Δn	The effective principal quantum number n_0^* 1.5	2.0	2.5	3.0	3.5	4.0	4.5	5.0
	Transition $p-p$; parameter A $\kappa=0$							
0.5	14 + 2	26 + 2	34 + 2	40 + 2	44 + 2	47 + 2	49 + 2	50 + 2
0.6	18 + 2	29 + 2	36 + 2	39 + 2	41 + 2	42 + 2	42 + 2	41 + 2
0.7	20 + 2	29 + 2	33 + 2	33 + 2	33 + 2	32 + 2	31 + 2	30 + 2
0.8	20 + 2	25 + 2	26 + 2	24 + 2	23 + 2	21 + 2	20 + 2	18 + 2
0.9	18 + 2	18 + 2	17 + 2	15 + 2	13 + 2	11 + 2	10 + 2	97 + 1
1.0	13 + 2	11 + 2	92 + 1	75 + 1	62 + 1	53 + 1	46 + 1	40 + 1
1.1	71 + 1	50 + 1	37 + 1	29 + 1	23 + 1	20 + 1	17 + 1	15 + 1
1.2	27 + 1	17 + 1	15 + 1	14 + 1	14 + 1	14 + 1	15 + 1	15 + 1
1.3	79 + 0	12 + 1	19 + 1	26 + 1	33 + 1	39 + 1	43 + 1	46 + 1
1.4	97 + 0	32 + 1	56 + 1	75 + 1	89 + 1	97 + 1	10 + 2	10 + 2
1.5	31 + 1	77 + 1	11 + 2	12 + 2	13 + 2	13 + 2	13 + 2	12 + 2
1.6	68 + 1	12 + 2	15 + 2	15 + 2	15 + 2	14 + 2	13 + 2	12 + 2
1.7	10 + 2	15 + 2	16 + 2	15 + 2	13 + 2	12 + 2	11 + 2	10 + 2
1.8	13 + 2	15 + 2	14 + 2	12 + 2	10 + 2	92 + 1	81 + 1	71 + 1
1.9	13 + 2	12 + 2	10 + 2	85 + 1	69 + 1	57 + 1	49 + 1	42 + 1
2.0	11 + 2	83 + 1	64 + 1	47 + 1	37 + 1	29 + 1	24 + 1	20 + 1
2.1	67 + 1	43 + 1	30 + 1	21 + 1	16 + 1	13 + 1	11 + 1	95 + 0
2.2	30 + 1	17 + 1	13 + 1	11 + 1	10 + 1	97 + 0	91 + 0	88 + 0
2.3	97 + 0	10 + 1	14 + 1	17 + 1	19 + 1	20 + 1	21 + 1	21 + 1
2.4	74 + 0	22 + 1	36 + 1	46 + 1	51 + 1	53 + 1	54 + 1	52 + 1
2.5	22 + 1	56 + 1	77 + 1	87 + 1	88 + 1	84 + 1	78 + 1	72 + 1
2.6	53 + 1	98 + 1	11 + 2	11 + 2	10 + 2	92 + 1	82 + 1	74 + 1
2.7	98 + 1	12 + 2	12 + 2	11 + 2	97 + 1	83 + 1	72 + 1	63 + 1
2.8	11 + 2	13 + 2	11 + 2	94 + 1	77 + 1	64 + 1	54 + 1	46 + 1
2.9	12 + 2	11 + 2	88 + 1	66 + 1	52 + 1	41 + 1	34 + 1	28 + 1
3.0	11 + 2	74 + 1	55 + 1	38 + 1	29 + 1	22 + 1	18 + 1	14 + 1
3.1	66 + 1	39 + 1	26 + 1	18 + 1	14 + 1	10 + 1	88 + 0	73 + 0
3.2	31 + 1	17 + 1	12 + 1	10 + 1	88 + 0	78 + 0	71 + 0	65 + 0
3.3	10 + 1	10 + 1	12 + 1	14 + 1	14 + 1	15 + 1	15 + 1	14 + 1
3.4	69 + 0	20 + 1	30 + 1	36 + 1	39 + 1	39 + 1	38 + 1	36 + 1
3.5	19 + 1	49 + 1	65 + 1	71 + 1	69 + 1	64 + 1	59 + 1	53 + 1
3.6	48 + 1	88 + 1	98 + 1	93 + 1	83 + 1	73 + 1	63 + 1	56 + 1
3.7	83 + 1	11 + 2	11 + 2	95 + 1	88 + 1	67 + 1	57 + 1	48 + 1
3.8	11 + 2	12 + 2	10 + 2	81 + 1	65 + 1	52 + 1	43 + 1	36 + 1
3.9	12 + 2	10 + 2	80 + 1	58 + 1	45 + 1	34 + 1	28 + 1	23 + 1
4.0	11 + 2	70 + 1	50 + 1	34 + 1	25 + 1	19 + 1	15 + 1	12 + 1
4.1	65 + 1	38 + 1	25 + 1	17 + 1	12 + 1	96 + 0	76 + 0	62 + 0
4.2	32 + 1	16 + 1	12 + 1	96 + 0	80 + 0	69 + 0	60 + 0	54 + 0
4.3	11 + 1	10 + 1	11 + 1	12 + 1	12 + 1	12 + 1	12 + 1	11 + 1
4.4	67 + 0	18 + 1	27 + 1	31 + 1	33 + 1	32 + 1	31 + 1	29 + 1

Table 6.3. (*continued*)

Δn	The effective principal quantum number n_0^*							
	1.5	2.0	2.5	3.0	3.5	4.0	4.5	5.0
	Transition $p - p$; parameter χ $\kappa = 0$							
0.5	12 + 1	16 + 1	17 + 1	18 + 1	17 + 1	17 + 1	16 + 1	15 + 1
0.6	14 + 1	17 + 1	18 + 1	17 + 1	16 + 1	15 + 1	14 + 1	13 + 1
0.7	15 + 1	17 + 1	17 + 1	16 + 1	14 + 1	13 + 1	11 + 1	10 + 1
0.8	16 + 1	17 + 1	16 + 1	14 + 1	12 + 1	10 + 1	94 + 0	82 + 0
0.9	17 + 1	16 + 1	14 + 1	11 + 1	96 + 0	80 + 0	68 + 0	58 + 0
1.0	17 + 1	14 + 1	11 + 1	85 + 0	67 + 0	53 + 0	43 − 0	35 − 0
1.1	15 + 1	10 + 1	73 + 0	54 − 0	41 − 0	32 − 0	26 − 0	21 − 0
1.2	10 + 1	59 + 0	48 − 0	44 − 0	44 − 0	45 − 0	45 − 0	46 − 0
1.3	45 − 0	55 + 0	88 + 0	11 + 1	14 + 1	16 + 1	17 + 1	18 + 1
1.4	49 − 0	12 + 1	20 + 1	25 + 1	28 + 1	30 + 1	29 + 1	28 + 1
1.5	10 + 1	20 + 1	27 + 1	29 + 1	29 + 1	28 + 1	26 + 1	24 + 1
1.6	14 + 1	23 + 1	27 + 1	26 + 1	25 + 1	22 + 1	20 + 1	18 + 1
1.7	17 + 1	24 + 1	24 + 1	22 + 1	20 + 1	17 + 1	15 + 1	13 + 1
1.8	19 + 1	23 + 1	21 + 1	18 + 1	15 + 1	13 + 1	11 + 1	97 + 0
1.9	19 + 1	20 + 1	17 + 1	14 + 1	11 + 1	94 + 0	78 + 0	65 + 0
2.0	19 + 1	17 + 1	13 + 1	10 + 1	78 + 0	61 + 0	49 − 0	40 − 0
2.1	17 + 1	12 + 1	87 + 0	64 + 0	49 − 0	38 − 0	31 − 0	25 − 0
2.2	12 + 1	73 + 0	58 + 0	51 + 0	7 − 0	45 − 0	43 − 0	41 − 0
2.3	61 + 0	58 + 0	83 + 0	10 + 1	12 + 1	13 + 1	13 + 1	14 + 1
2.4	44 − 0	11 + 1	19 + 1	24 + 1	27 + 1	28 + 1	28 + 1	27 + 1
2.5	93 + 0	20 + 1	28 + 1	32 + 1	32 + 1	30 + 1	28 + 1	26 + 1
2.6	14 + 1	25 + 1	29 + 1	29 + 1	27 + 1	25 + 1	22 + 1	20 + 1
2.7	17 + 1	26 + 1	27 + 1	25 + 1	22 + 1	19 + 1	16 + 1	14 + 1
2.8	19 + 1	24 + 1	23 + 1	19 + 1	16 + 1	14 + 1	12 + 1	10 + 1
2.9	20 + 1	22 + 1	18 + 1	15 + 1	12 + 1	10 + 1	82 + 0	69 + 0
3.0	20 + 1	18 + 1	14 + 1	10 + 1	83 + 0	65 + 0	52 + 0	42 − 0
3.1	18 + 1	13 + 1	94 + 0	69 + 0	53 + 0	41 − 0	33 − 0	27 − 0
3.2	13 + 1	79 + 0	63 + 0	54 + 0	49 − 0	45 − 0	43 − 0	40 − 0
3.3	68 + 0	61 + 0	84 + 0	10 + 1	11 + 1	12 + 1	12 + 1	12 + 1
3.4	44 − 0	11 + 1	19 + 1	23 + 1	26 + 1	27 + 1	27 + 1	26 + 1
3.5	89 + 0	20 + 1	29 + 1	32 + 1	33 + 1	31 + 1	29 + 1	26 + 1
3.6	14 + 1	25 + 1	31 + 1	31 + 1	29 + 1	26 + 1	23 + 1	20 + 1
3.7	17 + 1	27 + 1	28 + 1	26 + 1	23 + 1	20 + 1	17 + 1	15 + 1
3.8	19 + 1	25 + 1	24 + 1	20 + 1	17 + 1	14 + 1	12 + 1	10 + 1
3.9	20 + 1	23 + 1	19 + 1	15 + 1	12 + 1	10 + 1	85 + 0	71 + 0
4.0	20 + 1	18 + 1	14 + 1	11 + 1	86 + 0	67 + 0	54 + 0	44 − 0
4.1	19 + 1	13 + 1	97 + 0	72 + 0	55 + 0	43 − 0	35 − 0	28 − 0
4.2	14 + 1	82 + 0	66 + 0	56 + 0	50 + 0	46 − 0	43 − 0	40 − 0
4.3	72 + 0	62 + 0	85 + 0	99 + 0	11 + 1	11 + 1	11 + 1	12 + 1
4.4	44 − 0	11 + 1	18 + 1	23 + 1	25 + 1	26 + 1	26 + 1	25 + 1

Table 6.3. (*continued*)

	The effective principal quantum number n_0^*							
Δn	1.5	2.0	2.5	3.0	3.5	4.0	4.5	5.0
	Transition $p-p$, parameter A $\kappa = 2$							
0.1	13 − 0	42 − 0	77 + 0	11 + 1	16 + 1	21 + 1	25 + 1	30 + 1
0.2	63 + 0	16 + 1	27 + 1	38 + 1	47 + 1	56 + 1	63 + 1	69 + 1
0.3	15 + 1	33 + 1	49 + 1	61 + 1	69 + 1	75 + 1	79 + 1	81 + 1
0.4	25 + 1	48 + 1	63 + 1	71 + 1	75 + 1	76 + 1	75 + 1	74 + 1
0.5	36 + 1	57 + 1	65 + 1	68 + 1	67 + 1	64 + 1	61 + 1	58 + 1
0.6	42 + 1	56 + 1	58 + 1	56 + 1	52 + 1	48 + 1	45 + 1	41 + 1
0.7	42 + 1	48 + 1	45 + 1	40 + 1	36 + 1	32 + 1	29 + 1	27 + 1
0.8	37 + 1	35 + 1	30 + 1	26 + 1	22 + 1	20 + 1	18 + 1	16 + 1
0.9	28 + 1	22 + 1	18 + 1	15 + 1	13 + 1	12 + 1	11 + 1	10 + 1
1.0	17 + 1	12 + 1	10 + 1	95 + 0	91 + 0	90 + 0	89 + 0	88 + 0
1.1	78 + 0	67 + 0	73 + 0	83 + 0	92 + 0	99 + 0	10 + 1	10 + 1
1.2	36 − 0	61 + 0	95 + 0	12 + 1	15 + 1	16 + 1	17 + 1	18 + 1
1.3	34 − 0	11 + 1	19 + 1	26 + 1	30 + 1	31 + 1	31 + 1	29 + 1
1.4	82 + 0	24 + 1	36 + 1	41 + 1	41 + 1	38 + 1	34 + 1	31 + 1
1.5	18 + 1	37 + 1	43 + 1	42 + 1	37 + 1	33 + 1	29 + 1	25 + 1
1.6	28 + 1	41 + 1	40 + 1	35 + 1	29 + 1	25 + 1	21 + 1	19 + 1
1.7	33 + 1	37 + 1	31 + 1	26 + 1	21 + 1	18 + 1	15 + 1	13 + 1
1.8	32 + 1	28 + 1	22 + 1	17 + 1	14 + 1	12 + 1	10 + 1	94 + 0
1.9	26 + 1	18 + 1	14 + 1	11 + 1	99 + 0	87 + 0	77 + 0	70 + 0
2.0	17 + 1	11 + 1	92 + 0	82 + 0	75 + 0	70 + 0	66 + 0	63 + 0
2.1	83 + 0	68 + 0	70 + 0	74 + 0	77 + 0	78 + 0	77 + 0	76 + 0
2.2	41 − 0	60 + 0	85 + 0	10 + 1	11 + 1	11 + 1	11 + 1	11 + 1
2.3	33 − 0	98 + 0	15 + 1	19 + 1	20 + 1	20 + 1	19 + 1	17 + 1
2.4	66 + 0	20 + 1	28 + 1	31 + 1	29 + 1	26 + 1	23 + 1	19 + 1
2.5	15 + 1	32 + 1	36 + 1	33 + 1	28 + 1	24 + 1	20 + 1	17 + 1
2.6	25 + 1	37 + 1	34 + 1	28 + 1	23 + 1	19 + 1	15 + 1	13 + 1
2.7	31 + 1	33 + 1	27 + 1	21 + 1	17 + 1	14 + 1	11 + 1	98 + 0
2.8	31 + 1	26 + 1	19 + 1	15 + 1	12 + 1	10 + 1	84 + 0	72 + 0
2.8	26 + 1	17 + 1	13 + 1	10 + 1	86 + 0	73 + 0	63 + 0	56 + 0
3.0	18 + 1	10 + 1	87 + 0	75 + 0	67 + 0	61 + 0	56 + 0	52 + 0
3.1	86 + 0	67 + 0	68 + 0	70 + 0	69 + 0	67 + 0	65 + 0	61 + 0
3.2	43 − 0	60 − 0	81 + 0	94 + 0	99 + 0	98 + 0	95 + 0	90 + 0
3.3	32 − 0	93 + 0	14 + 1	16 + 1	17 + 1	16 + 1	14 + 1	13 + 1
3.4	60 + 0	18 + 1	25 + 1	26 + 1	24 + 1	21 + 1	18 + 1	15 + 1
3.5	14 + 1	30 + 1	33 + 1	30 + 1	24 + 1	20 + 1	16 + 1	13 + 1
3.6	24 + 1	35 + 1	31 + 1	25 + 1	20 + 1	16 + 1	13 + 1	10 + 1
3.7	30 + 1	32 + 1	25 + 1	19 + 1	15 + 1	12 + 1	99 + 0	82 + 0
3.8	31 + 1	25 + 1	18 + 1	14 + 1	11 + 1	88 + 0	73 + 0	61 + 0
3.9	26 + 1	16 + 1	12 + 1	97 + 0	79 + 0	66 + 0	56 + 0	49 − 0
4.0	18 + 1	10 + 1	84 + 0	72 + 0	62 + 0	56 + 0	50 + 0	46 − 0
4.1	87 + 0	67 + 0	66 + 0	67 + 0	65 + 0	61 + 0	58 + 0	54 + 0
4.2	44 − 0	59 + 0	78 + 0	88 + 0	90 + 0	88 + 0	83 + 0	77 + 0
4.3	32 − 0	90 + 0	13 + 1	15 + 1	15 + 1	14 + 1	12 + 1	11 + 1
4.4	58 + 0	17 + 1	23 + 1	24 + 1	22 + 1	18 + 1	15 + 1	13 + 1

Table 6.3. (*continued*)

Δn	The effective principal quantum number n_0^*							
	1.5	2.0	2.5	3.0	3.5	4.0	4.5	5.0
	Transition $p-p$, parameter χ $\kappa=2$							
0.1	63 + 0	84 + 0	10 + 1	12 + 1	14 + 1	16 + 1	17 + 1	18 + 1
0.2	11 + 1	14 + 1	17 + 1	19 + 1	20 + 1	21 + 1	21 + 1	21 + 1
0.3	14 + 1	18 + 1	20 + 1	21 + 1	21 + 1	20 + 1	19 + 1	17 + 1
0.4	17 + 1	20 + 1	20 + 1	19 + 1	18 + 1	16 + 1	15 + 1	13 + 1
0.5	18 + 1	20 + 1	18 + 1	16 + 1	14 + 1	12 + 1	11 + 1	98 + 0
0.6	19 + 1	18 + 1	16 + 1	13 + 1	11 + 1	96 + 0	82 + 0	70 + 0
0.7	18 + 1	16 + 1	13 + 1	10 + 1	85 + 0	70 + 0	58 + 0	49 − 0
0.8	17 + 1	13 + 1	10 + 1	78 + 0	61 + 0	50 + 0	41 − 0	34 − 0
0.9	15 + 1	10 + 1	73 + 0	56 + 0	46 − 0	38 − 0	33 − 0	29 − 0
1.0	12 + 1	71 + 0	54 + 0	47 − 0	42 − 0	40 − 0	38 − 0	37 − 0
1.1	88 + 0	54 + 0	56 + 0	62 + 0	67 + 0	71 + 0	73 + 0	73 + 0
1.2	52 + 0	71 + 0	10 + 1	13 + 1	15 + 1	16 + 1	17 + 1	16 + 1
1.3	66 + 0	16 + 1	25 + 1	31 + 1	33 + 1	33 + 1	31 + 1	28 + 1
1.4	15 + 1	31 + 1	40 + 1	42 + 1	39 + 1	35 + 1	30 + 1	25 + 1
1.5	24 + 1	37 + 1	39 + 1	35 + 1	30 + 1	25 + 1	20 + 1	17 + 1
1.6	28 + 1	33 + 1	30 + 1	24 + 1	20 + 1	16 + 1	13 + 1	11 + 1
1.7	27 + 1	26 + 1	21 + 1	16 + 1	13 + 1	10 + 1	89 + 0	74 + 0
1.8	24 + 1	19 + 1	14 + 1	11 + 1	91 + 0	74 + 0	62 + 0	52 + 0
1.9	20 + 1	13 + 1	10 + 1	81 + 0	67 + 0	57 + 0	50 + 0	44 − 0
2.0	15 + 1	94 + 0	76 + 0	67 + 0	62 + 0	58 + 0	48 − 0	52 + 0
2.1	10 + 1	71 + 0	74 + 0	80 + 0	85 + 0	86 + 0	86 + 0	84 + 0
2.2	65 + 0	82 + 0	11 + 1	14 + 1	16 + 1	16 + 1	16 + 1	16 + 1
2.3	66 + 0	15 + 1	24 + 1	29 + 1	31 + 1	30 + 1	28 + 1	26 + 1
2.4	13 + 1	30 + 1	41 + 1	43 + 1	40 + 1	35 + 1	30 + 1	26 + 1
2.5	24 + 1	40 + 1	43 + 1	39 + 1	33 + 1	27 + 1	22 + 1	18 + 1
2.6	29 + 1	37 + 1	34 + 1	28 + 1	23 + 1	18 + 1	15 + 1	12 + 1
2.7	29 + 1	29 + 1	24 + 1	19 + 1	15 + 1	12 + 1	10 + 1	84 + 0
2.8	26 + 1	21 + 1	16 + 1	13 + 1	10 + 1	86 + 0	71 + 0	60 + 0
2.9	22 + 1	15 + 1	11 + 1	93 + 0	78 + 0	67 + 0	58 + 0	51 + 0
3.0	17 + 1	10 + 1	86 + 0	77 + 0	71 + 0	67 + 0	62 + 0	58 + 0
3.1	11 + 1	78 + 0	83 + 0	89 + 0	93 + 0	94 + 0	92 + 0	89 + 0
3.2	71 + 0	87 + 0	12 + 1	15 + 1	16 + 1	16 + 1	16 + 1	15 + 1
3.3	67 + 0	15 + 1	24 + 1	29 + 1	30 + 1	29 + 1	27 + 1	24 + 1
3.4	13 + 1	30 + 1	41 + 1	43 + 1	40 + 1	35 + 1	30 + 1	26 + 1
3.5	23 + 1	40 + 1	45 + 1	41 + 1	34 + 1	28 + 1	23 + 1	19 + 1
3.6	30 + 1	39 + 1	36 + 1	30 + 1	24 + 1	19 + 1	16 + 1	13 + 1
3.7	30 + 1	31 + 1	25 + 1	20 + 1	16 + 1	13 + 1	10 + 1	90 + 0
3.8	27 + 1	23 + 1	17 + 1	14 + 1	11 + 1	92 + 0	77 + 0	65 + 0
3.9	22 + 1	16 + 1	12 + 1	10 + 1	84 + 0	72 + 0	63 + 0	56 + 0
4.0	17 + 1	10 + 1	91 + 0	83 + 0	76 + 0	72 + 0	66 + 0	62 + 0
4.1	11 + 1	82 + 0	88 + 0	94 + 0	98 + 0	98 + 0	96 + 0	92 + 0
4.2	74 + 0	90 + 0	12 + 1	15 + 1	16 + 1	16 + 1	16 + 1	15 + 1
4.3	67 + 0	16 + 1	24 + 1	28 + 1	30 + 1	29 + 1	26 + 1	24 + 1
4.4	12 + 1	30 + 1	41 + 1	43 + 1	40 + 1	35 + 1	30 + 1	25 + 1

Table 6.3. (*continued*)

Δn	The effective principal quantum number n_0^*							
	1.5	2.0	2.5	3.0	3.5	4.0	4.5	5.0
	Transition $p-d$, parameter A $\kappa=1$							
0.1			30 + 1	43 + 1	53 + 1	59 + 1	65 + 1	68 + 1
0.2			60 + 1	73 + 1	79 + 1	82 + 1	82 + 1	82 + 1
0.3			86 + 1	91 + 1	90 + 1	86 + 1	82 + 1	78 + 1
0.4			10 + 2	99 + 1	89 + 1	81 + 1	74 + 1	67 + 1
0.5			11 + 2	96 + 1	81 + 1	69 + 1	60 + 1	53 + 1
0.6		15 + 2	10 + 2	85 + 1	66 + 1	54 + 1	45 + 1	38 + 1*
0.7		16 + 2	96 + 1	68 + 1	49 + 1	38 + 1	30 + 1*	25 + 1*
0.8		15 + 2	76 + 1	49 + 1	32 + 1	24 + 1*	18 + 1*	15 + 1*
0.9		12 + 2	53 + 1	31 + 1	19 + 1*	13 + 1*	10 + 1*	84 + 0*
1.0		97 + 1	32 + 1	16 + 1*	10 + 1*	74 + 0*	61 + 0	54 + 0
1.1	19 + 2	59 + 1	17 + 1*	83 + 0*	60 + 0	54 + 0	54 + 0	55 + 0
1.2	15 + 2	29 + 1	79 + 0*	57 + 0	65 + 0*	78 + 0*	93 + 0*	10 + 1
1.3	10 + 2	11 + 1*	64 + 0	11 + 1	19 + 1	25 + 1	24 + 1	21 + 1
1.4	64 + 1	65 + 0	23 + 1	41 + 1	34 + 1	28 + 1	23 + 1	19 + 1
1.5	29 + 1	29 + 1	52 + 1	42 + 1	32 + 1	25 + 1	20 + 1	17 + 1
1.6	66 + 0*	68 + 1	53 + 1	39 + 1	28 + 1	21 + 1	17 + 1	13 + 1
1.7	38 − 0	79 + 1	49 + 1	34 + 1	23 + 1	17 + 1	13 + 1*	10 + 1*
1.8	34 + 1	83 + 1	43 + 1	28 + 1	17 + 1	12 + 1*	93 + 0*	73 + 0*
1.9	56 + 1	80 + 1	34 + 1	20 + 1	12 + 1*	83 + 0*	60 + 0*	48 − 0*
2.0	72 + 1	69 + 1	24 + 1	12 + 1*	73 + 0*	51 + 0*	39 − 0*	33 − 0
2.1	10 + 2	48 + 1	15 + 1	71 + 0*	46 − 0	36 − 0	33 − 0	31 − 0
2.2	98 + 1	27 + 1	79 + 0	45 − 0	41 − 0	42 − 0	45 − 0	48 − 0
2.3	79 + 1	12 + 1	50 + 0	57 + 0*	78 + 0*	10 + 1	11 + 1	11 + 1
2.4	54 + 1	61 + 0	87 + 0*	21 + 1	23 + 1	18 + 1	14 + 1	12 + 1
2.5	30 + 1	10 + 1	39 + 1	31 + 1	23 + 1	17 + 1	13 + 1	10 + 1
2.6	97 + 0	50 + 1	42 + 1	30 + 1	21 + 1	15 + 1	11 + 1	93 + 0
2.7	26 − 0	65 + 1	39 + 1	27 + 1	17 + 1	12 + 1	94 + 0*	73 + 0*
2.8	20 + 1	68 + 1	35 + 1	22 + 1	13 + 1	97 + 0*	69 + 0*	53 + 0*
2.9	41 + 1	67 + 1	29 + 1	17 + 1	98 + 0*	67 + 0*	47 − 0*	36 − 0*
3.0	54 + 1	61 + 1	21 + 1	11 + 1*	62 + 0*	43 − 0*	32 − 0*	26 − 0*
3.1	84 + 1	44 + 1	14 + 1	66 + 0*	40 − 0*	30 − 0	26 − 0	23 − 0
3.2	81 + 1	27 + 1	79 + 0*	41 − 0	33 − 0	32 − 0	33 − 0*	33 − 0
3.3	68 + 1	13 + 1	47 − 0	44 − 0*	54 + 0	66 + 0	75 + 0	75 + 0
3.4	50 + 1	63 + 0	65 + 0	13 + 1	17 + 1	14 + 1	11 + 1	92 + 0
3.5	30 + 1	73 + 0	32 + 1	27 + 1	19 + 1	14 + 1	11 + 1	86 + 0
3.6	11 + 1	42 + 1	37 + 1	26 + 1	17 + 1	13 + 1	96 + 0	74 + 0
3.7	29 − 0	58 + 1	35 + 1	24 + 1	15 + 1	10 + 1	78 + 0	60 + 0*
3.8	10 + 1	62 + 1	31 + 1	20 + 1	12 + 1	84 + 0*	59 + 0*	45 − 0*
3.9	35 + 1	61 + 1	26 + 1	15 + 1	87 + 0*	59 + 0*	41 − 0*	31 − 0*
4.0	47 + 1	57 + 1	19 + 1	10 + 1	57 + 0*	39 − 0*	28 − 0*	22 − 0*
4.1	74 + 1	42 + 1	13 + 1	63 + 0*	37 − 0	27 − 0	22 − 0	20 − 0
4.2	73 + 1	26 + 1	78 + 0*	38 − 0	30 − 0	28 − 0	27 − 0	26 − 0
4.3	63 + 1	13 + 1	46 − 0	39 − 0	45 − 0	52 + 0	58 + 0	58 + 0
4.4	48 + 1	65 + 0	56 + 0	10 + 1	14 + 1	12 + 1	99 + 0	78 + 0

Table 6.3. (*continued*)

Δn	The effective principal quantum number n_0^*							
	1.5	2.0	2.5	3.0	3.5	4.0	4.5	5.0
	Transition $p - d$, parameter χ $\kappa = 1$							
0.1			10 + 1	10 + 1	10 + 1	96 + 0	92 + 0	88 + 0
0.2			94 + 0	84 + 0	76 + 0	69 + 0	63 + 0	57 + 0
0.3			79 + 0	67 + 0	58 + 0	50 + 0	44 − 0	38 − 0
0.4			66 + 0	53 + 0	44 − 0	37 − 0	31 − 0	26 − 0
0.5			55 + 0	42 − 0	33 − 0	27 − 0	22 − 0	18 − 0
0.6		65 + 0	45 − 0	33 − 0	25 − 0	19 − 0	15 − 0	12 − 0*
0.7		57 + 0	36 − 0	25 − 0	18 − 0	13 − 0	10 − 0*	10 − 0*
0.8		49 − 0	28 − 0	19 − 0	12 − 0	10 − 0*	10 − 0*	10 − 0*
0.9		41 − 0	21 − 0	13 − 0	10 − 0*	10 − 0*	10 − 0*	10 − 0*
1.0		33 − 0	15 − 0	10 − 0*	10 − 0*	10 − 0*	10 − 0	10 − 0
1.1	56 + 0	25 − 0	10 − 0*	10 − 0*	10 − 0	10 − 0	11 − 0	14 − 0
1.2	48 − 0	16 − 0	10 − 0*	11 − 0	23 − 0*	40 − 0	59 + 0*	77 + 0
1.3	41 − 0	10 − 0*	19 − 0	76 + 0	18 + 1	24 + 1*	21 + 1	17 + 1
1.4	33 − 0	18 − 0	21 + 1	36 + 1	24 + 1	16 + 1	12 + 1	92 + 0
1.5	24 − 0	26 + 1	32 + 1	19 + 1	12 + 1	84 + 0	61 + 0	46 − 0
1.6	11 − 0*	32 + 1	16 + 1	95 + 0	63 + 0	45 − 0	33 − 0	25 − 0
1.7	44 − 0	16 + 1	87 + 0	55 + 0	37 − 0	26 − 0	19 − 0*	14 − 0*
1.8	30 + 1	10 + 1	54 + 0	34 − 0	22 − 0	15 − 0*	11 − 0*	10 − 0*
1.9	15 + 1	67 − 0	35 − 0	22 − 0*	13 − 0*	10 − 0*	10 − 0*	10 − 0*
2.0	95 + 0	48 − 0	23 − 0	13 − 0*	10 − 0*	10 − 0*	10 − 0*	10 − 0
2.1	73 + 0	34 − 0	15 − 0	10 − 0*	10 − 0	10 − 0	10 − 0	12 − 0
2.2	58 + 0	22 − 0	10 − 0*	10 − 0	16 − 0	25 − 0	36 − 0	48 − 0
2.3	47 − 0	13 − 0	13 − 0	37 − 0*	82 + 0*	13 + 1	17 + 1	16 + 1
2.4	36 − 0	12 − 0	76 + 0*	29 + 1	30 + 1	21 + 1	15 + 1	11 + 1
2.5	27 − 0	76 + 0	43 + 1	26 + 1	17 + 1	11 + 1	83 + 0	62 + 0
2.6	14 − 0	43 + 1	23 + 1	13 + 1	87 + 0	60 + 0	44 − 0	33 − 0
2.7	12 − 0	25 + 1	11 + 1	72 + 0	48 − 0	33 − 0	24 − 0*	18 − 0*
2.8	37 + 1	13 + 1	68 + 0	43 − 0	28 − 0	19 − 0*	14 − 0*	10 − 0*
2.9	22 + 1	81 + 0	42 − 0	26 − 0	17 − 0*	11 − 0*	10 − 0*	10 − 0*
3.0	12 + 1	55 + 0	27 − 0	16 − 0*	10 − 0*	10 − 0*	10 − 0*	10 − 0*
3.1	82 + 0	38 − 0	18 − 0	10 − 0*	10 − 0*	10 − 0	10 − 0	11 − 0
3.2	63 + 0	25 − 0	11 − 0*	10 − 0	14 − 0	21 − 0	30 − 0*	38 − 0
3.3	49 − 0	15 − 0	12 − 0	30 − 0*	61 + 0	10 + 1	13 + 1	14 + 1
3.4	38 − 0	12 − 0	53 + 0	20 + 1	29 + 1	23 + 1	17 + 1	13 + 1
3.5	28 − 0	47 − 0	44 + 1	30 + 1	19 + 1	13 + 1	96 + 0	71 + 0
3.6	16 − 0	45 + 1	28 + 1	15 + 1	10 + 1	70 + 0	51 + 0	38 − 0
3.7	10 − 0	30 + 1	13 + 1	82 + 0	55 + 0	38 − 0	28 − 0	21 − 0*
3.8	20 + 1	15 + 1	76 + 0	48 − 0	32 − 0	22 − 0*	16 − 0*	12 − 0*
3.9	28 + 1	90 + 0	47 − 0	29 − 0	19 − 0*	13 − 0*	10 − 0*	10 − 0*
4.0	14 + 1	59 + 0	30 − 0	18 − 0	11 − 0*	10 − 0*	10 − 0*	10 − 0*
4.1	88 + 0	41 − 0	19 − 0	11 − 0*	10 − 0*	10 − 0	10 − 0	10 − 0
4.2	66 + 0	27 − 0	12 − 0*	10 − 0	14 − 0	20 − 0	27 − 0	34 − 0
4.3	51 + 0	17 − 0	12 − 0	26 − 0	52 + 0	85 + 0	11 + 1	13 + 1
4.4	39 − 0	12 − 0	44 − 0	16 + 1	27 + 1	24 + 1	18 + 1	14 + 1

Table 6.3. (*continued*)

Δn	1.5	2.0	2.5	3.0	3.5	4.0	4.5	5.0
	The effective principal quantum number n_0^*							
	Transition $p-d$, parameter A $\kappa=3$							
0.1			27 − 1	66 − 1	11 − 0	18 − 0	26 − 0	36 − 0
0.2			19 − 0	42 − 0	72 + 0	10 + 1	14 + 1	19 + 1
0.3			63 + 0	12 + 1	19 + 1	27 + 1	34 + 1	42 + 1
0.4			14 + 1	25 + 1	36 + 1	46 + 1	54 + 1	61 + 1
0.5			25 + 1	40 + 1	51 + 1	60 + 1	65 + 1	68 + 1
0.6		19 + 1	38 + 1	53 + 1	61 + 1	65 + 1	65 + 1	63 + 1
0.7		29 + 1	49 + 1	59 + 1	62 + 1	59 + 1	56 + 1	51 + 1
0.8		40 + 1	56 + 1	58 + 1	54 + 1	48 + 1	42 + 1	37 + 1
0.9		48 + 1	55 + 1	49 + 1	41 + 1	34 + 1	29 + 1	25 + 1
1.0		51 + 1	47 + 1	36 + 1	28 + 1	22 + 1	18 + 1	15 + 1
1.1	31 + 1	48 + 1	35 + 1	23 + 1	17 + 1	13 + 1	11 + 1	10 + 1
1.2	34 + 1	38 + 1	22 + 1	14 + 1	10 + 1	95 + 0	91 + 0	92 + 0
1.3	33 + 1	25 + 1	12 + 1	86 + 0	83 + 0	92 + 0	10 + 1	11 + 1
1.4	28 + 1	13 + 1	69 + 0	77 + 0	10 + 1	13 + 1	16 + 1	19 + 1
1.5	21 + 1	59 + 0	65 + 0	11 + 1	17 + 1	23 + 1	27 + 1	29 + 1
1.6	11 + 1	38 − 0	10 + 1	21 + 1	29 + 1	34 + 1	34 + 1	33 + 1
1.7	40 − 0	61 + 0	20 + 1	32 + 1	38 + 1	37 + 1	34 + 1	30 + 1
1.8	13 − 0	13 + 1	31 + 1	39 + 1	38 + 1	33 + 1	28 + 1	24 + 1
1.9	22 − 0	23 + 1	39 + 1	39 + 1	32 + 1	26 + 1	21 + 1	17 + 1
2.0	62 + 0	33 + 1	39 + 1	32 + 1	24 + 1	18 + 1	15 + 1	12 + 1
2.1	13 + 1	38 + 1	33 + 1	23 + 1	16 + 1	12 + 1	10 + 1	88 + 0
2.2	20 + 1	36 + 1	23 + 1	14 + 1	11 + 1	92 + 0	80 + 0	74 + 0
2.3	24 + 1	27 + 1	14 + 1	96 + 0	83 + 0	81 + 0	81 + 0	83 + 0
2.4	25 + 1	17 + 1	84 + 0	76 + 0	85 + 0	10 + 1	10 + 1	11 + 1
2.5	21 + 1	86 + 0	65 + 0	92 + 0	12 + 1	15 + 1	16 + 1	17 + 1
2.6	14 + 1	46 − 0	85 + 0	15 + 1	20 + 1	23 + 1	23 + 1	22 + 1
2.7	65 + 0	48 − 0	15 + 1	24 + 1	28 + 1	28 + 1	25 + 1	22 + 1
2.8	21 − 0	92 + 0	24 + 1	32 + 1	31 + 1	27 + 1	22 + 1	19 + 1
2.9	15 − 0	17 + 1	33 + 1	34 + 1	28 + 1	22 + 1	17 + 1	15 + 1
3.0	37 − 0	28 + 1	36 + 1	29 + 1	22 + 1	16 + 1	13 + 1	10 + 1
3.1	99 + 0	34 + 1	32 + 1	22 + 1	15 + 1	11 + 1	95 + 0	78 + 0
3.2	16 + 1	34 + 1	23 + 1	14 + 1	10 + 1	88 + 0	71 + 0	68 + 1
3.3	21 + 1	28 + 1	15 + 1	98 + 0	80 + 0	76 + 0	71 + 0	69 + 0
3.4	23 + 1	18 + 1	91 + 0	75 + 0	78 + 0	87 + 0	89 + 0	93 + 0
3.5	21 + 1	99 + 0	66 + 0	84 + 0	10 + 1	12 + 1	13 + 1	14 + 1
3.6	15 + 1	52 + 0	78 + 0	12 + 1	16 + 1	19 + 1	19 + 1	19 + 1
3.7	78 + 0	45 − 0	13 + 1	21 + 1	24 + 1	24 + 1	21 + 1	19 + 1
3.8	28 − 0	79 + 0	22 + 1	28 + 1	27 + 1	24 + 1	19 + 1	16 + 1
3.9	14 − 0	15 + 1	30 + 1	31 + 1	26 + 1	20 + 1	15 + 1	15 + 1
4.0	29 − 0	26 + 1	34 + 1	27 + 1	20 + 1	15 + 1	11 + 1	10 + 1
4.1	83 + 0	33 + 1	31 + 1	21 + 1	15 + 1	11 + 1	87 + 0	79 + 0
4.2	14 + 1	34 + 1	23 + 1	14 + 1	10 + 1	86 + 0	66 + 0	69 + 0
4.3	20 + 1	28 + 1	15 + 1	98 + 0	79 + 0	75 + 0	62 + 0	63 + 0
4.4	22 + 1	19 + 1	94 + 0	75 + 0	74 + 0	85 + 0	76 + 0	73 + 0

Table 6.3. (*continued*)

	The effective principal quantum number n_0^*							
Δn	1.5	2.0	2.5	3.0	3.5	4.0	4.5	5.0
	Transition $p-d$, parameter χ $\kappa = 3$							
0.1			19 − 0	30 − 0	41 − 0	54 + 0	66 + 0	79 + 0
0.2			59 + 0	88 + 0	11 + 1	14 + 1	17 + 1	20 + 1
0.3			10 + 1	15 + 1	19 + 1	23 + 1	26 + 1	29 + 1
0.4			15 + 1	21 + 1	26 + 1	29 + 1	31 + 1	32 + 1
0.5			20 + 1	26 + 1	30 + 1	32 + 1	32 + 1	31 + 1
0.6		16 + 1	24 + 1	29 + 1	31 + 1	31 + 1	29 + 1	27 + 1
0.7		19 + 1	27 + 1	31 + 1	30 + 1	28 + 1	25 + 1	22 + 1
0.8		22 + 1	29 + 1	30 + 1	27 + 1	23 + 1	20 + 1	17 + 1
0.9		25 + 1	30 + 1	27 + 1	23 + 1	19 + 1	15 + 1	13 + 1
1.0		27 + 1	28 + 1	23 + 1	18 + 1	14 + 1	12 + 1	10 + 1
1.1	19 + 1	28 + 1	25 + 1	18 + 1	14 + 1	11 + 1	97 + 0	86 + 0
1.2	21 + 1	28 + 1	20 + 1	14 + 1	11 + 1	10 + 1	99 + 0	10 + 1
1.3	24 + 1	25 + 1	14 + 1	11 + 1	11 + 1	12 + 1	14 + 1	16 + 1
1.4	25 + 1	18 + 1	10 + 1	12 + 1	17 + 1	22 + 1	26 + 1	28 + 1
1.5	26 + 1	11 + 1	11 + 1	20 + 1	29 + 1	37 + 1	41 + 1	42 + 1
1.6	24 + 1	82 + 0	18 + 1	33 + 1	44 + 1	48 + 1	47 + 1	43 + 1
1.7	16 + 1	11 + 1	29 + 1	44 + 1	40 + 1	46 + 1	41 + 1	35 + 1
1.8	68 + 0	19 + 1	38 + 1	46 + 1	44 + 1	38 + 1	31 + 1	26 + 1
1.9	84 + 0	26 + 1	41 + 1	42 + 1	35 + 1	29 + 1	23 + 1	19 + 1
2.0	14 + 1	31 + 1	40 + 1	34 + 1	27 + 1	21 + 1	17 + 1	14 + 1
2.1	18 + 1	34 + 1	34 + 1	26 + 1	20 + 1	16 + 1	13 + 1	11 + 1
2.2	22 + 1	34 + 1	27 + 1	19 + 1	15 + 1	13 + 1	12 + 1	11 + 1
2.3	26 + 1	31 + 1	20 + 1	15 + 1	13 + 1	14 + 1	14 + 1	15 + 1
2.4	28 + 1	24 + 1	14 + 1	14 + 1	16 + 1	20 + 1	22 + 1	24 + 1
2.5	29 + 1	16 + 1	12 + 1	18 + 1	26 + 1	32 + 1	34 + 1	35 + 1
2.6	28 + 1	10 + 1	17 + 1	30 + 1	40 + 1	45 + 1	45 + 1	42 + 1
2.7	21 + 1	10 + 1	27 + 1	43 + 1	50 + 1	49 + 1	44 + 1	37 + 1
2.8	10 + 1	17 + 1	38 + 1	49 + 1	49 + 1	43 + 1	35 + 1	29 + 1
2.9	72 + 0	25 + 1	44 + 1	47 + 1	41 + 1	33 + 1	26 + 1	22 + 1
3.0	12 + 1	32 + 1	44 + 1	39 + 1	31 + 1	25 + 1	20 + 1	16 + 1
3.1	17 + 1	36 + 1	38 + 1	30 + 1	23 + 1	18 + 1	15 + 1	13 + 1
3.2	22 + 1	37 + 1	31 + 1	22 + 1	17 + 1	15 + 1	13 + 1	13 + 1
3.3	26 + 1	34 + 1	22 + 1	17 + 1	15 + 1	15 + 1	15 + 1	15 + 1
3.4	29 + 1	27 + 1	16 + 1	15 + 1	17 + 1	19 + 1	21 + 1	23 + 1
3.5	30 + 1	18 + 1	13 + 1	18 + 1	24 + 1	30 + 1	32 + 1	35 + 1
3.6	29 + 1	11 + 1	16 + 1	28 + 1	37 + 1	42 + 1	45 + 1	43 + 1
3.7	23 + 1	10 + 1	26 + 1	42 + 1	49 + 1	49 + 1	45 + 1	41 + 1
3.8	13 + 1	16 + 1	37 + 1	50 + 1	51 + 1	45 + 1	37 + 1	31 + 1
3.9	75 + 0	24 + 1	45 + 1	49 + 1	43 + 1	35 + 1	27 + 1	24 + 1
4.0	11 + 1	32 + 1	45 + 1	42 + 1	34 + 1	26 + 1	21 + 1	18 + 1
4.1	17 + 1	37 + 1	41 + 1	32 + 1	25 + 1	20 + 1	16 + 1	15 + 1
4.2	22 + 1	38 + 1	33 + 1	24 + 1	19 + 1	16 + 1	13 + 1	14 + 1
4.3	27 + 1	35 + 1	24 + 1	18 + 1	16 + 1	16 + 1	14 + 1	15 + 1
4.4	29 + 1	28 + 1	17 + 1	16 + 1	17 + 1	20 + 1	20 + 1	21 + 1

Table 6.3. (*continued*)

Δn	The effective principal quantum number n_0^*					
	2.5	3.0	3.5	4.0	4.5	5.0
	Transition $d - s$, parameter A					
0.1	24 − 0	54 + 0	90 + 0	12 + 1	16 + 1	20 + 1
0.2	71 + 0	15 + 1	23 + 1	30 + 1	36 + 1	42 + 1
0.3	10 + 1	20 + 1	28 + 1	35 + 1	40 + 1	44 + 1
0.4	10 + 1	19 + 1	25 + 1	30 + 1	33 + 1	35 + 1
0.5	83 + 0	14 + 1	18 + 1	21 + 1	23 + 1	24 + 1
0.6	56 + 0	96 + 0	12 + 1	13 + 1	14 + 1	15 + 1
0.7	34 − 0	59 + 0	75 + 0	86 + 0	92 + 0	96 + 0
0.8	21 − 0	38 − 0	51 + 0	59 + 0	65 + 0	68 + 0
0.9	17 − 0	35 − 0	47 − 0	56 + 0	62 + 0	66 + 0
1.0	11 − 0*	49 − 0	57 + 0	78 + 0	78 + 0	89 + 0
1.1	44 − 0	85 + 0	11 + 1	13 + 1	14 + 1	14 + 1
1.2	67 + 0	12 + 1	17 + 1	19 + 1	20 + 1	20 + 1
1.3	74 + 0	13 + 1	17 + 1	18 + 1	18 + 1	18 + 1
1.4	65 + 0	11 + 1	13 + 1	14 + 1	14 + 1	14 + 1
1.5	49 − 0	80 + 0	97 + 0	10 + 1	10 + 1	10 + 1
1.6	34 − 0	56 + 0	68 + 0	73 + 0	74 + 0	73 + 0
1.7	24 − 0	41 − 0	50 + 0	55 + 0	56 + 0	55 + 0
1.8	19 − 0	35 − 0	43 − 0	48 − 0	49 − 0	49 − 0
1.9	18 − 0	38 − 0	47 − 0	53 + 0	54 + 0	55 + 0
2.0	18 − 0	53 + 0	62 + 0	74 + 0	–	73 + 0
2.1	45 − 0	81 + 0	10 + 1	11 + 1	11 + 1	10 + 1
2.2	59 + 0	10 + 1	13 + 1	13 + 1	13 + 1	12 + 1
2.3	61 + 0	10 + 1	12 + 1	13 + 1	12 + 1	11 + 1
2.4	52 + 0	86 + 0	10 + 1	10 + 1	99 + 0	93 + 0
2.5	40 − 0	64 + 0	75 + 0	76 + 0	74 + 0	70 + 0
2.6	29 − 0	47 − 0	55 + 0	57 + 0	56 + 0	53 + 0
2.7	22 − 0	37 − 0	44 − 0	46 − 0	45 − 0	44 − 0
2.8	18 − 0	34 − 0	40 − 0	44 − 0	43 − 0	42 − 0
2.9	18 − 0	39 − 0	46 − 0	50 + 0	49 − 0	48 − 0
3.0	18 − 0	54 + 0	59 + 0	69 + 0	64 + 0	63 + 0
3.1	44 − 0	77 + 0	94 + 0	97 + 0	93 + 0	87 + 0
3.2	55 + 0	95 + 0	11 + 1	11 + 1	11 + 1	10 + 1
3.3	55 + 0	92 + 0	10 + 1	10 + 1	10 + 1	92 + 0
3.4	47 − 0	76 + 0	86 + 0	86 + 0	81 + 0	74 + 0
3.5	36 − 0	57 + 0	65 + 0	65 + 0	62 + 0	57 + 0
3.6	27 − 0	43 − 0	50 + 0	50 + 0	48 − 0	45 − 0
3.7	21 − 0	35 − 0	41 − 0	42 − 0	40 − 0	38 − 0
3.8	18 − 0	33 − 0	39 − 0	41 − 0	39 − 0	37 − 0
3.9	17 − 0	39 − 0	44 − 0	48 − 0	45 − 0	43 − 0
4.0	17 − 0	53 + 0	57 + 0	65 + 0	59 + 0	57 + 0
4.1	43 − 0	75 + 0	89 + 0	90 + 0	84 + 0	76 + 0
4.2	52 + 0	90 + 0	10 + 1	10 + 1	96 + 0	86 + 0
4.3	52 + 0	86 + 0	98 + 0	96 + 0	88 + 0	79 + 0
4.4	44 − 0	70 + 0	79 + 0	77 + 0	71 + 0	64 + 0

Table 6.3. (*continued*)

Δn	The effective principal quantum number n_0^*					
	2.5	3.0	3.5	4.0	4.5	5.0
	Transition $d-s$, parameter χ $\kappa=2$					
0.1	11 + 1	13 + 1	14 + 1	16 + 1	17 + 1	18 + 1
0.2	18 + 1	20 + 1	21 + 1	21 + 1	21 + 1	21 + 1
0.3	21 + 1	21 + 1	20 + 1	19 + 1	18 + 1	17 + 1
0.4	19 + 1	18 + 1	16 + 1	15 + 1	13 + 1	12 + 1
0.5	15 + 1	14 + 1	12 + 1	11 + 1	97 + 0	86 + 0
0.6	11 + 1	10 + 1	90 + 0	78 + 0	67 + 0	59 + 0
0.7	84 + 0	74 + 0	65 + 0	57 + 0	49 − 0	43 − 0
0.8	64 + 0	61 + 0	56 + 0	51 + 0	45 − 0	41 − 0
0.9	68 + 0	76 + 0	73 + 0	70 + 0	64 + 0	59 + 0
1.0	27 − 0*	14 + 1	11 + 1	13 + 1	10 + 1	11 + 1
1.1	23 + 1	27 + 1	28 + 1	27 + 1	25 + 1	23 + 1
1.2	34 + 1	39 + 1	41 + 1	39 + 1	36 + 1	32 + 1
1.3	34 + 1	37 + 1	36 + 1	34 + 1	30 + 1	27 + 1
1.4	27 + 1	28 + 1	26 + 1	23 + 1	21 + 1	18 + 1
1.5	20 + 1	19 + 1	18 + 1	15 + 1	13 + 1	12 + 1
1.6	14 + 1	13 + 1	12 + 1	11 + 1	98 + 0	85 + 0
1.7	10 + 1	10 + 1	99 + 0	90 + 0	79 + 0	70 + 0
1.8	93 + 0	10 + 1	98 + 0	91 + 0	82 + 0	73 + 0
1.9	10 + 1	13 + 1	12 + 1	12 + 1	11 + 1	10 + 1
2.0	10 + 1	21 + 1	19 + 1	20 + 1	–	16 + 1
2.1	28 + 1	33 + 1	34 + 1	32 + 1	29 + 1	26 + 1
2.2	35 + 1	41 + 1	42 + 1	39 + 1	35 + 1	31 + 1
2.3	34 + 1	38 + 1	37 + 1	34 + 1	30 + 1	26 + 1
2.4	27 + 1	28 + 1	27 + 1	24 + 1	21 + 1	18 + 1
2.5	20 + 1	20 + 1	19 + 1	17 + 1	15 + 1	13 + 1
2.6	14 + 1	15 + 1	14 + 1	12 + 1	11 + 1	97 + 0
2.7	11 + 1	12 + 1	11 + 1	10 + 1	95 + 0	84 + 0
2.8	10 + 1	12 + 1	12 + 1	11 + 1	10 + 1	91 + 0
2.9	11 + 1	16 + 1	15 + 1	15 + 1	13 + 1	12 + 1
3.0	12 + 1	24 + 1	22 + 1	23 + 1	20 + 1	18 + 1
3.1	30 + 1	36 + 1	37 + 1	34 + 1	31 + 1	27 + 1
3.2	35 + 1	42 + 1	43 + 1	40 + 1	35 + 1	31 + 1
3.3	33 + 1	38 + 1	37 + 1	34 + 1	30 + 1	26 + 1
3.4	27 + 1	29 + 1	27 + 1	24 + 1	21 + 1	18 + 1
3.5	20 + 1	21 + 1	19 + 1	17 + 1	15 + 1	13 + 1
3.6	15 + 1	15 + 1	15 + 1	13 + 1	11 + 1	10 + 1
3.7	12 + 1	13 + 1	12 + 1	11 + 1	10 + 1	93 + 0
3.8	11 + 1	13 + 1	13 + 1	12 + 1	11 + 1	10 + 1
3.9	12 + 1	18 + 1	17 + 1	17 + 1	15 + 1	13 + 1
4.0	13 + 1	26 + 1	24 + 1	25 + 1	21 + 1	20 + 1
4.1	30 + 1	37 + 1	38 + 1	35 + 1	32 + 1	28 + 1
4.2	35 + 1	42 + 1	43 + 1	40 + 1	35 + 1	30 + 1
4.3	33 + 1	38 + 1	37 + 1	34 + 1	29 + 1	25 + 1
4.4	26 + 1	29 + 1	27 + 1	25 + 1	22 + 1	19 + 1

Table 6.3. (*continued*)

Δn	The effective principal quantum number n_0^*					
	2.5	3.0	3.5	4.0	4.5	5.0
	Transition $d-p$, parameter A $\kappa = 1$					
0.1	19 + 1	31 + 1	41 + 1	48 + 1	53 + 1	58 + 1
0.2	24 + 1	37 + 1	46 + 1	51 + 1	55 + 1	57 + 1
0.3	21 + 1	32 + 1	38 + 1	41 + 1	43 + 1	43 + 1
0.4	15 + 1	23 + 1	27 + 1	29 + 1	29 + 1	29 + 1
0.5	94 + 0	15 + 1	17 + 1	18 + 1	18 + 1	18 + 1
0.6	52 + 0	87 + 0	99 + 0	10 + 1*	10 + 1*	10 + 1*
0.7	32 − 0	50 + 0	58 + 0	63 + 0	65 + 0	66 + 0
0.8	35 − 0	38 − 0	47 − 0	49 − 0	52 + 0	52 + 0
0.9	77 + 0	50 + 0*	63 + 0*	60 + 0*	63 + 0*	62 + 0*
1.0	18 + 1	12 + 1	15 + 1	13 + 1	13 + 1	11 + 1
1.1	16 + 1	20 + 1	23 + 1	21 + 1	20 + 1	18 + 1
1.2	14 + 1	17 + 1	18 + 1	17 + 1	16 + 1	14 + 1
1.3	99 + 0	12 + 1	13 + 1	12 + 1	11 + 1	11 + 1
1.4	63 + 0	87 + 0	92 + 0	89 + 0	85 + 0	79 + 0
1.5	39 − 0	59 − 0	62 + 0	62 + 0	59 + 0*	56 + 0*
1.6	28 − 0	40 − 0	43 − 0	43 − 0	42 − 0	40 − 0
1.7	31 − 0	31 − 0	35 − 0	35 − 0	34 − 0	32 − 0
1.8	57 + 0	33 − 0	39 − 0	35 − 0	35 − 0	33 − 0
1.9	13 + 1	51 + 0	64 + 0	50 + 0	50 + 0	44 − 0*
2.0	17 + 1	12 + 1	15 + 1	10 + 1	10 + 1	81 + 0
2.1	13 + 1	14 + 1	15 + 1	13 + 1	12 + 1	10 + 1
2.2	11 + 1	12 + 1	12 + 1	10 + 1	96 + 0	84 + 0
2.3	75 + 0	89 + 0	89 + 0	81 + 0	73 + 0	65 + 0
2.4	47 − 0	63 + 0	63 + 0	58 + 0	53 + 0	48 − 0
2.5	31 − 0	44 − 0	44 − 0	42 − 0	39 − 0	36 − 0
2.6	26 − 0	32 − 0	33 − 0	32 − 0	30 − 0	27 − 0
2.7	36 − 0	28 − 0	30 − 0	28 − 0	26 − 0	24 − 0
2.8	75 + 0	32 − 0	37 − 0	30 − 0	29 − 0	26 − 0
2.9	14 + 1	52 + 0	65 + 0	45 − 0	44 − 0	36 − 0
3.0	17 + 1	11 + 1	14 + 1	92 + 0	86 + 0	65 + 0
3.1	12 + 1	12 + 1	12 + 1	10 + 1	91 + 0	77 + 0
3.2	10 + 1	10 + 1	10 + 1	85 + 0	74 + 0	63 + 0
3.3	67 + 0	75 + 0	73 + 0	64 + 0	56 + 0	49 − 0
3.4	42 − 0	53 + 0	52 + 0	47 − 0	42 − 0	37 − 0
3.5	28 − 0	38 − 0	38 − 0	35 − 0	31 − 0	28 − 0
3.6	26 − 0	29 − 0	29 − 0	27 − 0	25 − 0	22 − 0
3.7	40 − 0	26 − 0	28 − 0	25 − 0	23 − 0	20 − 0
3.8	86 + 0	31 − 0	36 − 0	28 − 0	27 − 0	23 − 0
3.9	14 + 1	52 + 0	66 + 0	42 − 0	41 − 0	32 − 0
4.0	17 + 1	10 + 1	13 + 1	84 + 0	77 + 0	57 + 0
4.1	12 + 1	11 + 1	11 + 1	91 + 0	77 + 0	64 + 0
4.2	95 + 0	92 + 0	88 + 0	74 + 0	63 + 0	52 + 0
4.3	63 + 0	68 + 0	65 + 0	56 + 0	48 − 0	41 − 0
4.4	39 − 0	49 − 0	47 − 0	41 − 0	36 − 0	31 − 0

Table 6.3. (*continued*)

Δn	The effective principal quantum number n_0^*					
	2.5	3.0	3.5	4.0	4.5	5.0
	Transition $d-p$, parameter χ $\kappa = 1$					
0.1	10 + 1	10 + 1	99 + 0	95 + 0	90 + 0	86 + 0
0.2	88 + 0	79 + 0	71 + 0	64 + 0	59 + 0	53 + 0
0.3	67 + 0	57 + 0	49 − 0	43 − 0	38 − 0	33 − 0
0.4	48 − 0	40 − 0	33 − 0	28 − 0	24 − 0	20 − 0
0.5	32 − 0	27 − 0	21 − 0	18 − 0	15 − 0	12 − 0
0.6	21 − 0	17 − 0	14 − 0	11 − 0	10 − 0	10 − 0
0.7	19 − 0	13 − 0	11 − 0	10 − 0	10 − 0	10 − 0
0.8	42 − 0	19 − 0	18 − 0	14 − 0	12 − 0	10 − 0
0.9	16 + 1	57 = 0	55 + 0	42 − 0	38 − 0	32 − 0
1.0	44 + 1	26 + 1	25 + 1	18 + 1	15 + 1	12 + 1
1.1	40 + 1	41 + 1	36 + 1	31 + 1	26 + 1	22 + 1
1.2	27 + 1	25 + 1	21 + 1	17 + 1	15 + 1	12 + 1
1.3	15 + 1	13 + 1	11 + 1	93 + 0	78 + 0	66 + 0
1.4	88 + 0	74 + 0	61 + 0	50 + 0	42 − 0	35 − 0
1.5	54 + 0	45 − 0	36 − 0	30 − 0	25 − 0	20 − 0
1.6	44 − 0	31 − 0	26 − 0	21 − 0	17 − 0	14 − 0
1.7	65 + 0	30 − 0	27 − 0	21 − 0	18 − 0	15 − 0
1.8	16 + 1	49 − 0	48 − 0	34 − 0	31 − 0	25 − 0
1.9	41 + 1	12 + 1	12 + 1	84 + 0	75 + 0	60 + 0
2.0	46 + 1	40 + 1	39 + 1	26 + 1	23 + 1	17 + 1
2.1	42 + 1	42 + 1	36 + 1	31 + 1	26 + 1	21 + 1
2.2	28 + 1	26 + 1	22 + 1	18 + 1	15 + 1	12 + 1
2.3	16 + 1	14 + 1	12 + 1	10 + 1	85 + 0	71 + 0
2.4	98 + 0	85 + 0	70 + 0	57 + 0	47 − 0	40 − 0
2.5	65 + 0	53 + 0	44 − 0	36 − 0	29 − 0	24 − 0
2.6	63 + 0	40 − 0	34 − 0	27 − 0	23 − 0	19 − 0
2.7	11 + 1	42 − 0	39 − 0	29 − 0	25 − 0	21 − 0
2.8	27 + 1	70 + 0	70 + 0	47 − 0	43 − 0	34 − 0
2.9	47 + 1	16 + 1	18 + 1	11 + 1	10 + 1	76 + 0
3.0	45 + 1	45 + 1	45 + 1	29 + 1	26 + 1	19 + 1
3.1	42 + 1	42 + 1	36 + 1	30 + 1	25 + 1	21 + 1
3.2	29 + 1	27 + 1	22 + 1	18 + 1	15 + 1	13 + 1
3.3	17 + 1	15 + 1	12 + 1	10 + 1	88 + 0	73 + 0
3.4	10 + 1	91 + 0	74 + 0	61 + 0	51 + 0	42 − 0
3.5	72 + 0	58 + 0	48 − 0	39 − 0	33 − 0	27 − 0
3.6	77 + 0	46 − 0	39 − 0	31 − 0	26 − 0	22 − 0
3.7	14 + 1	50 + 1	47 − 0	34 − 0	30 − 0	24 − 0
3.8	33 + 1	84 + 0	86 + 1	56 + 0	51 + 0	40 − 0
3.9	47 + 1	19 + 1	21 + 1	12 + 1	11 + 1	87 + 0
4.0	45 + 1	48 + 1	47 + 1	31 + 1	27 + 1	20 + 1
4.1	42 + 1	42 + 1	36 + 1	30 + 1	25 + 1	21 + 1
4.2	29 + 1	27 + 1	22 + 1	19 + 1	15 + 1	13 + 1
4.3	17 + 1	16 + 1	13 + 1	10 + 1	90 + 0	75 + 0
4.4	10 + 1	94 + 0	77 + 0	64 + 0	53 + 0	44 − 0

Table 6.3. (*continued*)

Δn	The effective principal quantum number n_0^*					
	2.5	3.0	3.5	4.0	4.5	5.0
	Transition $d - p$, *parameter A* $\kappa = 3$					
0.1	20 − 1	55 − 1	10 − 0	16 − 0	24 − 0	33 − 0
0.2	11 − 0	30 − 0	55 + 0	85 + 0	12 + 1	15 + 1
0.3	29 − 0	74 + 0	12 + 1	19 + 1	25 + 1	31 + 1
0.4	53 + 0	12 + 1	20 + 1	27 + 1	34 + 1	39 + 1
0.5	73 + 0	15 + 1	23 + 1	29 + 1	34 + 1	37 + 1
0.6	80 + 0	15 + 1	21 + 1	25 + 1	27 + 1	29 + 1
0.7	70 + 0	12 + 1	16 + 1	18 + 1	19 + 1	20 + 1
0.8	51 + 0	88 + 0	11 + 1	12 + 1	12 + 1	13 + 1
0.9	32 − 0	55 + 0	71 + 0	80 + 0	87 + 0	90 + 0
1.0	20 − 0	37 − 0	51 + 0	62 + 0	70 + 0	76 + 0
1.1	15 − 0	32 − 0	49 − 0	65 + 0	78 + 0	89 + 0
1.2	17 − 0	41 − 0	67 + 0	92 + 0	11 + 1	13 + 1
1.3	26 − 0	62 + 0	10 + 1	14 + 1	16 + 1	19 + 1
1.4	40 − 0	91 + 0	14 + 1	18 + 1	21 + 1	22 + 1
1.5	53 + 0	11 + 1	16 + 1	19 + 1	20 + 1	21 + 1
1.6	58 + 0	11 + 1	15 + 1	16 + 1	17 + 1	17 + 1
1.7	51 + 0	93 + 0	11 + 1	12 + 1	12 + 1	12 + 1
1.8	39 − 0	69 + 0	85 + 0	93 + 0	94 + 0	92 + 0
1.9	27 − 0	49 − 0	62 + 0	70 + 0	73 + 0	73 + 0
2.0	20 − 0	38 − 0	52 + 0	61 + 0	66 + 0	67 + 0
2.1	17 − 0	36 − 0	53 + 0	66 + 0	74 + 0	76 + 0
2.2	20 − 0	44 − 0	67 + 0	87 + 0	96 + 0	10 + 1
2.3	27 − 0	60 + 0	93 + 0	11 + 1	13 + 1	13 + 1
2.4	38 − 0	82 + 0	12 + 1	15 + 1	15 + 1	16 + 1
2.5	48 − 0	97 + 0	13 + 1	15 + 1	15 + 1	15 + 1
2.6	51 + 0	96 + 0	12 + 1	13 + 1	13 + 1	12 + 1
2.7	45 − 0	81 + 0	10 + 1	10 + 1	10 + 1	99 + 0
2.8	34 − 0	62 + 0	76 + 0	82 + 0	81 + 0	76 + 0
2.9	25 − 0	46 − 0	58 + 0	64 + 0	65 + 0	63 + 0
3.0	19 − 0	38 − 0	51 + 0	58 + 0	62 + 0	59 + 0
3.1	18 − 0	37 − 0	52 + 0	64 + 0	69 + 0	65 + 0
3.2	28 − 0	44 − 0	65 + 0	83 + 0	84 + 0	92 + 0
3.3	28 − 0	59 + 0	87 + 0	11 + 1	11 + 1	11 + 1
3.4	38 − 0	77 + 0	11 + 1	13 + 1	13 + 1	13 + 1
3.5	46 − 0	90 + 0	12 + 1	13 + 1	13 + 1	13 + 1
3.6	48 − 0	88 + 0	11 + 1	12 + 1	11 + 1	11 + 1
3.7	42 − 0	75 + 0	92 + 0	98 + 0	93 + 0	85 + 0
3.8	32 − 0	58 + 0	71 + 0	76 + 0	74 + 0	69 + 0
3.9	24 − 0	45 − 0	56 + 0	62 + 0	62 + 0	58 + 0
4.0	19 − 0	37 − 0	50 + 0	56 + 0	61 + 0	50 + 0
4.1	18 − 0	37 − 0	52 + 0	64 + 0	68 + 0	57 + 0
4.2	21 − 0	44 − 0	63 + 0	83 + 0	75 + 0	82 + 0
4.3	28 − 0	58 + 0	83 + 0	11 + 1	98 + 0	98 + 0
4.4	37 − 0	75 + 0	10 + 1	12 + 1	12 + 1	11 + 1

Table 6.3. (*continued*)

Δn	The effective principal quantum number n_0^*					
	2.5	3.0	3.5	4.0	4.5	5.0
	Transition $d - p$, *parameter* χ $\kappa = 3$					
0.1	21 − 0	32 − 0	43 − 0	56 + 0	69 + 0	82 + 0
0.2	68 + 0	97 + 0	12 + 1	15 + 1	18 + 1	21 + 1
0.3	12 + 1	17 + 1	21 + 1	25 + 1	28 + 1	30 + 1
0.4	19 + 1	24 + 1	28 + 1	31 + 1	32 + 1	32 + 1
0.5	24 + 1	29 + 1	31 + 1	31 + 1	31 + 1	29 + 1
0.6	27 + 1	29 + 1	29 + 1	28 + 1	26 + 1	23 + 1
0.7	26 + 1	26 + 1	24 + 1	22 + 1	20 + 1	18 + 1
0.8	21 + 1	20 + 1	19 + 1	17 + 1	15 + 1	13 + 1
0.9	14 + 1	15 + 1	14 + 1	13 + 1	12 + 1	11 + 1
1.0	10 + 1	11 + 1	12 + 1	12 + 1	12 + 1	11 + 1
1.1	86 + 0	11 + 1	13 + 1	15 + 1	16 + 1	16 + 1
1.2	10 + 1	15 + 1	20 + 1	24 + 1	26 + 1	27 + 1
1.3	15 + 1	24 + 1	31 + 1	37 + 1	40 + 1	41 + 1
1.4	23 + 1	33 + 1	42 + 1	47 + 1	48 + 1	46 + 1
1.5	30 + 1	39 + 1	45 + 1	46 + 1	45 + 1	41 + 1
1.6	32 + 1	38 + 1	40 + 1	39 + 1	36 + 1	32 + 1
1.7	28 + 1	32 + 1	32 + 1	30 + 1	27 + 1	24 + 1
1.8	21 + 1	24 + 1	24 + 1	22 + 1	20 + 1	18 + 1
1.9	15 + 1	18 + 1	19 + 1	18 + 1	17 + 1	16 + 1
2.0	11 + 1	15 + 1	17 + 1	17 + 1	17 + 1	16 + 1
2.1	10 + 1	15 + 1	18 + 1	21 + 1	21 + 1	20 + 1
2.2	12 + 1	19 + 1	25 + 1	29 + 1	30 + 1	30 + 1
2.3	17 + 1	27 + 1	35 + 1	40 + 1	41 + 1	41 + 1
2.4	24 + 1	36 + 1	44 + 1	48 + 1	48 + 1	46 + 1
2.5	30 + 1	41 + 1	47 + 1	48 + 1	45 + 1	41 + 1
2.6	32 + 1	39 + 1	42 + 1	41 + 1	37 + 1	33 + 1
2.7	28 + 1	33 + 1	34 + 1	32 + 1	29 + 1	25 + 1
2.8	21 + 1	25 + 1	26 + 1	25 + 1	23 + 1	20 + 1
2.9	15 + 1	20 + 1	21 + 1	20 + 1	19 + 1	18 + 1
3.0	12 + 1	17 + 1	19 + 1	20 + 1	20 + 1	18 + 1
3.1	11 + 1	17 + 1	21 + 1	23 + 1	24 + 1	22 + 1
3.2	13 + 1	21 + 1	27 + 1	31 + 1	31 + 1	33 + 1
3.3	18 + 1	28 + 1	37 + 1	41 + 1	42 + 1	44 + 1
3.4	25 + 1	37 + 1	46 + 1	49 + 1	48 + 1	47 + 1
3.5	31 + 1	41 + 1	48 + 1	48 + 1	46 + 1	43 + 1
3.6	31 + 1	40 + 1	43 + 1	42 + 1	39 + 1	34 + 1
3.7	27 + 1	33 + 1	35 + 1	33 + 1	30 + 1	27 + 1
3.8	21 + 1	26 + 1	27 + 1	26 + 1	24 + 1	22 + 1
3.9	15 + 1	20 + 1	22 + 1	22 + 1	21 + 1	20 + 1
4.0	12 + 1	18 + 1	20 + 1	21 + 1	22 + 1	19 + 1
4.1	12 + 1	18 + 1	22 + 1	25 + 1	26 + 1	22 + 1
4.2	14 + 1	22 + 1	29 + 1	33 + 1	31 + 1	31 + 1
4.3	19 + 1	29 + 1	38 + 1	43 + 1	39 + 1	43 + 1
4.4	26 + 1	37 + 1	46 + 1	49 + 1	47 + 1	49 + 1

A summary of formulas defining the angular factors Q_κ for the actual cases most frequently met is given in Sect. 6.2. The functions $\Phi_\kappa(u)$ and $G_\kappa(\beta)$ are approximated by means of the following formulas:
for $\kappa = 1$:

$$\Phi_1(u) = C\left(\frac{u}{u+1}\right)^{1/2}\frac{\ln(16+u)}{u+\varphi}, \tag{6.1.9}$$

$$G_1(\beta) = A\frac{\sqrt{\beta(\beta+3)}}{\beta+\chi}\ln(16+1/\beta). \tag{6.1.10}$$

for $\kappa \neq 1$:

$$\Phi_\kappa(u) = C\left(\frac{u}{u+1}\right)^{1/2}\frac{1}{u+\varphi}, \tag{6.1.11}$$

$$G_\kappa(\beta) = A\frac{\sqrt{\beta(\beta+3)}}{\beta+\chi}. \tag{6.1.12}$$

The transition under consideration is characterized by the assignment of the effective principal quantum numbers of the lower level n_0^* and of the upper level n_1^*

$$n_0^* = \sqrt{z^2\mathrm{Ry}/|E_0|}, \quad n_1^* = \sqrt{z^2\mathrm{Ry}/|E_1|}$$

The quantities E_0 and E_1 are the ionization energies corresponding to a specified state of the atomic core; z is the spectroscopic symbol of an ion. For a neutral atom, $z = 1$, for a singly charged ion, $z = 2$, and so on.

The parameters C, φ and A, χ for transitions $s \to s$, $s \to p$, $s \to d$, $p \to s$, $p \to p$, $p \to d$, $d \to s$, $d \to p$ are given in Tables 6.2 and 6.3 as functions of n_0^* and $\Delta n = n_1^* - n_0^*$. The spacing with respect to n_0^* and Δn adopted in the tables ensures the possibility of linear interpolation almost everywhere. The tables give the order and the mantissa of the number; for example, 24 – 1, 47 – 0, 59+0, 42+1, and 12+2 denote respectively 0.024, 0.47, 0.59, 4.2, and 12.

The range of approximation for cross sections is $1 \leq u \leq 128$. The cases where the errors of approximation exceed 10 percent are indicated in Table 6.2 by asterisks. These errors, however, do not exceed a factor of 2. The rate coefficients are approximated in the range $1/32 \leq \beta \leq 4$. The asterisks in Tables 6.3 indicate the cases in which errors of approximation are greater than 25 percent.

6.1.4 Normalized Cross Sections for Specific Atoms and Ions (Tables 6.4–8)

(i) $\Delta S = 0$

For transitions with no change of spin ($\Delta S = 0$) the cross sections σ and rate coefficients $\langle v\sigma\rangle$ are fitted by

$$\sigma' = \frac{\pi a_0^2}{z^4}\left[\frac{E_1}{E_0}\right]^{3/2}\frac{Q'_\kappa(a_0 a_1)}{2l_0+1}\frac{C\Phi'(u)}{u+\varphi}, \tag{6.1.13}$$

$$u = (\mathscr{E} - \Delta E)/z^2 \mathrm{Ry}\,,$$

$$\langle v\sigma \rangle = 10^{-8} \frac{1}{z^3} \left[\frac{E_1}{E_0}\right]^{3/2} \frac{Q'_\kappa(a_0 a_1)}{2l_0+1} \cdot \frac{A G'(\beta)}{\beta+\chi} \exp(-p\beta), \quad \mathrm{cm^3 s^{-1}} \qquad (6.1.14)$$

$$\beta = z^2\ \mathrm{Ry}/T, \quad p = \Delta E/z^2\ \mathrm{Ry}\,.$$

Here, z is the charge of the parent ion (spectroscopic symbol of ion, $z = 1$ for a neutral atom, $z = 2$ for a singly charged ion, and so on), and $Q'_\kappa(a_0 a_1)$ is the angular factor. In fact, in all tables only the states with s core electrons (besides closed shells) are considered. In this case

$$L_p = 0, \quad G^{L_0 S_0}_{L_p S_p} = 1, \quad Q' = m$$

for excitation from the shells of equivalent electrons l_0^m, and

$$Q' = 1$$

for one electron outside closed electron shells. Since Q' are independent of κ we give the values, summed over κ, of σ and $\langle v\sigma \rangle$.

The formulas for $\Phi'(u)$ and $G'(\beta)$ are given by (5.1.12, 13 and 16).

(ii) $\Delta S = 1$

In the case of intercombination transitions

$$\sigma'' = \frac{\pi a_0^2}{z^4} \left[\frac{E_1}{E_0}\right]^{3/2} \frac{Q''_\kappa(a_0 a_1)}{2l_0+1} \frac{C\Phi''(u)}{u+\varphi}, \qquad (6.1.15)$$

$$u = (\mathscr{E} - \Delta E)/z^2\ \mathrm{Ry}\,,$$

$$\langle v\sigma'' \rangle = 10^{-8} \frac{1}{z^3} \left[\frac{E_1}{E_0}\right]^{3/2} \frac{Q''_\kappa(a_0 a_1)}{2l_0+1} \cdot \frac{A G''(\beta)}{\beta+\chi} \exp(-p\beta)\ [\mathrm{cm^3\ s^{-1}}], \qquad (6.1.16)$$

$$\beta = z^2\ \mathrm{Ry}/T, \quad p = \Delta E/z^2\ \mathrm{Ry}\,.$$

In the tables for intercombination transitions, as in those for $\Delta S = 0$ only the states with s core electrons (besides closed shells) are considered. In this case

$$L_p = 0, \quad G^{L_0 S_0}_{L_p S_p} = 1, \quad Q'' = m A_2 = m \frac{2S_1+1}{2(2S_p+1)} \qquad (6.1.17)$$

for excitation from the shells of m equivalent electrons. Since Q'' are independent of κ we give summed over κ values of σ'' and $\langle v\sigma'' \rangle$. The formulas for $\Phi''(u)$ and $G''(\beta)$ are given by (5.1.12, 13 and 16). The energy dependence of exchange cross sections varies from one transition to another, and the errors of fitting are rather large. For this reason only the rate coefficients are tabulated in most cases.

The set of parameters C, φ, D is adjusted for the range $0.02 < u < 16$, and the set A, χ, D, for $0.25 < \beta < 16$.

Table 6.4. Normalized Born and Coulomb-Born excitation cross sections. Transitions with no change of spin ($\Delta S = 0$)

Atom	Transition	C	φ	D	R	A	χ	D	R
H I	1s–2s	3.46	0.67	0.00	0.04	7.52	12.13	9.90	0.03
	1s–2p	70.32	4.27	0.00	0.03	24.12	0.34	0.50	0.02
	1s–3s	2.33	0.68	0.00	0.01	5.72	2.94	1.40	0.02
	1s–3p	38.11	3.64	0.00	0.05	18.59	0.28	0.00	0.02
	1s–3d	1.81	1.13	0.30	0.01	2.03	0.68	0.20	0.02
	1s–4s	2.06	0.69	0.00	0.01	5.24	2.53	1.00	0.02
	1s–4p	31.58	3.49	0.00	0.05	16.77	0.32	0.00	0.01
	1s–4d	2.07	1.29	0.20	0.01	2.40	0.65	0.10	0.02
	1s–4f	0.03	0.61	0.40	0.01	0.06	0.87	0.00	0.02
	2s–3s	18.27	0.26	0.00	0.02	20.15	1.64	2.10	0.02
	2s–3p	139.93	0.60	0.90	0.53	15.20	0.53	9.90	0.16
	2s–3d	57.52	0.18	0.00	0.02	89.18	1.14	0.60	0.02
	2s–4s	8.34	0.22	0.00	0.02	13.57	0.74	0.00	0.02
	2s–4p	59.82	0.60	0.90	0.58	7.72	0.68	9.90	0.12
	2s–4d	16.73	0.12	0.00	0.01	46.84	10.17	6.70	0.02
	2s–4f	8.94	0.13	0.00	0.03	21.24	4.10	2.60	0.02
	2p–3s	10.51	0.02	0.90	0.57	8.73	0.26	0.00	0.08
	2p–3p	60.96	0.41	0.00	0.06	42.94	1.75	3.90	0.02
	2p–3d	1014.92	1.59	0.20	0.06	147.18	0.67	6.70	0.02
	2p–4s	4.89	0.01	0.90	0.44	5.80	0.54	0.00	0.17
	2p–4p	27.75	0.24	0.00	0.02	42.45	0.84	0.20	0.02
	2p–4d	313.54	1.05	0.00	0.11	122.06	0.80	2.70	0.02
	2p–4f	38.55	0.35	0.00	0.08	41.11	1.74	2.30	0.02
	3s–4s	55.53	0.16	0.00	0.07	40.45	3.07	7.70	0.02
	3s–4p	359.31	0.26	0.90	0.25	44.43	0.57	9.90	0.22
	3s–4d	114.84	0.11	0.00	0.12	111.48	4.54	8.70	0.02
	3s–4f	56.65	0.03	0.00	0.01	121.39	0.99	0.00	0.02
	3p–4s	61.83	−0.01	1.00	0.44	25.01	0.63	2.90	0.02
	3p–4p	197.65	0.27	0.00	0.07	88.72	2.46	9.90	0.02
	3p–4d	1644.09	0.51	0.60	0.16	228.43	0.74	9.90	0.13
	3p–4f	394.47	0.13	0.00	0.10	332.30	4.44	9.90	0.02
	3d–4s	6.43	0.05	0.10	0.05	10.84	1.29	0.70	0.01
	3d–4p	52.52	0.02	0.70	0.25	49.05	0.50	0.30	0.02
	3d–4d	251.01	0.17	0.50	0.04	84.22	1.83	9.90	0.02
	3d–4f	5485.46	0.73	0.70	0.14	502.57	0.38	9.90	0.31
He I	$1^1S–2^1S$	1.85	1.26	0.00	0.01	4.13	1.61	0.40	0.02
	$1^1S–2^1P$	26.71	8.22	0.20	0.02	8.09	0.25	0.10	0.02
	$1^1S–3^1S$	1.43	1.32	0.00	0.01	3.43	1.23	0.00	0.02
	$1^1S–3^1P$	21.53	7.78	0.20	0.03	7.55	0.25	0.00	0.01
	$1^1S–3^1D$	0.36	2.90	0.20	0.02	0.33	0.52	0.00	0.02
	$1^1S–4^1S$	1.32	1.34	0.00	0.01	3.25	1.25	0.00	0.01
	$1^1S–4^1P$	21.77	8.99	0.10	0.02	7.35	0.27	0.00	0.01
	$1^1S–4^1D$	0.44	2.91	0.20	0.02	0.42	0.54	0.00	0.02
	$1^1S–4^1F$	0.00	1.48	0.30	0.02	0.00	0.97	0.20	0.02
	$2^1S–2^1P$	287.51	0.00	1.00	0.15	72.85	1.32	9.90	0.07
	$2^1S–3^1S$	14.84	0.28	0.00	0.02	17.95	1.32	1.30	0.02

Table 6.4. (*continued*)

Atom	Transition	C	φ	D	R	A	χ	D	R
He I	2^1S-3^1P	62.58	3.01	0.20	0.11	5.35	0.46	7.20	0.02
	2^1S-3^1D	45.04	0.23	0.00	0.01	70.67	0.79	0.10	0.02
	2^1S-4^1S	7.12	0.25	0.00	0.01	12.06	0.79	0.00	0.00
	2^1S-4^1P	34.69	2.21	0.50	0.09	3.39	0.53	7.60	0.03
	2^1S-4^1D	18.89	0.18	0.00	0.01	43.96	12.25	9.90	0.02
	2^1S-4^1F	5.27	0.17	0.00	0.03	12.09	2.77	1.50	0.02
	2^1P-3^1S	44.98	0.97	0.60	0.21	5.55	0.73	9.60	0.04
	2^1P-3^1P	60.81	0.41	0.00	0.06	42.80	1.75	3.90	0.02
	2^1P-3^1D	1010.04	1.59	0.20	0.06	147.66	0.67	6.60	0.02
	2^1P-4^1S	16.10	1.73	0.00	0.09	3.59	0.67	3.90	0.03
	2^1P-4^1P	27.69	0.24	0.00	0.02	42.42	0.84	0.20	0.02
	2^1P-4^1D	313.07	1.05	0.00	0.11	121.92	0.81	2.70	0.02
	2^1P-4^1F	38.15	0.36	0.00	0.08	40.64	1.74	2.30	0.02
	2^3S-2^3P	255.67	0.83	0.10	0.07	53.34	1.13	8.60	0.03
	2^3S-3^3S	11.64	0.31	0.00	0.03	15.61	0.94	0.50	0.02
	2^3S-3^3P	11.93	0.60	0.00	0.65	8.05	0.22	0.00	0.13
	2^3S-3^3D	31.92	0.30	0.00	0.01	48.95	0.71	0.00	0.01
	2^3S-4^3S	5.91	0.28	0.00	0.01	10.51	0.85	0.00	0.01
	2^3S-4^3P	10.22	1.02	0.00	0.53	5.23	0.16	0.10	0.09
	2^3S-4^3D	18.13	0.27	0.00	0.01	34.76	0.93	0.00	0.02
	2^3S-4^3F	2.72	0.23	0.00	0.04	5.97	1.51	0.40	0.02
	2^3P-3^3S	60.27	0.34	0.90	0.31	9.16	0.80	9.90	0.07
	2^3P-3^3P	55.97	0.43	0.00	0.04	41.32	1.60	3.30	0.02
	2^3P-3^3D	865.03	1.81	0.10	0.05	137.32	0.66	5.70	0.02
	2^3P-4^3S	16.22	0.34	0.90	0.42	3.45	0.93	7.80	0.06
	2^3P-4^3P	25.80	0.24	0.00	0.03	40.94	0.72	0.00	0.01
	2^3P-4^3D	309.41	1.21	0.00	0.11	112.13	0.77	2.70	0.02
	2^3P-4^3F					30.57	1.70	2.30	0.02
Li I	2s–2p	242.74	1.14	0.10	0.06	58.31	0.86	5.20	0.02
	2s–3s	9.92	0.26	0.20	0.04	14.43	0.66	0.00	0.01
	2s–3p	6.02	0.12	0.10	0.06	20.23	7.03	3.40	0.02
	2s–3d	24.23	0.37	0.00	0.02	36.32	0.70	0.00	0.01
	2p–3s	129.92	1.30	0.00	0.30	26.90	0.86	6.00	0.03
	2p–3p	57.91	0.42	0.00	0.04	46.49	1.22	2.10	0.02
	2p–3d	917.36	1.79	0.10	0.05	148.53	0.63	5.40	0.02
	3s–3p	1261.32	0.13	0.90	0.15	232.48	0.93	9.90	0.07
	3s–3d	75.59	0.06	0.00	0.04	115.65	2.17	2.10	0.02
	3p–3d	2994.65	0.10	0.20	0.23	1495.79	2.48	9.10	0.02
Li II	1^1S-2^1S	1.87	0.78	0.00	0.06	5.62	1.30	0.00	0.01
	1^1S-2^1P	43.32	7.77	0.20	0.02	11.31	0.21	0.20	0.02
	1^1S-3^1S	1.37	0.80	0.00	0.08	4.64	1.41	0.00	0.01
	1^1S-3^1P	31.48	7.63	0.20	0.02	9.38	0.22	0.10	0.02
	1^1S-3^1D	0.79	3.46	0.40	0.02	0.39	0.51	0.80	0.02

Table 6.4. (*continued*)

Atom	Transition	C	φ	D	R	A	χ	D	R
Li II	2^1S–2^1P	309.65	0.03	0.70	0.15	211.50	1.33	3.10	0.02
	2^1S–3^1S	14.84	0.19	0.00	0.02	27.82	1.02	0.20	0.02
	2^1S–3^1P	93.49	1.66	0.60	0.15	8.41	0.53	9.10	0.02
	2^1S–3^1D	50.10	0.32	0.00	0.03	61.23	1.18	1.10	0.02
	2^1P–3^1S	32.36	2.31	0.00	0.07	4.48	0.54	5.20	0.02
	2^1P–3^1P	59.47	0.21	0.00	0.01	93.50	1.21	0.70	0.02
	2^1P–3^1D	892.85	1.12	0.00	0.16	292.64	0.62	2.60	0.02
	2^3S–2^3P	249.92	0.03	0.80	0.18	164.06	1.11	2.60	0.02
	2^3S–3^3S	13.55	0.23	0.00	0.01	23.92	1.12	0.40	0.02
	2^3S–3^3P	57.24	1.87	0.70	0.15	4.15	0.45	9.90	0.02
	2^3S–3^3D	41.93	0.41	0.00	0.02	48.10	1.12	1.10	0.02
	2^3P–3^3S	43.37	2.15	0.10	0.17	5.70	0.60	6.20	0.03
	2^3P–3^3P	54.68	0.22	0.00	0.01	87.89	1.09	0.50	0.02
	2^3P–3^3D	780.14	1.21	0.00	0.16	268.62	0.51	1.90	0.02
Be I	2S–2P	151.61	2.11	0.00	0.09	58.21	0.40	0.90	0.02
Be II	2s–2p	222.45	0.02	0.90	0.24	141.37	0.92	2.20	0.02
	2s–3s	12.24	0.26	0.00	0.01	21.28	1.11	0.40	0.02
	2s–3p	29.67	4.26	0.00	0.03	3.48	0.30	3.00	0.02
	2s–3d	35.05	0.44	0.10	0.02	38.34	1.08	1.10	0.02
	2p–3s	70.98	1.72	0.00	0.10	13.17	0.65	4.80	0.03
	2p–3p	54.88	0.21	0.00	0.01	89.02	1.10	0.50	0.02
	2p–3d	790.48	1.20	0.00	0.16	273.34	0.51	1.90	0.02
	3s–3p	1517.48	0.03	0.80	0.15	615.05	1.51	6.60	0.02
	3s–3d	68.55	0.03	0.10	0.02	129.97	1.11	0.30	0.02
C I	$2p^2\ {}^3P$–$2p3s\ {}^3P$	18.47	2.92	0.50	0.08	4.14	0.41	2.00	0.03
	$2p^2\ {}^1D$–$2p3s\ {}^1P$	20.37	2.36	0.50	0.10	4.97	0.43	2.00	0.03
	$2p^2\ {}^1S$–$2p3s\ {}^1P$	30.78	2.09	0.20	0.08	9.70	0.46	1.50	0.02
O V	2^1S–2^1P	209.19	0.04	0.70	0.25	193.41	0.53	0.40	0.02
F VI	2^1S–2^1P	223.10	0.03	0.70	0.23	209.07	0.54	0.40	0.02
Na I	3s–3p	303.97	1.37	0.10	0.06	67.57	0.76	4.80	0.02
	3s–3d	27.94	0.34	0.00	0.01	42.88	0.72	0.00	0.01
	3s–4s	10.46	0.33	0.00	0.04	14.91	0.65	0.00	0.02
	3s–4p	19.82	0.11	0.60	0.09	31.17	0.74	0.00	0.02
	3p–3d	1392.49	1.44	0.30	0.05	159.32	0.68	9.10	0.02
	3p–4s	254.78	0.92	0.00	0.11	59.15	0.97	6.40	0.03
	3p–4p	70.07	0.38	0.00	0.10	53.67	1.39	2.70	0.02
	3p–4p	85.49	0.45	0.00	0.07	55.35	1.36	3.10	0.01
	3d–4p					676.14	1.62	9.90	0.04
	4s–3d	71.46	0.05	0.00	0.03	109.74	2.24	2.20	0.02
	4s–4p	1286.57	0.25	0.80	0.14	215.54	0.88	9.90	0.06
Na VIII	2^1S–2^1P	244.83	0.02	0.70	0.20	225.78	0.62	0.60	0.02
Mg I	3^1S–3^1P	214.17	2.32	0.00	0.06	62.81	0.45	1.60	0.02
	3^1S–4^1S	5.29	0.54	0.10	0.03	6.87	0.61	0.00	0.01

Table 6.4. (*continued*)

Atom	Transition	C	φ	D	R	A	χ	D	R
Mg I	3^1S-3^1D	26.04	0.44	0.00	0.01	45.35	0.85	0.00	0.01
	3^1S-4^1P	25.34	0.79	0.00	0.13	27.74	0.55	0.00	0.02
	3^1P-4^1S	281.84	0.84	0.00	0.10	68.56	1.02	6.50	0.02
	3^1P-3^1D	1140.87	1.06	0.50	0.07	130.28	0.68	9.90	0.01
	3^1P-4^1P	84.65	0.18	0.00	0.02	127.20	1.18	0.70	0.02
	3^3P-4^3S	103.99	1.56	0.10	0.18	23.65	0.73	4.30	0.03
	3^3P-4^3P	298.92	2.62	0.00	0.51	68.46	0.15	0.60	0.03
	3^3P-3^3D	706.60	2.22	0.10	0.05	141.69	0.53	3.30	0.02
Mg II	3s–3p	406.66	0.49	0.00	0.22	198.56	0.87	2.70	0.02
	3s–4s	15.14	0.21	0.00	0.02	25.99	1.33	0.70	0.02
	3s–3d	51.07	0.32	0.00	0.03	63.40	1.20	1.10	0.02
	3p–4s	208.84	0.74	0.00	0.21	66.65	0.76	3.70	0.02
	3p–3d	1604.38	0.79	0.00	0.17	475.20	0.81	4.30	0.02
	4s–3d	48.25	0.00	0.00	0.01	104.53	0.99	0.00	0.00
Mg IX	2S–2P	253.98	0.02	0.70	0.19	235.97	0.63	0.60	0.02
Mg X	2s–2p	319.36	0.01	0.70	0.14	278.24	0.90	1.30	0.02
	2s–3s	16.14	0.16	0.00	0.02	34.07	0.96	0.00	0.01
	2s–3p	119.68	2.26	0.00	0.11	22.37	0.38	2.70	0.02
	2s–3d	49.81	0.10	0.60	0.06	68.92	0.65	0.10	0.01
	2p–3s	23.35	1.69	0.00	0.17	5.40	0.39	2.30	0.02
	2p–3p	56.80	0.15	0.00	0.01	119.17	0.95	0.00	0.01
	2p–3d	745.42	0.68	0.00	0.29	439.66	0.40	0.60	0.02
Al I	3p–4s	77.72	1.68	0.20	0.13	18.76	0.66	3.50	0.02
Al II	3^1S-3^1P	329.94	0.81	0.00	0.19	157.76	0.57	1.50	0.02
K I	4s–4p	390.91	1.02	0.30	0.08	73.65	0.83	6.80	0.02
	4s–5s	12.79	0.30	0.00	0.02	16.82	0.92	0.50	0.02
	4s–3d	40.96	0.22	0.00	0.01	70.01	0.79	0.00	0.00
	4s–5p	23.27	0.13	0.50	0.08	35.02	0.70	0.00	0.01
	4p–5s	294.73	0.95	0.00	0.13	61.19	0.99	7.30	0.03
	4p–3d	1522.11	0.98	0.50	0.06	171.45	0.64	9.90	0.05
	4p–5p	80.13	0.36	0.00	0.13	59.54	1.50	3.10	0.02
	5s–3d	33.26	0.02	0.00	0.06	45.52	3.70	5.00	0.02
	5s–5p	1586.55	0.14	0.90	0.16	281.97	0.90	9.90	0.07
	3d–5p	858.63	0.53	0.00	0.11	196.56	1.34	9.90	0.03
Ca I	4^1S-3^1D	12.19	0.16	0.00	0.02	25.12	0.99	0.00	0.02
	4^1S-4^1P	302.55	1.82	0.10	0.06	68.44	0.62	3.50	0.02
	3^1D-4^1P	304.09	0.19	0.20	0.10	116.87	1.95	9.10	0.02
	4^3P-3^3D	283.09	0.59	0.00	0.09	69.84	1.40	9.40	0.02
Ca II	4s–3d	10.89	0.03	0.00	0.03	24.65	1.05	0.00	0.00
	4s–4p	596.95	0.03	0.90	0.27	178.03	0.99	3.70	0.02
	4s–5s	20.11	0.17	0.00	0.01	33.95	1.54	1.00	0.02
	3d–4p	369.95	0.02	0.80	0.15	236.50	1.23	3.00	0.02
	3d–5s	1.75	0.06	0.40	0.03	2.88	0.73	0.00	0.01
	4p–5s	294.60	0.68	0.00	0.22	85.36	0.82	4.60	0.02

Table 6.4. (*continued*)

Atom	Transition	C	φ	D	R	A	χ	D	R
Cu I	4s–4p	167.62	1.71	0.00	0.08	58.81	0.51	1.60	0.02
Zn I	4^1S–4^1P	163.76	2.58	0.00	0.06	56.38	0.38	0.90	0.02
	4^1S–5^1S	3.66	0.77	0.00	0.04	4.67	0.62	0.00	0.00
	4^1S–4^1D	15.12	0.67	0.00	0.00	23.90	0.79	0.00	0.01
	4^1S–5^1P	30.46	1.22	0.00	0.13	27.88	0.48	0.00	0.02
	4^1P–5^1S	240.58	0.83	0.00	0.10	62.92	1.02	6.00	0.02
	4^1P–4^1D	1082.46	1.73	0.20	0.04	152.82	0.63	6.40	0.02
	4^1P–5^1P	66.48	0.24	0.00	0.03	88.16	1.24	1.00	0.02
	4^3P–5^3S	88.10	1.62	0.00	0.10	22.88	0.71	3.50	0.02
	4^3P–5^3P	753.48	5.43	0.00	0.47	85.75	0.08	1.00	0.02
	4^3P–4^3D	470.12	2.25	0.10	0.07	114.73	0.44	2.10	0.02
Zn II	4s–4p	315.14	0.62	0.00	0.21	156.02	0.78	2.20	0.02
	4s–5s	10.72	0.24	0.00	0.02	20.05	0.84	0.00	0.02
	4s–4d	35.49	0.39	0.20	0.02	38.25	1.05	1.10	0.02
	4s–5p	13.28	0.19	0.30	0.05	24.07	0.83	0.00	0.00
	4s–4f	4.17	0.38	0.00	0.05	6.70	1.41	0.90	0.02
	4p–5s	177.20	0.80	0.00	0.20	57.89	0.74	3.40	0.02
	4p–4d	1253.98	0.96	0.00	0.17	388.52	0.70	3.30	0.02
	4p–5p	243.28	0.69	0.00	0.67	133.56	0.26	0.20	0.03
	4p–4f	71.52	0.15	0.60	0.03	69.26	1.19	1.70	0.02
Ga I	4p–5s	78.66	1.83	0.00	0.09	20.93	0.66	3.00	0.02
	4p–5p	22.83	0.00	1.00	0.04	30.47	0.70	0.20	0.02
	4p–4d	479.96	2.64	0.00	0.05	123.45	0.42	1.70	0.02
	4p–6s	21.00	1.47	0.30	0.12	8.21	0.54	1.50	0.03
	5s–5p	700.64	0.69	0.50	0.11	105.03	0.88	9.90	0.02
	5s–4d	57.69	0.11	0.00	0.04	96.59	0.99	0.30	0.02
	5s–6s	21.33	0.24	0.00	0.06	23.53	1.52	1.90	0.02
Ga II	4^1S–4^1P	299.06	0.94	0.00	0.18	138.94	0.54	1.40	0.02
	4^1S–5^1S	7.91	0.26	0.00	0.04	16.22	0.93	0.00	0.01
	4^1S–4^1D	40.57	0.48	0.00	0.02	48.84	1.07	0.90	0.02
	4^1P–5^1S	230.70	0.63	0.00	0.23	90.02	0.76	3.00	0.02
	4^1P–4^1D	1285.02	0.74	0.00	0.17	417.23	0.83	4.00	0.02
	4^3P–5^3S	124.23	1.23	0.00	0.15	33.28	0.67	3.50	0.02
	4^3P–4^3D	986.67	1.20	0.00	0.15	336.24	0.52	2.00	0.02
RbI	5s–5p	421.39	1.03	0.30	0.08	76.70	0.83	7.00	0.02
	5s–4d	39.60	0.19	0.00	0.01	70.04	0.82	0.00	0.00
	5s–6s	13.38	0.30	0.00	0.02	17.21	1.03	0.70	0.02
	5s–6p	29.20	0.12	0.60	0.11	38.59	0.60	0.00	0.00
	5s–5d	8.52	0.11	0.00	0.01	27.98	5.07	2.30	0.01
	5p–4d	1330.95	0.66	0.60	0.07	174.03	0.71	9.90	0.06
	5p–6s	338.26	0.92	0.00	0.11	68.96	1.01	7.70	0.03
	5p–6p	85.10	0.34	0.00	0.13	62.76	1.52	3.20	0.02
	5p–5d	42.03	0.03	0.80	0.37	42.62	0.41	0.00	0.04
	4d–6s	22.90	0.02	0.00	0.04	37.65	2.89	2.90	0.02
	4d–6p	395.52	0.55	0.00	0.15	103.37	1.20	7.60	0.03
	4d–5d	194.77	0.03	0.90	0.17	83.95	0.79	3.40	0.02

Table 6.4. (*continued*)

Atom	Transition	C	φ	D	R	A	χ	D	R
Sr I	5^1S-4^1D	21.03	0.16	0.00	0.02	41.28	0.93	0.00	0.01
	5^1S-5^1P	338.53	1.66	0.20	0.07	70.01	0.63	4.10	0.02
	5^1S-6^1S					11.65	1.01	0.80	0.02
	5^1S-6^1P	8.23	0.20	0.00	0.01	21.27	3.58	1.90	0.01
	4^1D-5^1P	552.67	0.29	0.00	0.08	181.86	1.81	9.90	0.03
	4^1D-6^1S	6.46	0.10	0.00	0.02	11.79	0.82	0.00	0.01
	4^1D-6^1P	31.88	0.73	0.00	0.32	13.04	0.46	1.60	0.02
	5^3P-4^3D	448.12	0.00	1.00	0.13	112.45	1.32	9.90	0.03
	5^3P-6^3S	181.20	0.89	0.50	0.24	33.94	0.79	6.60	0.03
	5^3P-6^3P					62.69	1.38	1.70	0.02
	4^3D-6^3S	6.11	0.09	0.00	0.03	11.59	0.86	0.00	0.00
	4^3D-6^3P					12.21	0.31	0.40	0.01
Sr II	5s–4d	18.13	0.03	0.00	0.01	39.07	0.99	0.00	0.01
	5s–5p	726.97	0.51	0.00	0.22	288.11	0.90	3.70	0.02
	5s–6s	21.95	0.15	0.00	0.01	36.83	1.61	1.10	0.02
	5s–5d	33.20	0.16	0.50	0.05	29.87	1.32	2.20	0.02
	5s–6p	25.65	0.11	0.30	0.04	43.48	0.77	0.00	0.01
	5s–4f	20.16	0.16	0.00	0.03	38.52	0.86	0.00	0.02
	5p–6s	503.13	0.03	0.90	0.32	199.26	0.93	4.30	0.02
	5p–5d	2168.30	0.69	0.00	0.19	558.87	0.88	5.70	0.02
	5p–6p	128.12	0.11	0.10	0.02	183.35	1.70	1.60	0.02
	5p–4f	281.14	0.12	0.30	0.04	308.03	1.88	2.70	0.02
	4d–5p	489.30	0.03	0.80	0.19	277.21	1.21	3.50	0.02
	4d–6s	2.98	0.06	0.40	0.04	4.56	0.96	0.40	0.01
	4d–5d	108.52	0.13	0.10	0.02	162.86	1.36	1.00	0.02
	4d–5d	50.72	0.01	0.70	0.85	82.76	0.81	0.00	0.23
	4d–5d	50.72	0.01	0.70	0.85	82.76	0.81	0.00	0.23
	4d–6p	22.17	0.07	0.40	0.05	35.77	0.74	0.00	0.00
	4d–4f	978.39	1.02	0.00	0.18	280.00	0.66	3.40	0.02
Ag I	5s–5p	189.97	1.82	0.00	0.07	62.12	0.50	1.70	0.02
	5s–6s	6.10	0.47	0.00	0.03	9.29	0.73	0.00	0.00
	5s–6p	20.24	0.62	0.00	0.12	26.49	0.65	0.00	0.03
	5s–5d	13.98	0.56	0.10	0.02	19.52	0.67	0.00	0.01
	5p–6s	162.11	1.06	0.00	0.10	42.21	0.87	4.80	0.03
	5p–6p	48.75	0.38	0.10	0.04	43.26	1.21	1.80	0.02
	5p–5d	820.00	1.89	0.10	0.05	140.91	0.61	4.80	0.02
	6s–6p	986.91	0.39	0.70	0.13	156.80	0.86	9.90	0.04
	6s–5d	77.56	0.08	0.00	0.06	119.28	1.84	1.60	0.02
Cd I	5^1S-5^1P	194.59	2.81	0.00	0.05	57.54	0.42	1.30	0.01
	5^1S-6^1S	4.44	0.66	0.10	0.02	5.78	0.63	0.00	0.01
	5^1S-5^1D	17.94	0.69	0.00	0.01	26.70	0.74	0.00	0.01
	5^1S-6^1P	29.17	1.20	0.00	0.12	26.20	0.46	0.00	0.01
	5^1P-6^1S	269.74	0.87	0.00	0.09	66.26	1.01	6.30	0.02
	5^1P-5^1D	1162.16	1.54	0.30	0.06	154.64	0.63	7.10	0.02
	5^1P-6^1P	76.03	0.19	0.00	0.03	114.68	1.12	0.60	0.02

Table 6.4. (*continued*)

Atom	Transition	C	φ	D	R	A	χ	D	R
Cd I	$5^3P–6^3S$	109.64	1.71	0.00	0.14	25.88	0.71	3.90	0.02
	$5^3P–6^3P$	233.46	2.37	0.00	0.50	61.24	0.17	0.50	0.03
	$5^3P–5^3D$	587.17	2.33	0.10	0.06	130.10	0.45	2.40	0.02
In I	5p–6s	66.44	1.18	0.60	0.17	13.28	0.67	4.90	0.03
	5p–6p	25.49	0.30	0.40	0.03	30.80	0.85	0.50	0.02
	5p–5d	340.34	2.38	0.30	0.08	71.91	0.42	2.40	0.03
In II	$5^1S–5^1P$	362.32	0.90	0.00	0.18	162.58	0.54	1.50	0.02
	$5^1S–6^1S$	9.61	0.28	0.00	0.02	17.44	1.08	0.30	0.02
	$5^3P–6^3S$	162.86	1.16	0.00	0.16	42.80	0.68	3.70	0.02
Cs I	6s–6p	482.05	1.03	0.30	0.07	80.69	0.85	7.90	0.02
	6s–5d	29.52	0.13	0.00	0.02	55.13	0.87	0.00	0.00
	6s–7s	14.48	0.29	0.00	0.02	18.03	1.11	0.90	0.02
	6s–7p	33.30	0.12	0.60	0.11	41.66	0.55	0.00	0.01
	6s–6d	4.59	0.19	0.30	0.02	6.68	0.66	0.00	0.01
	6s–4f	6.84	0.15	0.00	0.02	15.99	2.13	0.90	0.02
	6p–5d	812.31	0.00	1.00	0.15	191.12	1.19	9.90	0.04
	6p–7s	370.22	0.92	0.00	0.14	72.16	1.03	8.20	0.03
	6p–7p	90.97	0.33	0.00	0.13	66.24	1.57	3.40	0.02
	6p–6d	466.25	0.87	0.60	0.16	57.17	0.74	9.90	0.03
	6p–4f	136.19	0.21	0.00	0.04	170.92	1.60	1.70	0.02
	5d–7s	10.35	0.03	0.00	0.02	20.70	0.90	0.00	0.01
	5d–7p	39.98	0.02	0.90	0.35	21.00	0.77	2.40	0.02
	5d–6d	128.76	0.03	0.90	0.11	63.48	0.90	3.20	0.02
	5d–4f	1253.22	1.12	0.30	0.09	169.13	0.80	9.20	0.02
Ba I	$6^1S–5^1D$	6.24	0.11	0.00	0.04	11.63	0.86	0.00	0.01
	$6^1S–6^1P$	319.72	1.32	0.20	0.07	69.95	0.72	4.70	0.02
	$5^1D–6^1P$	194.93	0.57	0.00	0.09	60.67	1.31	6.80	0.02
	$5^1D–7^1S$					0.81	3.84	6.70	0.02
	$5^3D–6^3P$	191.96	0.37	0.00	0.09	67.23	1.63	8.00	0.02
	$5^3D–7^3S$					1.24	2.31	3.00	0.02
	$6^3P–7^3S$	124.76	0.78	0.50	0.27	25.11	0.84	6.60	0.03
Ba II	6s–5d	18.71	0.01	0.00	0.04	42.83	1.08	0.00	0.00
	6s–6p	836.47	0.03	0.90	0.27	355.84	1.02	4.30	0.02
	6s–7s	24.63	0.14	0.00	0.01	41.03	1.60	1.10	0.02
	6s–6d	41.98	0.17	0.40	0.07	38.18	1.42	2.40	0.02
	5d–6p	597.78	0.04	0.80	0.21	308.59	1.16	3.80	0.02
	5d–7s	2.60	0.07	0.30	0.03	4.53	0.78	0.00	0.01
	5d–6d	102.46	0.13	0.10	0.01	158.29	1.41	1.00	0.02
	6p–7s	439.45	0.54	0.00	0.24	134.20	0.88	4.90	0.02
	6p–6d	2135.33	0.70	0.00	0.20	538.64	0.86	5.70	0.02
Hg I	$6^1S–6^1P$	195.00	2.95	0.40	0.08	44.18	0.39	1.80	0.02
	$6^1S–7^1S$	10.02	0.29	0.00	0.02	14.91	0.82	0.20	0.02
	$6^1P–7^1S$	309.30	1.01	0.00	0.08	68.78	0.96	6.50	0.02
	$6^3P–7^3S$	168.16	1.70	0.10	0.14	36.45	0.69	4.20	0.02

Table 6.5. Excitation of multiply charged ions. Normalized Coulomb-Born-exchange cross sections for hydrogenlike ion Ne X. Parameters C, φ, D and A, χ, D given below can be used for any hydrogenlike ions with $z > 3$.

Transition	C	φ	D	R	A	χ	D	R
1s–2s	3.42	0.94	0.00	0.02	5.90	0.82	0.00	0.00
1s–2p	65.74	4.63	0.00	0.08	23.92	0.20	0.00	0.01
1s–3s	2.25	1.02	0.00	0.03	4.20	0.89	0.00	0.01
1s–3p	34.74	4.07	0.00	0.12	16.34	0.28	0.00	0.02
1s–3d	1.33	0.89	0.00	0.38	2.65	5.52	4.40	0.09
1s–4s	1.97	1.05	0.00	0.04	3.79	0.91	0.00	0.01
1s–4p	28.24	3.86	0.00	0.14	14.60	0.31	0.00	0.03
1s–4d	1.46	0.84	0.00	0.40	3.55	8.45	6.00	0.08
1s–4f	0.02	0.14	0.10	0.39	0.29	4.56	0.00	0.01
2s–3s	17.64	0.17	0.00	0.01	32.45	1.08	0.30	0.02
2s–3p	168.98	1.87	0.00	0.13	34.82	0.41	2.70	0.02
2s–3d	55.08	0.07	0.70	0.08	64.69	0.75	0.50	0.01
2s–4s	8.12	0.23	0.00	0.01	15.15	0.84	0.00	0.02
2s–4p	71.75	1.87	0.00	0.13	19.11	0.39	1.80	0.02
2s–4d	16.06	0.07	0.70	0.09	24.72	0.68	0.00	0.01
2s–4f	8.70	0.03	0.80	0.09	14.46	0.74	0.00	0.01
2p–3s	11.25	0.65	0.00	0.63	5.90	0.23	0.40	0.02
2p–3p	57.98	0.10	0.20	0.01	122.67	0.98	0.00	0.01
2p–3d	789.52	0.72	0.00	0.29	431.23	0.39	0.70	0.02
2p–4s	4.23	0.03	0.90	0.66	4.10	0.34	0.00	0.10
2p–4p	26.75	0.09	0.30	0.02	66.44	2.20	0.90	0.02
2p–4d	261.84	0.64	0.00	0.29	200.66	0.29	0.00	0.02
2p–4f	32.88	0.02	0.90	0.21	39.02	0.67	0.40	0.02
3s–4s	52.74	0.06	0.00	0.03	105.23	0.91	0.00	0.01
3s–4p	405.74	1.15	0.00	0.17	49.19	0.72	9.90	0.02
3s–4d	122.48	0.02	0.80	0.22	80.69	2.81	7.50	0.02
3s–4f	57.08	0.05	0.30	0.14	85.15	1.77	1.60	0.02
3p–4s	70.33	0.79	0.00	0.56	10.50	0.72	9.40	0.02
3p–4p	190.45	0.05	0.10	0.03	364.70	0.87	0.00	0.01
3p–4d	1274.21	0.00	1.00	0.45	453.05	0.72	4.10	0.02
3p–4f	396.20	0.04	0.60	0.15	393.76	2.22	3.70	0.02
3d–4s	6.47	0.02	0.50	0.11	10.54	0.73	0.00	0.01
3d–4p	50.47	0.00	0.90	0.30	42.37	0.41	0.30	0.01
3d–4d	250.15	0.04	0.10	0.03	531.28	0.99	0.00	0.00
3d–4f	4204.00	0.02	0.90	0.35	2319.00	0.62	1.80	0.02

Table 6.6. Excitation of multiply charged ions. Normalized Coulomb-Born-exchange cross sections for heliumlike ion Na x. Transitions with no change of spin ($\Delta S = 0$) Parameters C, φ, D and A, χ, D can be used for any heliumlike ion with $z > 3$

Transition	C	φ	D	R	A	χ	D	R
1^1S-2^1S	3.31	1.43	0.10	0.02	3.77	0.60	0.10	0.02
1^1S-2^1P	62.45	5.56	0.50	0.04	10.57	0.31	1.70	0.01
1^1S-3^1S	2.25	1.64	0.20	0.01	2.38	0.56	0.10	0.02

Table 6.6. (*continued*)

Transition	C	φ	D	R	A	χ	D	R
1^1S-3^1P	41.37	6.90	0.50	0.04	6.37	0.32	1.80	0.01
1^1S-3^1D	1.70	3.76	0.70	0.07	0.31	0.44	3.20	0.02
1^1S-4^1S	2.01	1.78	0.20	0.01	2.05	0.54	0.10	0.02
1^1S-4^1P	36.13	7.33	0.50	0.06	5.46	0.32	1.80	0.02
1^1S-4^1D	1.96	3.99	0.70	0.08	0.35	0.43	3.10	0.02
1^1S-4^1F	0.02	1.69	0.00	0.33	0.02	0.55	0.00	0.08
2^1S-2^1P	409.42	0.03	0.50	0.10	290.52	1.99	4.80	0.02
2^1S-3^1S	17.37	0.20	0.00	0.02	29.40	1.63	1.10	0.02
2^1S-3^1P	166.30	2.46	0.00	0.09	23.32	0.50	4.70	0.02
2^1S-3^1D	55.29	0.20	0.50	0.07	47.55	1.08	1.80	0.02
2^1S-4^1S	8.14	0.28	0.00	0.03	12.86	1.76	1.40	0.02
2^1S-4^1P	70.82	2.16	0.20	0.12	11.51	0.55	4.50	0.03
2^1S-4^1D	17.46	0.32	0,40	0.04	15.60	1.03	1.50	0.02
2^1S-4^1F	8.20	0.00	1.00	0.08	6.44	1.83	4.20	0.01
2^1P-3^1S	21.23	2.84	0.00	0.04	2.59	0.38	4.30	0.02
2^1P-3^1P	58.14	0.18	0.00	0.02	103.91	0.87	0.10	0.02
2^1P-3^1D	836.51	0.89	0.00	0.23	373.36	0.48	1.30	0.02
2^1P-4^1S	8.47	2.56	0.00	0.08	1.54	0.36	2.60	0.02
2^1P-4^1P	27.18	0.24	0.00	0.02	48.39	0.79	0.00	0.02
2^1P-4^1D	291.36	0.95	0.00	0.19	155.65	0.48	0.90	0.02
2^1P-4^1F	38.25	0.51	0.30	0.07	23.51	0.95	2.30	0.02
3^1S-4^1S	51.98	0.06	0.00	0.04	102.46	0.89	0.00	0.01
3^1S-4^1P	387.56	1.41	0.00	0.11	41.23	0.65	9.90	0.02
3^1S-4^1D	116.56	0.03	0.80	0.22	68.13	2.91	8.90	0.02
3^1S-4^1F	56.59	0.10	0.10	0.16	64.90	2.85	4.40	0.02
3^1P-4^1S	99.72	1.41	0.00	0.27	10.43	0.61	9.90	0.05
3^1P-4^1P	190.96	0.06	0.10	0.04	354.90	0.92	0.10	0.02
3^1P-4^1D	1463.19	0.56	0.00	0.31	427.74	0.74	4.50	0.02
3^1P-4^1F	380.73	0.90	0.30	0.19	341.30	2.51	5.10	0.01
3^1D-4^1S	6.80	0.13	0.20	0.27	5.48	2.54	5.80	0.02
3^1D-4^1P	50.17	0.02	0.90	0.32	25.18	0.90	3.20	0.02
3^1D-4^1D	247.32	0.05	0.00	0.02	505.88	0.93	0.00	0.01
3^1D-4^1F	4214.	0.02	0.90	0.35	2237.	0.64	2.00	0.02
2^3S-2^3P	361.31	0.01	0.70	0.10	270.49	1.48	3.10	0.01
2^3S-3^3S	16.59	0.19	0.00	0.02	29.62	1.22	0.50	0.02
2^3S-3^3P	146.76	2.41	0.00	0.08	22.56	0.46	3.90	0.02
2^3S-3^3D	51.27	0.09	0.70	0.09	54.73	0.82	0.80	0.01
2^3S-4^3S	7.82	0.26	0.00	0.01	13.75	1.12	0.40	0.02
2^3S-4^3P	68.41	2.43	0.00	0.08	14.29	0.39	2.30	0.03
2^3S-4^3D	16.82	0.10	0.70	0.10	22.59	0.56	0.00	0.01
2^3S-4^3F	7.36	0.05	0.80	0.11	10.68	0.81	0.30	0.02
2^3P-3^3S	16.66	1.27	0.00	0.37	4.91	0.27	1.40	0.01

Table 6.6. (*continued*)

Transition	C	φ	D	R	A	χ	D	R
2^3P-3^3P	56.23	0.13	0.10	0.02	114.22	0.92	0.00	0.00
2^3P-3^3D	772.12	0.79	0.00	0.27	390.74	0.42	0.90	0.02
2^3P-4^3S	5.64	0.73	0.00	0.54	3.40	0.21	0.10	0.03
2^3P-4^3P	26.31	0.11	0.30	0.02	58.49	1.04	0.00	0.01
2^3P-4^3D	269.86	0.75	0.00	0.26	181.43	0.37	0.30	0.02
2^3P-4^3F	30.56	0.03	0.90	0.22	33.07	0.58	0.40	0.02
3^3S-4^3S	50.54	0.06	0.00	0.04	100.15	0.90	0.00	0.01
3^3S-4^3P	350.20	1.41	0.00	0.13	37.00	0.64	9.90	0.02
3^3S-4^3D	106.33	0.02	0.80	0.22	66.88	2.72	7.70	0.02
3^3S-4^3F	56.00	0.06	0.30	0.14	76.69	2.15	2.40	0.02
3^3P-4^3S	98.32	1.15	0.00	0.37	11.50	0.65	9.90	0.03
3^3P-4^3P	187.59	0.05	0.10	0.03	356.10	0.86	0.00	0.02
3^3P-4^3D	1351.28	0.55	0.00	0.34	403.69	0.72	4.30	0.02
3^3P-4^3F	391.17	0.04	0.60	0.15	371.17	2.26	4.00	0.02
3^3D-4^3S	7.65	0.01	0.70	0.20	9.49	1.44	1.40	0.02
3^3D-4^3P	53.06	0.00	0.90	0.35	37.24	0.50	0.90	0.02
3^3D-4^3D	250.60	0.04	0.10	0.03	522.72	0.97	0.00	0.00
3^3D-4^3F	4199.	0.02	0.90	0.35	2302.	0.61	1.80	0.02

Table 6.7. Normalized cross sections for intercombination transitions ($\Delta S = 1$)

Atom	Transition	C	φ	D	R	A	χ	D	R
He I	1^1S-2^1S					1.10	2.25	0.20	0.03
	1^1S-2^1P	22.66	0.69	2.60	0.31	9.92	1.88	0.20	0.04
	1^1S-3^1S					1.91	3.68	0.20	0.03
	1^1S-3^1P	27.63	0.37	3.70	0.38	12.18	2.37	0.20	0.04
	1^1S-3^1D					0.75	0.73	0.00	0.07
	1^1S-4^1S					2.14	4.06	0.20	0.03
	1^1S-4^1P	28.54	0.34	3.90	0.38	12.94	2.52	0.20	0.04
	1^1S-4^1D					0.98	0.74	0.00	0.07
	1^1S-4^1F					0.03	0.93	0.00	0.14
	2^3S-2^1S					9.38	1.77	0.00	0.04
	2^3S-2^1P					22.20	1.78	0.00	0.02
	2^3S-3^1S					13.66	3.89	0.00	0.02
	2^3S-3^1P					15.76	2.16	0.00	0.02
	2^3S-2^3D					88.17	4.80	0.00	0.01
	2^1S-2^3P					13.13	1.69	0.00	0.04
	2^1S-3^3S					4.93	2.59	0.00	0.04
	2^1S-3^3P					19.10	4.26	0.00	0.03
	2^1S-3^3D					108.73	5.50	0.00	0.01
	2^3P-2^1P					19.58	1.04	0.00	0.05
	2^3P-3^1S					17.22	3.70	0.00	0.02

Table 6.7. (*continued*)

Atom	Transition	C	φ	D	R	A	χ	D	R
	2^3P-3^1P					36.07	3.89	0.10	0.05
	2^3P-3^1D					166.02	5.17	0.00	0.04
	2^1P-3^3S					14.15	3.22	0.00	0.01
	2^1P-3^3P					40.87	4.57	0.10	0.06
	2^1P-3^3D					159.60	4.81	0.00	0.03
Li II	1^1S-2^3S					0.04	0.87	9.90	0.33
	1^1S-2^3P	11.43	1.20	1.20	0.42	7.20	0.45	0.00	0.05
	1^1S-3^3S					0.04	0.85	9.90	0.34
	1^1S-3^3P	11.64	1.23	1.40	0.48	6.35	0.40	0.00	0.06
	1^1S-3^3D					1.53	0.90	0.00	0.02
	2^3S-2^1S					12.95	3.64	0.00	0.04
	2^3S-2^1P					25.83	3.60	0.00	0.03
	2^3S-3^1S					8.42	5.78	0.10	0.08
	2^3S-3^1P					9.80	3.20	0.00	0.04
	2^3S-3^1D					52.71	4.79	0.00	0.03
	2^1S-2^3P					14.93	3.27	0.00	0.03
	2^1S-3^3S					6.10	4.59	0.00	0.04
	2^1S-3^3P					6.54	2.50	0.00	0.02
	2^1S-3^3D					50.03	4.77	0.00	0.02
	2^3P-2^1P					41.35	1.87	0.00	0.02
	2^3P-3^1S					9.70	3.28	0.00	0.03
	2^3P-3^1P					22.67	1.64	0.00	0.05
	2^3P-3^1D					107.21	3.63	0.00	0.01
	2^1P-3^3S					14.98	4.25	0.00	0.02
	2^1P-3^3P					19.92	1.69	0.00	0.05
	2^1P-3^3D					123.15	3.94	0.00	0.01
Be I	2^1S-2^3P					10.55	0.80	0.00	0.04
B II	2^1S-2^3P					14.67	1.76	0.00	0.01
C III	2^1S-2^3P					10.16	2.37	0.00	0.04
O V	2^1S-2^3P					6.83	2.05	0.00	0.02
F VI	2^1S-2^3P					7.55	2.09	0.00	0.02
Na VIII	2^1S-2^3P					7.05	2.15	0.00	0.02
Mg I	3^1S-3^3P					15.38	0.95	0.00	0.05
	3^1S-4^3S					2.36	0.92	0.00	0.08
	3^1S-4^3P					11.54	1.31	0.00	0.02
	3^1S-3^3D					64.55	2.36	0.00	0.02
	3^3P-3^1P					38.63	1.66	0.00	0.03
	3^3P-4^1S					5.74	1.59	0.00	0.04
	3^3P-3^1D					185.98	4.44	0.10	0.06
	3^3P-4^1P					37.61	3.46	0.10	0.05
Mg IX	2^1S-2^3P					7.01	2.21	0.00	0.02
Al II	3^1S-3^3P					35.75	3.87	0.00	0.02

Table 6.7. (*continued*)

Atom	Transition	C	φ	D	R	A	χ	D	R
Ca I	4^1S-4^3P					20.73	1.54	0.00	0.04
	4^1S-3^3D					34.57	2.12	0.00	0.02
	4^3P-4^1P					37.19	1.78	0.00	0.00
	4^3P-3^1D					31.84	1.42	0.00	0.01
Zn I	4^1S-4^3P					13.08	0.68	0.00	0.06
	4^1S-5^3S					1.64	0.73	0.00	0.03
	4^1S-5^3P					10.63	1.04	0.00	0.02
	4^1S-4^3D					36.60	1.90	0.00	0.02
	4^3P-4^1P					32.85	1.40	0.00	0.04
	4^3P-5^1S					8.37	3.75	0.10	0.08
	4^3P-4^1D					222.55	6.75	0.10	0.06
	4^3P-5^1D					35.37	3.27	0.10	0.04
Ga II	4^1S-4^3P					37.95	3.53	0.00	0.02
	4^1S-5^3S					7.66	4.37	0.00	0.04
	4^1S-4^3D					62.19	4.24	0.00	0.03
	4^3P-4^1P					72.04	3.02	0.00	0.01
	4^3P-5^1S					16.86	4.45	0.00	0.01
	4^3P-4^1D					127.73	2.81	0.00	0.01
Sr I	5^1S-5^3P					25.27	1.99	0.00	0.03
	5^1S-4^3D					43.30	2.38	0.00	0.01
	5^1S-6^3S					8.71	3.08	0.00	0.03
	5^1S-6^3P					23.66	3.13	0.00	0.02
	5^3P-4^1D					36.50	1.47	0.00	0.02
	5^3P-5^1P					39.25	2.07	0.00	0.01
	5^3P-6^1S					8.87	2.14	0.00	0.05
	5^3P-6^1P					29.75	2.89	0.00	0.02
	4^3D-5^1P					22.87	1.40	0.00	0.07
Cd I	5^1S-5^3P					15.11	0.80	0.00	0.08
	5^1S-6^3S					2.13	0.91	0.00	0.09
	5^1S-6^3P					11.86	1.27	0.00	0.02
	5^1S-5^3D					45.80	2.32	0.00	0.02
	5^3P-5^1P					35.17	1.49	0.00	0.03
	5^3P-6^1S					8.47	3.70	0.10	0.06
	5^3P-5^1D					241.55	7.05	0.10	0.08
	5^3P-6^1P					33.72	3.41	0.10	0.05
In II	5^1S-5^3P					40.08	4.32	0.00	0.03
	5^1S-6^3S					6.08	4.49	0.00	0.04
	5^3P-5^1P					89.74	4.11	0.00	0.01
	5^3P-6^1S					20.63	5.93	0.00	0.02
Ba I	6^1S-5^3D					12.56	1.30	0.00	0.04
	6^1S-6^3P					22.01	1.71	0.00	0.03
	6^1S-7^3S					5.62	2.48	0.00	0.03

Table 6.7. (*continued*)

Atom	Transition	C	φ	D	R	A	χ	D	R
	5^3D-6^1P					16.62	1.12	0.00	0.08
	5^3D-7^1S					6.85	1.90	0.00	0.05
	5^1D-6^3P					11.02	0.65	0.00	0.13
Hg I	6^1S-6^3P					19.83	1.17	0.00	0.08
	6^1S-7^3S					3.84	1.34	0.00	0.12
	6^3P-6^1P					37.16	1.61	0.00	0.01
	6^3P-7^1S					12.30	4.18	0.10	0.06
	6^1P-7^3S					21.52	4.36	0.00	0.04
	7^3S-7^1S					13.63	3.94	0.00	0.01

Table 6.8. Excitation of multipy charged ions. Normalized exchange cross section for heliumlike ion Na X. Intercombination transitions ($\Delta S = 1$). Parameters C, φ, D and A, χ, D can be used for any heliumlike ion with $z > 3$.

Transition	C	φ	D	R	A	χ	D	R
1S–2S	2.45	1.31	1.00	0.09	1.59	0.69	0.00	0.02
1S–2P	4.61	0.81	0.50	0.11	10.64	1.12	0.00	0.03
1S–3S	2.27	1.52	1.00	0.08	1.49	0.69	0.00	0.02
1S–3P	4.50	0.84	0.60	0.10	9.57	1.09	0.00	0.03
1S–3D	0.36	0.52	0.30	0.21	2.29	1.60	0.00	0.02
1S–4S	2.30	1.37	1.10	0.08	1.53	0.70	0.00	0.02
1S–4P	4.35	0.90	0.60	0.10	9.13	1.08	0.00	0.03
1S–4D	0.46	0.57	0.30	0.21	2.83	1.57	0.00	0.02
1S–4F	0.01	0.36	0.10	0.35	0.17	2.14	0.00	0.02
2S–3S					2.10	3.33	0.00	0.03
2S–3P					5.42	3.46	0.00	0.03
2S–3D					16.26	3.47	0.00	0.03
2S–4S					1.92	3.35	0.00	0.04
2S–4P					4.90	3.44	0.00	0.03
2S–4D					11.33	3.23	0.00	0.03
2S–4F					8.76	4.51	0.00	0.03
2P–3S					4.02	3.06	0.00	0.03
2P–3P					26.37	3.64	0.00	0.03
2P–3D					65.05	4.42	0.00	0.02
2P–4S					3.65	3.05	0.00	0.03
2P–4P					23.65	3.65	0.00	0.03
2P–4D					54.02	4.25	0.00	0.03
2P–4F					22.88	5.53	0.00	0.02
3S–4S					2.01	6.90	0.00	0.02
3S–4S					2.01	6.90	0.00	0.02
3S–4P					4.54	7.02	0.00	0.02
3S–4D					8.38	7.17	0.00	0.02
3S–4F					20.05	6.76	0.00	0.02
3P–4S					3.87	6.44	0.00	0.03

Table 6.8. (*continued*)

Transition	C	φ	D	R	A	χ	D	R
3P–4P					17.28	6.79	0.00	0.02
3P–4D					32.97	8.34	0.00	0.01
3P–4F					58.74	7.55	0.00	0.02
3D–4S					6.48	6.77	0.00	0.02
3D–4P					24.22	7.53	0.00	0.02
3D–4D					50.84	7.86	0.00	0.02
3D–4F					121.00	8.86	0.00	0.02

6.1.5 Transitions between Closely Spaced Levels (Tables 6.9–10)

In the case of transitions between closely spaced levels under the conditions $\Delta E \ll E_0, E_1$ and $\Delta E \ll \mathcal{E}$, the dependence of multipole and exchange cross sections calculated by means of the first-order methods on ΔE is almost absent. For the optically allowed transitions ($\Delta l = 1$) a weak logarithmic dependence exists. The calculations for multiply-charged ions have been made using the Coulomb–Born approximation for transitions with no change of spin ($\Delta S = 0$) and using the orthogonalized functions method for intercombination transitions. The data of Tables 6.9 and 6.10 were obtained for a set of values of ΔE and can be applied to arbitrary multiply-charged ions with $z > 3$. For quadrupole and intercombination transitions the value of ΔE is not important. For dipole transitions one has to interpolate data for particular values of ΔE. The fitting formulas and the range of analytic approximation are quite the same as in Sect. 6.1.4.

6.1.6 Ionization Cross Sections (Table 6.11 and 6.12)

The ionization cross sections have been calculated in accordance with (3.1.38) in the partial wave representation. In cases of ions the Coulomb–Born approximation has been used:

$$\sigma_i(a_0) = Q_i\sigma(l_0), \quad \mathrm{DE} = z^2 \ \mathrm{Ry}\,, \tag{6.1.18}$$

$$\langle v\sigma_i(a_0)\rangle = Q_i\langle v\sigma_i(l_0)\rangle\,. \tag{6.1.19}$$

$$u = (\mathcal{E} - E_0)/z^2 \ \mathrm{Ry}\,,$$

$$\beta = z^2 \ \mathrm{Ry}/T, \quad p = E_0/z^2 \ \mathrm{Ry}\,.$$

The fitting formulas and the angular factors Q_i are given by (5.1.21–25). For the total cross section of ionization from a shell l_0^m,

$$Q_i = m\,, \tag{6.1.20}$$

where m is a number of equivalent electrons.

The set of parameters C, φ, D is adjusted for the range $0.0625 < u < 64$, and the set A, χ, D, for $0.125 < \beta < 8$.

Table 6.9. Transitions between the closely spaced levels with no change of spin ($\Delta S = 0$) The Coulomb-Born-exchange cross sections for multiply charged ions

Transition	$\Delta E/z^2$ [cm^{-1}]	C	φ	D	R	A	χ	D	R
				Dipole transitions					
2s–2p	1480.	321.	0.01	0.70	0.15	265.	0.96	1.60	0.02
2s–2p	740.	389.	0.01	0.70	0.08	269.	1.72	4.00	0.02
2s–2p	370.	433.	0.02	0.50	0.09	315.	2.06	4.80	0.02
2s–2p	185.	491.	0.01	0.50	0.06	406.	2.16	4.30	0.01
2s–2p	93.	531.	0.01	0.40	0.08	524.	1.96	3.20	0.02
2s–2p	46.	589.	0.01	0.40	0.06	650.	1.72	2.30	0.02
2s–2p	23.	647.	0.01	0.40	0.05	773.	1.52	1.70	0.02
2s–2p	12.	704.	0.00	0.40	0.04	883.	1.42	1.40	0.02
3s–3p	856.	1849.	0.01	0.70	0.12	1232.	1.66	4.10	0.02
3s–3p	428.	2163.	0.03	0.50	0.10	1492.	2.10	5.20	0.02
3s–3p	214.	2439.	0.02	0.40	0.10	2024.	2.24	4.60	0.02
3s–3p	107.	2762.	0.01	0.40	0.08	2671.	2.19	3.70	0.01
3s–3p	54.	3092.	0.01	0.40	0.06	3403.	1.89	2.60	0.02
3s–3p	27.	3436.	0.01	0.40	0.05	4173.	1.55	1.70	0.02
3s–3p	13.	3677.	0.01	0.30	0.07	4890.	1.47	1.40	0.02
3p–3d	856.	2356.	0.01	0.70	0.11	1588.	1.68	4.10	0.02
3p–3d	428.	2738.	0.03	0.50	0.10	1904.	2.11	5.20	0.02
3p–3d	214.	3080.	0.02	0.40	0.10	2565.	2.24	4.60	0.02
3p–3d	107.	3482.	0.01	0.40	0.08	3387.	2.15	3.60	0.01
3p–3d	54.	3892.	0.01	0.40	0.06	4304.	1.84	2.50	0.02
3p–3d	27.	4319.	0.01	0.40	0.05	5242.	1.55	1.70	0.02
3p–3d	13.	4617.	0.01	0.30	0.07	6132.	1.46	1.40	0.02
				Quadrupole transitions					
2p–2p	100.	13.53	0.00	0.30	0.03	20.46	1.15	0.70	0.02
2p–2p	10.	13.51	0.00	0.30	0.04	20.37	1.14	0.70	0.02
3s–3d	1000.	51.26	0.01	0.20	0.05	86.06	1.69	1.20	0.02
3s–3d	100.	51.64	0.00	0.20	0.04	96.24	0.85	0.00	0.02
3s–3d	10.	51.53	0.00	0.20	0.04	95.54	0.84	0.00	0.02
3p–3p	100.	91.07	0.00	0.10	0.03	178.25	0.87	0.00	0.01
3p–3p	10.	91.17	0.00	0.10	0.03	178.31	0.87	0.00	0.02
3d–3d	100.	75.06	0.00	0.10	0.03	144.24	0.85	0.00	0.02
3d–3d	10.	75.09	0.00	0.10	0.03	143.78	0.93	0.10	0.02

Table 6.10. Intercombination transitions between the closely spaced levels ($\Delta S = 1$). The summed over κ exchange rate coefficients for multiply charged ions

Transition	$\Delta E/z^2$ [cm^{-1}]	A	χ	D	R
2s–2s	200.	6.00	3.07	0.00	0.03
2s–2p	747.	7.84	2.83	0.00	0.02
2p–2p	200.	62.32	3.51	0.00	0.03
3s–3s	20.	6.08	6.62	0.00	0.02
3s–3p	210.	9.00	7.80	0.00	0.02
3s–3d	252.	15.36	8.35	0.00	0.01
3p–3p	20.	41.08	7.15	0.00	0.02
3p–3d	42.	47.72	8.59	0.00	0.01

Table 6.11. Ionization cross sections for atoms and ions in Coulomb-Born approximation

Atom	Level	C	φ	D	R	A	χ	D	R
H I	1s	9.581	2.37	0.60	0.07	7.371	0.12	−0.70	0.01
	2s	32.6	0.13	0.00	0.11	198.5	21.45	5.10	0.05
	2p	86.9	0.33	−0.60	0.08	430.0	5.25	1.00	0.02
	3s	66.9	0.03	−0.50	0.11	746.3	15.23	1.60	0.03
	3p	175.0	0.11	−0.80	0.10	1961.0	9.00	0.60	0.02
	3d	294.0	0.08	−0.80	0.12	3462.0	7.11	0.20	0.01
	4s	106.0	0.00	−0.70	0.13	2084.0	19.66	1.00	0.02
	4p	283.0	0.01	−0.80	0.11	5919.0	15.61	0.50	0.02
	4d	483.0	0.00	−0.80	0.12	10416.0	14.02	0.30	0.02
	4f	680.0	−0.01	−0.80	0.21	17566.0	14.32	0.10	0.00
	5s	145.0	−0.01	−0.80	0.16	4445.0	26.75	0.80	0.02
	5p	396.0	−0.02	−0.80	0.12	13595.0	25.52	0.50	0.02
	5d	674.0	−0.03	−0.80	0.13	24815.0	25.57	0.40	0.02
	5f	956.0	−0.03	−0.80	0.19	35980.0	22.74	0.20	0.01
	5g	1296.0	−0.03	−0.80	0.28	57240.0	23.95	0.00	0.01
He I	1^1S	5.986	4.50	3.50	0.05	1.583	0.16	−0.10	0.02
	2^3S	19.67	0.21	0.20	0.11	86.14	17.47	5.60	0.06
	2^3P	88.1	0.44	−0.60	0.07	366.8	5.65	1.50	0.03
	2^1S	25.3	0.16	0.10	0.12	128.4	18.37	5.10	0.06
	2^1P	88.0	0.34	−0.60	0.08	426.8	5.42	1.10	0.02
He II	1s	9.570	2.41	1.10	0.07	4.500	0.25	−0.10	0.04
	2s	32.45	0.22	0.00	0.12	89.60	13.12	7.20	0.06
	2p	85.1	0.40	−0.60	0.05	295.6	3.81	1.10	0.02
	3s	66.2	0.07	−0.50	0.12	406.4	8.30	1.60	0.03
	3p	176.0	0.09	−0.70	0.09	1388.0	6.37	0.60	0.02
	3d	287.0	0.11	−0.80	0.03	2548.0	5.65	0.30	0.01
Li I	2s	13.80	0.24	0.00	0.06	77.41	16.35	4.10	0.04
	2p	88.7	0.40	−0.60	0.08	389.9	5.45	1.30	0.02
	3s	34.1	0.08	−0.70	0.03	386.5	9.60	0.70	0.02
	3p	179.0	0.13	−0.80	0.12	1913.0	9.65	0.80	0.02
	3d	294.0	0.08	−0.80	0.12	3458.0	7.12	0.20	0.01
Li II	1^1S	8.046	2.68	4.00	0.07	1.872	0.20	0.30	0.02
	2^3S	25.00	0.30	0.20	0.11	44.27	7.56	6.00	0.07
	2^1S	29.41	1.33	−0.80	0.09	60.58	9.83	6.80	0.07
	2^3P	85.3	0.49	−0.60	0.07	259.3	3.89	1.40	0.03
	2^1P	85.7	0.41	−0.60	0.05	294.2	3.95	1.20	0.02
Be I	2^1S	8.56	0.73	0.60	0.08	11.87	0.17	−0.80	0.09
	2^3P	64.51	1.05	0.10	0.09	77.73	0.13	−0.80	0.01
	2^1P	88.5	0.56	−0.50	0.08	285.4	7.85	3.30	0.04
B I	2p	53.08	1.70	0.60	0.08	36.56	0.29	−0.30	0.04
C I	2s	4.529	3.25	1.10	0.05	2.349	0.18	−0.40	0.01
	$2p^3P$	36.88	2.60	1.70	0.10	12.47	0.36	0.50	0.02
	$2p^1D$	41.11	2.30	1.30	0.08	16.79	0.36	0.30	0.02
	$2p^1S$	47.58	1.88	0.90	0.07	25.70	0.30	−0.10	0.02

Table 6.11. (*continued*)

Atom	Level	C	φ	D	R	A	χ	D	R
C II	2p	67.48	1.01	0.00	0.11	74.04	0.23	−0.60	0.04
C III	2^1S	21.63	0.31	0.50	0.10	31.72	0.17	−0.80	0.05
	2^3P	76.3	0.40	−0.10	0.09	165.9	11.48	8.20	0.05
	2^1P	84.6	0.34	−0.30	0.09	239.8	6.21	3.00	0.04
C IV	2s	25.86	0.23	0.30	0.11	53.42	9.73	6.70	0.07
	2p	84.4	0.42	−0.60	0.08	302.0	4.89	1.60	0.03
C V	1^1S	9.203	1.42	3.80	0.05	3.519	0.21	−0.10	0.02
	2^3S	29.04	0.19	0.20	0.11	76.64	12.60	7.00	0.06
	2^3P	83.4	0.40	−0.60	0.07	312.3	4.89	1.50	0.03
N I	2p	26.43	3.78	3.00	0.10	5.18	0.25	0.70	0.02
O I	2s	2.644	3.66	8.00	0.04	0.485	−0.11	−0.10	0.01
	2p	24.92	3.64	2.70	0.06	5.45	0.19	0.30	0.01
O II	2s	8.321	2.56	1.50	0.12	2.540	0.45	1.10	0.03
	2p	50.62	1.88	0.80	0.08	21.90	0.54	0.90	0.03
O III	2s	13.67	0.92	0.80	0.11	11.63	0.23	−0.50	0.04
	2p	66.66	0.74	0.20	0.11	80.48	0.11	−0.80	0.03
O IV	2s	19.41	0.40	0.70	0.11	24.52	0.13	−0.80	0.02
	2p	75.1	0.41	0.00	0.10	143.3	9.42	7.40	0.05
O V	2^1S	23.95	0.26	0.50	0.11	37.80	5.43	4.40	0.08
	2^3P	79.1	0.31	−0.20	0.08	226.1	7.11	3.50	0.04
	2^1P	83.9	0.33	−0.40	0.09	273.7	5.95	2.40	0.03
O VI	2S	28.03	0.19	0.30	0.10	66.22	11.34	6.90	0.07
F I	2p	17.07	5.88	5.00	0.06	1.98	0.10	0.40	0.01
Ne I	2p	13.17	5.82	6.50	0.03	1.39	0.09	0.30	0.01
Na I	3s	13.06	0.42	−0.60	0.02	83.04	8.29	1.50	0.02
	3p	85.2	0.26	−0.70	0.11	553.7	5.03	0.50	0.02
	3d	296.0	0.09	−0.80	0.11	3362.0	6.89	0.20	0.02
Mg I	3^1S	9.21	0.68	−0.40	0.03	39.92	9.37	2.90	0.03
	3^3P	68.4	0.57	−0.40	0.08	209.8	6.26	2.60	0.03
	3^1P	77.3	0.32	−0.70	0.10	464.8	4.99	0.60	0.02
Mg IX	2^1S	27.47	0.20	0.40	0.10	57.88	9.80	6.50	0.07
	2^3P	80.4	0.32	−0.40	0.08	272.0	5.79	2.20	0.03
	2^1P	83.1	0.35	−0.50	0.08	304.3	5.79	2.00	0.03
Mg X	2s	29.94	0.17	0.30	0.10	80.34	12.65	6.80	0.06
	2p	83.7	0.29	−0.50	0.07	331.4	5.37	1.60	0.03
Mg XI	1^1S	9.477	1.23	3.20	0.06	4.417	0.23	−0.20	0.02
Ar I	3p	30.62	1.95	1.70	0.41	18.03	0.73	0.70	0.06
K I	4s	16.36	0.36	−0.60	0.05	92.98	6.23	1.10	0.02
Ca I	4^1S	12.31	0.71	−0.40	0.11	34.93	5.37	2.40	0.03
	4^3P	88.3	0.53	−0.50	0.14	285.1	4.35	1.40	0.03
	3^3D	174.1	0.39	0.60	0.14	200.9	0.12	−0.80	0.02
	3^1D	188.4	0.39	0.40	0.15	234.6	0.13	−0.80	0.03

Table 6.11. (*continued*)

Atom	Level	C	φ	D	R	A	χ	D	R
	4^1P	99.4	0.42	−0.60	0.09	405.2	6.91	2.10	0.03
Cu I	4s	10.27	0.75	−0.30	0.04	37.25	13.17	5.40	0.04
Kr I	4p	39.98	2.47	1.70	0.31	16.19	0.60	1.00	0.05
Rb I	5s	16.32	0.42	−0.60	0.05	77.12	5.69	1.30	0.02
	5p	95.7	0.23	−0.70	0.05	624.6	6.07	0.80	0.02
	4d	306.0	0.30	−0.80	0.13	1926.0	7.56	1.30	0.03
Sr I	5^1S	13.22	0.59	−0.50	0.04	56.38	8.94	2.90	0.03
	5^3P	99.7	0.54	−0.50	0.11	331.0	7.97	3.20	0.04
	4^3D	236.3	0.46	0.70	0.14	393.7	6.67	5.40	0.07
	4^1D	39.77	1.59	1.40	0.09	36.03	0.20	−0.60	0.03
	5^1P	117.9	0.36	0.20	0.13	585.7	9.49	2.50	0.04
Sr II	5s	25.0	0.18	0.20	0.03	144.2	5.44	0.90	0.02
	4d	325.5	0.22	0.60	0.16	785.6	8.18	4.90	0.05
Ag I	5s	11.28	0.71	0.60	0.04	40.13	18.90	8.30	0.04
Xe I	5p	46.68	1.87	1.60	0.18	35.78	0.34	−0.30	0.03
Cs I	6s	19.2	0.38	0.10	0.05	122.0	6.53	1.00	0.02
	6p	125.3	0.22	0.10	0.08	869.6	6.49	0.80	0.02
	5d	297.0	0.23	0.40	0.16	1128.0	11.10	4.20	0.05
Ba I	6^1S	14.52	0.51	0.20	0.05	73.18	6.39	1.40	0.02
	5^3D	153.3	0.45	2.70	0.09	100.5	0.22	−0.40	0.01
	5^1D	163.3	0.38	2.60	0.09	118.8	0.20	−0.50	0.01
	6^3P	96.6	0.43	0.20	1.10	399.1	4.96	1.20	0.03
Ba II	6s	27.3	0.14	0.20	0.02	171.0	5.58	0.80	0.02
	5d	352.1	0.21	0.60	0.16	883.3	6.90	3.80	0.05
	6p	127.0	0.16	0.00	0.04	1018.0	5.52	0.40	0.02
Hg I	6^1S	9.67	0.24	2.60	0.12	18.60	12.23	8.50	0.16
	6^3P	89.2	0.22	1.80	0.13	129.9	0.18	−0.80	0.12
	6^1P	114.9	0.85	0.10	0.12	413.2	12.41	4.90	0.05

Table 6.12. Rate coefficients of dielectronic recombination in Coulomb-Born-exchange approximation. Parameters A_d and χ_d.

X_{z+1}	α_0		α_1	A_d	χ_d
	H-like ions				
He II	1s	–	2p	31.18	0.74
Be IV	1s	–	2p	36.69	0.73
C VI	1s	–	2p	31.34	0.71
O VIII	1s	–	2p	25.78	0.69
Ne X	1s	–	2p	21.06	0.67
Mg XII	1s	–	2p	17.20	0.66
Si XIV	1s	–	2p	14.08	0.64
S XVI	1s	–	2p	11.56	0.63

Table 6.12. (*continued*)

X_{z+1}	α_0 — α_1	A_d	χ_d
Ca XX	1s – 2p	7.828	0.61
Fe XVI	1s – 2p	4.420	0.60
	He-like ions		
Li II	$1^1S - 2^1P$	17.43	1.13
C V	$1^1S - 2^1P$	22.21	0.85
C V	$1^1S - 2^3P$	0.024	0.90
O VII	$1^1S - 2^1P$	18.21	0.79
O VII	$1^1S - 2^3P$	0.078	0.85
Ne IX	$1^1S - 2^1P$	14.37	0.75
Ne IX	$1^1S - 2^3P$	0.170	0.82
Mg XI	$1^1S - 2^1P$	11.16	0.72
Mg XI	$1^1S - 2^3P$	0.309	0.80
Si XIII	$1^1S - 2^1P$	8.659	0.69
Si XIII	$1^1S - 2^3P$	0.495	0.78
S XV	$1^1S - 2^1P$	6.738	0.68
S XV	$1^1S - 2^3P$	0.714	0.75
Ca XIX	$1^1S - 2^1P$	4.144	0.66
Ca XIX	$1^1S - 2^3P$	1.151	0.71
Fe XXV	$1^1S - 2^1P$	2.138	0.64
Fe XXV	$1^1S - 2^3P$	1.500	0.65
	Li-like ions		
Be II	2s – 2p	10.81	0.07
C IV	2s – 2p	6.405	0.04
O VI	2s – 2p	3.706	0.02
Ne VIII	2s – 2p	2.400	0.02
Mg X	2s – 2p	1.690	0.01
Si XII	2s – 2p	1.182	0.01
S XIV	2s – 2p	.9387	0.01
Ca XVIII	2s – 2p	.6229	0.01
Fe XXIV	2s – 2p	.5012	0.01
	Be-like ions		
C III	$2^1S - 2^1P$	21.07	0.10
O V	$2^1S - 2^1P$	11.62	0.06
Ne VII	$2^1S - 2^1P$	7.010	0.04
Na VIII	$2^1S - 2^1P$	5.665	0.03
Mg IX	$2^1S - 2^1P$	4.663	0.03
Fe XXIII	$2^1S - 2^1P$	1.003	0.01

6.1.7 Dielectronic Recombination Rate Coefficients (Table 6.12)

The methods of calculations of the rate coefficients for the dielectronic recombination process,

$$X_{z+1}(\alpha_0) + e \rightarrow X_z(\alpha_1 nl) \rightarrow X_z(\alpha_0 nl) + \hbar\omega ,$$

are described in Sect. 5.2. The simplified model (5.2.12) with its modification (5.2.29) for $s-p$ transitions was used. The excitation cross section for the transition $\alpha_0-\alpha_1$ of an ion X_{z+1} has been calculated in the Coulomb–Born approximation with exchange whenever it has been substantial. The rate coefficient for dielectronic recombination connected with the transition $\alpha_0-\alpha_1$ is expressed in the form

$$\kappa_\mathrm{d}(\alpha_0,\alpha_1)=10^{-13}\,Q_\mathrm{d}(\alpha_0,\alpha_1)\frac{A_\mathrm{d}}{2l_0+1}\beta^{3/2}\exp(-\beta\chi_\mathrm{d}),\quad \beta=\frac{(z+1)^2\mathrm{Ry}}{T}\,. \tag{6.1.21}$$

Parameters A_d and χ_d for the most important actual cases are given in Table 6.12 and the angular factors Q_d for these cases are given by

$$\begin{aligned}&Q_\mathrm{d}(n_0l_0^m,\,n_0l_0^{m-1}n_1l_1)=m\,,\\&Q_\mathrm{d}(n_0l_0^Nn_1l_1^m,\,n_0l_0^{N-1}n_1l_1^{m+1})=N\left(1-\frac{m}{2(2l_1+1)}\right).\end{aligned} \tag{6.1.22}$$

In the case of heliumlike ions the total rate coefficient for dielectronic recombination is the sum of contributions from excitation of both singlet and triplet P levels.

6.2 Formulas Defining the Angular Factors

6.2.1 Rules for the Addition of Cross Sections

In various applications, cross sections are required for transitions between separate levels, between two groups of closely spaced levels, for transitions from a given level to a group of levels, and for transitions from the whole group of levels to a given level. For example, one may be interested in transitions between separate fine structure components $L_0S_0J_0-L_1S_1J_1$ of two terms or in the transition between the terms $L_0S_0-L_1S_1$ as a whole.

The cross section for transition from a given level a of the group A to the group B of levels b is, clearly,

$$\sigma(Aa,B)=\sum_b\sigma(Aa,Bb)\,, \tag{6.2.1}$$

where $\sigma(Aa,Bb)$ is the cross section for the transition $a-b$. If every level a of the group A is populated proportionally to its statistical weight, then the cross section for the transition $A-Bb$ is defined by

$$\sigma(A,Bb)=\frac{1}{g(A)}\sum_a g(a)\,\sigma(Aa,Bb)\,, \tag{6.2.2}$$

and the cross section for the transition $A-B$, by

$$\sigma(A,B)=\frac{1}{g(A)}\sum_{ab} g(a)\,\sigma(Aa,Bb)\,. \tag{6.2.3}$$

Here $g(a)$ is the statistical weight of level a, and $g(A) = \sum_a g(a)$ is the statistical weight of the group of levels A.

The tabulated cross sections are given by formulas (6.1.1,2) where the dependence of effective cross sections on angular momenta is determined by the factors Q'_κ and Q''_κ. Therefore the summation of the cross sections over the fine-structure components of terms and over the terms belonging to a single electronic configuration is equivalent to the summation of these angular factors. The next subsections give a summary of formulas defining the factors Q'_κ and Q''_κ for the cases which can be met when using the tables of cross sections given in Sect. 6.1.

6.2.2 *LS*-Coupling; Q_κ for transitions between levels *LSJ*

In this and the following subsections we give general formulas for Q-factors in the LS-coupling. The derivation for transitions not involving the shells of equivalent electrons was given in Sect. 2.3. We consider also some most important particular cases. $Q_\kappa^{(p)}$ means Q'_κ or Q''_κ. To simplify the notation, we denote by γ the whole set of quantum numbers defining the term, specifying if necessary the spin S_p and orbital angular momentum L_p of the atomic core, the orbital momentum of an electron l, the total spin S, and the total orbital momentum of an atom L. The unnecessary quantum numbers will be omitted in formulas.

The multipole order κ in general can vary between

$$\kappa_{\min} = |l_0 - l_1| \,, \quad \kappa_{\max} = l_0 + l_1 \,,$$

and Q'_κ is not zero only if

$$\kappa = \kappa_{\min}, \kappa_{\min} + 2, \ldots, l_0 + l_1 \,,$$

For transitions between LSJ levels Q-factor can be written as follows, compare (2.3.3–5)

$$Q_\kappa^{(p)}(L_0 S_0 J_0, L_1 S_1 J_1) = \frac{2l_0+1}{2J_0+1} \sum_{qv} B^2_{q\kappa v}(J)\, C^{(p)}(q) \,, \tag{6.2.4}$$

$$C'(q) = 2\delta(q), \quad C''(q) = \frac{1}{2}[q]^2 \,, \tag{6.2.5}$$

$$B_{q\kappa v}(J) = B_{q\kappa}(SL)\, M_{q\kappa v}(SLJ) \tag{6.2.6}$$

The factor M according to (2.2.23) is equal to

$$M_{q\kappa v}(SLJ) = [J_0 J_1 v] \begin{Bmatrix} S_0 & S_1 & q \\ L_0 & L_1 & \kappa \\ J_0 & J_1 & v \end{Bmatrix} . \tag{6.2.7}$$

$B_{q\kappa}(SL) \equiv B_{q\kappa}(S_0 L_0, S_1 L_1)$ does not depend on J, but depends on the type of transition. It is discussed in the next subsection.

The sum over fine-structure components J_1 is independent of J_0:

$$Q_\kappa^{(p)}(L_0S_0J_0,\, L_1S_1) = Q_\kappa^{(p)}(L_0S_0,\, L_1S_1) = \frac{[l_0]^2}{[L_0S_0]^2}\sum_q B_{q\kappa}^2(SL)\, C^{(p)}(q)\,, \tag{6.2.8}$$

Q_κ averaged over J_0 of the initial levels is

$$Q_\kappa^{(p)}(L_0S_0,\, L_1S_1J_1) = \frac{[J_1]^2}{[L_1S_1]^2} Q_\kappa^{(p)}(L_0S_0,\, L_1S_1)\,, \tag{6.2.9}$$

i.e., it is proportional to the upper level statistical weight.

6.2.3 *LS*-Coupling; Q_κ for transitions between Terms *LS*

Q-factors for transition L_0S_0–L_1S_1 are defined by (6.2.8). On substitution of $C^{(p)}$ we obtain

$$\begin{aligned} Q'_\kappa(L_0\,S_0,\, L_1\,S_1) &= \frac{2[l_0]^2}{[L_0\,S_0]^2} B_{0\kappa}^2\,(S_0L_0,\, S_1L_1)\,,\\ Q''_\kappa(L_0\,S_0,\, L_1\,S_1) &= \frac{2[l_0]^2}{[L_0\,S_0]^2} \sum_q B_{q\kappa}^2\,(S_0\,L_0,\, S_1\,L_1)\,[q]^2\,. \end{aligned} \tag{6.2.10}$$

a) Transitions not involving the shells of equivalent electrons

$$\begin{aligned} &a_0 = [L_pS_p]l_0L_0\,S_0, \quad a_1 = [L_pS_p]l_1L_1\,S_1,\\ &B_{q\kappa}(SL) = M_{0qq}(S_p\,sS)\,M_{0\kappa\kappa}(L_plL)\,[S_pL_p]\,. \end{aligned} \tag{6.2.11}$$

Therefore

$$\begin{aligned} Q'_\kappa(L_0S_0,\, L_1S_1) &= Q_\kappa(L_0,\, L_1)\cdot A_0\,, \qquad A_0 = \delta(S_0,\, S_1)\,,\\ Q''_\kappa(L_0S_0,\, L_1S_1) &= Q_\kappa(L_0,\, L_1)\cdot A_2\,, \qquad A_2 = (2S_1+1)/2(2S_p+1) \end{aligned} \tag{6.2.12}$$

where

$$Q_\kappa(L_0,\, L_1) = [L_pS_p]^2M_{0\kappa\kappa}^2(L_plL) = [l_0L_0]^2 \begin{Bmatrix} L_0 & L_1 & \kappa \\ l_1 & l_0 & L_p \end{Bmatrix}^2 . \tag{6.2.13}$$

The sum over L_1 and average over L_0 are independent on L_0:

$$Q_\kappa(L_0,\, L_pl_1) = Q_\kappa(L_pl_0,\, L_pl_1) = 1\,, \tag{6.2.14}$$

$$Q_\kappa(L_pl_0,\, L_1) = [L_1]^2/[L_pl_1]^2\,. \tag{6.2.15}$$

Similar sums take place for spin factors

$$\frac{1}{2[S_p]^2}\sum_{S_0}[S_0]^2A_0 = \sum_{S_1}A_0 = \sum_{S_1}A_2 = 1\,, \quad \frac{1}{2[S_p]^2}\sum_{S_0}[S_0]^2A_2 = A_2\,. \tag{6.2.16}$$

b) Transitions from the shell of equivalent electrons

$$a_0 = l_0^m L_0 S_0\,, \quad a_1 = l_0^{m-1}[L_p S_p] l_1 L_1 S_1\,,$$
$$B_{q\kappa}(SL) = \sqrt{m} G^{L_0 S_0}_{L_p S_p} \cdot M_{0qq}(S_p s S) M_{0\kappa\kappa}(L_p l L)[S_p L_p]\,, \tag{6.2.17}$$

where $G^{L_1 S_1}_{L_p S_p}$ is the fractional parentage coefficient (Sect. 6.2.5). Q-factors are defined by (6.2.12) with

$$Q_\kappa(L_0, L_1) = m\left(G^{L_0 S_0}_{L_p S_p}\right)^2 [l_0 L_0]^2 \begin{Bmatrix} L_0 & L_1 & \kappa \\ l_1 & l_0 & L_p \end{Bmatrix}^2 . \tag{6.2.18}$$

We see that $Q_\kappa(L_0, L_1)$ depends in fact on S_0 through G. The sums over L_1 and $L_p S_p$ are

$$Q_\kappa(a_0, l_0^{m-1}[L_p S_p] l_1) = m\left(G^{L_0 S_0}_{L_p S_p}\right)^2 , \tag{6.2.19}$$

$$Q_\kappa(l_0^m L_0, l_0^{m-1} l_1) = Q_\kappa(l_0^m, l_0^{m-1} l_1) = m\,. \tag{6.2.20}$$

c) Transitions between shells of equivalent electrons

$$a_0 = l_0^N l_1^m L_0 S_0\,, \quad a_1 = l_0^{N-1} l_1^{m+1}[L_p S_p] L_1 S_1 \quad N = 4l_0 + 2$$

We assume here for simplicity that l_0^N is the closed shell. Here, $L_p S_p$ are momenta of the shell l_1^{m+1}. For such transition one can use (6.2.18) for the transition $l_1^{M-m} - l_1^{M-m-1} l_0$, $M = 2(2l_1 + 1)$.

$$Q_\kappa(L_0, L_1) = (M - m)\left(G^{L_0 S_0}_{L_p S_p}(l_1^{M-m})\right)^2 [l_0 L_1]^2 \begin{Bmatrix} L_0 & L_1 & \kappa \\ l_0 & l_1 & L_p \end{Bmatrix}^2 \tag{6.2.21}$$

and corresponding substitutions into (6.2.19, 20). We note also that for any value of $n \le 4l_0 + 2$

$$Q_\kappa(l_0^n l_1^m, l_0^{n-1} l_1^{m+1}) = \frac{n(M - m)}{M} . \tag{6.2.22}$$

We used here for the total statistical weight the expression

$$g(l^m) = \frac{(4l + 2)!}{m!(M - m)!} \tag{6.2.23}$$

d) Transitions inside the shell of equivalent electrons

$$a_0 = l^m S_0 L_0, \quad a_1 = l^m S_1 L_1,$$
$$B_{q\kappa}(SL) = (l^m S_0 L_0 \| V^{q\kappa} \| l^m S_1 L_1) \tag{6.2.24}$$
$$= \sum_{S_p L_p} \sqrt{m} G^{L_0 S_0}_{L_p S_p} \cdot M_{0qq}(S_p s S) M_{0\kappa\kappa}(L_p l L)[S_p L_p]\,.$$

Table 6.13. Reduced matrix elements

$(p^2L_0S_0\|U^2\|p^2L_1S_1)$				$(p^3L_0S_0\|U^2\|p^3L_1S_1)$			
	^1S	^3P	^1D		^4S	^2P	^2D
^1S	0	0	$2/\sqrt{3}$	^4S	0	0	0
^3P	0	-1	0	^2P	0	0	$-\sqrt{3}$
^1D	$2/\sqrt{3}$	0	$\sqrt{7/3}$	^2D	0	$\sqrt{3}$	0

Thus we obtain

$$\begin{aligned} Q'_\kappa(a_0, a_1) &= \frac{2[l]^2}{[S_0L_0]^2}(l^mS_0L_0\|U^\kappa\|l^mS_1L_1)^2 \\ Q''_\kappa(a_0, a_1) &= \frac{1}{4}Q'_\kappa(a_0, a_1) + \frac{3[l]^2}{[S_0L_0]^2}(l^mS_0L_0\|V^{1\kappa}\|l^mS_1L_1)^2 \end{aligned} \tag{6.2.25}$$

where

$$\begin{aligned} (l^mS_0L_0\|U^\kappa\|l^mS_1L_1) &= \frac{2^{1/2}}{[S_0]}(l^mS_0L_0\|V^{0\kappa}\|(l^mS_1L_1), \\ (l^mS_0L_0\|U^\kappa\|l^mS_1L_1) &= \sum_{L_pS_p} G^{L_0S_0}_{L_pS_p}G^{L_1S_1}_{L_pS_p}[L_p]\begin{Bmatrix} L_p & L_p & 0 \\ l & l & \kappa \\ L_0 & L_1 & \kappa \end{Bmatrix}, \\ (l^mS_0L_0\|V^{q\kappa}\|l^mS_1L_1) &= \sum_{L_pS_p} G^{L_0S_0}_{L_pS_p}G^{L_1S_1}_{L_pS_p}[L_pS_p] \\ &\quad \times \begin{Bmatrix} S_p & S_p & 0 \\ s & s & q \\ S_0 & S_1 & q \end{Bmatrix} \cdot \begin{Bmatrix} L_p & L_p & 0 \\ l & l & \kappa \\ L_0 & L_1 & \kappa \end{Bmatrix}. \end{aligned} \tag{6.2.26}$$

The reduced matrix elements $(a_0\|U^\kappa\|a_1)$ and $(a_0\|V^{q\kappa}\|a_1)$ were defined and partly tabulated in [6.1]. More detailed tables are given in [6.3]. For $\kappa = 0$

$$(l^mS_0L_0\|U^0\|l^mS_1L_1) = \delta(L_0, L_1)m[L_0]/[l] \tag{6.2.27}$$

Therefore, transitions with a change of L are possible only with $\kappa \geq 2$ or due to exchange. At $\Delta L = 0$, $\Delta S \neq 0$ also exchange is necessary; at $\Delta L = \Delta S = 0$ there is no transition. In the Table 6.13 reduced elements for $\kappa = 2$ are given for $m = 2, 3$; for $m > M/2$ one can use relation:

$$(l^{M-m}S_0L_0\|U^2\|l^{M-m}S_1L_1) = -(l^mS_0L_0\|U^2\|l^mS_1L_1). \tag{6.2.28}$$

6.2.4 *jl* Coupling

Below we shall give formulas for two cases: (i) both initial and final levels are described by jl coupling; (ii) the initial term L_0S_0 is described by LS coupling, and the final term K is described by jl coupling.

(i) For transitions between the fine structure components

$J_0 = K_0 \pm \frac{1}{2}, \quad J_1 = K_1 \pm \frac{1}{2}$, we have

$$Q_\kappa(K_0J_0, K_1J_1) = (2K_0+1)(2J_1+1)\begin{Bmatrix} K_0 & J_0 & \frac{1}{2} \\ J_1 & K_1 & \kappa \end{Bmatrix}^2 Q_\kappa(K_0, K_1). \tag{6.2.29}$$

Summation over J_1 gives

$$Q_\kappa(K_0J_0, K_1) = Q_\kappa(K_0, K_1). \tag{6.2.30}$$

Equation (6.2.29), being averaged over J_0, yields

$$Q_\kappa(K_0, K_1J_1) = \frac{2J_1+1}{2(2K_1+1)} Q_\kappa(K_0, K_1). \tag{6.2.31}$$

The jl-coupling scheme cannot be used for a shell with equivalent electrons. Therefore only the case of transitions which do not involve groups of equivalent electrons should be considered. For transitions between the terms $\gamma_0 = [L_pS_pj]l_0K_0$, $\gamma_1 = [L_pS_pj]l_1K_1$ as a whole the factor Q_κ is

$$Q_\kappa(\gamma_0, \gamma_1) = (2l_0+1)(2K_1+1)\begin{Bmatrix} l_0 & K_0 & j \\ K_1 & l_1 & \kappa \end{Bmatrix}^2. \tag{6.2.32}$$

Summing (6.2.32) over K_1 gives

$$Q_\kappa([L_pS_pj]\,l_0K_0, [L_pS_pj]\,l_1) = Q_\kappa(l_0, l_1) = 1, \tag{6.2.33}$$

and averaging over K_0 provides

$$Q_\kappa(l_0, l_1K_1) = \frac{2K_1+1}{(2j+1)(2l_1+1)}. \tag{6.2.34}$$

(ii) For transitions from the level $L_0S_0J_0$ described by LS coupling to the level $[L_pS_pj]l_1K_1J_1$ described by the jl-coupling scheme, we have

$$\gamma_0 = [L_pS_p]l_0L_0S_0, \quad \gamma_1 = [L_pS_pj]l_1K_1,$$

$$Q_\kappa(\gamma_0J_0, \gamma_1J_1) \tag{6.2.35}$$

$$= (2l_0+1)(2J_1+1)(2S_0+1)(2L_0+1)(2j+1)(2K_1+1)\begin{bmatrix} L_0 & S_p & \kappa & K_1 \\ l_0 & J_0 & j & \frac{1}{2} \\ L_p & l_1 & S_0 & J_1 \end{bmatrix}^2.$$

The definition of the $12j$ symbol used here is given by (6.3.23). Averaging over J_0 and summing after that over J_1 gives for transition between the terms,

$$Q_\kappa(\gamma_0\gamma_1)$$

$$= (2l_0+1)\frac{(2j+1)(2K_1+1)}{2S_p+1}\sum_r(2r+1)\begin{Bmatrix} L_p & l_1 & r \\ \kappa & L_0 & l_0 \end{Bmatrix}^2\begin{Bmatrix} L_p & l_1 & r \\ K_1 & S_p & j \end{Bmatrix}^2. \tag{6.2.36}$$

Summing (6.2.36) over K_1, we obtain

$$Q_\kappa(\gamma_0, [L_pS_pj]l_1) = \frac{2j+1}{(2L_p+1)(2S_p+1)} . \tag{6.2.37}$$

By summing further over j, one has

$$Q_\kappa(\gamma_0, [L_pS_p]l_1) = Q_\kappa(l_0, l_1) = 1 . \tag{6.2.38}$$

Averaging (6.2.36) over L_0, we have

$$Q_\kappa([L_pS_p]l_0S_0, \gamma_1) = \frac{2K_1+1}{(2L_p+1)(2S_p+1)(2l_1+1)} . \tag{6.2.39}$$

For transitions from the shell of equivalent electrons $l_0^m - l_0^{m-1}l_1$, the formulas (6.2.35–38) should be multiplied by $m(G_{L_pS_p}^{L_0S_0})^2$.

6.3 3*nj* Symbols and Fractional Parentage Coefficients

Formulas for the angular factors Q'_κ and Q'' contain 6*j* symbols, 9*j* symbols, 12*j* symbols, and the fractional parentage coefficients $G_{L_pS_p}^{LS}$. The detailed description of their invariance properties, and formulas, sum rules, and numerical values can be found in [6.2–7]. Here we give only those which are necessary for this book.

I) The 6*j* symbol $\begin{Bmatrix} a_1 & a_2 & a_3 \\ b_1 & b_2 & b_3 \end{Bmatrix}$ obeys the following symmetry relations: it remains invariant under any permutation of its columns and also on transposing the lower and upper arguments in each of any two columns. For example,

$$\begin{Bmatrix} a_1 & a_2 & a_3 \\ b_1 & b_2 & b_3 \end{Bmatrix} = \begin{Bmatrix} a_2 & a_1 & a_3 \\ b_2 & b_1 & b_3 \end{Bmatrix} = \begin{Bmatrix} a_3 & a_2 & a_1 \\ b_3 & b_2 & b_1 \end{Bmatrix}, \begin{Bmatrix} a_1 & a_2 & a_3 \\ b_1 & b_2 & b_3 \end{Bmatrix} = \begin{Bmatrix} b_1 & b_2 & a_3 \\ a_1 & a_2 & b_3 \end{Bmatrix} .$$

The 6*j* symbol is nonzero if the following triangular conditions are fulfilled:

$$\Delta(a_1a_2a_3), \quad \Delta(a_1b_2b_3), \quad \Delta(b_1a_2b_3), \quad \Delta(b_1b_2a_3) .$$

The triangular condition $\Delta(abc)$ means that the sum of any two arguments is greater than or equal to the third argument and the modulus of the difference of any two arguments is less than or equal to the third one.

The 9*j* symbol remains invariant under any permutation of its rows or columns and also under transposition (change of rows to columns). Triangular conditions are fulfilled for every row and column. Here we give summation formulas for *j*-symbols, including some formulas that are missing in most books. $\{abc\}$ means "0*j*-symbol": the set of triangle rules $\Delta(abc)$ and condition that $a+b+c$ is integer. Sums of one j-symbol.

$$\sum_x [x]^2 \{j_2 j_1 x\} = [j_2 j_1]^2 \tag{6.3.1a}$$

$$\sum_x (-1)^x [x]^2 \{j_2 j_1 x\} = (-1)^{j_1+j_2}[j_m]^2, \quad j_m = \max(j_2 j_1) \tag{6.3.1b}$$

$$\sum_x [x]^2 \begin{Bmatrix} j_1 & j_2 & j_3 \\ j_1 & j_2 & x \end{Bmatrix} = (-1)^{2j_3}\{j_1 j_2 j_3\} \tag{6.3.2a}$$

$$\sum_x (-1)^x [x]^2 \begin{Bmatrix} j_1 & j_1 & j_3 \\ l_1 & l_1 & x \end{Bmatrix} = (-1)^{-j_1-l_1}[j_1 l_1]\delta(j_3, 0) \tag{6.3.2b}$$

$$\sum_x [x]^2 \begin{Bmatrix} j_1 & j_2 & j_3 \\ l_1 & l_2 & l_3 \\ j_3 & l_3 & x \end{Bmatrix} = [l_1]^{-2}\delta(j_2, l_1) \tag{6.3.3a}$$

$$\sum_x (-1)^x [x]^2 \begin{Bmatrix} j_1 & j_2 & j_3 \\ l_1 & l_2 & l_3 \\ l_3 & j_3 & x \end{Bmatrix} = (-1)^{2l_2+j_2-l_1}[j_1]^{-2}\delta(j_1, l_2) \tag{6.3.3b}$$

Sums of two j-symbols.

$$\sum_x [x]^2 \begin{Bmatrix} j_1 & j_2 & j_3 \\ l_1 & l_2 & x \end{Bmatrix} \begin{Bmatrix} j_1 & j_2 & j_3' \\ l_1 & l_2 & x \end{Bmatrix} = [j_3]^{-2}\delta(j_3, j_3') \tag{6.3.4a}$$

$$\sum_x (-1)^x [x]^2 \begin{Bmatrix} j_1 & j_2 & j_3 \\ l_1 & l_2 & x \end{Bmatrix} \begin{Bmatrix} j_1 & l_1 & j_3' \\ j_2 & l_2 & x \end{Bmatrix} = (-1)^{-j_3-j_3'} \begin{Bmatrix} j_1 & j_2 & j_3 \\ l_2 & l_1 & j_3' \end{Bmatrix} \tag{6.3.4b}$$

$$\sum_x [x]^2 \begin{Bmatrix} j_1 & j_2 & j_3 \\ l_1 & l_2 & l_3 \\ k_1 & k_2 & x \end{Bmatrix} \begin{Bmatrix} k_1 & k_2 & x \\ j_3 & l_3 & v \end{Bmatrix} = \begin{Bmatrix} j_1 & j_2 & j_3 \\ k_2 & v & l_2 \end{Bmatrix} \begin{Bmatrix} l_1 & l_2 & l_3 \\ v & k_1 & j_1 \end{Bmatrix} (-1)^{2v} \tag{6.3.4c}$$

$$\sum_x [x]^2 \begin{Bmatrix} j_1 & j_2 & j_3 \\ l_1 & l_2 & l_3 \\ k_1 & k_2 & x \end{Bmatrix} \begin{Bmatrix} p_1 & p_2 & j_3 \\ q_1 & q_2 & l_3 \\ k_1 & k_2 & x \end{Bmatrix} (-1)^{j_3+l_3-k_1-k_2}$$

$$= \sum_v [v]^2 \begin{Bmatrix} j_1 & j_3 & j_2 \\ k_1 & p_1 & q_1 \\ l_1 & p_2 & v \end{Bmatrix} \begin{Bmatrix} l_2 & k_2 & j_2 \\ l_3 & q_2 & q_1 \\ l_1 & p_2 & v \end{Bmatrix} (-1)^{j_2+q_1-l_1-p_2} \tag{6.3.5}$$

$$\sum_{x_1 x_2} [x_1 x_2]^2 \begin{Bmatrix} j_1 & j_2 & j_3 \\ l_1 & l_2 & l_3 \\ x_1 & x_2 & k_3 \end{Bmatrix} \begin{Bmatrix} j_1 & j_2 & j_3' \\ l_1 & l_2 & l_3' \\ x_1 & x_2 & k_3 \end{Bmatrix} = [j_3 l_3]^{-2}\delta(j_3 l_3, j_3' l_3') \tag{6.3.6a}$$

$$\sum_{x_1 x_2} (-1)^{x_2} [x_1 x_2]^2 \begin{Bmatrix} j_1 & j_2 & j_3 \\ l_1 & l_2 & l_3 \\ x_1 & x_2 & k_3 \end{Bmatrix} \begin{Bmatrix} j_1 & l_2 & j_3' \\ l_1 & j_2 & l_3' \\ x_1 & x_2 & k_3 \end{Bmatrix} = (-1)^{2j_2+l_3-l_3'} \begin{Bmatrix} j_1 & j_2 & j_3 \\ l_2 & l_1 & l_3 \\ j_3' & l_3' & k_3 \end{Bmatrix} \tag{6.3.6b}$$

Sums of three j-symbols.

$$\sum_x (-1)^{2x} [x]^2 \begin{Bmatrix} j_1 & j_2 & j_3 \\ l_3 & k_3 & x \end{Bmatrix} \begin{Bmatrix} l_1 & l_2 & l_3 \\ j_2 & x & k_2 \end{Bmatrix} \begin{Bmatrix} k_1 & k_2 & k_3 \\ x & j_1 & l_1 \end{Bmatrix} = \begin{Bmatrix} j_1 & j_2 & j_3 \\ l_1 & l_2 & l_3 \\ k_1 & k_2 & k_3 \end{Bmatrix} \tag{6.3.7a}$$

$$\sum_x (-1)^x [x]^2 \begin{Bmatrix} j_1 & j_2 & k_3 \\ l_2 & l_1 & x \end{Bmatrix} \begin{Bmatrix} j_1 & k_2 & j_3 \\ l_3 & x & l_1 \end{Bmatrix} \begin{Bmatrix} k_1 & j_2 & j_3 \\ x & l_3 & l_2 \end{Bmatrix}$$

$$= \sum_x (-1)^x [x]^2 \begin{Bmatrix} j_1 & x & l_1 \\ l_2 & k_3 & j_2 \end{Bmatrix} \begin{Bmatrix} j_2 & x & l_2 \\ l_3 & k_1 & j_3 \end{Bmatrix} \begin{Bmatrix} j_3 & x & l_3 \\ l_1 & k_2 & j_1 \end{Bmatrix} = (-1)^a \begin{Bmatrix} k_1 & k_2 & k_3 \\ j_1 & j_2 & j_3 \end{Bmatrix} \begin{Bmatrix} k_1 & k_2 & k_3 \\ l_1 & l_2 & l_3 \end{Bmatrix} \quad (6.3.7b)$$

$$a = j_1 + j_2 + j_3 + l_1 + l_2 + l_3 + k_1 + k_2 + k_3$$

$$\sum_x [x]^2 \begin{Bmatrix} j_1 & j_2 & j_3 \\ l_1 & l_2 & l_3 \\ k_1 & k_2 & x \end{Bmatrix} \begin{Bmatrix} u_1 & l_3 & u_3 \\ j_3 & u_2 & x \end{Bmatrix} \begin{Bmatrix} u_1 & k_2 & v_3 \\ k_1 & u_2 & x \end{Bmatrix} (-1)^{j_3+l_3+k_1+k_2}$$

$$= \sum_y [y]^2 \begin{Bmatrix} u_1 & l_3 & u_3 \\ k_2 & l_2 & j_2 \\ v_3 & l_1 & y \end{Bmatrix} \begin{Bmatrix} j_1 & j_2 & j_3 \\ u_3 & u_2 & y \end{Bmatrix} \begin{Bmatrix} j_1 & l_1 & k_1 \\ v_3 & u_2 & y \end{Bmatrix} (-1)^{j_2+u_3+l_1+v_3} . \quad (6.3.8)$$

Sums of four j-symbol.

$$\sum_x [x]^2 \begin{Bmatrix} j_1 & j_2 & x \\ l_1 & l_2 & k_1 \end{Bmatrix} \begin{Bmatrix} l_1 & l_2 & x \\ j_3 & j_4 & k_2 \end{Bmatrix} \begin{Bmatrix} j_3 & j_4 & x \\ l_3 & l_4 & k_3 \end{Bmatrix} \begin{Bmatrix} l_3 & l_4 & x \\ j_1 & j_2 & k_4 \end{Bmatrix} (-1)^{k_1-k_2+k_3-k_4}$$

$$= \sum_x [y]^2 \begin{Bmatrix} l_2 & l_4 & y \\ k_4 & k_1 & j_1 \end{Bmatrix} \begin{Bmatrix} k_4 & k_1 & y \\ l_1 & l_3 & j_2 \end{Bmatrix} \begin{Bmatrix} l_1 & l_3 & y \\ k_3 & k_2 & j_4 \end{Bmatrix} \begin{Bmatrix} k_3 & k_2 & y \\ l_2 & l_4 & j_3 \end{Bmatrix} (-1)^{j_1-j_2+j_4-j_3} \quad (6.3.9a)$$

$$\sum_x (-1)^x [x]^2 \begin{Bmatrix} j_1 & j_2 & x \\ l_1 & l_2 & k_1 \end{Bmatrix} \begin{Bmatrix} l_1 & l_2 & x \\ j_3 & j_4 & k_2 \end{Bmatrix} \begin{Bmatrix} j_3 & j_4 & x \\ l_3 & l_4 & k_3 \end{Bmatrix} \begin{Bmatrix} l_3 & l_4 & x \\ j_2 & j_1 & k_4 \end{Bmatrix}$$

$$= \sum_y (-1)^y [y]^2 \begin{Bmatrix} k_4 & l_4 & j_2 \\ l_3 & k_3 & j_4 \\ j_1 & j_3 & y \end{Bmatrix} \begin{Bmatrix} j_1 & j_3 & y \\ k_2 & k_1 & l_2 \end{Bmatrix} \begin{Bmatrix} j_2 & j_4 & y \\ k_2 & k_1 & l_1 \end{Bmatrix} (-1)^a \quad (6.3.9b)$$

$$a = l_1 + l_2 + l_3 + k_1 + k_2 + k_3$$

$$\sum_{xy} [x]^2 \begin{Bmatrix} a & x & y \\ j_1 & j_2 & j_3 \end{Bmatrix} \begin{Bmatrix} a & x & y \\ l_1 & l_2 & l_3 \end{Bmatrix} \begin{Bmatrix} j_1 & j_3 & x \\ l_3 & l_1 & p \end{Bmatrix} \begin{Bmatrix} j_1 & j_2 & y \\ l_2 & l_1 & q \end{Bmatrix} (-1)^{x+y}$$

$$= \delta(p, q) \frac{1}{[p]^2} \begin{Bmatrix} a & j_2 & j_3 \\ p & l_3 & l_2 \end{Bmatrix} (-1)^{j_1+j_2+j_3+l_1+l_2+l_3+a+p} . \quad (6.3.10)$$

For what follows it is convenient to adopt the designations:

$$f(\mathbf{abk}) = \begin{Bmatrix} a_2 & b_2 & k_2 \\ a_1 & b_1 & k_1 \\ a_0 & b_0 & k_0 \end{Bmatrix} = \begin{Bmatrix} a_2 & a_1 & a_0 \\ b_2 & b_1 & b_0 \\ k_2 & k_1 & k_0 \end{Bmatrix}, \quad \begin{Bmatrix} a & b & k \\ p & q & l \end{Bmatrix}_i = \begin{Bmatrix} a_i & b_i & k_i \\ p_i & q_i & l_i \end{Bmatrix}, \quad \mathbf{a} = (a_2 a_1 a_0) \quad (6.3.11)$$

Then

$$\sum_{k_1 k_2} [k_2 k_1]^2 f(\mathbf{abk}) f(\mathbf{pqk}) \begin{Bmatrix} a & b & k \\ p & q & l \end{Bmatrix}_2 \begin{Bmatrix} a & b & k \\ p & q & l \end{Bmatrix}_1$$

$$= \sum_{l_0} [l_0]^2 f(\mathbf{aql}) f(\mathbf{pbl}) \begin{Bmatrix} a & q & l \\ p & b & k \end{Bmatrix}_0 \quad (6.3.12)$$

using the definition of the M by (2.2.23) this equation can be written as

$$\sum_{k_1 k_2} M_k(\mathbf{a},\mathbf{b}) M_k(\mathbf{p},\mathbf{q}) \begin{Bmatrix} a\, b\, k \\ p\, q\, l \end{Bmatrix}_2 \begin{Bmatrix} a\, b\, k \\ p\, q\, l \end{Bmatrix}_1 \cdot [k_2 k_1]^2/[k_0]^2$$
$$= \sum_{l_0} M_l(\mathbf{a},\mathbf{q}) M_l(\mathbf{p},\mathbf{b}) \begin{Bmatrix} a\, q\, l \\ p\, b\, k \end{Bmatrix}_0 = \sum_{l_0} M_q(\mathbf{l},\mathbf{a}) M_p(\mathbf{b},\mathbf{l}) \begin{Bmatrix} a\, q\, l \\ p\, b\, k \end{Bmatrix}_0 . \tag{6.3.13}$$

We have also

$$\sum_{k_0 k_1 k_2} [k_2 k_1 k_0]^2 f(\mathbf{abk}) f(\mathbf{pqk}) \begin{Bmatrix} a\, b\, k \\ p\, q\, l \end{Bmatrix}_2 \begin{Bmatrix} a\, b\, k \\ p\, q\, l \end{Bmatrix}_1 \begin{Bmatrix} a\, b\, k \\ u\, v\, s \end{Bmatrix}_0 \begin{Bmatrix} p\, q\, k \\ u\, v\, t \end{Bmatrix}_0$$
$$= \sum_{l_0} [l_0]^2 f(\mathbf{aql}) f(\mathbf{pbl}) \begin{Bmatrix} v\, t\, p \\ s\, u\, b \\ a\, q\, l \end{Bmatrix}_0 , \tag{6.3.14}$$

$$\sum_{\substack{a_2 a_1 \\ b_2 b_1}} (-1)^{a_1+b_1} [a_2 b_2 a_1 b_1]^2 f(\mathbf{abc}) f(\mathbf{mna}) f(\mathbf{pqb}) \begin{Bmatrix} a\, b\, c \\ q\, m\, p \end{Bmatrix}_2 \begin{Bmatrix} a\, b\, k \\ n\, p\, q \end{Bmatrix}_1 \cdot$$
$$\cdot\, \delta(p_2 q_1, n_2 m_1) \tag{6.3.15}$$
$$= \begin{Bmatrix} a_2\, b_2\, k_2 \\ a_1\, b_1\, k_1 \\ a_0\, b_0\, k_0 \end{Bmatrix} \begin{Bmatrix} m_0\, q_0\, c_2 \\ q_2\, m_2\, m_1 \end{Bmatrix} \begin{Bmatrix} p_0\, n_0\, l_1 \\ n_1\, p_1\, n_2 \end{Bmatrix} (-1)^m ; \; m = n_2+q_2+q_1+n_0+p_0+b_0 ,$$

$$\sum_{\substack{k_0 k_1 k_2 \\ k_1' k_2'}} [k_2' k_1' k_2 k_1 k_0]^2 f(\mathbf{a'b'k'}) f(\mathbf{p'q'k'}) f(\mathbf{abk}) f(\mathbf{pqk}) \cdot$$
$$\cdot \begin{Bmatrix} a'\, b'\, k' \\ p'\, q'\, l' \end{Bmatrix}_2 \begin{Bmatrix} a'\, b'\, k' \\ p'\, q'\, l' \end{Bmatrix}_1 \begin{Bmatrix} a\, b\, k \\ p\, q\, l \end{Bmatrix}_1 \begin{Bmatrix} a\, b\, k \\ p\, q\, l \end{Bmatrix}_0 \delta(k', k)\, \delta(0', 2)$$
$$= \sum_{l_2} [l_2]^2 f(\mathbf{a'q'l'}) f(\mathbf{p'b'l'}) f(\mathbf{aql}) f(\mathbf{pbl}) \delta(0', 2) \tag{6.3.16}$$
$$\delta(0', 2) = \delta(a_0' b_0' p_0' q_0' l_0', a_2 b_2 p_2 q_2 l_2) .$$

Sums including $3jm$-symbols.

$$\sum_m (-1)^{j-m} \begin{pmatrix} j & j & j' \\ m & -m & m' \end{pmatrix} = [j]\delta(j'm', 00) , \tag{6.3.17}$$

$$\sum_{jm} (-1)^{j-m} [j]^2 \begin{pmatrix} j_1 & j_2 & j \\ m_1 & m_2 & m \end{pmatrix} \begin{pmatrix} j_1 & j_2 & j \\ -m_1' & -m_2' & -m \end{pmatrix}$$
$$= (-1)^{j_1-m_1} \delta(m_1 m_2, m_1' m_2') , \tag{6.3.18}$$

$$\sum_{m_1 m_2} (-1)^{j_1-m_1-j_2-m_2} [j]^2 \begin{pmatrix} j_1 & j_2 & j \\ m_1 & m_2 & m \end{pmatrix} \begin{pmatrix} j_1 & j_2 & j \\ -m_1 & -m_2 & -m' \end{pmatrix}$$
$$= (-1)^{j-m} \delta(jm, j'm') , \tag{6.3.19}$$

$$\begin{aligned}&\sum_{x\mu}(-1)^{x-\mu}[x]^2\begin{Bmatrix}x & j_1 j_2\\ l & l_1 l_2\end{Bmatrix}\begin{pmatrix}x & j_1 & j_2\\ \mu & m_1 & m_2\end{pmatrix}\begin{pmatrix}x & l_1 & l_2\\ -\mu & m_1' & m_2'\end{pmatrix}\\ &\quad=\sum_{\nu}(-1)^{l-\nu}\begin{pmatrix}l & j_1 & l_2\\ \nu & m_1 & m_2'\end{pmatrix}\begin{pmatrix}l & l_1 & j_2\\ -\nu & m_1' & m_2\end{pmatrix}.\end{aligned} \tag{6.3.20}$$

In particular, if $m_1 = m_2 = m_1' = m_2' = 0$,

$$\begin{aligned}&\sum_{x}(-1)^{x}[x]^2\begin{Bmatrix}x & j_1 j_2\\ l & l_1 l_2\end{Bmatrix}\begin{pmatrix}x & j_1 j_2\\ 0 & 0\ 0\end{pmatrix}\begin{pmatrix}x & l_1 l_2\\ 0 & 0\ 0\end{pmatrix}\\ &\quad=(-1)^{l}\begin{pmatrix}l & j_1 l_2\\ 0 & 0\ 0\end{pmatrix}\begin{pmatrix}l & l_1 j_2\\ 0 & 0\ 0\end{pmatrix}.\end{aligned} \tag{6.3.21}$$

A summary of formulas for $6j$ symbols in which one of the arguments does not exceed unity is given in Table 6.14.

II) $9j$ symbols and $12j$ symbols are defined in terms of the $6j$ symbols in the following way:

$$\begin{Bmatrix}a & b & c\\ d & e & f\\ p & q & r\end{Bmatrix}=\sum_{x}(-1)^{2x}(2x+1)\begin{Bmatrix}a & b & c\\ f & r & x\end{Bmatrix}\begin{Bmatrix}d & e & f\\ b & x & q\end{Bmatrix}\begin{Bmatrix}p & q & r\\ x & a & d\end{Bmatrix}, \tag{6.3.22}$$

Table 6.14. Formulas for $6j$ symbols

$s = a + b + c$
$\begin{Bmatrix}a & b & c\\ o & c & b\end{Bmatrix}=(-1)^s[(2b+1)(2c+1)]^{-1/2}$
$\begin{Bmatrix}a & b & c\\ 1/2 & c-1/2 & b+1/2\end{Bmatrix}=(-1)^s\left[\dfrac{(a+c-b)(a+b-c+1)}{(2b+1)(2b+2)2c(2c+1)}\right]^{-1/2}$
$\begin{Bmatrix}a & b & c\\ 1/2 & c-1/2 & b-1/2\end{Bmatrix}=(-1)^s\left[\dfrac{(a+b+c+1)(b+c-a)}{2b(2b+1)2c(2c+1)}\right]^{-1/2}$
$\begin{Bmatrix}a & b & c\\ 1 & c & b\end{Bmatrix}=(-1)^s\dfrac{2[a(a+1)-b(b+1)-c(c+1)]}{[2b(2b+1)(2b+2)2c(2c+1)(2c+2)]^{1/2}}$
$\begin{Bmatrix}a & b & c\\ 1 & c-1 & b-1\end{Bmatrix}=(-1)^s\left[\dfrac{s(s+1)(s-2a-1)(s-2a)}{(2b-1)2b(2b+1)(2c-1)2c(2c+1)}\right]^{1/2}$
$\begin{Bmatrix}a & b & c\\ 1 & c-1 & b\end{Bmatrix}=(-1)^s\left[\dfrac{2(s+1)(s-2a)(s-2b)(s-2c+1)}{2b(2b+1)(2b+2)(2c-1)2c(2c+1)}\right]^{1/2}$
$\begin{Bmatrix}a & b & c\\ 1 & c-1 & b+1\end{Bmatrix}=(-1)^s\left[\dfrac{(s-2b)(s-2b-1)(s-2c+1)(s-2c+2)}{(2b+1)(2b+2)(2b+3)(2c-1)2c(2c+1)}\right]^{1/2}$

$$\begin{bmatrix} a_1\, a_2\, a_3\, a_4 \\ b_1\, b_2\, b_3\, b_4 \\ c_1\, c_2\, c_3\, c_4 \end{bmatrix} = (-1)^{b_1-b_2-b_3+b_4} \sum_x (2x+1) \begin{Bmatrix} c_1 & c_2 & x \\ a_3 & a_1 & b_1 \end{Bmatrix} \begin{Bmatrix} c_3 & c_4 & x \\ a_3 & a_1 & b_2 \end{Bmatrix} \begin{Bmatrix} c_1 & c_2 & x \\ a_4 & a_2 & b_3 \end{Bmatrix} \begin{Bmatrix} c_3 & c_4 & x \\ a_4 & a_2 & b_4 \end{Bmatrix} . \tag{6.3.23}$$

III) Our notation for the fractional parentage coefficients, $G^{LS}_{L_pS_p}$ agrees with that of [6.2]. The Racah notation is related to this by

$$(l^{m-1}[L_pS_p]\, lLS\}\, l^m LS) = G^{LS}_{L_pS_p} .$$

The values of fractional parentage coefficients for electron configurations p^m with $m = 3, 4, 5$ are given by Tables 6.15–17. For configurations s^2, p^2, p^6, the fractional parentage coefficients are equal to unity.

Table 6.15. Fractional parentage coefficients $(p^2[L_pS_p]\, pLS\}\, p^3LS)$

p^2 \ p^3	^4S	^2P	^2D
^1S	0	$\frac{\sqrt{2}}{3}$	0
^3P	1	$-\frac{1}{\sqrt{2}}$	$\frac{1}{\sqrt{2}}$
^1D	0	$-\sqrt{\frac{5}{18}}$	$-\frac{1}{\sqrt{2}}$

Table 6.16. Fractional parentage coefficients $(p^3[L_pS_p]pLS\}\, p^4LS)$

p^3 \ p^4	^1S	^3P	^1D
^4S	0	$-\frac{1}{\sqrt{3}}$	0
^2P	1	$-\frac{1}{2}$	$-\frac{1}{2}$
^2D	0	$\sqrt{\frac{5}{12}}$	$\sqrt{\frac{3}{4}}$

Table 6.17. Fractional parentage coefficients $(p^4[L_pS_p]p^2P\} p^5\,{}^2P)$

p^5 \ p^4	^2P
^1S	$\sqrt{\frac{1}{15}}$
^3P	$\sqrt{\frac{3}{5}}$
^1D	$\sqrt{\frac{1}{3}}$

7 Broadening of Spectral Lines

Various phenomena of spectral line broadening connected with the most interesting applications of atomic spectroscopy to plasma diagnostics, astrophysics, laser physics, and other areas are considered in this chapter. The presentation of the general theory of impact broadening is based on the density-matrix and quantum kinetic equation methods. These methods permit not only the line shape to be described in the case of spontaneous emission or linear absorption, but also allow nonlinear effects arising in laser spectroscopy to be considered. There are many books and review articles discussing the progress in theoretical and experimental work on the problem of spectral lines broadening [7.1–16]. For an extensive bibliography on line shapes see [7.17–19]. For a brief review of recent developments in the theory with stress to applications to nonlinear laser spectroscopy see [7.20].

7.1 Model of a Classical Oscillator

7.1.1 Formulation of the Problem

The theory of spectral line broadening caused by the interaction of an atom with surrounding particles is closely connected with the general theory of atomic collisions. Moreover in the region of not very high pressure, when the impact approximation is valid, the calculation of the profile of a spectral line includes calculation of the scattering amplitudes or scattering phases. Nevertheless it is useful to begin the study of pressure effects by considering a model simplified to the maximum extent. We shall make the following assumptions:

i) the relative motion of the atom and the perturbing particle is quasi-classical, which enables one to use the concept of the trajectory of the perturbing particle;
ii) this trajectory is rectilinear;
iii) interactions with the nearest perturbing particle (binary interactions) play the principal role in the broadening, therefore multiparticle interactions can be neglected;
iv) the perturbation is adiabatic, i.e., does not induce transitions between different states of the atom.

Within these assumptions, the picture of broadening is outlined as follows. The perturbing particle produces an external field

$$V(R) = V\left[\sqrt{\rho^2 + v^2(t - t_0)^2}\right] , \qquad (7.1.1)$$

where R is the distance between the atom and perturbing particle at a given time t, ρ is the impact parameter, t_0 is the time of nearest approach, and v is the relative velocity. As a result the energy levels of the atom and, consequently, the frequency of oscillations of the atomic oscillator vary in time. Therefore the oscillation of the atomic oscillator can be described in the form

$$f(t) = \exp\left[i\omega_0 t + i \int_{-\infty}^{t} \kappa(t')\, dt'\right] , \tag{7.1.2}$$

where ω_0 is the unperturbed frequency and $\kappa(t)$ is the frequency shift due to the interaction. Perturbation of the monochromaticity of the oscillations leads to broadening of the corresponding spectral line. The line shape is given by the expansion of the function $f(t)$ in a Fourier integral,

$$\begin{aligned} I(\omega) &= \lim_{T\to\infty} \left| \frac{1}{\sqrt{2\pi T}} \int_{-T/2}^{T/2} f(t) \exp(-i\omega t)\, dt \right|^2 \\ &= \lim_{T\to\infty} \frac{1}{2\pi T} \left| \int_{-T/2}^{T/2} \exp[-i(\omega - \omega_0)t + i\eta(t)]\, dt \right|^2 , \end{aligned} \tag{7.1.3}$$

$$\eta(t) = \int_{-\infty}^{t} \kappa(t')\, dt' , \tag{7.1.4}$$

where $\eta(t)$ is the phase of the oscillation caused by interaction. If the frequency ω is measured from the unperturbed frequency ω_0, then the exponent $\exp(i\omega_0 t)$ must be omitted. In this case,

$$I(\omega) = \lim_{T\to\infty} \frac{1}{2\pi T} \left| \int_{-T/2}^{T/2} \exp[-i\omega t + i\eta(t)]\, dt \right|^2 . \tag{7.1.5}$$

In the theory of spectral line broadening, conditions are usually considered when gas pressure and temperature, state of ionization, and so on, do not vary with time. This means that the functions $\eta(t)$ and $f(t) = \exp[i\eta(t)]$ are stationary random processes, and (7.1.3) can be rewritten in the following way:

$$I(\omega) = \frac{1}{\pi} \mathrm{Re}\left\{ \int_{0}^{\infty} \Phi(\tau) \exp(-i\omega\tau)\, d\tau \right\} , \tag{7.1.6}$$

$$\Phi(\tau) = \lim_{T\to\infty} \frac{1}{T} \int_{-T/2}^{T/2} f^*(t) f(t+\tau)\, dt = \overline{f^*(t) f(t+\tau)} , \tag{7.1.7}$$

where $\Phi(\tau)$ is the correlation function.

Time averaging can be replaced by averaging over the statistical assembly of quantities defining the function $f(t)$. We shall denote such averaging by angle brackets,

$$\Phi(\tau) = \langle f^*(0) f(\tau) \rangle . \tag{7.1.8}$$

For $f(t) = \exp[\mathrm{i}\eta(t)]$, we have

$$\Phi(\tau) = \exp\{\overline{\mathrm{i}[\eta(t+\tau) - \eta(t)]}\} = \langle \exp[\mathrm{i}\eta(\tau)] \rangle \ . \tag{7.1.9}$$

7.1.2 Impact Broadening

We shall consider in this section an approximation which is called the impact approximation. This approximation is based on the assumption that the decisive factor in the broadening of a line is the disruption of the coherence of the oscillations of an atomic oscillator during collisions. In other words if the duration of collision is small as compared with the mean time between collisions, then one can neglect radiation during collisions and consider the collisions to be instantaneous. Therefore the collisions are manifested only in phase shifts η.

Using this assumption of instantaneous collision, it is possible to calculate the correlation function $\Phi(\tau)$ in the following way [7.21]. In accordance with (7.1.9) the difference $\Delta\Phi = \Phi(\tau + \Delta\tau) - \Phi(\tau)$ can be written in the form

$$\begin{aligned}\Delta\Phi &= \langle \exp[\mathrm{i}\eta(\tau+\Delta\tau)] \rangle - \langle \exp[\mathrm{i}\eta(\tau)] \rangle \\ &= \langle \exp[\mathrm{i}\eta(\tau)] \exp(\mathrm{i}\Delta\eta) \rangle - \langle \exp[\mathrm{i}\eta(\tau)] \rangle \ ,\end{aligned}$$

where $\Delta\eta$ is the phase shift produced by collisions during time interval $\Delta\tau$. Since collisions are instantaneous, the phase shift $\Delta\eta$ does not depend on $\eta(\tau)$. Therefore $\eta(\tau)$ and $\Delta\eta$ are statistically independent, and consequently

$$\Delta\Phi = \langle \exp[\mathrm{i}\eta(\tau)] \rangle [\langle \exp(\mathrm{i}\Delta\eta) \rangle - 1] = -\Phi(\tau) \langle 1 - \exp(\mathrm{i}\Delta\eta) \rangle \ . \tag{7.1.10}$$

We shall denote the number of collisions per second with parameters ρ and v as $P(\rho, v) d\rho dv$. The number of collisions during time interval $\Delta\tau$ is equal $P(\rho, v) d\rho dv\, \Delta\tau$. Therefore

$$\langle 1 - \exp(\mathrm{i}\Delta\eta) \rangle = \vartheta \Delta\tau, \quad \vartheta = \int [1 - \exp(\mathrm{i}\eta)] P(\rho, v)\, d\rho dv \ , \tag{7.1.11}$$

where η is the phase shift produced by collision with parameters ρ, v.

If the density of perturbing particles is N and their distribution over velocities v is given by the distribution function $\mathscr{F}(v)$, then

$$\theta = N \int_0^\infty v \mathscr{F}(v)\, dv\, 2\pi \int_0^\infty \rho d\rho [1 - \exp(\mathrm{i}\eta)] \ . \tag{7.1.12}$$

Denoting

$$\sigma' = 2\pi \int_0^\infty (1 - \cos\eta)\, \rho d\rho \ , \tag{7.1.13}$$

$$\sigma'' = 2\pi \int_0^\infty \sin\eta\, \rho d\rho \ , \tag{7.1.14}$$

we have

$$\theta = N\langle v(\sigma' - \mathrm{i}\sigma'')\rangle \ . \tag{7.1.15}$$

From (7.1.10, 11) it follows that

$$\frac{d\Phi}{d\tau} = -\theta\Phi \ , \tag{7.1.16}$$

$$\Phi = \exp(-\theta\tau) \ . \tag{7.1.17}$$

By substituting (7.1.17) into (7.1.6), we obtain

$$I(\omega) = \frac{\gamma}{2\pi} \cdot \frac{1}{(\omega - \Delta)^2 + (\gamma/2)^2} \ , \tag{7.1.18}$$

$$\gamma = 2N\langle v\sigma'\rangle ; \quad \Delta = N\langle v\sigma''\rangle \ . \tag{7.1.19}$$

The spectral distribution given by (7.1.18) is usually called the Lorentzian distribution. The width of the distribution (the distance between symmetrical points ω_1 and ω_2, for which $I(\omega_1) = I(\omega_2) = I_{\max}/2$) is γ. The shift of the line peak from ω_0 is Δ. The quantities σ' and σ'' are called the width and shift effective cross sections.

Let us assume that the perturbing particle at a distance R produces the frequency shift $\kappa = C_n R^{-n}$. Then

$$\kappa(t) = C_n \cdot R^{-n} = C_n[\rho^2 + v^2(t - t_0)^2]^{-n/2} \ , \tag{7.1.20}$$

$$\eta(\rho, v) = C_n \int_{-\infty}^{\infty} [\rho^2 + v^2(t - t_0)^2]^{-n/2}\, dt = \alpha_n \frac{C_n}{v\rho^{n-1}} \ , \tag{7.1.21}$$

$$\alpha_n = \sqrt{\pi}\, \frac{\Gamma\left(\dfrac{n-1}{2}\right)}{\Gamma\left(\dfrac{n}{2}\right)} \ . \tag{7.1.22}$$

For $n = 2, 3, 4, 5, 6$, we have $\alpha_n = \pi, 2, \pi/2, 4/3, 3\pi/8$. Substituting (7.1.21) in (7.1.13, 14), it is not difficult to obtain the following formulas for γ and Δ which we shall use below in estimations of the width and shift:

$$\begin{aligned}
&n = 3 \quad \gamma = 2\pi^2 C_3 N \ , \\
&n = 4 \quad \gamma = \left(\frac{\pi^5}{4}\right)^{1/3} \Gamma\left(\frac{1}{3}\right) C_4^{2/3} \langle v^{1/3}\rangle N \simeq 11.4\, C_4^{2/3} \langle v^{1/3}\rangle N \ , \\
&\Delta = \frac{\sqrt{3}}{2}\gamma \ , \\
&n = 6 \quad \gamma \simeq 8.16\, C_6^{2/5} \langle v^{3/5}\rangle N \ , \\
&\Delta \simeq 0.36\,\gamma \ .
\end{aligned} \tag{7.1.23}$$

It is not difficult to show that the main contribution in σ' is given by the strong collisions for which $\eta \gtrsim 1$ and $\rho < \rho_0$, where ρ_0 is defined by condition $\eta(\rho_0) = 1$:

$$\rho_0 = \left(\frac{\alpha_n C_n}{v}\right)^{1/(n-1)} . \tag{7.1.24}$$

The impact parameter ρ_0 is usually called the Weisskopf radius. Therefore to an order to magnitude,

$$\sigma' \simeq \pi\rho_0^2 \tag{7.1.25}$$

The shift cross section σ'', see (7.1.14), is determined by more distant collisions $\rho \gtrsim \rho_0$.

In the case of $n = 2$, the phase $\eta(\rho) \propto \rho^{-1}$. Thus σ' diverges as $\ln \rho_m$ and σ'' diverges as ρ_m, where ρ_m is the upper limit of integration in (7.1.13) and (7.1.14). The divergence of the integrals (7.1.13) and (7.1.14) means that the approximation of binary collision is not valid. It is evident that in this case broadening is determined by distant (weak) collisions with $\rho > \rho_0$.

7.1.3 Quasi-Static Broadening

If the external field varies sufficiently slowly, i.e., if it is quasi-static, it is possible to assume that $I(\omega)d\omega$ is simply proportional to the statistical weight of the configuration of perturbing particles for which the frequency of the atomic oscillator is included in the interval $\omega, \omega + d\omega$. In the binary approximation the frequency shift is produced by the nearest particle. Consequently, to calculate $I(\omega)$, it is necessary to find the probability $W(R)dR$ of the nearest particle being within the range of distance $(R, R + dR)$ from the atom. For R much larger than the atomic dimensions the interaction potential could be neglected and this probability is

$$W(R)dR = 4\pi R^2 N \exp\left(-\frac{4\pi}{3} NR^3\right) dR = \exp\left[-\left(\frac{R}{R_0}\right)^3\right] d\left(\frac{R}{R_0}\right)^3 , \tag{7.1.26}$$

where $R_0 = (3/4\pi N)^{1/3}$. Substituting $R = (C_n/\kappa)^{1/n} = [C_n/(\omega - \omega_0)]^{1/n}$ in (7.1.26), we obtain the probability distribution for a frequency shift of an atomic oscillator. In accordance with the basic assumption of the quasi-static approximation, the shape of the spectral line is also determined by this distribution. If the notation $\overline{\Delta\omega} = C_n R_0^{-n}$ is introduced it follows from (7.1.26) that

$$I(\omega)\, d\omega = \frac{4\pi}{n} N C_n^{3/n} (\omega - \omega_0)^{-(3+n)/n} \exp\left[-\left(\frac{\overline{\Delta\omega}}{\omega - \omega_0}\right)^{3/n}\right] d\omega . \tag{7.1.27}$$

This distribution is valid only for sufficiently large values of $\omega - \omega_0$ for which

$R = C_n^{1/n}(\omega - \omega_0)^{-1/n} \ll R_0$ For $R \gtrsim R_0$, the binary approximation is not valid. Thus (7.1.27) describes only the outer part of a line. The condition $R \ll R_0$ means that $\Delta\omega \ll \omega - \omega_0$. Thus the exponential factor can be omitted in (7.1.27), after which we obtain

$$I(\omega)\,d\omega = \frac{4\pi}{n} N C_n^{3/n} (\omega - \omega_0)^{-(3+n)/n}\,d\omega\,. \tag{7.1.28}$$

7.1.4 Relationship and Limits of Applicability of the Impact and Quasi-Static Approximations

Let us return to the general relations (7.1.3, 4). We shall first of all consider (7.1.3) for high values of $\Delta\omega = \omega - \omega_0$. If $\Delta\omega$ is large, the integrand in (7.1.3) oscillates strongly everywhere, except in the vicinity of the points t_k for which

$$\left(\frac{d\eta}{dt}\right)_{t_k} = \kappa(t_k) = \Delta\omega\,.$$

Thus the principal contribution to (7.1.3) give small regions $\Delta\tau_k$ around these points and instead of (7.1.3) one can write

$$I(\omega) = \lim_{T\to\infty} \frac{1}{2\pi T} \left| \sum_k \int_{\Delta\tau_k} \exp\{\mathrm{i}[\eta(t) + (\omega_0 - \omega)(t - t_k) + (\omega_0 - \omega)\,t_k]\}\,dt \right|^2\,. \tag{7.1.29}$$

We shall expand the function $\eta(t)$ in a series near t_k in powers of $t - t_k$. Since $(d\eta/dt)_{t_k} = \Delta\omega$, the linear terms in the exponent in (7.1.29) cancel, and the series begins with the term

$$\frac{1}{2}\left(\frac{d^2\eta}{dt^2}\right)_{t_k} (t - t_k)^2\,.$$

Only the region $\Delta\tau_k$, where this term is less than unity, is significant in the integration (beyond this region strong oscillations begin). Hence

$$\Delta\tau_k \simeq \sqrt{2}\left|\left(\frac{d^2\eta}{dt^2}\right)_{t_k}\right|^{-\frac{1}{2}} = \sqrt{2}\left|\left(\frac{d\kappa}{dt}\right)_{t_k}\right|^{-\frac{1}{2}}\,. \tag{7.1.30}$$

If within the limits of this region the next term of the expansion

$$\frac{1}{6}\left|\left(\frac{d^3\eta}{dt^3}\right)(t - t_k)^3\right| \ll 1\,, \quad \left(\frac{d^3\eta}{dt^3}\right)_{t_k} = \left(\frac{d^2\kappa}{dt^2}\right)_{t_k}\,, \tag{7.1.31}$$

which must fulfill the following inequality:

$$\left|\left(\frac{d^2\kappa}{dt^2}\right)_{t_k}\right| \cdot \left|\left(\frac{d\kappa}{dt}\right)_{t_k}\right|^{-3/2} \ll 1\,, \tag{7.1.32}$$

the series can be broken of at a term proportional to $(t - t_k)^2$, and in each term of the sum (7.1.29), the limits of integration can be extended from $-\infty$ to ∞. In this case[1]

$$I(\omega) = \lim_{T\to\infty} \frac{1}{2\pi T} \left| \sum_k \exp\{\mathrm{i}[\eta(t_k) + (\omega_0 - \omega)\, t_k]\} \right.$$

$$\left. \times \int_{-\infty}^{\infty} \exp\left\{ \mathrm{i}\left[\frac{1}{2}\left(\frac{d^2\eta}{dt^2}\right)_{t_k} (t - t_k)^2 \right] \right\} dt \right|^2 \qquad (7.1.33)$$

$$= \lim_{T\to\infty} \frac{1}{T} \left| \sum_k \exp\{\mathrm{i}[\eta(t_k) + (\omega_0 - \omega)\, t_k + \pi/4]\} \left(\frac{d^2\eta}{dt^2}\right)_{t_k}^{-1/2} \right|^2$$

$$= \lim_{T\to\infty} \frac{1}{T} \sum_k \left(\frac{d\kappa}{dt}\right)_{t_k}^{-1} .$$

It is easy to see that $\sum_k (d\kappa/dt)_{t_k}^{-1} d\omega$ is the time during which $\kappa(t)$ is included in the interval $\omega - \omega_0$, $\omega - \omega_0 + d\omega$. Since $d\tau_k$ and $d\omega$ in Fig. 7.1 are connected by the relation $(d\kappa/dt)_{t_k} d\tau_k = d\omega$, (7.1.33) gives the quasi-static intensity distribution $W(\omega - \omega_0)d\omega$. We shall replace the summation in (7.1.33) by integration. The number of particles incident on the annular element $2\pi\rho\; d\rho$ in the time T is $2\pi\rho d\rho\; NvT$, where N is the density of perturbing particles. Taking into account that each collision with $\rho \leq \rho_{\Delta\omega} = (C_n/\Delta\omega)^{1/n}$, $\kappa_{\max} = C_n\rho^{-n} \gg \Delta\omega$ gives two points t_k and t_{k+1} (Fig. 7.1), we obtain

$$I(\omega)\, d\omega = d\omega \int_0^{\rho_{\Delta\omega}} 4\pi\rho \left(\frac{d\kappa}{dt}\right)^{-1} Nv d\rho = \frac{4\pi}{n} N C_n^{3/n} \frac{d\omega}{\Delta\omega^{1+3/n}} , \qquad (7.1.34)$$

i.e., the quasi-statical distribution in the wing of the line.

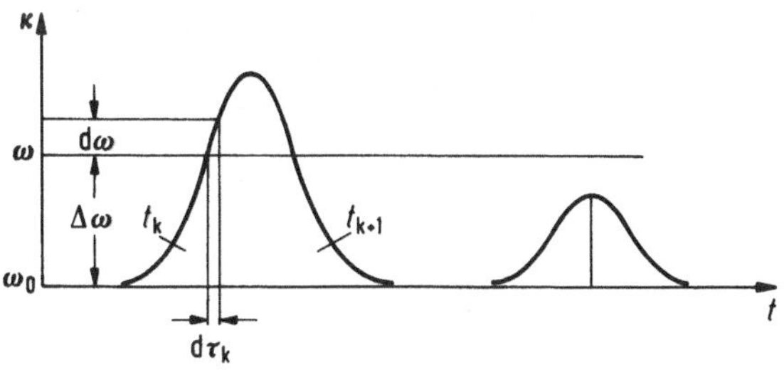

Fig. 7.1 Instantaneous frequency shift $\kappa(t)$

[1] It is assumed that the phases $\alpha_k = [\eta(t_k) + (\omega - \omega_0)t_k]$ are independent. This assumption will be discussed below.

If a small neighbourhood around the instant of closest approach is not considered, then

$$\frac{d\kappa}{dt} \simeq \frac{C_n v}{\rho^{n+1}}, \quad \frac{d^2\kappa}{dt^2} \simeq \frac{C_n v^2}{\rho^{n+2}}, \tag{7.1.35}$$

and relation (7.1.32) takes the form

$$\frac{C_n}{v\rho^{n-1}} \gg 1 . \tag{7.1.36}$$

Only collisions with $\rho \leq \rho_{\Delta\omega} = (C_n/\Delta\omega)^{1/n}$ give points t_k and t_{k+1}. Therefore (7.1.36) can be rewritten in another form

$$\Delta\omega \gg \frac{v^{n/(n-1)}}{C_n^{1/(n-1)}} = \Omega . \tag{7.1.37}$$

According to condition (7.1.37) the quasi-static distribution is valid for large $\Delta\omega$, i.e., in the wing of a line.

We shall now consider (7.1.3) in the limiting case of small $\Delta\omega$. If $\Delta\omega$ is so small that $1/\Delta\omega$ is much greater than the duration of the collision

$$\frac{1}{\Delta\omega} \gg \frac{\rho}{v}, \tag{7.1.38}$$

the change of phase in the collision can be considered to be instantaneous. Hence it follows that the impact approximation can be used. The main contribution in the impact broadening of a line is given by collisions with $\rho \backsim \rho_0 = (\alpha_n C_n/v)^{1/(n-1)}$. Substituting ρ_0 in (7.1.38), we obtain a relation opposite to (7.1.37):

$$\Delta\omega \ll \frac{v^{n/(n-1)}}{C_n^{1/(n-1)}} = \Omega .$$

Thus in the center of a line, $\Delta\omega \ll \Omega$, the impact (Lorentzian) distribution of intensity is valid. For high values of $\Delta\omega$, $\Delta\omega \gg \Omega$, the impact distribution is replaced by the quasi-static one. The quasi-static wing can appear both on the long-wave and on the short-wave side depending on the direction of shift of the terms.

If Ω considerably exceeds the impact width γ, then the greater part of the integral intensity of a line is concentrated in the impact region. Taking into account that

$$\gamma \simeq 2\pi\rho_0^2 N v = 2\pi N v(\alpha C_n/v)^{2/(n-1)} , \text{ we obtain}$$

$$2\pi\rho_0^2 N v \ll \Omega = \frac{v}{\rho_0} ;$$

whence

$$h = \rho_0^3 N \ll 1 , \tag{7.1.39}$$

where the dimensionless parameter h determines the number of perturbers in the sphere of the Weisskopf radius. Thus, for low pressures and high velocities, so long as the inequality (7.1.39) is fulfilled, the impact mechanism of broadening plays a decisive role. A relatively negligible part of the total intensity is concentrated in the quasi-static wing. At high pressures and low velocities, when

$$h = \rho_0^3 N \backsim 1 , \tag{7.1.40}$$

the impact approximation is inapplicable even to the inner part of a line.

Let us note that if condition (7.1.39) is not fulfilled then the binary approximation is violated. In fact relation (7.1.40) means that the effective radius ρ_0 is approximately equal to the mean distance between perturbers. Although when $\rho_0^3 N \gtrsim 1$ the quasi-static distribution is applicable practically to the whole profile of a line, the expressions (7.1.27, 28) obtained above in the approximation of binary interactions are valid only in the wing of a line.

The assumption of the independence of the phases α_k was made above in the derivation of the formula (7.1.33). Since only strong collisions for which $\eta \simeq C_n/\rho^n \cdot \rho/v \gg 1$ are responsible for the quasi-static wing, the difference $\alpha_{k+1} - \alpha_k \gg 1$. In a nonpublished work [7.4] Anderson and Talman investigated in detail the limiting expressions for $I(\omega)$ valid for the central part of the line and for the wings, and obtained also an interpolation expression for the intermediate part. The same problem is discussed also in [7.22, 23].

7.1.5 Doppler Effect

The frequency of an oscillator whose velocity component in the direction of observation is v is displaced in accordance with the Doppler principle by an amount $\omega_0 v/c$. Let the distribution of the radiating atoms with respect to v be defined by the function $W(v)$. Then $\omega = \omega_0 + \omega_0 v/c$, $v = c(\omega - \omega_0)/\omega_0$, and

$$I(\omega)\,d\omega = W\left(c\frac{\omega - \omega_0}{\omega_0}\right)\frac{c}{\omega_0}\,d\omega . \tag{7.1.41}$$

With a Maxwellian distribution

$$W(v)\,dv = \frac{1}{\sqrt{\pi}} \exp\left[-\left(\frac{v}{v_0}\right)^2\right]\frac{dv}{v_0} , \tag{7.1.42}$$

where $v_0 = \sqrt{2kT/m}$, we obtain

$$I(\omega)\,d\omega = \frac{1}{\sqrt{\pi}} \exp\left[-\left(\frac{\omega - \omega_0}{\Delta\omega_\mathrm{D}}\right)^2\right]\frac{d\omega}{\Delta\omega_\mathrm{D}}, \quad \Delta\omega_\mathrm{D} = \omega_0\frac{v_0}{c} . \tag{7.1.43}$$

The intensity distribution (7.1.43) is symmetrical. The magnitude of the broadening is defined by the parameter $\Delta\omega_\mathrm{D}$. The width of the line, which we shall denote by δ, and the peak density $I(\omega_0)$ are expressed in terms of the parameter

$\Delta\omega_D$:

$$\delta = 2\sqrt{\ln 2}\,\Delta\omega_D \tag{7.1.44}$$

$$I(\omega_0) = 1/\sqrt{\pi}\,\Delta\omega_D \tag{7.1.45}$$

Here δ is defined as a difference between the symmetrical frequencies ω_1 and ω_2 for which $I(\omega_1) = I(\omega_2) = I(\omega_0)/2$. The parameter $\Delta\omega_D$ is usually called the Doppler width of a line.

When deriving (7.1.41, 43), it is assumed that there is only one frequency $\omega_0(1 + v/c)$ in the spectrum of the oscillator with velocity v. This assumption is valid if v does not vary in time or remains a constant quantity during a sufficiently long time. If velocity is constant only during time interval τ, then this interval contributes to the intensity of radiation in a spectral interval with width $1/\tau$ around the frequency $\omega_0 + \omega_0 v/c$. Formula (7.1.41) is valid if $\omega_0 v/c \gg 1/\tau$. Substituting for τ the free path time $\tau_0 = L/v$, where L is the mean free path, we have

$$\omega_0 \frac{v}{c} \gg \frac{v}{L}, \quad 2\pi \frac{L}{\lambda} \gg 1 . \tag{7.1.46}$$

In the general case, the Doppler broadening is determined by Fourier transform of the function

$$f(t) = \exp\left[\mathrm{i}\frac{\omega_0}{c}x(t)\right], \quad x(t) = \int_{-\infty}^{t} v(t')\,dt' \tag{7.1.47}$$

Substituting (7.1.47) in (7.1.6, 7), we have

$$I(\omega) = \frac{1}{\pi}\,\mathrm{Re}\left\{\int_0^{\infty} \Phi(\tau)\exp(-\mathrm{i}\omega\tau)\,d\tau\right\}, \quad \Phi(\tau) = \left\langle \exp\left[\mathrm{i}\frac{\omega_0}{c}x(\tau)\right]\right\rangle . \tag{7.1.48}$$

The function $\Phi(\tau)$ can be calculated using the kinetic equation method [7.24].

We shall write $\Phi(\tau)$ in the form

$$\Phi(\tau) = \langle \exp[\mathrm{i}\boldsymbol{k}\cdot\boldsymbol{r}(\tau)]\rangle, \quad \boldsymbol{r}(\tau) = \int_0^{\tau} \boldsymbol{v}(t)\,dt , \tag{7.1.49}$$

and introduce the distribution function $f(\boldsymbol{r}, \boldsymbol{v}, t)$ for the oscillator coordinate $\boldsymbol{r}$ and velocity $\boldsymbol{v}$. This distribution function satisfies the Boltzmann equation

$$\frac{\partial f}{\partial t} + \boldsymbol{v}\cdot\nabla f = \left(\frac{\partial f}{\partial t}\right)_{\mathrm{coll.}}, \quad \left(\frac{\partial f}{\partial t}\right)_{\mathrm{coll.}} = \hat{G}f , \tag{7.1.50}$$

and the initial condition

$$f(\boldsymbol{r}, \boldsymbol{v}, 0) = W(\boldsymbol{v})\,\delta(\boldsymbol{r}) . \tag{7.1.51}$$

Here $(\partial f/\partial t)_{\mathrm{coll.}}$ is the collisional integral or collisional term, $\hat{G}$ is the linear operator of collisions, and $W(\boldsymbol{v})$ is the distribution function for $\boldsymbol{v}$. The correlation function $\Phi(\tau)$ and the intensity distribution $I(\omega)$ can be defined as

$$\Phi(\tau) = \int d\boldsymbol{v} \int f(\boldsymbol{r}, \boldsymbol{v}, \tau) \exp(\mathrm{i}\boldsymbol{k} \cdot \boldsymbol{r})\, d\boldsymbol{r}\ , \tag{7.1.52}$$

$$I(\omega) = \frac{1}{\pi} \operatorname{Re}\{\int d\boldsymbol{v} \int f(\boldsymbol{r}, \boldsymbol{v}, \tau) \exp[-\mathrm{i}(\omega\tau - \boldsymbol{k} \cdot \boldsymbol{r})]\, d\boldsymbol{r}\, d\tau\}\ . \tag{7.1.53}$$

Here the frequency difference $\omega - \omega_0$ is replaced by ω. According to (7.1.53), $I(\omega)$ is expressed in terms of the Fourier component $F_\omega(\boldsymbol{v}, \boldsymbol{k})$ of the distribution function:

$$I(\omega) = \frac{1}{\pi} \operatorname{Re}\{\int F_\omega(\boldsymbol{v}, \boldsymbol{k})\, d\boldsymbol{v}\}\ , \tag{7.1.54}$$

$$F_\omega(\boldsymbol{v}, \boldsymbol{k}) = \int f(\boldsymbol{r}, \boldsymbol{v}, \tau) \exp[-\mathrm{i}(\omega\tau - \boldsymbol{k} \cdot \boldsymbol{r})]\, d\boldsymbol{r}\, d\tau. \tag{7.1.55}$$

Carrying out the Fourier transform of (7.1.50) and taking into consideration the initial condition (7.1.51), we obtain the following equation for $F_\omega(\boldsymbol{v}, \boldsymbol{k})$:

$$-W(\boldsymbol{v}) + \mathrm{i}(\omega - \boldsymbol{k} \cdot \boldsymbol{v}) F_\omega(\boldsymbol{v}, \boldsymbol{k}) = (\hat{G}f)_{\omega k}\ . \tag{7.1.56}$$

For simplicity, the functions F_ω and $(\hat{G}f)_{\omega k}$ will be written below without indices ω.

In the absence of collisions $(\partial f/\partial t)_{\mathrm{coll.}} = 0, (\hat{G}f)_k = 0$, and

$$F(\boldsymbol{v}, \boldsymbol{k}) = \frac{W(\boldsymbol{v})}{\mathrm{i}(\omega - \boldsymbol{k} \cdot \boldsymbol{v})}\ .$$

For the Maxwellian $W(\boldsymbol{v})$ this gives the usual Doppler distribution (7.1.43). In fact, $W(\boldsymbol{v})d\boldsymbol{v} = W(v)w(\boldsymbol{v}_\perp)dv\,d\boldsymbol{v}_\perp$, where v is the velocity component in the direction of vector $\boldsymbol{k}, \boldsymbol{v}_\perp$ is perpendicular to $\boldsymbol{k}$. Integrating in (7.1.54) over $\boldsymbol{v}_\perp$ gives unity. When integrating over v, it is necessary to replace $\omega - \boldsymbol{k}\boldsymbol{v}$ by $\omega - kv - i\varepsilon$, considering $\varepsilon \to 0$. We find

$$I(\omega) = \lim \int \frac{\varepsilon}{\pi} \frac{W(v)\, dv}{(\omega - kv)^2 + \varepsilon^2} = \int \delta(\omega - kv)\, W(v)\, dv$$

$$= \frac{1}{\sqrt{\pi}\Delta\omega_{\mathrm{D}}} \exp\left[-\left(\frac{\omega}{\Delta\omega_{\mathrm{D}}}\right)^2\right]\ . \tag{7.1.57}$$

We shall consider now the influence of collisions assuming the model of Brownian motion [7.25]. This model can be used in the case of so-called weak collisions. In the framework of the model of Brownian motion, the collisional term in (7.1.50) has the form

$$\left(\frac{\partial f}{\partial t}\right)_{\mathrm{coll.}} = \nu\, \mathrm{div}_{\boldsymbol{v}}(\boldsymbol{v}f) + \frac{v_0^2}{2} \nu\, \Delta_{\boldsymbol{v}} f\ . \tag{7.1.58}$$

The effective frequency of collision ν is assumed not to depend on velocity and $v_0^2 = 2kT/m$. Solving (7.1.50) with the collisional term (7.1.58) and taking into account the initial condition (7.1.51), it is possible to obtain

$$\Phi(\tau) = \exp\left[-\frac{\Delta\omega_D^2}{2\nu^2}\left(\nu\tau - 1 + e^{-\nu\tau}\right)\right] , \quad \Delta\omega_D = kv_0 , \tag{7.1.59}$$

$$I(\omega) = \frac{1}{\pi}\,\mathrm{Re}\left\{\frac{2\nu}{\Delta\omega_D^2 - 2i\nu\omega}\,\Phi\left(1, 1 + \frac{\Delta\omega_D^2}{2\nu^2} - i\frac{\omega}{\nu}; \frac{\Delta\omega_D^2}{2\nu^2}\right)\right\} , \tag{7.1.60}$$

where $\Phi(\alpha, \gamma; z)$ is a confluent hypergeometric function

$$\Phi(\alpha, \gamma; z) = 1 + \frac{\alpha z}{\gamma 1!} + \frac{\alpha(\alpha+1) z^2}{\gamma(\gamma+1)\, 2!} + \frac{\alpha(\alpha+1)(\alpha+2) z^3}{\gamma(\gamma+1)(\gamma+2)\, 3!} + \dots . \tag{7.1.61}$$

For $\nu = 0$, (7.1.60) gives the usual Doppler distribution. When $\nu \neq 0$, but $\nu \ll \Delta\omega_D$, for the central part of a profile $\omega = 0$ and for the far wing $\omega \gg \Delta\omega_D$, we have, respectively,

$$I(0) = \frac{1}{\sqrt{\pi}\Delta\omega_D}\left(1 + \frac{2}{3\sqrt{\pi}}\frac{\nu}{\Delta\omega_D}\right) , \quad I(\omega) \simeq \frac{1}{\sqrt{\pi}\Delta\omega_D}\frac{\nu\Delta\omega_D^3}{2\sqrt{\pi}\omega^4} . \tag{7.1.62}$$

Thus due to collisions with $\nu \ll \Delta\omega_D$ the intensity in the central part of a profile is increased and the wing with intensity distribution $\propto \omega^{-4}$ appears. In the limiting case of high densities when $\nu \gg \Delta\omega_D$,

$$I(\omega) \simeq \frac{1}{\pi}\frac{\gamma_d}{\omega^2 + \gamma_d^2} \cdot \frac{\nu^2}{\omega^2 + \nu^2}, \quad \nu_d = \frac{1}{2}\frac{\Delta\omega_D^2}{\nu} . \tag{7.1.63}$$

The central part of a line is described by the Lorentzian distribution

$$I(\omega) = \frac{1}{\pi}\frac{\gamma_d}{\omega^2 + \gamma_d^2} \tag{7.1.64}$$

with width $2\gamma_d = \Delta\omega_D^2/\nu$. Since $\nu \simeq v_0/L$, $2\gamma_d = \Delta\omega_D\, 2\pi L/\lambda$ i.e., the width decreases $\propto L$ with increase of density. This result was first obtained by Dicke [7.26]. For $\omega \gg \nu$, (7.1.63) coincides with the intensity distribution in the wing from (7.1.62).

The qualitative picture of modification of the Doppler distribution due to collisions does not depend on the specific model of Brownian motion used above.

We shall consider now the model of strong collisions assuming that after every collision the distribution of velocities does not depend on the velocity before collision and is Maxwellian. In this case, the collisional term in (7.1.50) can be written in the form

$$\left(\frac{\partial f}{\partial t}\right)_{\mathrm{coll.}} = -\nu f + \nu W(\boldsymbol{v}) \int f(\boldsymbol{r}, \boldsymbol{v}', t)\, d\boldsymbol{v}' . \tag{7.1.65}$$

For $(\hat{G}f)_{\boldsymbol{k}}$ in (7.1.56), we have

$$(\hat{G}f)_{\boldsymbol{k}} = -\nu F(\boldsymbol{v}, \boldsymbol{k}) + \nu W(\boldsymbol{v}) \int F(\boldsymbol{v}', \boldsymbol{k})\, d\boldsymbol{v}' . \tag{7.1.66}$$

Substituting (7.1.66) in (7.1.56) we obtain

$$F(\boldsymbol{v},\boldsymbol{k}) = \frac{\nu W(\boldsymbol{v})}{\nu + \mathrm{i}(\omega - \boldsymbol{k}\cdot\boldsymbol{v})} \int F(\boldsymbol{v}',\boldsymbol{k})\, d\boldsymbol{v}' + \frac{W(\boldsymbol{v})}{\nu + \mathrm{i}(\omega - \boldsymbol{k}\cdot\boldsymbol{v})} \,. \tag{7.1.67}$$

Integrating the right-hand side and left-hand side of this equation over $\boldsymbol{v}$ we have

$$\int F(\boldsymbol{v},\boldsymbol{k})\, d\boldsymbol{v} = \int \frac{W(\boldsymbol{v})\, d\boldsymbol{v}}{\nu + \mathrm{i}(\omega\boldsymbol{k}\cdot\boldsymbol{v})} \left(\nu \int F(\boldsymbol{v}',\boldsymbol{k})\, d\boldsymbol{v}' + 1\right)$$

and after replacing in the right-hand side $\boldsymbol{v}'$ by $\boldsymbol{v}$

$$\int F(\boldsymbol{v},\boldsymbol{k})\, d\boldsymbol{v} \left(1 - \nu \int \frac{W(\boldsymbol{v})\, d\boldsymbol{v}}{\nu + \mathrm{i}(\omega - \boldsymbol{k}\cdot\boldsymbol{v})}\right) = \int \frac{W(\boldsymbol{v})\, d\boldsymbol{v}}{\nu + \mathrm{i}(\omega - \boldsymbol{k}\cdot\boldsymbol{v})}$$

It is possible now to find $F(\boldsymbol{v},\boldsymbol{k})$. Then using (7.1.54) we obtain

$$I(\omega) = \mathrm{Re}\left\{ \frac{\frac{1}{\pi} \int \frac{W(\boldsymbol{v})\, d\boldsymbol{v}}{\nu + \mathrm{i}(\omega - \boldsymbol{k}\cdot\boldsymbol{v})}}{1 - \nu \cdot \int \frac{W(\boldsymbol{v})\, d\boldsymbol{v}}{\nu + \mathrm{i}(\omega - \boldsymbol{k}\cdot\boldsymbol{v})}} \right\} . \tag{7.1.68}$$

When $\nu \ll \Delta\omega_{\mathrm{D}}$, the second term in the denominator of (7.1.68) has the order of magnitude of $\nu/\Delta\omega_{\mathrm{D}}$. If this term is neglected, (7.1.68) gives the usual Doppler distribution. In the general case, the intensity distribution (7.1.68) is similar to that given by (7.1.60). Instead of (7.1.62), it follows from (7.1.68) for $I(0)$ and $\omega \gg \Delta\omega_{\mathrm{D}}$,

$$I(0) \simeq \frac{1}{\sqrt{\pi}\,\Delta\omega_{\mathrm{D}}} \left(1 + \frac{\pi - 2}{\sqrt{\pi}} \frac{\nu}{\Delta\omega_{\mathrm{D}}}\right), \quad I(\omega) \simeq \frac{1}{\sqrt{\pi}\Delta\omega_{\mathrm{D}}} \frac{\nu \Delta\omega_{\mathrm{D}}^3}{2\sqrt{\pi}\omega^4} \,. \tag{7.1.69}$$

In the limiting case $\nu \gg \Delta\omega_{\mathrm{D}}$, (7.1.68) leads to a Lorentzian distribution in the central part of a line with width $\Delta\omega_{\mathrm{D}}^2/\nu$. In the region of high frequencies $\omega > \nu$, the Lorentzian distribution is replaced by a wing $I(\omega) \propto \omega^{-4}$.

Thus in both cases considered above (weak and strong collisions), with an increase of the density a narrowing of the central part of the line occurs. At high densities, when $L < \lambda/2\pi$, the narrowing of a line is proportional to $L \propto N^{-1}$, and in the limiting case of $L \ll \lambda/2\pi$, the central part of a line is described by a Lorentzian distribution with width $2\pi L/\lambda$ times less than the usual Doppler broadening $\Delta\omega_{\mathrm{D}}$. Such a narrowing of the Doppler profile caused by collisions can be observed only in cases when there is no broadening due to interaction with perturbers. In the general case there are no grounds for separating the effect of interaction and the Doppler effect. The same collisions can produce both a phase shift and a change of velocity of the atom. This means the statistical dependence of both effects. It must be noted that Doppler broadening is usually of interest just under the condition $L > \lambda/2\pi$. In fact, the condition $\Delta\omega_{\mathrm{D}} \gtrsim \gamma$, where γ is the impact width can be rewritten in the form $2\pi v/\lambda \gtrsim Nv\sigma' = Nv\sigma\sigma'/\sigma = v\sigma'/L\sigma$, where σ is the gas-kinetic effective cross section of the atom. As a rule, $\sigma' \gtrsim \sigma$ and, consequently, $\Delta\omega_{\mathrm{D}} \gtrsim \gamma$ when $L > \lambda/2\pi$.

Nevertheless the statistical dependence of Doppler and impact broadening in some cases must be taken into account. This problem will be considered below in the framework of the quantum theory of broadening. A bibliography on Dicke narrowing may be found in [7.27].

7.1.6 Convolution of the Doppler and Lorentzian Distributions

If $L \gg \lambda/2\pi$, the combined treatment of impact and Doppler broadenings (statistically independent) leads to the convolution of Doppler and Lorentzian distributions. The Lorentzian intensity distribution with width γ and shift Δ, corresponding to the atom with velocity component v in the direction of observation, is given by

$$I_v(\omega) = \frac{\gamma}{2\pi} \frac{1}{(\omega - \Delta - \omega_0 v/c)^2 + (\gamma/2)^2} . \tag{7.1.70}$$

To obtain the intensity distribution for an assembly of atoms, it is necessary to average (7.1.70) over the velocity distribution $W(v)$. Thus

$$I(\omega) = \frac{\gamma}{2\pi} \int \frac{W(v)\, dv}{(\omega - \Delta - \omega_0 v/c)^2 + (\gamma/2)^2} . \tag{7.1.71}$$

For a Maxwellian distribution

$$I(\omega) = \frac{\gamma}{2\pi} \frac{1}{\sqrt{\pi} v_0} \int \frac{\exp[-(v/v_0)^2]\, dv}{(\omega - \Delta - \omega_0 v/c)^2 + (\gamma/2)^2} . \tag{7.1.72}$$

When $\Delta\omega_{\mathrm{D}} \ll \gamma/2$ the term $\omega_0 v/c$ can be neglected in the denominator in (7.1.72), after which the integration over v gives a Lorentzian distribution with width γ. Consequently, when $\Delta\omega_{\mathrm{D}} \ll \gamma/2$ Doppler broadening can be neglected.

When $\Delta\omega_{\mathrm{D}} \gg \gamma/2$ a significant contribution to the integral (7.1.72) can be given by two ranges of values of v: $v \sim 0$ and $v \sim c(\omega - \Delta)/\omega_0$. In the first of those ranges, the term $\omega_0 v/c$ in the denominator can be neglected and in the second v can be replaced in the numerator by $c(\omega - \Delta)/\omega_0$. After this it is easy to obtain two approximate expressions for $I(\omega)$ valid for the center of a line $\omega - \Delta \ll \Omega_{\mathrm{D}}$ and for the wing $\omega - \Delta \gg \Omega_{\mathrm{D}}$, where Ω_{D} is determined by the relation

$$\Omega_{\mathrm{D}}^2 = \Delta\omega_{\mathrm{D}}^2 \ln \left[2\pi^{3/2} \frac{\Delta\omega_{\mathrm{D}}}{\gamma} \left(\frac{\Omega_{\mathrm{D}}}{\Delta\omega_{\mathrm{D}}} \right)^2 \right] . \tag{7.1.73}$$

In the center of a line $\omega - \Delta \ll \Omega_{\mathrm{D}}$, $I(\omega)$ coincides with the usual Doppler distribution. In the wing of a line, $I(\omega) \propto \gamma/2\pi\omega^2$. Thus for any relation between $\Delta\omega_{\mathrm{D}}$ and $\gamma/2$ at sufficiently high values of ω, the Doppler distribution is replaced by the Lorentzian wing.

We shall write (7.1.72) in the form

$$I(\omega) = \frac{1}{\sqrt{\pi}\,\Delta\omega_{\mathrm{D}}} \mathrm{Re} \left\{ W\left(\frac{\omega - \Delta}{\Delta\omega_{\mathrm{D}}} \right), \left(\frac{\gamma}{2\Delta\omega_{\mathrm{D}}} \right) \right\} , \tag{7.1.74}$$

$$W(x,y) = \frac{\mathrm{i}}{\pi}\int_{-\infty}^{\infty}\frac{\exp(-t^2)dt}{x+\mathrm{i}y-t} = \frac{1}{\sqrt{\pi}}\int_0^{\infty}\exp[-z^2+\mathrm{i}(x+\mathrm{i}y)z]\,dz\ , \qquad (7.1.75)$$

where $x = (\omega - \Delta)/\Delta\omega_D$, $y = \gamma/2\Delta\omega_D$, $t = (v/c)(\omega_0/\Delta\omega_D)$. The function $W(x,y)$ can be expressed in terms of the probability integral with complex argument[1]

$$W(x,y) = \exp[-(x+\mathrm{i}y)^2]\left[1 - \frac{2}{\sqrt{\pi}}\int_0^{\mathrm{i}(x+\mathrm{i}y)}\exp(-t^2)\,dt\right]\ . \qquad (7.1.76)$$

The intensity distribution $I(\omega)$ for any relation between parameters γ and $\Delta\omega_D$ can be calculated using (7.1.76).

7.2 General Theory of Impact Broadening

7.2.1 Density Matrix Method in the Quasi-Classical Approximation

In the quasi-classical approximation the interaction of the atom with the surrounding particles can be described by the time-dependent perturbation $V(t)$. In this case the coordinates of the perturbing particles can be considered not as dynamic variables but as assigned functions of time, which enables one to introduce the perturbation $V(t)$ instead of the perturbation $V(\boldsymbol{R})$. It will be shown in this section how the shape of a line is calculated when an atom undergoes an arbitrary perturbation $V(t)$. From the theory of the interaction of a quantum system with electromagnetic radiation we know that for dipole transition $\alpha \rightarrow \beta$ [7.2]

$$I(\omega) \propto |\int P_{\alpha\beta}(t)\exp(-\mathrm{i}\omega t)\,dt|^2\ , \qquad (7.2.1)$$

where $P_{\alpha\beta}(t)$ is the matrix element of the dipole moment of an atom calculated by means of the perturbed wave functions $\Psi_\alpha(t)$ and $\Psi_\beta(t)$. These functions are the solutions of the Schrödinger equation for the Hamiltonian

$$H = H_0 + V(t)\ . \qquad (7.2.2)$$

Formula (7.2.1) is the natural generalization of the classical formula (7.1.3). It is helpful to write this formula in a form similar to (7.1.6),

$$I(\omega) = \frac{1}{\pi}\mathrm{Re}\left\{\int_0^{\infty}\Phi(\tau)\exp(-\mathrm{i}\omega\tau)\,d\tau\right\}\ , \qquad (7.2.3)$$

where

$$\Phi(\tau) = \overline{P_{\alpha\beta}(t+\tau)P^*_{\alpha\beta}(t)} = \overline{P_{\alpha\beta}(t+\tau)P_{\beta\alpha}(t)}\ , \qquad (7.2.4)$$

[1] Tables of function $W(x,y)$ are given in [7.28, 29]

or

$$\Phi(\tau) = \langle P_{\alpha\beta}(\tau) P_{\beta\alpha}(0) \rangle \ . \tag{7.2.5}$$

We shall consider further a transition between the two degenerate levels a and b, the indices α and β numbering the states belonging respectively to the initial and final levels. We shall assume that all states α are populated with equal probability. In this case

$$I(\omega) = \sum_{\alpha\beta} I_{\alpha\beta}(\omega);$$

therefore instead of (7.2.4, 5) it is necessary to assume

$$\Phi(\tau) = \sum_{\alpha\beta} \overline{P_{\alpha\beta}(t+\tau) P^*_{\alpha\beta}(t)} \ , \tag{7.2.6}$$

$$\Phi(\tau) = \sum_{\alpha\beta} \langle P_{\alpha\beta}(\tau) P^*_{\alpha\beta}(0) \rangle \ . \tag{7.2.7}$$

Equations (7.2.6, 7) are easily generalized to the case when a line is formed by a set of transitions between two groups of closely spaced levels. We shall indicate by the indices α states belonging to initial levels and by the indices β, those belonging to final levels, and we shall denote by W_α the population of the state α, $\sum_\alpha W_\alpha = 1$. Then

$$\Phi(\tau) = \sum_{\alpha\beta} W_\alpha \overline{P_{\alpha\beta}(t+\tau) P_{\beta\alpha}(t)} = \sum_{\alpha\beta} W_\alpha \langle P_{\alpha\beta}(\tau) P_{\beta\alpha}(0) \rangle . \tag{7.2.8}$$

The perturbed functions $\Psi_\alpha(t)$ and $\Psi_\beta(t)$ can be expanded in terms of time-independent functions of the isolated atom

$$\Psi_\alpha(t) = \sum_{\alpha'} a_{\alpha'\alpha}(t) \Psi_{\alpha'} \exp\left(-\frac{\mathrm{i}}{\hbar} E_{\alpha'} t\right) ,$$

$$\Psi_\beta(t) = \sum_{\beta'} a_{\beta'\beta}(t) \Psi_{\beta'} \exp\left(-\frac{\mathrm{i}}{\hbar} E_{\beta'} t\right) ,$$

$$\begin{aligned} \langle P_{\alpha\beta}(\tau) P_{\beta\alpha}(0) \rangle &= \sum_{\alpha'\beta'} \langle a^*_{\alpha'\alpha}(\tau) a_{\beta'\beta}(\tau) \rangle \ P_{\alpha'\beta'} P_{\beta\alpha} \exp(\mathrm{i}\omega_{\alpha'\beta'}\tau) \\ &= \sum_{\alpha'\beta'} \rho^{(\alpha\beta)}_{\alpha'\beta'}(\tau) \ P_{\alpha'\beta'} P_{\beta\alpha} \ , \end{aligned}$$

where $\rho^{(\alpha\beta)}_{\alpha'\beta'} = \langle a^*_{\alpha'\alpha} a_{\beta'\beta} \rangle$ is the density matrix of an atom, the matrix elements $P_{\alpha'\beta'}$ and $P_{\beta\alpha}$ do not depend on t. The upper indices $(\alpha\beta)$ define the initial conditions $\rho^{(\alpha\beta)}_{\alpha'\beta'}(0) = \delta_{\alpha\alpha'}\delta_{\beta\beta'}$. The evolution of the density matrix with time is given by the following equation

$$\frac{d\rho}{dt} = \frac{\mathrm{i}}{\hbar}(H\rho - \rho H) \ , \quad H = H_0 + V(t) \ . \tag{7.2.9}$$

For correlation function $\Phi(\tau)$ we have

$$\Phi(\tau) = \sum_{\alpha\beta\alpha'\beta'} W_\alpha \rho^{(\alpha\beta)}_{\alpha'\beta'}(\tau)\, P_{\alpha'\beta'} P_{\beta\alpha} \, . \tag{7.2.10}$$

In the framework of impact approximation, the equation for the density matrix ρ can be written in the form

$$\frac{d\rho}{dt} = \frac{\mathrm{i}}{\hbar}(H_0\rho - \rho H_0) + \left(\frac{d\rho}{dt}\right)_{\text{coll.}} \, . \tag{7.2.11}$$

The last term in right-hand side of this equation describes the evolution of the operator ρ caused by collisions.

We shall assume that the wave functions before collision Ψ and after collision Ψ' are connected by the relation

$$\Psi' = S\Psi \, , \tag{7.2.12}$$

where S is the collision S matrix. According to (7.2.12) the corresponding transformation of the operator ρ is

$$\rho \to S^\dagger \rho S \, .$$

The increase of ρ caused by collision with parameters v (impact parameter, velocity, and so on) is

$$\Delta\rho = S^\dagger \rho S - \rho \, , \quad S = S(v) \, , \tag{7.2.13}$$

where $S^\dagger_{ik} = S^*_{ki}$. Therefore

$$\left(\frac{d\rho}{dt}\right)_{\text{coll.}} = \int [S^\dagger(v)\rho S(v) - \rho]\, P(v)\, dv \tag{7.2.14}$$

where $P(v)dv$ is the number of collisions with parameters v in the interval v, $v+dv$ per second. If by v are understood the impact parameter ρ and velocity v, then $P(v)dv = Nvf(v)\, 2\pi\rho d\rho dv$, where N is the density of the perturbing particles, and $f(v)$ is the distribution function for v.

Substituting (7.2.14) in (7.2.11) and carrying out the Fourier transform, it is not difficult to obtain the system of algebraic equations for matrix elements $\rho^{(\alpha\beta)}_{\alpha'\beta'}(\omega)$ with different indices $\alpha'\beta'$ and the same indices $\alpha\beta$,

$$\begin{aligned} &\delta_{\alpha\alpha'}\delta_{\beta\beta'} - \mathrm{i}(\omega - \omega_{\alpha'\beta'})\rho^{(\alpha\beta)}_{\alpha'\beta'}(\omega) \\ &\quad = \sum_{\alpha''\beta''} \rho^{(\alpha\beta)}_{\alpha''\beta''}(\omega) \int P(v)\, dv\, [\delta_{\alpha'\alpha''}\delta_{\beta'\beta''} - S^*_{\alpha''\alpha'} S_{\beta''\beta'}] \, . \end{aligned} \tag{7.2.15}$$

From (7.2.3) and (7.2.10), we have

$$I(\omega) = \frac{1}{\pi} \operatorname{Re} \left\{ \sum_{\alpha\beta\alpha'\beta'} W_\alpha P_{\alpha'\beta'} P_{\beta\alpha} \rho^{(\alpha\beta)}_{\alpha'\beta'}(\omega) \right\} . \tag{7.2.16}$$

It is supposed in (7.2.16) that radiation of definite polarization is of interest. In the case of arbitrary polarization, $P_{\alpha'\beta'} P^*_{\alpha\beta}$ must be replaced by $\boldsymbol{P}_{\alpha'\beta'} \boldsymbol{P}^*_{\alpha\beta}$.

Summing with respect to $\alpha'\beta'$ and $\alpha''\beta''$ in (7.2.15) and (7.2.16) in the general case means summing over all stationary states of an atom. However, in calculation of the intensity distribution in a narrow spectral range corresponding to the transitions between the two groups of closely spaced states α and β, one can assume that the indices α', α'' take the same values as α, and the indices β', β'', the same values as β.

Therefore, if the S matrix is known, the spectrum $I(\omega)$ can be calculated by solving the system of (7.2.15) and using (7.2.16).

For an isolated line $\alpha \to \beta$, we have only one equation:

$$1 - \mathrm{i}(\omega - \omega_0)\rho_{\alpha\beta} = \rho_{\alpha\beta} \int [1 - S^*_{\alpha\alpha} S_{\beta\beta}]\, P(v)\, dv\,.$$

Thus

$$I(\omega) = W|P_{\alpha\beta}|^2 \frac{\gamma_{\alpha\beta}}{2\pi} \frac{1}{(\omega - \omega_0 - \Delta_{\alpha\beta})^2 + (\gamma_{\alpha\beta}/2)^2}\,, \tag{7.2.17}$$

$$\frac{\gamma_{\alpha\beta}}{2} - \mathrm{i}\Delta_{\alpha\beta} = \int [1 - S^*_{\alpha\alpha} S_{\beta\beta}]\, P(v)\, dv\,, \tag{7.2.18}$$

$$\sigma' - \mathrm{i}\sigma'' = \int [1 - S^*_{\alpha\alpha} S_{\beta\beta}]\, 2\pi\rho d\rho\,. \tag{7.2.19}$$

In the general case, the S matrix is complex: $S_{\alpha\alpha} = \exp(-\Gamma_\alpha + \mathrm{i}\eta_\alpha)$, $S_{\beta\beta} = \exp(-\Gamma_\beta + \mathrm{i}\eta_\beta)$. Therefore the width $\gamma_{\alpha\beta}$ and shift $\Delta_{\alpha\beta}$ are determined by the expressions

$$\begin{aligned} \gamma_{\alpha\beta} &= \int \{1 - \exp[-(\Gamma_\alpha + \Gamma_\beta)]\, \cos(\eta_\alpha - \eta_\beta)\}\, P(v)\, dv\,, \\ \Delta_{\alpha\beta} &= \int \exp[-(\Gamma_\alpha + \Gamma_\beta)]\, \sin(\eta_\alpha - \eta_\beta)\, P(v)\, dv\,. \end{aligned} \tag{7.2.20}$$

We shall consider now the intensity distribution in the wing of a line corresponding to the set of transitions $\alpha \to \beta$. If the differences $\omega - \omega_{\alpha'\beta'}$ are high enough, one can substitute in the right-hand side of (7.2.15) the zeroth approximation $\rho^{(\alpha\beta)}_{\alpha''\beta''} = -\delta_{\alpha\alpha''}\delta_{\beta\beta''}(\omega - \omega_0)^{-1}$, where ω_0 is the mean value of the frequencies $\omega_{\alpha'\beta'}$. After this it is not difficult to obtain

$$I(\omega) = \sum_{\alpha\beta} W_\alpha \mid P_{\alpha\beta} \mid^2 \frac{\gamma}{2\pi(\omega - \omega_0)^2}\,, \tag{7.2.21}$$

where

$$\gamma = \Big(\sum_{\alpha\beta} W_\alpha |P_{\alpha\beta}|^2\Big)^{-1} 2\,\mathrm{Re}\left\{ \sum_{\alpha\beta\alpha'\beta'} W_\alpha P_{\alpha'\beta'} P_{\beta\alpha} \int [\delta_{\alpha\alpha'}\delta_{\beta\beta'} - S^*_{\alpha\alpha'} S_{\beta\beta'}]\, P(v)\, dv \right\}. \tag{7.2.22}$$

The intensity distribution in the central part of a line is more complex than in the case of a Lorentzian profile.

If the S matrix is diagonal (for example, when the perturbation $V(t)$ is adiabatic), the spectrum $I(\omega)$ can be easily calculated in the general form. In this case the system of equations (7.2.15) splits into independent equations.

Solving these equations, we obtain

$$I(\omega) = \sum_{\alpha\beta} W_\alpha \mid P_{\alpha\beta} \mid^2 \frac{\gamma_{\alpha\beta}}{2\pi} \frac{1}{(\omega - \omega_{\alpha\beta} - \Delta_{\alpha\beta})^2 + (\gamma_{\alpha\beta}/2)^2} . \tag{7.2.23}$$

This intensity distribution has the form of the sum of independent Lorentzian distributions.

In general case in order to calculate $I(\omega)$ one must determine the S matrix and then solve the system of (7.2.15). In calculating the S matrix, it is possible to use different approximate methods and results of the general theory of atomic collisions. In particular, one can define the S matrix using the method of successive approximations of perturbation theory.

The eigenfunction $\Psi(t)$ of the Hamiltonian $H_0 + V$ satisfying the initial condition $\Psi(t_0) = \Phi_n(t_0) \exp\left(-\frac{\mathrm{i}}{\hbar} E_n t_0\right)$ (where Φ is the unperturbed wave function) can be written in the form

$$\Psi(t) = \sum_m S_{mn}(t) \Phi_m \exp\left(-\frac{\mathrm{i}}{\hbar} E_m t\right), \quad S_{mn}(t_0) = \delta_{mn} ,$$

where the coefficients of expansion $S_{mn}(t)$ satisfy the system of equations

$$\mathrm{i}\hbar \dot{S}_{mn} = \sum_k V_{mk} S_{kn} \exp\left[\frac{\mathrm{i}}{\hbar}(E_m - E_k)t\right] .$$

We shall introduce the operator

$$\tilde{V} = \exp\left(\frac{\mathrm{i}}{\hbar} H_0 t\right) V \exp\left(-\frac{\mathrm{i}}{\hbar} H_0 t\right) ,$$

for which

$$\tilde{V}_{mk} = V_{mk} \exp\left[\frac{\mathrm{i}}{\hbar}(E_m - E_k)t\right] .$$

Therefore the system of equations for S_{mn} can be written in operator form as

$$\mathrm{i}\hbar \dot{S}_{mn} = \sum_k \tilde{V}_{mk} S_{kn} ,$$

$$\mathrm{i}\hbar \dot{S} = \tilde{V} S .$$

We shall seek the solution of this equation by the method of successive approximations:

$$S(t) = 1 + \left(-\frac{\mathrm{i}}{\hbar}\right) \int_{t_0}^{t} \tilde{V}(t')\, dt' + \left(-\frac{\mathrm{i}}{\hbar}\right)^2 \int_{t_0}^{t} \tilde{V}(t')\, dt' \int_{t_0}^{t'} \tilde{V}(t'')\, dt'' + \ldots . \tag{7.2.24}$$

In order to determine the S matrix corresponding to the collision with parameters v, it is necessary to replace $\tilde{V}(t)$ by $\tilde{V}_v(t)$, where $\tilde{V}_v(t) \to 0$ at $t \to \pm\infty$, and assume $t_0 = -\infty, \quad t = \infty$:

$$S(v) = 1 + \left(-\frac{i}{\hbar}\right) \int_{-\infty}^{\infty} \tilde{V}_v(t)\,dt + \left(-\frac{i}{\hbar}\right)^2 \int_{-\infty}^{\infty} \tilde{V}_v(t)\,dt \int_{-\infty}^{t} \tilde{V}_v(t')\,dt' + \dots \tag{7.2.25}$$

Let us restrict ourselves to the first two terms of (7.2.25). In this approximation it is easy to obtain for the quantities Γ and η in (7.2.20) the following expressions:

$$\Gamma_\alpha = \frac{1}{2\hbar^2} \sum_s \left| \int_{-\infty}^{\infty} \langle\alpha|V_v|s\rangle \exp(i\omega_{\alpha s}t)\,dt \right|^2$$

$$\eta_\alpha = -\frac{1}{\hbar} \int_{-\infty}^{\infty} \langle\alpha|V_v(t)|\alpha\rangle\,dt - \frac{1}{\hbar^2} \mathrm{Im} \left\{ \sum_s{}' \int_{-\infty}^{\infty} \langle\alpha|V_v(t)|s\rangle \exp(i\omega_{\alpha s}t)\,dt \right.$$
$$\left. \times \int_{-\infty}^{t} \langle s|V_v(t')|\alpha\rangle \exp(i\omega_{s\alpha}t')\,dt' \right\}. \tag{7.2.26}$$

The quantity $2\Gamma_\alpha$ is the total probability of transitions from the state α into all other states of the atom. Such transitions shorten the lifetime of an atom in the state α, which is equivalent to broadening of the corresponding level. The quantity η_α is the phase shift caused by the shift of the level during the collision. The increase of Γ_α and Γ_β increases the width and reduces the shift of the line, see (7.2.20). The broadening caused by collisional transitions is usually called the broadening due to inelastic collisions.

7.2.2 Degeneracy of Levels

We shall consider the isolated spectral line corresponding to the radiative transition between levels with angular momenta j_1 and j_2. Such line is produced by the set of transitions between degenerated states $j_1m_1 - j_2m_2$, where m_1 and m_2 are the magnetic quantum numbers. We shall write down the equation (7.2.11) for density matrix in the form

$$\frac{d}{dt}\rho_{m_1m_2} = i\omega_0\rho_{m_1m_2} + \sum_{m_1'm_2'} \rho_{m_1'm_2'} G_{m_1'm_2',\,m_1m_2}\,, \tag{7.2.27}$$

where $\omega_0 = \omega_{j_1j_2}$, and in accordance with (7.2.14) and (7.2.15),

$$G_{m_1'm_2'm_1m_2} = -\int P(v)\,dv\,[\delta_{m_1'm_1}\delta_{m_2'm_2} - S^*_{m_1'm_1}S_{m_2'm_2}]\,.$$

Let us introduce now linear combinations of $\rho_{m_1m_2}$ which have the same transformation properties as the irreducible tensor operators T^s_σ see [Ref. 7.30, Sect. 4.3]. Since

$$\psi^*_{j_1m_1}\psi_{j_2m_2} = (-1)^{j_1+m_1}\psi_{j_1-m_1}\psi_{j_2m_2}\,,$$

the required transformation has the form

$$\rho_{s\sigma} = \sum_{m_1 m_2} (-1)^{-j_1-m_1}(-m_1 m_2|s\sigma)\rho_{m_1 m_2}$$

$$= \sum_{m_1 m_2} (-1)^{-j_2-m_1-\sigma}\sqrt{2s+1}\begin{pmatrix} j_1 & j_2 & s \\ -m_1 & m_2 & -\sigma \end{pmatrix}\rho_{m_1 m_2}\,, \qquad (7.2.28)$$

$$\rho_{m_1 m_2} = \sum_{s\sigma}(-1)^{j_1+m_1}(-m_1 m_2|s\sigma)\rho_{s\sigma}$$

$$= \sum_{s\sigma}(-1)^{j_2+m_1+\sigma}\sqrt{2s+1}\begin{pmatrix} j_1 & j_2 & s \\ -m_1 & m_2 & -\sigma \end{pmatrix}\rho_{s\sigma}\,,$$

where $(-m_1 m_2|s\sigma)$ are the Clebsch–Gordan coefficients [Ref. 7.30, Sect. 4.2]. Multiplying (7.2.27) by $(-1)^{-j_1-m_1}(-m_1 m_2|s\sigma)$, summing with respect to m_1 and m_2, and expressing $\rho_{m_1' m_2'}$ in the collisional term in terms of $\rho_{s\sigma}$, we obtain

$$\frac{d}{dt}\rho_{s\sigma} = \mathrm{i}\omega_0\rho_{s\sigma} + \sum_{s'\sigma'}\rho_{s'\sigma'}G_{s'\sigma',\,s\sigma}\,,$$

$$G_{s'\sigma',\,s\sigma} = \sum_{m_1 m_2 m_1' m_2'}(-1)^{-m_1+m_1'}(-m_1' m_2'|s'\sigma')G_{m_1' m_2' m_1 m_2}(-m_1 m_2|s\sigma)\,.$$

The collisional term in (7.2.14) must be averaged over different orientation of the vectors $\boldsymbol{v}$ and $\boldsymbol{\rho}$ in space. The interaction of an atom with perturbing particles averaged in such a way is isotropic. This means that the equations defining the irreducible tensor quantities $\rho_{s\sigma}$ are invariant under rotations and the matrix G is diagonal with respect to s and σ. Moreover the matrix G does not depend on σ:

$$G_{s'\sigma',\,s\sigma} = \delta_{ss'}\delta_{\sigma\sigma'}G_s$$

Thus, the system of equations for $\rho_{s\sigma}$ has the form

$$\frac{d}{dt}\rho_{s\sigma} = \mathrm{i}\omega_0\rho_{s\sigma} + G_s\rho_{s\sigma}\,. \qquad (7.2.29)$$

This system splits into independent equations for each of the quantities $\rho_{s\sigma}$. The similar approach to the density matrix equation based on the expansion of the operators $\rho_{m_1 m_2}$ in terms of the irreducible tensor operators has been used in [7.17, 18].

That the matrix G is diagonal with respect to s and σ can be verified by direct calculation. The integrand in (7.2.14) averaged over directions of $\boldsymbol{v}$ and ρ must have the form

$$\sum_{\kappa q}\langle m_1' m_2'|(-1)^q U_q^\kappa V_{-q}^\kappa|m_1 m_2\rangle$$

$$= \sum_{\kappa q}(-1)^{q+j_1-m_1'+j_2-m_2'}(j_1\|U^\kappa\|j_1)(j_2\|V^\kappa\|j_2)\begin{pmatrix} j_1 & \kappa & j_1 \\ -m_1' & -q & m_1 \end{pmatrix}$$

$$\times\begin{pmatrix} j_2 & \kappa & j_2 \\ -m_2' & -q & m_2 \end{pmatrix}\,,$$

where U_q^κ and V_q^κ are the irreducible tensor operators acting on the functions $\psi_{j_1 m_1}$ and $\psi_{j_2 m_2}$. Substituting this expression in the formula defining the matrix $G_{m_1' m_2' m_1 m_2}$ and carrying out the summation of the product of the four $3j$ symbols in the expression for $G_{s'\sigma', s\sigma}$ with respect to m_1, m_2, m_1', m_2', we obtain

$$G_{s'\sigma', s\sigma} = \delta_{ss'}\delta_{\sigma\sigma'}G_s, \quad G_s = \sum_\kappa G_s(\kappa)\,.$$

Let us proceed to the calculation of the spectrum $I(\omega)$. From (7.2.16), we have

$$I(\omega) = \frac{1}{\pi}\mathrm{Re}\left\{ \sum_{m_1 m_2 m_1' m_2' \lambda} W_{m_1}(P_\lambda)_{m_1' m_2'}(P_\lambda)^*_{m_1 m_2}\rho^{(m_1 m_2)}_{m_1' m_2'}(\omega)\right\},$$

where P_λ is the spherical component of the vector P. We shall assume that the population of the state m_1 is $(2j_1+1)^{-1}$. Transforming (7.2.15) in the same way as (7.2.29), we have

$$\rho^{(m_1 m_2)}_{s\sigma}(\omega) = (-1)^{-j_2 - m_1 - \sigma}\sqrt{2s+1}\begin{pmatrix} j_1 & j_2 & s \\ -m_1 & m_2 & -\sigma \end{pmatrix}\frac{1}{-\mathrm{i}(\omega-\omega_0)+G_s}\,.$$

The quantities $\rho^{(m_1 m_2)}_{m_1' m_2'}(\omega)$ are expressed in terms of $\rho^{(m_1 m_2)}_{s\sigma}$ in accordance with the second of equation (7.2.28).

For the matrix elements of P_λ, we have [Ref. 7.30, Sect. 4.3].

$$(P_\lambda)_{m_1' m_2'} = (-1)^{j_1 - m_1'}\begin{pmatrix} j_1 & 1 & j_2 \\ -m_1' & \lambda & m_2' \end{pmatrix}(j_1\|P\|j_2)\,,$$

$$(P_\lambda)^*_{m_1 m_2} = (-1)^{j_1 - m_1}\begin{pmatrix} j_1 & 1 & j_2 \\ -m_1 & \lambda & m_2 \end{pmatrix}(j_1\|P\|j_2)^*\,.$$

Therefore, it is necessary to calculate the following sum of the product of the four $3j$ symbols:

$$\begin{aligned}
&\sum_{m_1 m_2 m_1' m_2' \lambda}(-1)^{2j_1 - 2m_1}\begin{pmatrix} j_1 & j_2 & s \\ -m_1 & m_2 & -\sigma \end{pmatrix}\begin{pmatrix} j_1 & j_2 & s \\ -m_1' & m_2' & -\sigma \end{pmatrix}\begin{pmatrix} j_1 & 1 & j_2 \\ -m_1' & \lambda & m_2' \end{pmatrix}\\
&\quad\times\begin{pmatrix} j_1 & 1 & j_2 \\ -m_1 & \lambda & m_2 \end{pmatrix}\\
&=\sum\begin{pmatrix} j_1 & j_2 & s \\ -m_1 & m_2 & -\sigma \end{pmatrix}\begin{pmatrix} j_1 & j_2 & 1 \\ -m_1 & m_2 & \lambda \end{pmatrix}\begin{pmatrix} j_1 & j_2 & s \\ -m_1' & m_2' & -\sigma \end{pmatrix}\\
&\quad\times\begin{pmatrix} j_1 & j_2 & 1 \\ -m_1' & m_2' & \lambda \end{pmatrix}\\
&=\frac{1}{9}\delta_{s1}\sum_\lambda \delta_{\lambda,-\sigma} = \frac{1}{3}\delta_{s1}\,.
\end{aligned}$$

Let us recall that $2j_1 - 2m_1$ is even.

Finally we obtain

$$I(\omega) = \frac{|(j_1\|P\|j_2)|^2}{2j_1+1} \cdot \frac{1}{\pi}\mathrm{Re}\left\{\frac{1}{-\mathrm{i}(\omega-\omega_0)+G_1}\right\}$$

$$= \frac{|(j_1\|P\|j_2)|^2}{2j_1+1} \frac{\mathrm{Re}\{G_1\}}{\pi} \frac{1}{(\omega-\omega_0-\mathrm{Im}\{G_1\})^2+(\mathrm{Re}\{G_1\})^2} \,. \tag{7.2.30}$$

For width γ and shift Δ, we have

$$\gamma = -2\mathrm{Re}\{G_1\}, \quad \Delta = \mathrm{Im}\{G_1\} \,. \tag{7.2.31}$$

Although the possible values of the numbers s are $s = j_1 + j_2, j_1 + j_2 - 1, \ldots .$ $|j_1 - j_2|$, the broadening is completely determined by the density-matrix element $\rho_{1\sigma}$ and the width and shift of a line are expressed in terms of the real and imaginary parts of G_1.

As shown in Sect. 7.2.1 in the general cases of a line corresponding to the set of transitions α — β between the two groups of closely spaced levels, the spectrum $I(\omega)$ is not described by Lorentzian distribution. Only in the wings of a line is the intensity distribution determined by the simple formula (7.2.21).

Nevertheless in the case of a line formed by a set of transitions between degenerate states m_1 and m_2 belonging to the levels j_1 and j_2, (7.2.16) for $I(\omega)$ gives the Lorentzian distribution. In order to determine the width γ and shift Δ of this distribution, it is necessary to calculate only one quantity G_1. Therefore the broadening of a line corresponding to dipole radiative transitions is completely determined by one of the equations (7.2.29), namely by the equation corresponding to the irreducible representation $s = 1$. Repeating the derivation of (7.2.30) for quadrupole radiative transitions, it is not difficult to show that in this case the broadening is determined by the density-matrix element $\rho_{s\sigma}$ with $s = 2$, the width and shift being $\gamma = -2\mathrm{Re}\{G_2\}$, $\Delta = \mathrm{Im}\{G_2\}$.

7.2.3 Quantum Theory

It has been shown above that in cases when the relative motion of an atom and perturbing particles can be described in the framework of classical mechanics, the theory of the broadening of spectral lines is a natural generalization of the classical oscillator model. We shall consider now the quantum theory of pressure effects in which not only the motion of the atomic electrons but also the relative motion of the atom and perturbing particles is described by the Schrödinger equation. Such a quantum theory enables one to take into consideration the broadening caused by light perturbing particles, especially by electrons. Let us begin with a relatively simplified problem, assuming that the mass of the atom is large as compared with the mass of the perturbing particle, so that the atom is at rest. We also assume that only elastic scattering take place.

We shall define the density matrix of the atom using the general rule [7.31]:

$$\rho(\boldsymbol{r}, \boldsymbol{r}') = \int \Psi^*(\boldsymbol{r}, \boldsymbol{R})\, \Psi(\boldsymbol{r}', \boldsymbol{R})\, d\boldsymbol{R}\,, \tag{7.2.32}$$

where $\boldsymbol{R}$ is the coordinate of the perturbing particle. The stationary states of the system consisting of the atom and perturbing particle are described by the wave functions

$$\Psi_{\alpha k}(\boldsymbol{r}, \boldsymbol{R}) = \Phi_\alpha(\boldsymbol{r}) \frac{1}{\sqrt{V}} \left[\exp(\mathrm{i}\boldsymbol{k}\cdot\boldsymbol{R}) + f_\alpha(\vartheta)\frac{\exp(\mathrm{i}kR)}{R}\right], \tag{7.2.33}$$

where $\Phi_\alpha(\boldsymbol{r})$ is the atomic wavefunction, $f_\alpha(\vartheta)$ is the scattering amplitude, $\boldsymbol{k}$ is the wave vector of the perturbing particle, and V is the normalization volume.

Let us expand the wave functions on the right-hand side of (7.2.32) in terms of the wave functions (7.2.33):

$$\Psi(\boldsymbol{r}, \boldsymbol{R}) = \sum_\alpha a_\alpha \Psi_{\alpha k}(\boldsymbol{r}, \boldsymbol{R})\,.$$

We shall expand the density matrix $\rho(\boldsymbol{r}, \boldsymbol{r}')$ in a sum over atomic stationary states

$$\rho(\boldsymbol{r}, \boldsymbol{r}') = \sum_{\alpha\beta} \rho_{\alpha\beta} \Phi_\alpha(\boldsymbol{r}) \Phi_\beta(\boldsymbol{r}')\,. \tag{7.2.34}$$

As a result, we obtain

$$\begin{aligned}\rho_{\alpha\beta} &= a_\alpha^* a_\beta \frac{1}{V} \int_V \left[\exp(-\mathrm{i}\boldsymbol{k}\cdot\boldsymbol{R}) + f_\alpha^* \frac{\exp(-\mathrm{i}kR)}{R}\right]\left[\exp(\mathrm{i}\boldsymbol{k}\cdot\boldsymbol{R})\right.\\ &\quad \left. + f_\beta \frac{\exp(\mathrm{i}kR)}{R}\right] d\boldsymbol{R} = a_\alpha^* a_\beta \left\{1 + \frac{1}{V}\int\left[f_\beta \frac{\exp(-\mathrm{i}\boldsymbol{k}\cdot\boldsymbol{R} + \mathrm{i}kR)}{R}\right.\right.\\ &\quad \left.\left. + f_\alpha^* \frac{\exp(\mathrm{i}\boldsymbol{k}\cdot\boldsymbol{R} - \mathrm{i}kR)}{R} + \frac{f_\alpha^* f_\beta}{R^2}\right] d\boldsymbol{R}\right\}\end{aligned} \tag{7.2.35}$$

The exponential factor $\exp\,(-\mathrm{i}\boldsymbol{k}\cdot\boldsymbol{R} + \mathrm{i}kR) = \exp(-\mathrm{i}kR\cos\vartheta + \mathrm{i}kR)$ rapidly oscillates everywhere with the exception of the small region $\cos\vartheta \simeq 1$, $\vartheta = 0$. Therefore in calculating the first two integrals in (7.2.35), we can use the approximation

$$\begin{aligned}\frac{1}{V}\int \exp(-\mathrm{i}\boldsymbol{k}\cdot\boldsymbol{R} + \mathrm{i}kR) f(\vartheta)\frac{d\boldsymbol{R}}{R} &\simeq \frac{f(0)}{V}\int \exp(-\mathrm{i}\boldsymbol{k}\cdot\boldsymbol{R} + \mathrm{i}kR)\frac{d\boldsymbol{R}}{R}\\ &= f(0)\frac{2\pi\mathrm{i}}{k}\frac{R}{V}\,,\end{aligned}$$

where R is the radius of the spherical volume V and $f(0)$ is the amplitude of forward scattering. Thus

$$\rho_{\alpha\beta} = a_\alpha^* a_\beta \left\{1 - \frac{2\pi\mathrm{i}}{k}\frac{R}{V}[f_\alpha^*(0) - f_\beta(0)] + \frac{R}{V}\int f_\alpha^*(\vartheta) f_\beta(\vartheta)\, dO\right\}. \tag{7.2.36}$$

If one neglects interaction and assumes that in (7.2.33) and (7.2.35) $f_\alpha = f_\beta = 0$, then $\rho_{\alpha\beta} = a_\alpha^* a_\beta$. Therefore the last two terms in (7.2.36) give the collision contribution to the density matrix:

$$\Delta\rho_{\alpha\beta} = \rho_{\alpha\beta}\frac{R}{V}\left\{-\frac{2\pi i}{k}[f_\alpha^*(0) - f_\beta(0)] + \int f_\alpha^* f_\beta\, dO\right\} . \tag{7.2.37}$$

This expression is the quantum generalization of the quasi-classical expression (7.2.13).

In order to obtain the collisional term $(d\rho/dt)_{\mathrm{coll.}}$ in the general equation for the density matrix in the impact approximation (7.2.11), it is necessary to take into consideration that the wave functions (7.2.33) are normalized in such a way that volume V contains only one perturbing particle and the time during which one collision occurs is R/v, where $v = \hbar k/m$ is the velocity and m is the mass of the perturbing particle.

After multiplying $\Delta\rho_{\alpha\beta}$ by NVv/R and averaging over velocities v, we have

$$\left(\frac{d\rho_{\alpha\beta}}{dt}\right)_{\mathrm{coll.}} = -N\langle v(\sigma' - i\sigma'')\rangle\, \rho_{\alpha\beta} , \tag{7.2.38}$$

$$\sigma' - i\sigma'' = i\frac{2\pi}{k}[f_\alpha^*(0) - f_\beta(0)] - \int f_\alpha^* f_\beta\, dO . \tag{7.2.39}$$

We shall express the scattering amplitudes f_α and f_β in the form of partial-wave expansions

$$f = \frac{1}{2ik}\sum_l (2l+1)[S^l - 1]\, P_l(\cos\vartheta) , \quad S^l = \exp(i2\eta_l) , \tag{7.2.40}$$

$$\sigma' - i\sigma'' = \frac{\pi}{k^2}\sum(2l+1)(1 - S_\alpha^{l*}S_\beta^l) , \tag{7.2.41}$$

$$\sigma' = \frac{\pi}{k^2}\sum(2l+1)[1 - \cos 2(\eta_\alpha^l - \eta_\beta^l)] , \tag{7.2.42}$$

$$\sigma'' = -\frac{\pi}{k^2}\sum(2l+1)\sin 2(\eta_\alpha^l - \eta_\beta^l) . \tag{7.2.43}$$

Thus the width and shift cross sections σ' and σ'' are expressed in the terms of scattering amplitudes or scattering phases. [Equations (7.2.39, 42, and 43) were obtained by a somewhat different way in [7.32, 33].

Expressions (7.2.41–43) establish the connection between the broadening of lines and the elastic scattering of perturber. In those cases when the perturbation of one of the levels can be neglected, $\sigma' = \sigma/2$, where σ is the elastic cross section (let us recall that $\gamma = 2Nv\sigma'$).

In the quasi-classical limit,

$$\hbar l \to mv\rho, \frac{\pi}{k^2}\sum(2l+1) \to 2\pi\int \rho\, d\rho, \quad 2(\eta_\alpha^l - \eta_\beta^l) \to \eta(\rho) ,$$

(7.2.42, 43) coincide with (7.1.13, 14). Equation (7.1.21) for $\eta(\rho)$ is the limiting expression of the general quasi-classical formula for the phase $2(\eta_\alpha^l - \eta_\beta^l)$, which is valid (in the case of the field $\hbar C_n R^{-n}$) providing

$$\frac{mC_n k^{n-2}}{\hbar} \gg 1 .$$

This condition can be rewritten in the form

$$\rho_0 \gg \lambda = \hbar/mv ,$$

where ρ_0 is the Weisskopf radius (7.1.24). For heavy perturbing particles, this condition is always fulfilled. But in the case of light perturbing particles, in particular for electrons, it is violated.

We shall consider the dependence of σ' on the perturber velocity v at high v, i.e., in the Born approximation. The cross section σ' of (7.2.42) can be written in the form

$$\sigma' = \frac{1}{2} \int |f_\alpha(\vartheta) - f_\beta(\vartheta)|^2 dO . \tag{7.2.44}$$

For electron scattering in the Born approximation,

$$f(\vartheta) = \frac{2me^2}{\hbar^2 q^2} [\mathscr{Z} - F(q)] , \tag{7.2.45}$$

where $\mathscr{Z}$ is the total number of electrons in the atom, $F(q)$ is the atomic form factor (scattering factor) and $q = 2k\hbar^{-1} \sin \vartheta/2$. Substituting this expression for $f(\vartheta)$ in (7.2.45), we obtain

$$\sigma' = \frac{4\pi m^2 e^4}{\hbar^4 k^2} \int_0^{2k} \frac{[F_\alpha(q) - F_\beta(q)]^2}{q^4} q dq . \tag{7.2.46}$$

At high velocities, the scattering amplitude has a very sharp maximum in the region of small angles. This means that the integral (7.2.46) does not depend on the upper limit ($q = 2k$, when $\vartheta = \pi/2$). Thus integration in (7.2.46) can be extended to $q = \infty$. After this the integral in (7.2.46) no longer depends on k and

$$\sigma' \propto k^{-2} \propto v^{-2}, \quad \gamma \propto v^{-1} .$$

Thus at high velocities, the width of a line is inversely proportional to the velocity for any type of interaction. It is necessary only that the integral (7.2.46) converges.

The formulas obtained above are easy to generalize so that they also include inelastic collisions. By repeating the derivation of the expression for the collisional term $(d\rho/dt)_{\mathrm{coll}}$, it is not difficult to show that (7.2.39) and (7.2.41) preserve their form. But the quantities S^l are now $S^l = \exp(-2\Gamma^l - \mathrm{i}2\eta^l)$. Therefore

$$\sigma' = \frac{\pi}{k^2} \sum (2l+1)\{1 - \exp[-2(\Gamma_\alpha^l + \Gamma_\beta^l)] \cos 2(\eta_\alpha^l - \eta_\beta^l)\} , \tag{7.2.47}$$

$$\sigma'' = -\frac{\pi}{k^2} \sum (2l+1) \exp[-2(\Gamma_\alpha^l + \Gamma_\beta^l)] \sin 2(\eta_\alpha^l - \eta_\beta^l) . \tag{7.2.48}$$

If the perturbation of one of the states can be neglected, then

$$\sigma' = \frac{\pi}{k^2}\sum(2l+1)\{1-\exp(-2\Gamma^l)\cos 2\eta^l\}$$
$$= \frac{1}{2}(\sigma_{\text{el.}} + \sigma_{\text{inel.}}) \,, \tag{7.2.49}$$

where $\sigma_{\text{el.}}$ and $\sigma_{\text{inel.}}$ are, respectively, the cross sections of elastic and inelastic scattering.

7.2.4 Quantum Kinetic Equation Method

In this section we shall generalize the density-matrix equation discussed above taking into consideration not only the motion of the perturbing particle but also the motion of the atom. Such a generalization is necessary for understanding the very important problems connected with the statistical dependence of the impact and Doppler broadening [7.20, 34–38]. In many applications of the theory of spectral line broadening such as plasma diagnostics, effects of the statistical dependence of the impact and Doppler broadening do not play a significant role. They are of special interest for nonlinear laser spectroscopy [7.39]. The density-matrix equation describing this general case has the same structure as (7.2.11),

$$\frac{d\rho}{dt} = \frac{\mathrm{i}}{\hbar}(H_0\rho - \rho H_0) + \left(\frac{d\rho}{dt}\right)_{\text{coll.}} \,. \tag{7.2.50}$$

But now the unperturbed wave functions must be taken in the form

$$\Phi_{m\boldsymbol{p}} = \Phi_m(\boldsymbol{r})\exp(\mathrm{i}\boldsymbol{p}\cdot\boldsymbol{R}) \,, \tag{7.2.51}$$

where $\boldsymbol{R}$ is the coordinate of the center of mass of the atom, m is the quantum number describing the atomic state, $\boldsymbol{p}$ is the wave vector of the atom as a whole, and $\hbar\boldsymbol{p}$ is the momentum of the atom. In (7.2.16) for the intensity distribution $I(\omega)$ in the case of an isolated spectral line $m \rightarrow n$, the following substitutions must be made:

$$\alpha \rightarrow m\boldsymbol{p}_0,\ \beta \rightarrow n\boldsymbol{\kappa}_0,\ \alpha' \rightarrow m\boldsymbol{p},\ \beta' \rightarrow n\boldsymbol{\kappa},\ W_\alpha = W(p_0),\ \boldsymbol{P} \rightarrow \boldsymbol{d}\exp(\mathrm{i}\boldsymbol{k}\cdot\boldsymbol{R}) \,,$$

where $\boldsymbol{k}$ is the wave vector of the photon, and $\boldsymbol{d}$ is the electric dipole moment. In addition, the sum over $\alpha\beta\alpha'\beta'$ must be replaced by an integral over $\boldsymbol{p}_0$, $\boldsymbol{p}$, κ_0, κ. As

$$\langle m\boldsymbol{p}|d\exp(\mathrm{i}\boldsymbol{k}\cdot\boldsymbol{R})|n\boldsymbol{\kappa}\rangle \propto d_{mn}\delta(-\boldsymbol{p}+\boldsymbol{k}+\boldsymbol{\kappa}) \,,$$

$$\langle n\boldsymbol{\kappa}_0|d\exp(-\mathrm{i}\boldsymbol{k}\cdot\boldsymbol{R})|m\boldsymbol{p}_0\rangle \propto d^*_{mn}\delta(-\boldsymbol{\kappa}_0-\boldsymbol{k}+\boldsymbol{p}_0) \,,$$

it is not difficult to obtain

$$I(\omega) = \frac{|d_{mn}|^2}{\pi}\mathrm{Re}\{\int d\boldsymbol{p}_0\, d\boldsymbol{p}\, W(\boldsymbol{p}_0)\, \rho^{m\boldsymbol{p}_0,\, n(\boldsymbol{p}_0-\boldsymbol{k})}_{m\boldsymbol{p},\, n(\boldsymbol{p}-\boldsymbol{k})}\}$$
$$= \frac{|d_{mn}|^2}{\pi}\mathrm{Re}\{\int d\boldsymbol{p}\, F(\boldsymbol{p}, \boldsymbol{p}-\boldsymbol{k})\}\,, \tag{7.2.52}$$

where $F(\boldsymbol{p}, \boldsymbol{p}-\boldsymbol{k}) = \int d\boldsymbol{p}_0\, W(\boldsymbol{p}_0)\, \rho^{m\boldsymbol{p}_0,\, n(\boldsymbol{p}-\boldsymbol{k})}_{m\boldsymbol{p},\, n(\boldsymbol{p}-\boldsymbol{k})}$ is the corresponding density matrix element, averaged over the initial distribution of $\boldsymbol{p}_0$. We shall omit the factor $|d_{mn}|^2$ everywhere below. Let us introduce the notation $(d\rho/dt)_{\mathrm{coll.}} = \hat{G}\rho$, where $\hat{G}$ is the operator describing the evolution of ρ due to collisions. From (7.2.50), we obtain

$$-W(\boldsymbol{p}) + \mathrm{i}\left(\omega - \frac{\hbar}{m}\boldsymbol{p}\cdot\boldsymbol{k}\right)F(\boldsymbol{p}, \boldsymbol{p}-\boldsymbol{k}) = (\hat{G}f)_{\boldsymbol{p},\boldsymbol{p}-\boldsymbol{k}}\,, \tag{7.2.53}$$

where m is the mass of an atom[1].

Equations (7.2.52, 53) have the same form as (7.1.54) and (7.1.56). Now, however, it is possible to express the collisional term $(\hat{G}f)$ in terms of the exact scattering amplitudes[2]. The general expression of $(\hat{G}f)$ is very complicated due to the fact that density matrix is defined in the laboratory system of coordinates and the scattering amplitudes are defined in the system of the center of mass of the colliding particles. We shall give below the expressions for $(\hat{G}F)$ in two limiting cases of light and heavy perturbing particles, $m_{\mathrm{p}} \ll m$ and $m_{\mathrm{p}} \gg m$, where m_{p} is the mass of the perturbing particle, and m is the mass of the atom.

In the first case, in the limit $m_{\mathrm{p}}/m \to 0$,

$$(\hat{G}F)_{\boldsymbol{p},\boldsymbol{p}-\boldsymbol{k}} = -\left(\frac{\gamma}{2} - \mathrm{i}\Delta\right)F_{\boldsymbol{p},\boldsymbol{p}-\boldsymbol{k}}\,,$$
$$\gamma/2 - \mathrm{i}\Delta = N\langle v_{\mathrm{p}}(\sigma' - \mathrm{i}\sigma'')\rangle\,, \tag{7.2.54}$$

where σ' and σ'' are respectively the width and shift cross sections, determined by (7.2.39–43), and v_{p} is the velocity of the perturbing particle. Substituting (7.2.54) in (7.2.53), it is not difficult to show that the spectrum $I(\omega)$ has the form of a convolution of the Doppler and Lorentzian distributions:

$$I(\omega) = \frac{1}{\pi}\mathrm{Re}\left\{\int \frac{W(\boldsymbol{p})\, d\boldsymbol{p}}{\mathrm{i}(\omega - \Delta - \hbar\boldsymbol{p}\cdot\boldsymbol{k}/m) + \gamma/2}\right\}\,, \quad \frac{\hbar\boldsymbol{p}\cdot\boldsymbol{k}}{m} = \boldsymbol{v}\cdot\boldsymbol{k}\,, \tag{7.2.55}$$

where $\boldsymbol{v}$ is the velocity of an atom.

---

[1] A term proportional to k^2 is omitted in this equation: $(H_0\rho - \rho H_0)_{m\boldsymbol{p},n\boldsymbol{p}-\boldsymbol{k}} = \rho_{m\boldsymbol{p},n\boldsymbol{p}-\boldsymbol{k}}$ $\left\{E_m - E_n + \frac{\hbar^2}{2m}[p^2 - (\boldsymbol{p}-\boldsymbol{k})^2]\right\} \simeq \rho_{m\boldsymbol{p},n\boldsymbol{p}-\boldsymbol{k}}\left(\hbar\omega_0 + \frac{\hbar^2}{m}\boldsymbol{p}\cdot\boldsymbol{k}\right)$.

[2] The general expression for $(\hat{G}f)$ is given in [7.37].

In the opposite limiting case of light atom and heavy perturbing particle $m_p \gg m, v_p \ll v$,

$$(\hat{G}F)_{p,p-k} = -N\frac{\hbar p}{m}[\sigma'(p) - i\sigma''(p)]\,F(\mathbf{p},\ \mathbf{p}-\mathbf{k})$$
$$-N\frac{\hbar p}{m}\tilde{\sigma}(p)\,F(\mathbf{p},\ \mathbf{p}-\mathbf{k}) + N\frac{\hbar p}{m}\int dO_q f_m^*(\mathbf{q},\ \mathbf{p}) f_n(\mathbf{q},\ \mathbf{p})\,F(\mathbf{q},\ \mathbf{q}-\mathbf{k})\,, \tag{7.2.56}$$

where σ' and σ'' are, as before, the width and shift cross sections, but now they are dependent on the wave vector p of the atom:

$$\tilde{\sigma}(p) = \int dO_q f_m^*(\mathbf{q},\ \mathbf{p}) f_n(\mathbf{q},\ \mathbf{p}) = \frac{\pi}{p^2}\sum(2l+1)(S_n^l - 1)(S_m^l - 1)^*\,. \tag{7.2.57}$$

The equation (7.2.57) coincides with the expression defining the elastic cross section σ if $f_m^* f_n$ is replaced by $|f_n|^2$ or $|f_m|^2$. In the collisional term (7.2.56), the cross section $\tilde{\sigma}$ plays the role of the elastic cross section σ. The cross section $\tilde{\sigma}$ differs from σ, being in the general case complex.

If the scattering is isotropic, or almost isotropic, then $\tilde{\sigma}(p) \simeq 4\pi\overline{f_m^* f_n}$. Taking this into consideration we obtain from (7.2.56)

$$\begin{aligned}(\hat{G}F)_{p,p-k} &= -Nv[\sigma'(v) - i\sigma''(v)]\,F(\mathbf{p},\ \mathbf{p}-\mathbf{k}) \\ &\quad - Nv\tilde{\sigma}(v)\,F(\mathbf{p},\ \mathbf{p}-\mathbf{k}) + Nv\tilde{\sigma}(v)\int\frac{dO_q}{4\pi}F(\mathbf{q},\ \mathbf{q}-\mathbf{k})\,.\end{aligned} \tag{7.2.58}$$

Equation (7.2.53) takes the form [cf.(7.1.67)]

$$\frac{W(\mathbf{p})}{i(\omega - \mathbf{k}\cdot\mathbf{v}) + Nv(\sigma' - i\sigma'' + \tilde{\sigma})} = F(\mathbf{p},\ \mathbf{p}-\mathbf{k})$$
$$-\frac{Nv\tilde{\sigma}}{i(\omega - \mathbf{k}\cdot\mathbf{v}) + Nv(\sigma' - i\sigma'' + \tilde{\sigma})}\int\frac{dO_q}{4\pi}F(\mathbf{q},\ \mathbf{q}-\mathbf{k})\,.$$

Integrating the left- and right-hand side of this equation over dO_p and taking into consideration that for isotropic distribution

$$W(\mathbf{p}) = \frac{w(p)}{4\pi p^2},\quad w(p)\,dp = w(v)\,dv,\quad dO_p = dO_v\,,$$

it is easy to find $\int F(\mathbf{p},\ \mathbf{p}-\mathbf{k})dO_p$ and then using (7.2.52) to obtain the spectrum

$$I(\omega) = \frac{1}{\pi}\mathrm{Re}\left\{\int w(v)\,dv \frac{\displaystyle\int\frac{dO_v}{4\pi}\frac{1}{i(\omega - \mathbf{k}\cdot\mathbf{v}) + Nv(\sigma' - i\sigma'' + \tilde{\sigma})}}{\displaystyle 1 - Nv\tilde{\sigma}\int\frac{dO_v}{4\pi}\frac{1}{i(\omega - \mathbf{k}\cdot\mathbf{v}) + Nv(\sigma' - i\sigma'' + \tilde{\sigma})}}\right\}\,. \tag{7.2.59}$$

This equation differs from (7.1.68), taking account not only of the perturbation of the internal atomic state, but also of the perturbation of the motion of the atom as a whole.

All the cross sections contained in (7.2.59), σ', σ'', and $\tilde{\sigma}$, are dependent on v. When $N \to 0$, (7.2.59) coincides with the usual Doppler distribution (7.1.57). At high densities when one can neglect the term $\boldsymbol{k} \cdot \boldsymbol{v}$, (7.2.59) takes the form

$$I(\omega) = \frac{1}{\pi} \int w(v)\, dv \frac{Nv\sigma'}{(\omega - Nv\sigma'')^2 + (Nv\sigma')^2} \,. \tag{7.2.60}$$

Instead of the usual Lorentzian distribution, (7.2.60) describes a superposition of Lorentzian distributions, each having width $2Nv\sigma'$, shift $Nv\sigma''$, and relative weight $w(v)$. If one assumes that the scattering in one of the states m and n can be neglected, when $\tilde{\sigma} = 0$, then (7.2.59) takes the form

$$I(\omega) = \frac{1}{\pi} \mathrm{Re} \left\{ \int \frac{W(\boldsymbol{v})\, d\boldsymbol{v}}{\mathrm{i}(\omega - \boldsymbol{k} \cdot \boldsymbol{v}) + Nv\sigma' - \mathrm{i}Nv\sigma''} \right\} \,. \tag{7.2.61}$$

This expression differs from the usual convolution of the Doppler and Lorentzian distributions containing $2Nv\sigma'(v)$ and $Nv\sigma''(v)$ instead of the averaged quantities $\gamma = 2N\langle v\sigma' \rangle$ and $\Delta = N\langle v\sigma'' \rangle$. The intensity distributions of (7.2.60, 61) are asymmetric.

In the case $\sigma' = \sigma'' = 0$ (the scattering is purely elastic and $f_m = f_n$), (7.2.59) describes a narrowing of the Doppler profile similar to that given by (7.1.68). The parameter $Nv\tilde{\sigma}$ plays the same role as the frequency of collisions in (7.1.68).

We shall consider now the case of high densities ($\sigma' = \sigma'' = 0$, $Nv\tilde{\sigma} \gg \boldsymbol{k} \cdot \boldsymbol{v}$):

$$\int \frac{dO_v}{4\pi} \frac{1}{\mathrm{i}(\omega - \boldsymbol{k} \cdot \boldsymbol{v}) + Nv\tilde{\sigma}} \simeq \frac{1}{\mathrm{i}\omega + Nv\tilde{\sigma}} - \frac{k^2 v^2}{3} \frac{1}{(\mathrm{i}\omega + Nv\tilde{\sigma})^3} \,.$$

For frequencies $\omega \ll Nv\tilde{\sigma}$ (7.2.59) gives

$$I(\omega) = \frac{1}{\pi} \mathrm{Re} \left\{ \int w(v)\, dv \frac{1}{\mathrm{i}\omega + \frac{1}{3} k^2 v^2 (Nv\tilde{\sigma})^{-1}} \right\} \,.$$

We shall introduce the notation $\tilde{D} = v/3N\tilde{\sigma}$. Then

$$I(\omega) = \int w(v)\, dv \frac{k^2 \mathrm{Re}\{\tilde{D}\}}{\pi} \cdot \frac{1}{(\omega + k^2 \mathrm{Im}\{\tilde{D}\})^2 + (k^2 \mathrm{Re}\{\tilde{D}\})^2} \,. \tag{7.2.62}$$

If $f_m = f_n$, and $\tilde{\sigma}$ coincides with the elastic cross section σ, then $\mathrm{Im}\{\tilde{D}\} = 0$ and $\mathrm{Re}\{\tilde{D}\} = \tilde{D}$ is the diffusion coefficient depending on v instead of $\langle v \rangle$. In the general case of complex $\tilde{\sigma}$ and $\tilde{D}$, the real part of $\tilde{D}$ determines the width and the imaginary part of $\tilde{D}$ determines the shift. The width $2k^2\mathrm{Re}\{\tilde{D}\}$ is proportional to N^{-1}. Thus, at $Nv\tilde{\sigma} \gg \boldsymbol{k} \cdot \boldsymbol{v}$, the Doppler distribution narrows due to collisions. The resulting line profile has the form of a superposition of Lorentzian distributions with widths $2k^2\mathrm{Re}\{\tilde{D}\}$ and shifts $k^2\mathrm{Im}\{\tilde{D}\}$. This profile is asymmetric.

If σ' and σ'' are less than $\tilde{\sigma}$ but are not equal to zero, then on increasing N, there is first a narrowing of the Doppler contour to a width $\sim Nv\sigma'$, and then a broadening.

The results of the calculation of the spectrum $I(\omega)$ by means of the quantum kinetic equation contain a number of new elements, the most interesting of which are the following. Even in the treatment of the simplest example, purely impact broadening, qualitative differences from the formulas usually used arise. Only in the case of broadening by light particles, such as electrons, does a single Lorentzian contour arise, with width $2N\langle v_p\sigma'\rangle$ and shift $N\langle v_p\sigma''\rangle$, where v_p is the velocity of the electrons and the angle brackets denote averaging over v_p. In the general case of $m_p \sim m$, after averaging over the velocities of the perturbing particles v_p, the cross sections $\sigma'(v)$ and $\sigma''(v)$ contained in the collisional term retain their dependence on the velocity v of the atom. As a result, the following intensity distribution arises:

$$I(\omega) = \int w(v)\, dv \frac{Nv\sigma'(v)}{\pi} \frac{1}{[\omega - Nv\sigma''(v)]^2 + [Nv\sigma'(v)]^2} . \tag{7.2.63}$$

This distribution is asymmetric.

The greatest difference arises in the case of scattering of a light atom by heavy (almost at rest) perturbing particles. The perturbations due to different perturbing particles combine in completely different ways, depending on the masses of these particles. If the perturbing particles of type 1 and type 2 are light, then the sum of the corresponding widths and shifts arises:

$$\frac{\gamma}{2} - \mathrm{i}\Delta = N_1\langle v_1\sigma_1'\rangle + N_2\langle v_2\sigma_2'\rangle + \mathrm{i}\,N_1\langle v_1\sigma_1''\rangle + \mathrm{i}\,N_2\langle v_2\sigma_2''\rangle .$$

But if the perturbation is created by heavy particles (type 1) and electrons (type 2), then (at $\tilde{\sigma} = 0$)

$$I(\omega) = \frac{1}{\pi}\mathrm{Re}\left\{\int \frac{w(v)\,dv}{\mathrm{i}(\omega - \boldsymbol{k}\cdot\boldsymbol{v}) + N_1 v_1\sigma_1'(v) - \mathrm{i}N_1 v_1\sigma_1''(v) + N_2\langle v_2\sigma_2'\rangle - \mathrm{i}N_2\langle v_2\sigma_2''\rangle}\right\} . \tag{7.2.64}$$

All this is a reflection of the statistical dependence of the Doppler and impact broadenings.

The second characteristic feature is the fact that the cross section $\tilde{\sigma}$ responsible for the collisional compensation of the Doppler broadening is complex. Let us recall that $\tilde{\sigma} \neq 0$ if both the scattering amplitudes f_m and f_n are nonzero.

It is not difficult to show that in the examples treated, asymmetry arises for two reasons – the dependence on the atomic velocity of the parameters of the equation, and the fact that the cross section $\tilde{\sigma}$ is complex.

In calculation of the width of the resulting spectrum, all characteristic features of collisional broadening connected with the effect of the statistical dependence of the Doppler and impact broadenings are usually not very important. Nevertheless they can be of interest for some other problems, for example, those arising in the theory of nonlinear resonances in the spectra of gas lasers [7.38, 39] (see also the bibliography given in [7.20].

7.2.5 Absorption Spectrum

The energy absorbed in one second by a system of electric charges interacting with an electric field

$$\mathcal{E}(\boldsymbol{R},t) = \frac{1}{2}\{\mathcal{E}_0 \exp[\mathrm{i}(\omega t - \boldsymbol{k}\cdot\boldsymbol{R}) + \mathcal{E}_0^* \exp[-\mathrm{i}(\omega t - \boldsymbol{k}\cdot\boldsymbol{R})]\}$$

is (in the electric dipole approximation)

$$Q = -\mathcal{E}\cdot\dot{\boldsymbol{d}} = \mathrm{Re}\,\{\mathrm{i}\omega\mathcal{E}_0^*\cdot\boldsymbol{d}_\omega \exp(\mathrm{i}\boldsymbol{k}\cdot\boldsymbol{R})\}\ ,$$

where

$$\boldsymbol{d} = \boldsymbol{d}_\omega(\boldsymbol{R})\exp(\mathrm{i}\omega t) + \boldsymbol{d}_\omega^*(\boldsymbol{R})\exp(-\mathrm{i}\omega t)$$

is the electric dipole moment induced by the field $\mathcal{E}(\boldsymbol{R},t)$. The quantum generalization of this expression for Q has the form

$$Q = \mathrm{Re}\,[\mathrm{i}\omega\,\mathcal{E}_0^*\,\mathrm{Trace}\,\{\boldsymbol{d}\exp(\mathrm{i}\boldsymbol{k}\cdot\boldsymbol{R})\,\rho(\omega)\}]\ . \tag{7.2.65}$$

Here $\boldsymbol{d}$ is the electric dipole operator, $\boldsymbol{R}$ is the coordinate of the center of mass of the atom, $\boldsymbol{k}$ is the wave vector of the photon, and $\rho(\omega)$ is the Fourier component of the density matrix, satisfying the equation

$$\frac{d\rho}{dt} - \frac{\mathrm{i}}{\hbar}(H_0\rho - \rho H_0) - \hat{G}\rho = \frac{\mathrm{i}}{\hbar}(\mathcal{E}\cdot\boldsymbol{d}\rho - \rho\mathcal{E}\cdot\boldsymbol{d})\ . \tag{7.2.66}$$

This equation contains an additional term describing the interaction of the atom with the electric field. Solving this equation by the method of successive approximation, it is possible by means of (7.2.65) to determine the absorption (or emission) power Q.

If a set of transitions $\alpha \to \beta$ are perturbed by the field $\mathcal{E}$ (the frequencies $\omega_{\alpha\beta} \simeq \omega$), then using (7.2.65) we obtain

$$Q = \mathrm{Re}\,\{\mathrm{i}\omega\,\mathcal{E}_0^*\sum_{\alpha\beta}\boldsymbol{d}_{\alpha\beta}\int d\boldsymbol{p}\,\rho_{\beta\boldsymbol{p},\alpha\boldsymbol{p}+\boldsymbol{k}}(\omega)\}\ , \tag{7.2.67}$$

where $\boldsymbol{p}$ is the momentum of the atom.

We shall neglect the collisional term simplicity and solve (7.2.66) in the linear approximation according to the field. We shall substitute in the right-hand side of (7.2.66) the zeroth-order density matrix, which is diagonal in indices α, β and

$\boldsymbol{p},\boldsymbol{p}'$:

$$\rho^{(0)}_{\beta\boldsymbol{p},\beta\boldsymbol{p}} = N_\beta W(\boldsymbol{p}), \quad \rho^{(0)}_{\alpha\boldsymbol{p}+\boldsymbol{k},\alpha\boldsymbol{p}+\boldsymbol{k}} = N_\alpha W(\boldsymbol{p}+\boldsymbol{k}) \simeq N_\alpha W(\boldsymbol{p}),$$

where N_β and N_α are respectively the populations of the states β and α, and $W(\boldsymbol{p})$ is the distribution function for momentum $\boldsymbol{p}$. Since $\boldsymbol{k} \ll \boldsymbol{p}$, we can assume that $W(\boldsymbol{p}+\boldsymbol{k}) \simeq W(\boldsymbol{p})$.

In this approximation we have [cf. (7.1.57)]

$$\rho_{\beta\boldsymbol{p},\alpha\boldsymbol{p}+\boldsymbol{k}} = \frac{\mathrm{i}}{2\hbar}\mathcal{E}_0\cdot\boldsymbol{d}_{\beta\alpha}(N_\beta - N_\alpha)\,W(\boldsymbol{p})\frac{1}{\mathrm{i}(\omega-\omega_0+\hbar\boldsymbol{p}\cdot\boldsymbol{k}/m)},$$

$$Q = \frac{\pi}{2}\frac{\omega}{\hbar}\sum_{\alpha\beta}(N_\beta - N_\alpha)|\mathcal{E}_0\cdot\boldsymbol{d}_{\alpha\beta}|^2 I(\omega),$$

$$I(\omega) = \frac{1}{\pi}\mathrm{Re}\left\{\lim_{\varepsilon\to 0}\int\frac{W(\boldsymbol{p})d\boldsymbol{p}}{\mathrm{i}(\omega-\omega_0-\hbar\boldsymbol{p}\cdot\boldsymbol{k}/m)+\varepsilon}\right\}.$$

Therefore the absorbed power Q is proportional to the difference of the populations $(N_\beta - N_\alpha)$ and the function $I(\omega)$, which describes the usual Doppler distribution in the spectrum of spontaneous emission. In the general case, when (7.2.66) contains the collisional term, the expression for Q remains the same, but the function $I(\omega)$ has a more complex form describing the Doppler and impact broadenings. All results obtained above for the spectrum of spontaneous emission can be obtained also from (7.2.65, 66). By solving (7.2.65) in the next approximations according to the field, it is possible to calculate the power of nonlinear absorption [7.38].

7.2.6 Interference Effects: Narrowing of Spectral Lines

In cases when the frequencies of some atomic transitions coincide or are so closely spaced that the corresponding spectral lines overlap, specific interference effects can arise [7.40]. In some particular cases, these interference effects are so important that they alter the entire picture of the broadening. We shall illustrate this by considering as an example the four-level system shown in Fig. 7.2. We shall assume that the transition frequencies $\omega_{kl} = \omega_1$ and $\omega_{mn} = \omega_2$ are almost

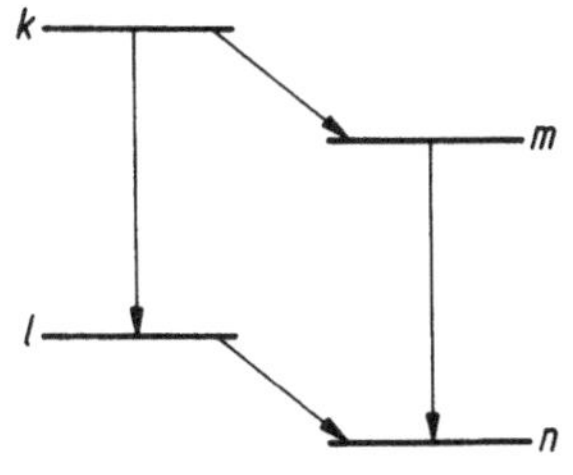

Fig. 7.2. Levels scheme and the radiative transitions which are considered

the same, $\omega_1 \simeq \omega_2$, but that all other transition frequencies differ very much from ω_1 and ω_2. Therefore in (7.2.16) and (7.2.15), the indices $\alpha, \alpha', \alpha''$ stand for k, m and the indices β, β', β'' stand for l, n. We shall denote the pair of indices k, l by 1 and the pair of indices m, n by 2. We shall use the notation $W_k = W_1$ and $W_m = W_2$. By solving the system of equations (7.2.15), we obtain

$$I(\omega) = \frac{1}{\pi}\mathrm{Re}\left\{\frac{W_1|P_1|^2 G_{22} + W_2|P_2|^2 G_{11} - W_2 P_1 P_2^* G_{12} - W_1 P_1^* P_2 G_{21}}{G_{11}G_{22} - G_{21}G_{12}}\right\}, \tag{7.2.68}$$

where

$$\begin{aligned} G_{11} &= \mathrm{i}(\omega - \omega_1) + \int (1 - S_{kk}^* S_{ll}) P(v)\, dv, \\ G_{22} &= \mathrm{i}(\omega - \omega_2) + \int (1 - S_{mm}^* S_{nn}) P(v)\, dv, \\ G_{12} &= -\int S_{mk}^* S_{nl}\, P(v)\, dv, \\ G_{21} &= -\int S_{km}^* S_{ln}\, P(v)\, dv. \end{aligned} \tag{7.2.69}$$

We shall assume that S matrix elements obey the relations

$$\begin{aligned} &S_{kk} = S_{ll}, \quad S_{mm} = S_{nn}, \quad S_{km} = S_{ln}, \quad S_{mk} = S_{nl}, \\ &S_{kl} = S_{mn} = S_{kn} = S_{ml} = 0. \end{aligned} \tag{7.2.70}$$

This means that collisions produce only the mutual perturbation of the states l and n and also the states k and m. It will be shown that such a situation can arise in a number of systems.

Using (7.2.70) and also the unitarity properties of the S matrix,

$$\sum_b |S_{ab}|^2 = 1,$$

we have

$$\begin{aligned} S_{mk}^* S_{nl} &= |S_{mk}|^2 = 1 - |S_{mm}|^2, \\ S_{km}^* S_{ln} &= |S_{km}|^2 = 1 - |S_{kk}|^2. \end{aligned} \tag{7.2.71}$$

From the general definition of the inelastic cross section σ, it follows that

$$\int |S_{km}|^2 P(v)\, dv = N\langle v\sigma_{km}\rangle, \quad \int |S_{mk}|^2 P(v)\, dv = N\langle v\sigma_{mk}\rangle, \tag{7.2.72}$$

where N is the density of the perturbing particles, v is the relative velocity, and the angle brackets denote averaging over velocities. In Boltzmann equilibrium the level populations W_1, W_2 and transition frequencies are connected by the relations

$$N\langle v\sigma_{km}\rangle = \frac{\gamma_1}{2}, \quad N\langle v\sigma_{mk}\rangle = \frac{\gamma_2}{2}, \quad W_1\gamma_1 = W_2\gamma_2. \tag{7.2.73}$$

Thus

$$\begin{aligned} G_{11} &= \mathrm{i}(\omega - \omega_1) + \gamma_1/2, \quad & G_{22} &= \mathrm{i}(\omega - \omega_2) + (W_1/W_2)\gamma_1/2, \\ G_{12} &= -(W_1/W_2)\gamma_1/2, \quad & G_{21} &= -\gamma_1/2. \end{aligned} \tag{7.2.74}$$

We shall introduce the notation

$$\omega_1 = \omega_0 - \delta, \quad \omega_2 = \omega_0 + \delta, \quad 2\delta = \Delta . \tag{7.2.75}$$

After this (7.2.68) gives

$$I(\omega) = \frac{1}{2\pi} \frac{(W_1\gamma_1 + W_2\gamma_2)|(\omega - \delta)P_1 - (\omega + \delta)P_2|^2}{[(\omega + \delta)(\omega - \delta)]^2 + [\gamma_1(\omega - \delta) + \gamma_2(\omega + \delta)]^2} . \tag{7.2.76}$$

We shall consider now the limiting cases of small and high values of γ.

In this first case, when $\gamma_{1,2}/\delta \ll 1$, the second term in the denominator of (7.2.76) is small, and the function $I(\omega)$ has two sharp maxima at $\omega = \pm\delta$. Equation (7.2.76) can be rewritten in the form

$$I(\omega) \simeq \frac{1}{\pi}\left(\frac{W_1\gamma_1|P_1|^2}{(\omega + \delta)^2 + \gamma_1^2} + \frac{W_2\gamma_2|P_2|^2}{(\omega - \delta)^2 + \gamma_2^2}\right) . \tag{7.2.77}$$

In the other limiting case, when $\gamma_{1,2}/\delta \gg 1$, (7.2.76) has one sharp maximum at

$$\omega \simeq \omega_M = \frac{\gamma_1 - \gamma_2}{\gamma_1 + \gamma_2}\delta .$$

In the vicinity of this frequency, (7.2.76) gives

$$I(\omega) \simeq \frac{(W_1\gamma_1 + W_2\gamma_2)|\gamma_2 P_1 + \gamma_1 P_2|^2}{2\gamma_1\gamma_2(\gamma_1 + \gamma_2)} \frac{1}{\pi} \frac{\Gamma}{(\omega - \omega_M)^2 + \Gamma^2} , \tag{7.2.78}$$

where

$$\Gamma = \frac{4\gamma_1\gamma_2\delta^2}{(\gamma_1 + \gamma_2)^3} .$$

Equation (7.2.78) describes a Lorentzian distribution with width $\Gamma \propto N^{-1}$.

Therefore at low pressures, when $\gamma_{1,2}/\delta \ll 1$, the two components of a line are independent and their widths γ_1 and γ_2 are proportional to N. On further increase of N when the components of the line begin to overlap, the picture of the broadening alters completely, and in the limiting case of $\gamma_{1,2}/\delta \gg 1$ a single Lorentzian distribution with width proportional to N^{-1} arises.

In the far wing of a line $\omega \gg \gamma_1, \gamma_2, \delta$, in accordance with (7.2.76), the spectrum $I(\omega)$ has the form

$$I(\omega) \simeq \frac{W_1\gamma_1 + W_2\gamma_2}{2\pi} \frac{|\omega(P_1 - P_2) - \delta(P_1 + P_2)|^2}{\omega^4} . \tag{7.2.79}$$

Note that in the particular case $P_1 = P_2$, $I(\omega) \propto \omega^{-4}$.

All qualitative features of the example considered above are connected with the conditions (7.2.70) for the S matrix. There are a number of systems for which the S matrix obeys such conditions. We shall consider two subsystems I and II with the levels A, B and a, b, respectively, shown in Fig. 7.3a. Let us assume that subsystem I is perturbed by collisions, but they do not act on subsystem II.

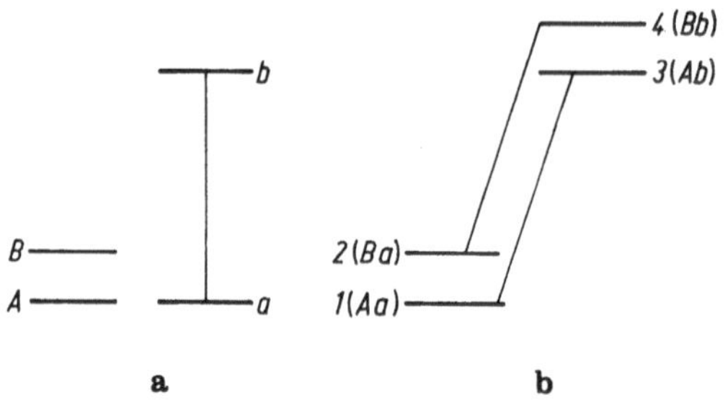

Fig. 7.3. (a) Levels scheme of subsystems I and II (b) Levels scheme of the whole system.

The system as a whole has four levels: $1(Aa)$, $2(Ba)$, $3(Ab)$, and $4(Bb)$, shown in Fig. 7.3.

Two transitions $3 \to 1$ and $4 \to 2$ correspond to the transition $a \to b$ of subsystem II. If the interaction between subsystems I and II is not great, the interaction produces a splitting of the frequencies $\omega_{31} - \omega_{42} = \Delta \neq 0$, but does not influence the probability of the collisional transition A—B in subsystem I.

In this case the S matrix has the form

$$\begin{bmatrix} S_{11} & S_{12} & S_{13} & S_{14} \\ S_{21} & S_{22} & S_{23} & S_{24} \\ S_{31} & S_{32} & S_{33} & S_{34} \\ S_{41} & S_{42} & S_{43} & S_{44} \end{bmatrix} = \begin{bmatrix} S_{AA} & S_{AB} & 0 & 0 \\ S_{BA} & S_{BB} & 0 & 0 \\ 0 & 0 & S_{AA} & S_{AB} \\ 0 & 0 & S_{BA} & S_{BB} \end{bmatrix}, \qquad (7.2.80)$$

i.e., is in full agreement with the conditions (7.2.70).

Thus the example considered above describes the rather typical situation when the relaxation processes in subsystem I perturb the spectrum of subsystem II. We assume of course that subsystems I and II interact. If this interaction, and consequently also the frequency splitting Δ, is great enough, i.e., $\gamma/\Delta \ll 1$, then relaxation processes in subsystem I produce broadening of the line, corresponding to the radiative transition is subsystem II. In the other limiting case of weak interaction ($\gamma/\Delta \to \infty$), the spectrum of subsystem II is not sensitive to the relaxation processes in subsystem I. For a given value of Δ, the inequalities $\gamma/\Delta \ll 1$ and $\gamma/\Delta \gg 1$ correspond respectively to low and high densities N. Therefore, on increase of N, there is first a broadening of the spectrum of subsystem II to a width $\sim \gamma$, and then a narrowing proportional to N^{-1}. The spectrum narrowing with increase of N is due to the interference of the amplitudes of the radiative transitions when the corresponding line components begin to overlap.

Let us consider for example the splitting of the term ^2P in a strong magnetic field H [Ref. 7.30, Sect. 8.2]:

$$\Delta E = \mu_0 H(M_L + 2M_S) + AM_LM_S\,, \qquad (7.2.81)$$

where μ_0 is the Bohr magneton, A is the fine structure constant, and M_L, M_S are the magnetic quantum numbers of the orbital and spin angular momenta. We shall consider the magnetic dipole transition $M_S = \frac{1}{2} \to M'_S = -\frac{1}{2}$, which gives three components of the line corresponding to the three possible values of $M_L = 0, \pm 1$. The frequency splitting of these components is of the order of A. The cross section of the spin reorientation σ_S usually is less than the cross section of the orbital angular momentum reorientation σ_L. Therefore one can assume

$$\gamma = N\langle v\sigma_L\rangle \ .$$

If $\gamma \ll A$, collisional reorientation of the orbital angular momentum produces broadening of the transitions $M_S = \frac{1}{2} \to M'_S = -\frac{1}{2}\,(M_L = 0, \pm 1)$. If density N is so high that $\gamma > A$, then the relaxation transitions $M_L \to M'_L$ begins to be ineffective in broadening the spectrum of the transition $M_S = \frac{1}{2} - M'_S = -\frac{1}{2}$. Moreover, on increase of N, a narrowing of the spectra must be observed. In the limiting case of $N\langle v\sigma_L\rangle \gg A$, the width of the spectrum is less than the initial splitting A. It must be noted that the condition $N\langle v\sigma_L\rangle \gg A$ can be fulfilled only for light atoms for which the fine splitting is not too great. For example, the fine splitting of the ground level of Li atom is $0.34\,\mathrm{cm}^{-1}$.

As will be shown in Sect. 7.3 the interference narrowing of a line can take place in the spectra of highly excited hydrogen atoms.

7.3 Broadening of Lines of the Hydrogen Spectrum in a Plasma

7.3.1 Preliminary Estimates

The main contribution to the broadening of lines of the hydrogen spectrum in a plasma is due to the linear Stark effect in the fields of electrons and ions. The perturbing particle with charge Ze produces the electric field $\mathcal{E} = \mathrm{Ze}\mathrm{R}^{-2}$. Using the well-known formula for the linear Stark effect [7.30] $\Delta\omega = 3/2n\,(n_1 - n_2)ea_0\mathcal{E}/\hbar$, where n, n_1, n_2 are the principal and parabolic quantum numbers, we can assume $\Delta\omega = C_2/R^2$. The constant C_2 for the level with principal quantum number n has the order of magnitude $Zn(n-1)e^2a_0/\hbar \simeq Zn(n-1)\,[\mathrm{cm}^2\mathrm{s}^{-1}]$. We shall estimate the magnitude of the dimensionless parameters (7.1.39)

$$h_\mathrm{e} \sim N_\mathrm{e}\left(\frac{n(n-1)e^2a_0}{\hbar v_\mathrm{e}}\right)^3 , \quad h_\mathrm{i} \sim N_\mathrm{i}\left(\frac{Zn(n-1)e^2a_0}{\hbar v_\mathrm{i}}\right)^3 \tag{7.3.1}$$

where v_e and v_i are the velocities of electrons and ions, respectively. The range of temperatures and densities for which $h_\mathrm{i} \gg 1$ and $h_\mathrm{e} \ll 1$ is usually of greatest interest. This means that the field of the ions is quasi-static and the electrons cause impact broadening.

7.3.2 Ion Broadening: Holtsmark Theory

For $h_i \gg 1$, number of ions in the sphere of Weisskopf radius is large and the binary approximation is inapplicable. Thus the main problem which arises in considering ion broadening is to find the quasi-static intensity distribution taking into account the simultaneous action on the atom of a large number of ions.

We shall consider the component α—β of a line and denote the shift of this component in the field $\mathscr{E}$ by [Ref. 7.30, Sect. 7.2]

$$\kappa = B_{\alpha\beta}\mathscr{E}, \quad B_{\alpha\beta} = \frac{(C_2)_{\alpha\beta}}{Ze} = \frac{3}{2}[n(n_1 - n_2) - n'(n_1' - n_2')]\frac{ea_0}{\hbar}, \tag{7.3.2}$$

where α and β are the set of parabolic quantum numbers $n_1 n_2 m$ and $n_1' n_2' m'$. The intensity distribution at a given field $\mathscr{E}$ is given by

$$I_{\alpha\beta}(\mathscr{E}, \Delta\omega)d\omega = \frac{I_{\alpha\beta}}{B_{\alpha\beta}}\delta\left(\frac{\Delta\omega - \kappa}{B_{\alpha\beta}}\right)d\omega . \tag{7.3.3}$$

Averaging this expression by means of the distribution function $W(\mathscr{E})$, we obtain

$$I_{\alpha\beta}(\omega)d\omega = \frac{I_{\alpha\beta}}{B_{\alpha\beta}}W\left(\frac{\omega - \omega_0}{B_{\alpha\beta}}\right)d\omega . \tag{7.3.4}$$

The resulting ionic field $\mathscr{E}$ is equal to the vector sum over all ions,

$$\mathscr{E} = \sum_k \mathscr{E}_k = Ze\sum_k \frac{\boldsymbol{R}_k}{R_k^3} .$$

The function $W(\mathscr{E})$ determines the probability of a given magnitude of the absolute value of $\mathscr{E}$. This function was calculated by Holtsmark in the ideal gas approximation. (A detailed discussion is given in [7.25]). In this approximation, one assumes that each of the ions can with equal probability be located at any point of the volume independently of how all the other ions are located. Therefore the function $W(\mathscr{E})$ can be calculated in the following way:

$$\begin{aligned} W_H(\mathscr{E})\,d\mathscr{E} &= \langle\delta\left(\mathscr{E} - Ze\sum_{k=1}^{NV}\frac{\boldsymbol{R}_k}{R_k^3}\right)d\mathscr{E}\rangle \\ &= \int\frac{d\boldsymbol{R}_1}{V}\int\frac{d\boldsymbol{R}_2}{V}\dots\delta\left(\mathscr{E} - Ze\sum_{k=1}^{NV}\frac{\boldsymbol{R}_k}{R_k^3}\right)d\mathscr{E} \\ &= \frac{1}{(2\pi)^3}\int\frac{d\boldsymbol{R}_1}{V}\int\frac{d\boldsymbol{R}_2}{V}\dots\int d\boldsymbol{\rho}\exp(\mathrm{i}\boldsymbol{\rho}\cdot\mathscr{E})\sum_{k=1}^{NV}\exp\left(-\mathrm{i}\frac{Ze\boldsymbol{\rho}\cdot\boldsymbol{R}_k}{R_k^3}\right)d\mathscr{E} . \end{aligned}$$

Here we use the well-known representation of the δ-function, introducing the additional integration over $\boldsymbol{\rho}$. By changing the order of integration over $\boldsymbol{R}$ and

$\boldsymbol{\rho}$, we obtain in the limit $V \to \infty$

$$W_{\mathrm{H}}(\mathscr{E}) = \frac{1}{(2\pi)^3} \int d\boldsymbol{\rho} \exp(\mathrm{i}\boldsymbol{\rho}\cdot\mathscr{E}) \left\{1 - \frac{1}{V}\int d\boldsymbol{R}_k \left[1 - \exp\left(-\mathrm{i}\frac{Ze\boldsymbol{\rho}\cdot\boldsymbol{R}_k}{R_k^3}\right)\right]\right\}^{NV}$$

$$= \frac{1}{(2\pi)^3} \int d\boldsymbol{\rho} \exp(\mathrm{i}\boldsymbol{\rho}\cdot\mathscr{E}) \exp\left\{-N\int d\boldsymbol{R}\left[1 - \exp\left(-\mathrm{i}\frac{Ze\boldsymbol{\rho}\cdot\boldsymbol{R}}{R^3}\right)\right]\right\} .$$

Integration over $\boldsymbol{R}$ gives

$$\int d\boldsymbol{R}\left[1 - \exp\left(-\mathrm{i}\frac{Ze\boldsymbol{\rho}\cdot\boldsymbol{R}}{R^3}\right)\right] = \frac{4}{15}(2\pi Ze\rho)^{3/2} .$$

Then it is possible to carry out the integration over the angular variables of the vector $\boldsymbol{\rho}$. As a result we have

$$W_H(\mathscr{E})d\mathscr{E} = \frac{d\mathscr{E}}{\mathscr{E}}\mathscr{H}\left(\frac{\mathscr{E}}{\mathscr{E}_0}\right) , \tag{7.3.5}$$

where

$$\mathscr{H}(\beta) = \frac{2}{\pi\beta}\int_0^\infty x \sin x \exp\left[-\left(\frac{x}{\beta}\right)^{3/2}\right] dx , \tag{7.3.6}$$

$$\mathscr{E}_0 = 2\pi\left(\frac{4}{15}\right)^{2/3} ZeN^{2/3} = 2.6031\, ZeN^{2/3} . \tag{7.3.7}$$

Values of the function $\mathscr{H}(\beta)$ for a wide range of values of the parameter β are given in Table 7.1. In addition the function $\mathscr{H}(\beta)$ is shown in Fig. 7.4. The maximum of the function $\mathscr{H}(\beta)$ corresponds to the point $\beta = 1.607$. In the two limiting cases, high and low values of β, the function $\mathscr{H}(\beta)$ can be approximated by the series

$$\mathscr{H}(\beta) \simeq \begin{cases} 1.496\,\beta^{-5/2}(1 + 5.107\beta^{-3/2} + 14.93\beta^{-3} + \cdots) \quad (\beta \gg 1) , & (7.3.8) \\ \frac{4}{3\pi}\beta^2(1 - 0.463\beta^2 + 0.1227\beta^4 + \cdots) \quad (\beta \ll 1) . & (7.3.9) \end{cases}$$

If in the expression for $\mathscr{H}(\beta)$, the field $\mathscr{E}_0$ is redefined by putting $\mathscr{E}_0 = ZeR_0^{-2}$, where $R_0 = (3/4\pi N)^{1/3}$, then instead of (7.3.8) we have $\mathscr{H}(\beta) \simeq 1.5\beta^{-5/2}$, which coincides with the binary distribution (7.1.26). We note that from the practical point of view the difference between the two definitions of $\mathscr{E}_0$ is unimportant.

In accordance with (7.3.8), in the wing of the line,

$$I(\omega) \simeq (\omega - \omega_0)^{-5/2} 1.5 \sum_{\alpha\beta} I_{\alpha\beta}(B_{\alpha\beta})^{3/2}\mathscr{E}_0^{3/2} , \tag{7.3.10}$$

in full agreement with the binary distribution (7.1.28). This is due to the fact that the strongest fields are created mainly by the nearest ion. It must be noted that the distribution function of the binary approximation is fairly close to $\mathscr{H}(\beta)$ everywhere, with the exception of the range of low values of β. Weak fields,

Table 7.1. Holtsmark distribution function

β	$\mathscr{H}(\beta)$	β	$\mathscr{H}(\beta)$	β	$\mathscr{H}(\beta)$
0.0	0.00000	2.4	0.272746	6.8	0.016494
0.1	0.0042245	2.6	0.238221	7.0	0.015165
0.2	0.016665	2.8	0.205563	7.2	0.013981
0.3	0.036643	3.0	0.176063	7.4	0.012922
0.4	0.063082	3.2	0.150242	7.6	0.011974
0.5	0.094596	3.4	0.128118	7.8	0.011120
0.6	0.129587	3.6	0.109422	8.0	0.010350
0.7	0.166360	3.8	0.093753	9.0	0.007438
0.8	0.203233	4.0	0.080674	10.0	0.005561
0.9	0.238641	4.2	0.069765	11.0	0.004289
1.0	0.271221	4.4	0.060654	12.0	0.003392
1.1	0.299870	4.6	0.053023	13.0	0.002739
1.2	0.323782	4.8	0.046604	14.0	0.002249
1.3	0.342461	5.0	0.041180	15.0	0.001875
1.4	0.355702	5.2	0.036573	17.0	0.001351
1.5	0.363566	5.4	0.032640	20.0	0.0008856
1.6	0.366334	5.6	0.029263	24.0	0.0005537
1.7	0.364456	5.8	0.026349	28.0	0.0003733
1.8	0.358502	6.0	0.023822	33.0	0.0002457
1.9	0.349109	6.2	0.021619	38.0	0.0001718
2.0	0.336939	6.4	0.019690	43.0	0.0001256
2.2	0.306821	6.6	0.017993	48.0	0.0000952
				53.0	0.0000741

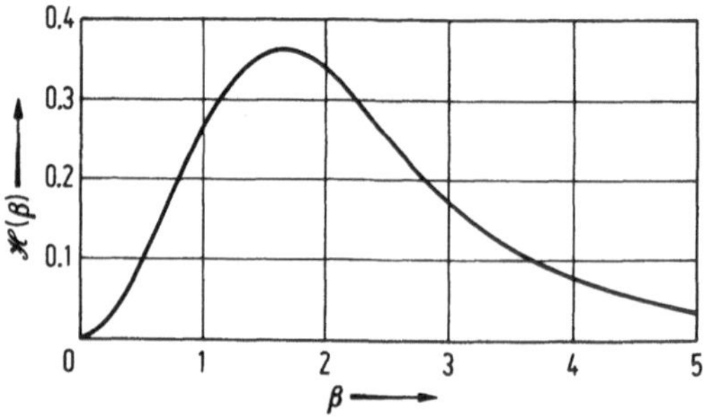

Fig. 7.4. Holtsmark distribution $\mathscr{H}(\beta)$

obviously, are produced by a large number of comparatively distant ions. The calculations of Holtsmark intensity distribution for a number of hydrogen spectral lines were carried out in [7.41]. It is convenient to rewrite (7.3.10) introducing the effective Stark-effect constant B for a line as a whole [7.42]

$$I(\omega) \simeq I_0 1.5(\omega - \omega_0)^{-5/2}(B\mathscr{E}_0)^{3/2}, \quad I_0 = \sum_{\alpha\beta} I_{\alpha\beta}\,, \tag{7.3.11}$$

where in accordance with (7.3.2)

$$B^{3/2} = I_0^{-1} \sum_{\alpha\beta} I_{\alpha\beta} \left(\frac{e}{\hbar}\right)^{3/2} (z_{\alpha\alpha} - z_{\beta\beta})^{3/2} . \tag{7.3.12}$$

Here z is the coordinate of the atomic electron. Comparison with the results of accurate numerical calculations shows that for a hydrogenlike ion with nuclear charge $\mathscr{Z}e$, the constant B can be approximated by the expression

$$B = \left(\frac{3}{8}\right)^{2/3} \frac{\hbar}{\mathscr{Z}me}(n^2 - n'^2) , \tag{7.3.13}$$

where n and n' are the principal quantum numbers of the initial and final levels. Similarly for the contour of the line $I(\omega) = \sum_{\alpha\beta} I_{\alpha\beta}(\omega)$, one can also use the approximate expression [7.42]

$$I(\omega) = \frac{I_0}{B\mathscr{E}_0} T_{\mathrm{H}} \left(\frac{\omega - \omega_0}{B\mathscr{E}_0}\right) . \tag{7.3.14}$$

The dependence of $T_{\mathrm{H}}(\beta)$ on β is given in Table 7.2. At high values of β, $T_{\mathrm{H}}(\beta) \to 1.5\, \beta^{-5/2}$. Since the contour of the line (7.3.4), and also (7.3.14), is symmetrical with respect to ω_0, the Holtsmark width of a line $\Delta\omega_{\mathrm{H}}$ is approximately equal to $8B\mathscr{E}_0$. Using (7.3.13), we obtain for the hydrogen spectrum,

$$\Delta\omega_{\mathrm{H}} \simeq 12.5(n^2 - n'^2) N^{2/3} . \tag{7.3.15}$$

Formula (7.3.14) describes sufficiently well the contour of the line everywhere apart from the central region.

In order to improve the Holtsmark theory one must take into consideration the mutual correlation of ion positions. In the Holtsmark theory, the exponential factor

$$\exp[-V(\boldsymbol{R}_1, \boldsymbol{R}_2, \dots)/kT]$$

in the expression for probability of an ion configuration $\boldsymbol{R}_1, \boldsymbol{R}_2, \dots$ with potential V is neglected. Thus, the relative probability of such configurations to which high positive values of V correspond are overestimated. In particular, the Holtsmark theory overestimates the probabilities of large frequency shifts κ, i.e., of high values of $\mathscr{E}$, and underestimates the probabilities of low κ. The simplest way of introducing the corresponding corrections to the Holtsmark theory is to take the Debye–Hückel screening into account. The field of the ion surrounded by a cloud of other ions and electrons of the plasma decreases at large distances as

Table 7.2. Function $T_{\mathrm{H}}(\beta)$

β	0	0.5	1	2	3	5	7	10	15	20
$T_{\mathrm{H}}(\beta)$	0.1	0.1	0.098	0.086	0.070	0.039	0.02	0.0072	0.0023	0.00099

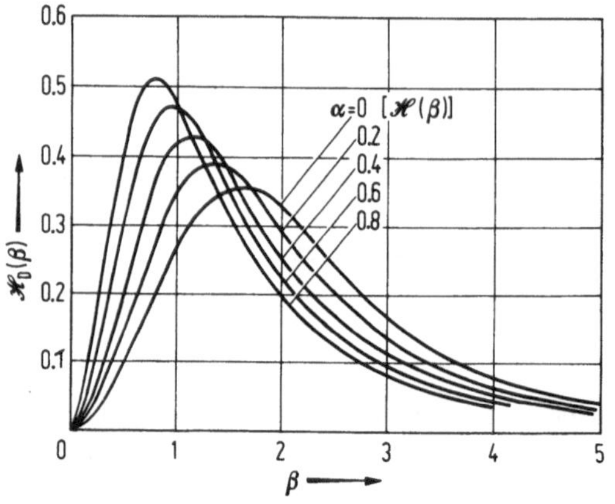

Fig. 7.5. Hooper distribution $\mathscr{H}_{\mathrm{D}}(\beta)$ for the neutral point [7.43]

exp $(-R/R_{\mathrm{D}})$, where R_{D} is the Debye radius

$$R_{\mathrm{D}} = \sqrt{\frac{kT}{4\pi Ne^2(1+Z^2)}} \,. \tag{7.3.16}$$

The difference between the distribution function $\mathscr{H}_{\mathrm{D}}(\beta)$ taking this screening into account and the Holtsmark one depends on the magnitude of the parameter

$$\alpha = \frac{R_0}{R_{\mathrm{D}}} \propto N^{1/6}\,T^{-1/2} \,. \tag{7.3.17}$$

The quantity α^{-3} is equal to the number of ions inside the Debye sphere N_{D}. As $\alpha \to 0$ ($R_{\mathrm{D}} \to \infty$), the function $\mathscr{H}_{\mathrm{D}}(\beta)$ coincides with $\mathscr{H}(\beta)$. The dependence of the function $\mathscr{H}_{\mathrm{D}}$ on β is shown in Fig. 7.5. [7.43].[1] The difference between $\mathscr{H}_{\mathrm{D}}(\beta)$ and $\mathscr{H}(\beta)$ becomes noticeable at $\alpha \simeq 0.4$($N_{\mathrm{D}} \simeq 16$). Thus, the condition of applicability of the Holtsmark distribution is the inequality

$$\alpha^{-3} = N_{\mathrm{D}} = \frac{4\pi}{3}\left(\frac{kT}{4\pi e^2(1+Z^2)}\right)^{3/2} N^{-1/2} \gg 1 \,. \tag{7.3.18}$$

A special situation arises in those cases when the broadening of lines of hydrogenlike ions He^{+}, Li^{2+}, and so on is considered. In calculating $W(\mathscr{E})$ it is necessary to take into account the repulsion between emitting and perturbing ions.

[1] Calculations in [7.43] are not based on the simple picture of Debye shielding and are more rigorous and rather complicated.

7.3.3 Correction for Thermal Motion

Due to the thermal motion of ions, the amplitude and direction of the vector $\mathcal{E}$ varies with time. Changing the amplitude of $\mathcal{E}$ causes the changing of the level splitting and leads to a phase modulation. If the vector $\mathcal{E}$ rotates very slowly, the angular momentum of the atom follows $\mathcal{E}$ adiabatically. This results in the change of the dipole moment component in the direction of a photon wave vector $\boldsymbol{k}$. Therefore the slow reorientation of vector $\mathcal{E}$ produces amplitude modulation.

The level splitting with respect to M components is of the order of $B\mathcal{E}$. Reorientation of the vector $\mathcal{E}$ can be considered to be slow when $B\mathcal{E} \gg 1/\tau$, where τ is the typical time of reorientation of $\mathcal{E}$. In the opposite limiting case $B\mathcal{E} \ll 1/\tau$, the magnetic field arising in the rotating coordinate system induces transitions between sublevels M. This means that perturbation is no longer adiabatic.

The corrections for thermal motion of ions taking into consideration all above-mentioned effects were calculated in [7.44, 45]. According to these calculations, the central part of a line is the most sensitive to the thermal motion of ions. The experiment [7.46] revealed that the dynamic effects of the ion influence substantially the shapes of the hydrogen lines. These effects can be taken into account by a modified relaxation theory of line broadening [7.47, 48] and the "Model Microfield Method" [7.49–52]. The dynamic effects can be very important in case of multiply-charged ions [7.51].

7.3.4 Electron Broadening[1]

A theory of broadening of lines of the hydrogen spectrum by electrons must take into account the nonadiabatic nature of the perturbation. Collisional transitions between the states of the same level play the main role in the broadening. We shall consider the spectral line corresponding to a radiative transition between the level a (the states α, $\alpha' \dots$) and the level b (the states $\beta, \beta' \dots$). The intensity distribution $I(\omega)$ can be calculated by means of (7.2.16). It is necessary to determine the elements of the S matrix and solve the system of equations (7.2.15). The collisional term in (7.2.15) can be written in the form

$$\sum_{\alpha''\beta''} \rho^{(\alpha\beta)}_{\alpha''\beta''} \langle\langle \alpha''\beta''|\Phi|\alpha'\beta' \rangle\rangle \,, \tag{7.3.19}$$

where Φ is the collision operator,

$$\Phi = N \int_0^\infty v f(v)\, dv \int_0^\infty 2\pi\rho\, d\rho \{1 - S_a^* S_b\} \,. \tag{7.3.20}$$

[1] A detailed treatment of the different aspects of electron broadening in hydrogen and hydrogenlike spectra using different approaches is given in [7.6, 53–64].

Here $f(v)$ is the Maxwellian distribution function, the braces { } denote the averaging over directions of vectors $\boldsymbol{\rho}$ and $\boldsymbol{v}$, and the notation

$$\langle\!\langle \alpha''\beta''|S_a^* S_b|\alpha'\beta'\rangle\!\rangle = \langle \alpha''|S^*|\alpha'\rangle\langle \beta''|S|\beta'\rangle \tag{7.3.21}$$

is used.

We shall determine the S matrix using the perturbation theory expansion in series (7.2.24).

The perturbation $V(t)$ produced by collision with the electron has the form

$$V(t) = e^2 \frac{\boldsymbol{r}\cdot(\boldsymbol{\rho}+\boldsymbol{v}t)}{(\rho^2+v^2t^2)^{3/2}}, \tag{7.3.22}$$

where $\boldsymbol{r}$ is the radius vector of an atomic electron.

As already noted above in the case of the interaction $V \propto R^{-2}$, the principal role in the broadening is played by collisions with high values of $\rho \gg \rho_0$, where ρ_0 is the Weisskopf radius. This means that we can restrict ourselves to the first nonvanishing term in (7.2.24). After averaging (7.2.25) over all directions of the vectors $\boldsymbol{\rho}$ and $\boldsymbol{v}$, we have $\langle \tilde{V}_v \rangle = 0$ and

$$[V(t)V(t')] = \frac{\boldsymbol{r}\cdot\boldsymbol{r}}{3}\frac{\rho^2+v^2tt'}{(\rho^2+v^2t^2)^{3/2}(\rho^2+v^2t'^2)^{3/2}}. \tag{7.3.23}$$

Therefore

$$(1-S_a^*S_b) = \frac{1}{\hbar^2}\left[-\int_{-\infty}^{\infty}\tilde{V}_a(t)\,dt\int_{-\infty}^{\infty}\tilde{V}_b(t)\,dt + \int_{-\infty}^{\infty}\tilde{V}_a(t)\,dt\int_{-\infty}^{t}\tilde{V}_a(t')\,dt' + \int_{-\infty}^{\infty}\tilde{V}_b(t)\,dt\int_{-\infty}^{t}\tilde{V}_b(t')\,dt'\right]. \tag{7.3.24}$$

The first term of this expression is

$$-\frac{e^4}{3}\boldsymbol{r}_a\cdot\boldsymbol{r}_b\int_{-\infty}^{\infty}\int_{-\infty}^{\infty}dt\,dt'\frac{\rho^2+v^2tt'}{(\rho^2+v^2t^2)^{3/2}(\rho^2+v^2t'^2)^{3/2}} = -\frac{2e^4}{3}\boldsymbol{r}_a\cdot\boldsymbol{r}_b\frac{1}{\rho^2v^2}, \tag{7.3.25}$$

The second and the third terms can be calculated as follows:

$$\langle\!\langle \alpha''\beta''|\int_{-\infty}^{\infty}\tilde{V}_a(t)\,dt\int_{-\infty}^{t}\tilde{V}_a(t')\,dt'\,|\alpha'\beta'\rangle\!\rangle \tag{7.3.26}$$

$$= \frac{e^4}{3}\sum_a \boldsymbol{r}_{\alpha''\alpha}\cdot\boldsymbol{r}_{\alpha\alpha'}\int_{-\infty}^{\infty}dt\int_{-\infty}^{\infty}dt'\frac{(\rho^2+v^2tt')\exp(\mathrm{i}\omega_{\alpha''\alpha}t)\exp(\mathrm{i}\omega_{\alpha\alpha'}t')}{(\rho^2+v^2t^2)^{3/2}(\rho^2+v^2t'^2)^{3/2}}.$$

We shall introduce the dimensionless variables

$$z_1 = \frac{\rho}{v}\omega_{\alpha''\alpha},\quad z_2 = \frac{\rho}{v}\omega_{\alpha'\alpha},\quad x_1 = \frac{vt}{\rho},\quad x_2 = \frac{vt'}{\rho}. \tag{7.3.27}$$

Then (7.3.26) can be expressed in terms of the integrals

$$A(z_1,z_2)+iB(z_1,z_2)$$
$$=\frac{1}{2}\int_{-\infty}^{\infty}dx_1\int_{-\infty}^{x_1}dx_2\frac{1+x_1x_2}{(1+x_1^2)^{3/2}(1+x_2^2)^{3/2}}\exp\left[i(z_1x_1-z_2x_2)\right]. \quad (7.3.28)$$

The summation over α is restricted to the states of the level a. The perturbation due to all the other levels is neglected. The matrix elements $\boldsymbol{r}_{\alpha''\alpha}$ are nonzero only for neighboring Stark components α'',α. At $\alpha''=\alpha', z_1=z_2=z$, and at $\alpha''\neq\alpha'$, $z_1=-z_2=z$. Let us denote the corresponding integrals by $A_+(z), B_+(z)$ and $A_-(z), B_-(z)$. All these integrals can be expressed in terms of Bessel functions

$$A_\pm(z)=z^2[K_1^2(z)\pm K_0^2(z)]\simeq\begin{cases}\pi|z|\exp(-2|z|) & (z\gg 1),\\ 1 & (z\ll 1),\end{cases} \quad (7.3.29)$$

$$B_\pm(z)=\frac{2|z|}{\pi}P\int_{-\infty}^{\infty}\frac{A_\pm(z')}{z^2-z'^2}dz'\simeq\begin{cases}\frac{\pi}{4z} & (z\gg 1),\\ \to 0 & (z\ll 1).\end{cases} \quad (7.3.30)$$

For low values of ρ, the approximation (7.3.23) is not valid. The region of $\rho<\rho_0$ gives a comparatively small contribution to the broadening. Therefore it is possible to assume

$$\int_0^\infty 2\pi\rho\,d\rho(1-S_a^*S_b)\simeq\pi\rho_0^2+\int_{\rho_0}^\infty 2\pi\rho\,d\rho(1-S_a^*S_b).$$

The upper limit of integration must be taken as $\rho=R_D$, see (7.3.16). It is not difficult to prove that for typical plasma parameters, the quantity $z_{\max}=R_D\omega_{\alpha''\alpha}/v$, where $\omega_{\alpha''\alpha}$ is the Stark splitting in the field $\mathscr{E}_0=2.6\,ZeN^{2/3}$, can be neglected as compared to unity. Therefore one can assume $A=1$ and $B=0$[1]. As a result

$$\Phi=\frac{4\pi N}{3}\frac{e^4}{\hbar^2}[\boldsymbol{r}_a\cdot\boldsymbol{r}_a+\boldsymbol{r}_b\cdot\boldsymbol{r}_b-2\boldsymbol{r}_a\cdot\boldsymbol{r}_b]\int\frac{dv}{v}f(v)\left(1+\ln\frac{R_D}{\rho_0}\right). \quad (7.3.31)$$

It must be noted that in some cases the upper limit of integration over ρ must be chosen not as R_D but as $v/\Delta\omega$. The spectrum $I(\omega)$ in the region of the frequencies $\Delta\omega$ is determined by the correlation function $\Phi(\tau)$ at $\tau\sim 1/\Delta\omega$. The impact approximation can be used only if the duration of the collision $\rho/v>1/\Delta\omega$, or $\rho<\rho_m=v/\Delta\omega$. Therefore the upper limit must be chosen as the minimum value of the two values $R_D, v/\Delta\omega$.

When integrating over v in (7.3.31) the lower limit of integration must be determined from the condition $\rho_0(v_{\min})\lesssim R_D$. As a result we obtain the following

[1] The same integrals are encountered in the problem of the broadening of nonhydrogenlike spectral lines in plasma (Sect. 7.4).

expression for the operator Φ:

$$\Phi = \frac{16}{3} N \left(\frac{e^2}{\hbar}\right)^2 \left(\frac{1}{\langle v \rangle}\right) [\boldsymbol{r}_a \cdot \boldsymbol{r}_a + \boldsymbol{r}_b \cdot \boldsymbol{r}_b - 2\boldsymbol{r}_a \cdot \boldsymbol{r}_b] \left(\ln \frac{R_D}{\rho_0} + 0.215\right) \quad (7.3.32)$$

Here $\langle v^{-1} \rangle = 4/\pi \langle v \rangle^{-1}$, and the constant 0.215 gives the relative contribution of the strong collisions $\rho < \rho_0$. We shall determine the Weisskopf radius ρ_0 in accordance with that of Sect. 7.1 as

$$\rho_0^2 = \frac{2}{3} \left(\frac{e^2}{\hbar \langle v \rangle}\right)^2 I(n, n') a_0^2 , \quad (7.3.33)$$

$$I(n, n') = \Big(\sum_{\alpha\beta} |P_{\alpha\beta}|^2\Big)^{-1}$$

$$\sum_{\alpha\beta\alpha'\beta'} P_{\alpha'\beta'} P_{\beta\alpha} \ll \alpha\beta | \boldsymbol{r}_a \cdot \boldsymbol{r}_a + \boldsymbol{r}_b \cdot \boldsymbol{r}_b - 2\boldsymbol{r}_a \cdot \boldsymbol{r}_b | \alpha'\beta' \gg a_0^{-2} . \quad (7.3.34)$$

where n and n' are the principal quantum numbers of the levels a and b.

The intensity distribution in the wing of a line is given by (7.2.21) and (7.2.22), where in accordance with (7.3.32–34), the effective constant of the impact broadening γ is

$$\gamma = \frac{32}{3} N \langle v \rangle^{-1} \frac{\hbar^2}{m^2} \left(\ln \frac{R_D}{\rho_0} + 0.215\right) I(n, n') , \quad (7.3.35)$$

The same quantity gives the order of magnitude of the broadening by electrons.

The quantity $I(n, n')$ can be written in the form

$$I(n, n') = I(n) + I(n') - 2K(n, n') . \quad (7.3.36)$$

As $\langle \boldsymbol{r}_a \cdot \boldsymbol{r}_a \rangle \simeq a_0^2 n^4$ we have $I(n) \sim n^4$ and $I(n') \sim n'^4$. (For hydrogenlike ion with a nuclear charge $\mathscr{Z} I(n) \sim n^4 \mathscr{Z}^{-2}$ and $I(n') \sim n'^4 \mathscr{Z}^{-2}$). If $n \simeq n'$, then to an order of magnitude $K(n, n') \sim I(n)$, $I(n')$. For example, in the case of radiative transition $n = 3 \to n' = 2$, we have $I(n) = 56$, $I(n) + I(n') = 68$, and $I(n, n') = 27$. In the case of the transition $n = 4 \to n' = 2$, we have $I(n) = 359$ and $I(n, n') = 291$. Nevertheless using (7.3.36) for estimates of the broadening in the case of transitions between lower excited levels, one can assume $I(n, n') \simeq n^4$, where n is the maximum value of n, n', or $I(n, n') \simeq (n^4 + n'^4)/2$.

In the case of transitions between neighbouring highly excited levels $n \gg 1$, $n' \gg 1$, and $|n - n'| \ll n, n'$, the term $2K(n, n')$ leads to a significant reduction of the broadening. This effect will be considered in Sect 7.3.6.

In order to calculate the total intensity distribution produced by electrons for a given ion field, one must solve the system of equations (7.2.15). The Hamiltonian of an atom in the electric field being diagonal in parabolic coordinates, the calculation of the matrix Φ must be carried out using parabolic atomic wave functions.

The resulting effect of impact broadening by electrons in the presence of a fixed ion field can be described in the following way. The ion field splits the line into the Stark components, each of these components being broadened by electrons. If the ion splitting is less than the impact broadening produced by the electrons $B\mathscr{E} \lesssim \gamma$, then the Stark components overlap. In this case, nondiagonal matrix elements of Φ are of importance. But the main contribution to the line shape is given by the ion fields $\mathscr{E} \sim \mathscr{E}_0$ for which $B\mathscr{E}_0 > \gamma$.

The transition from the impact distribution to the quasistatic wing produced by electrons in the case of a hydrogen spectral line must be considered taking into account nonadiabatic effects. This problem is very complicated. For a discussion see [7.59–65].

7.3.5 Combined Effect of Electrons and Ions

The intensity distribution resulting from the combined effect of electrons and ions can be obtained by calculating the electron broadening with a fixed ion field $\mathscr{E}$ and averaging the result over all possible values of $\mathscr{E}$.

If the electron broadening is described by the Lorentzian distribution with width γ from (7.3.35) and ion broadening in the same approximation as in (7.3.14), then for the normalized distribution of intensity it is easy to obtain [7.42]

$$I(\omega) = \frac{1}{B\mathscr{E}_0} T\left(\frac{\omega - \omega_0}{B\mathscr{E}_0}, \frac{\gamma}{B\mathscr{E}_0}\right) . \tag{7.3.37}$$

The function $T(x, y)$ is defined by the expression

$$T(x, y) = \frac{y}{2\pi} \int \frac{T_{\mathrm{H}}(x')\,dx'}{(x - x')^2 + (y/2)^2} . \tag{7.3.38}$$

Values of the function $\log T(x, y)$ are given in Table 7.3. When $y \to 0, T(x, y) \to T_{\mathrm{H}}(x)$.

In the line wing (high values of x),

$$T(x, y) \simeq 1.5\, x^{-5/2} + \frac{y}{2\pi} x^{-2} . \tag{7.3.39}$$

In order to improve the description of a line shape one must take into consideration two effects: (i) in the far wing of a line impact distribution is replaced by quasistatic one; and (ii) in the central part of a line, ion broadening must be calculated taking into consideration the thermal motion of ions; moreover the distribution function $W_{\mathrm{D}}(\mathscr{E})$ instead of $W_{\mathrm{H}}(\mathscr{E})$ must be used (see Fig. 7.5).

Detailed calculations of the hydrogen line shapes have been carried out in [7.66–69]. The results of these calculations are in many cases in a reasonable agreement with the experimental data [7.70–73]. The difference between experiment and calculations is, however, noticeable in the central part of the lines, especially for the lines having an unshifted Stark component (L_α, H_α). When plasma density is not high the discrepancy of experimental [7.74, 75] and theoretical linewidths for L_α and H_α can exceed a factor of 2, see also [7.48]. In

Table 7.3. log $T(x, y)$

y	x: 0	0.5	1	2	3	5	7	10	15	20
0.0	$\bar{1}.004$	$\bar{1}.000$	$\bar{2}.992$	$\bar{2}.936$	$\bar{2}.846$	$\bar{2}.251$	$\bar{2}.292$	$\bar{3}.860$	$\bar{3}.370$	$\bar{4}.996$
0.5	$\bar{2}.963$	$\bar{2}.958$	$\bar{2}.943$	$\bar{2}.912$	$\bar{2}.828$	$\bar{2}.579$	$\bar{2}.313$	$\bar{3}.975$	$\bar{3}.455$	$\bar{3}.159$
1.0	$\bar{2}.912$	$\bar{2}.911$	$\bar{2}.905$	$\bar{2}.862$	$\bar{2}.794$	$\bar{2}.588$	$\bar{2}.356$	$\bar{2}.050$	$\bar{3}.580$	$\bar{3}.252$
1.5	$\bar{2}.874$	$\bar{2}.872$	$\bar{2}.864$	$\bar{2}.826$	$\bar{2}.768$	$\bar{2}.592$	$\bar{2}.385$	$\bar{2}.092$	$\bar{3}.666$	$\bar{3}.354$
2.0	$\bar{2}.837$	$\bar{2}.834$	$\bar{2}.826$	$\bar{2}.794$	$\bar{2}.742$	$\bar{2}.588$	$\bar{2}.402$	$\bar{2}.123$	$\bar{3}.729$	$\bar{3}.434$
3.0	$\bar{2}.768$	$\bar{2}.766$	$\bar{2}.759$	$\bar{2}.734$	$\bar{2}.692$	$\bar{2}.570$	$\bar{2}.415$	$\bar{2}.175$	$\bar{3}.822$	$\bar{3}.550$
4.0	$\bar{2}.706$	$\bar{2}.704$	$\bar{2}.699$	$\bar{2}.678$	$\bar{2}.645$	$\bar{2}.545$	$\bar{2}.417$	$\bar{2}.208$	$\bar{3}.888$	$\bar{3}.633$
5.0	$\bar{2}.651$	$\bar{2}.649$	$\bar{2}.646$	$\bar{2}.628$	$\bar{2}.600$	$\bar{2}.518$	$\bar{2}.410$	$\bar{2}.228$	$\bar{3}.938$	$\bar{3}.698$

case of hydrogenlike ions the accuracy of methods used in [7.66–68] can even be poorer for such lines [7.51]. As noted above (Sect. 7.3.3) the dynamic effects of the ion must be taken into account.

7.3.6 New Approaches to the Theory of Stark Broadening

The new, very promising approaches to the problem of spectral line broadening by charged particles have been given in [7.47–52, 76–79]. In [7.76, 77] the study of Stark profiles for hydrogen lines is carried out by means of direct computer simulation. The radiating atom is assumed to be placed in some volume together with a fixed number of charged particles, electrons and ions, the motion of which is described by classical orbits. Numerical solution of the Schrödinger equation for a radiating atom perturbed by Debye-screened fields of charged particles is averaged over the initial coordinates and the velocities of the perturbers. The Stark profiles are obtained as a result of such a procedure.

A new version of relaxation theory proposed in [7.47, 48] showed good agreement with experiment and with the results of computer simulation [7.76, 77].

In the so called "Model Microfield Method" proposed in [7.78, 79] the intensity distribution of a spectral line is expressed in terms of a correlation function for the plasma microfield. This method proved to be very successful in taking into account the "ion-radiator dynamic effects" [7.49–52], and also showed good agreement with experiment and with the results of computer simulation.

The above-mentioned theoretical methods can be applied to the nonhydrogenic spectra, too.

7.3.7 Highly Excited States

The reduction of electron broadening due to the term $K(n, n')$ in (7.3.36) is connected with the interference effects discussed in Sect. 7.2.

In the case of high n and $n' = n-1$ the matrix elements $S_{nl,nl\pm1}$ and $S_{n'l,n'l\pm1}$ for large values of l are almost the same:

$$S_{nl,nl\pm1} - S_{n'l,n'l\pm1} \ll S_{nl,nl\pm1} .$$

Moreover the matrix elements $P_{nl,n'l\pm1}$ and $P_{nl',n'l'\pm1}$, where $l' = l \pm 1$, are also almost the same. Therefore there arises a very close similarity with the four-level system considered in Sect. 7.2.6. It can be shown [7.6, 80, 81] that in the case of radiative transitions between neighboring highly excited levels, $n \gg 1, n' \gg 1$, and $|n-n'| \ll n, n'$,

$$I(n,n') \ll I(n), I(n') .$$

Thus for transitions $n \to n-1, n \gg 1$ we have

$$I(n,n-1) \sim \frac{1}{n^2} I(n) . \tag{7.3.40}$$

According to (7.3.40) for such transitions the dependence $\gamma_e \propto n^4$ is replaced by the dependence $\gamma_e \propto n^2$. This means that one cannot neglect the perturbation due to the all other atomic levels. As a result the broadening due to the quadratic stark effect as in the case of nonhydrogenlike spectra begins to be of importance. Broadening of this kind is considered in Sect. 7.4.

As a rule the lines corresponding to radiative transitions between highly excited levels are observed in plasmas with very low electron density, when $h_e \ll 1$ and $h_i \ll 1$. Expression for the width of a line $n - n'$ is given by

$$\begin{aligned} \gamma &= \gamma_e + \gamma_i , \\ \gamma_e &= N_e[\langle v\sigma_e(n)\rangle + \langle v\sigma_e(n')\rangle] , \\ \gamma_i &= N_i[\langle v\sigma_i(n)\rangle + \langle v\sigma_i(n')\rangle] , \end{aligned} \tag{7.3.41}$$

where $\sigma(n) = \sum_{|k|>0} \sigma(n, n+k)$ are the effective cross sections of inelastic scattering. Using the quasi-classical calculations of cross sections for highly excited level described in Sect. 3.5, one can write for the quantity $\langle v\sigma(n)\rangle$ the following approximate formula:

$$\langle v\sigma_e(n)\rangle \simeq 10^{-8} \frac{n^4}{z^3 \Theta^{1/2}} \Phi(x) f(\Theta) \ [\mathrm{cm^3 s^{-1}}], \tag{7.3.42}$$

where

$$\begin{aligned} &\Theta = T/z^2\mathrm{Ry}, \quad x = E_n/T = 1/n^2\Theta , \\ &\Phi(x) = 2.18\{0,82\varphi(x) + 1.47[1 - x\varphi(x)]\}, \\ &\varphi(x) = -\exp(x) E_i(-x) \simeq \ln\left[1 + \frac{1+1.4\gamma x}{\gamma \cdot x \cdot (1+1.4x)}\right], \quad \gamma = 1.78 \end{aligned} \tag{7.3.43}$$

$$f(\Theta) = \ln\left[1 + \frac{n\sqrt{\Theta}}{z(1+2.5/z\sqrt{\Theta})}\right] \ln(1 + n\sqrt{\Theta}/z) \tag{7.3.44}$$

when $z \to \infty$ or $\Theta \to \infty, f(\Theta) \to 1$. If perturbing particles are protons, the temperature T in (7.3.42–44) should be replaced by the quantity Tm/M, where m is the mass of electron and M is the reduced mass of the colliding particles. Stark broadening of the lines in the far-infrared solar spectrum corresponding to the transitions between highly excited hydrogenlike states is of interest for diagnosing the structure of the solar atomosphere [7.82].

7.4 Line Broadening of Nonhydrogenlike Spectra in a Plasma

7.4.1 Preliminary Estimates

The spectral lines of nonhydrogenlike atoms in the presence of a constant and homogeneous electric field undergo a shift and also a splitting proportional to $\mathscr{E}^2$ – the quadratic Stark effect. We shall assume that the field $\mathscr{E} = QR^{-2}$, produced by a charge Q, varies little for atomic dimensions (this is valid for sufficiently large values of R). Then in (7.1.20) for the shift of oscillator frequency, $n = 4$ and $\kappa = C_4 R^{-4}$. The constant C_4 for a transition $n \to k$ is defined as

$$C_4 = (C_4)_n - (C_4)_k; \quad (C_4)_n = Q^2/\hbar \sum_m \frac{|(D_z)_{nm}|^2}{\Delta E_{nm}} ,$$

where $(D_z)_{nm}$ are the matrix elements of the z component of the electric dipole operator, $\Delta E_{nm} = E_n - E_m$ (see [1.1]). The parameters h_e (electron broadening) and h_i (ion broadening) are

$$h_e = N\left(\frac{\pi}{2}\frac{C_4}{v_e}\right), \quad h_i = N\left(\frac{\pi}{2}\frac{C_4}{v_i}\right). \tag{7.4.1}$$

The quadratic Stark-effect constants C_4, as a rule, have the order of magnitude $10^{-12} - 10^{-15}$ cm^4 s^{-1}, although values of $C_4 < 10^{-15}$ and $C_4 \sim 10^{-11} - 10^{-10}$ are also encountered. For $C_4 = 10^{-12} - 10^{-15}, v_e = 5.10^7$ cm s^{-1} and $v_i = 2.10^5$cm s^{-1}, we have

$$h_e = 3\cdot(10^{-19} - 10^{-22})N, \quad h_i = 0.75(10^{-17} - 10^{-20})N.$$

At not very high values of the density of charged particles $N < 10^{15}$ cm^{-3}, $h_e \ll 1$, and $h_i \ll 1$. This means that both electrons and ions produce impact broadening.

According to (7.1.23) $\gamma_4, \Delta_4 \propto v^{1/3}$. Thus electrons play the principal role in the broadening of a line. The interaction with ions only slightly increases the impact width and shift of a line, by approximately 15–20%, because $(v_e/v_i)^{1/3} \simeq (M/m)^{1/6} \simeq$ 5–6. Since $\kappa \propto Q^2$ the sign of the shift of a line is the same for electrons and for ions.

7.4.2 Electron Broadening

We shall describe electron broadening of lines of a nonhydrogenlike atom in the framework of quasi-classical theory discussed in Sect. 7.2. In the case of the isolated spectral line the spectrum $I(\omega)$ is described by Lorentzian distribution (7.2.17). The width and shift are given by (7.2.18) and (7.2.20). The elements of the S matrix averaged over M components of atomic levels must be substituted in these equations. As a rule the main contribution to the broadening is given by collisions with relatively large values of the impact parameter ρ. Therefore in calculating the S matrix we can restrict ourselves to the first terms of the expansion in powers of Γ and η. Therefore one can average over M directly the quantities Γ and η.

In the case of the dipole interaction between a neutral atom and a charged particle $V = -\boldsymbol{d}\cdot\mathcal{E}$ the linear term in (7.2.25) for the S matrix $S(v)$, being averaged over M (or over directions of the vectors $\boldsymbol{\rho}$ and $\boldsymbol{v}$), is equal to zero. Only the next term containing $\tilde{V}_v(t)\tilde{V}_v(t')$ in the integrand is nonvanishing.

We shall assume now that the perturbation of one of the levels (initial or final) can be neglected. Then for the radiative transition n—k, assuming that the level k is not perturbed, we can obtain

$$\eta = \sum_s{}' \eta_s = -\sum_s{}' 2\left(\frac{\hbar}{mv}\right)^2 \left(f_{ns}\frac{\mathrm{Ry}}{\Delta E_{ns}}\right)\frac{1}{\rho^2}B\left(\frac{\omega_{ns}\rho}{v}\right)\frac{\Delta E_{ns}}{|\Delta E_{ns}|}, \tag{7.4.2}$$

$$\Gamma = \sum_s{}' \Gamma_s = \sum_s{}' 2\left(\frac{\hbar}{mv}\right)^2 \left(f_{ns}\frac{\mathrm{Ry}}{\Delta E_{ns}}\right)\frac{1}{\rho^2}A\left(\frac{\omega_{ns}\rho}{v}\right). \tag{7.4.3}$$

Here $\Delta E_{ns} = E_s - E_n$, f_{ns} is the oscillator strength of the transition $n \to s$ (see [Ref. 1.1, Sect. 9.2]), and the sum over s extends over all atomic levels, for which $f_{ns} \neq 0$.

The functions A and B coincide with the functions A_+ and B_+ defined by (7.3.29, 30). Values of the functions A and B are given in Table 7.4.

As a rule the principal contribution to (7.4.2, 3) is provided by the nearest perturbing levels and, in some cases, by only one of them. In the approximation of one perturbing level,

$$\gamma = 2N\langle v\rangle\sigma_0'(\langle v\rangle)J'(\beta), \tag{7.4.4}$$

$$\Delta = N\langle v\rangle\sigma_0''(\langle v\rangle)J''(\beta), \tag{7.4.5}$$

Table 7.4. Functions $A(z)$ and $B(z)$.

z	$A(z)$	$B(z)$	z	$A(z)$	$B(z)$
0.0	1.000	0.000	1.8	0.177	0.507
0.2	1.035	0.160	2.0	0.130	0.467
0.4	0.962	0.359	2.4	0.0688	0.393
0.6	0.829	0.498	2.8	0.0355	0.331
0.8	0.680	0.576	3.2	0.0181	0.283
1.0	0.540	0.606	3.6	0.0090	0.245
1.2	0.418	0.603	4.0	0.0045	0.216
1.4	0.318	0.580	5.0	0.00075	0.166
1.6	0.239	0.546	$\to\infty$	$\pi z e^{-2z}$	$\pi/4z$

where σ_0' and σ_0'' are the width and shift cross sections defined by (7.1.13, 14),

$$\sigma_0' = \left(\frac{\pi}{2}\right)^{5/3} \Gamma\left(\frac{1}{3}\right) C_4^{2/3} \langle v \rangle^{-2/3} \simeq 5.7 C_4^{2/3} \langle v \rangle^{-2/3} ,$$

$$\sigma_0'' = \sqrt{3}\sigma_0' . \qquad (7.4.6)$$

$\langle v \rangle$ is the mean value of the electron velocity, and the constant C_4 is defined as

$$C_4 = \frac{2e^2 a_0^3}{\hbar} f \left(\frac{\mathrm{Ry}}{\Delta E}\right)^2 .$$

The functions $J'(\beta)$ and $J''(\beta)$ depend on the dimensionless parameter

$$\beta = \left| f \frac{\mathrm{Ry}}{\Delta E} \right|^{1/2} \frac{|\Delta E|}{m\langle v \rangle^2} . \qquad (7.4.7)$$

As $\beta \to \infty$, J', $J'' \to 0.97$. This case corresponds to adiabatic perturbation when $\Gamma = 0$, and the broadening is determined by the phase shift η. For $\beta \ll 1$, on the contrary, inelastic collisions play the main role. In this case,

$$\langle v\sigma' \rangle = 4\pi f \frac{\mathrm{Ry}}{\Delta E} \langle v \left(\frac{\hbar}{mv}\right)^2 \ln \left(\frac{mv^2}{2|\Delta E|} \left| f \frac{\mathrm{Ry}}{\Delta E} \right|^{1/2} \right) \rangle , \qquad (7.4.8)$$

$$\langle v\sigma'' \rangle = -2\pi^2 f \frac{\mathrm{Ry}}{\Delta E} \langle v \left(\frac{\hbar}{mv}\right)^2 \rangle \frac{\Delta E}{|\Delta E|} . \qquad (7.4.9)$$

The results of numerical calculations of the functions $J'(\beta)$ and $J''(\beta)$ are given in Table 7.5.

We shall now consider to what extent the results obtained above can be generalized to the case of several perturbing levels. This problem obviously arises

Table 7.5. Factors $J'(\beta)$ and $J''(\beta)$.

β	J'	J''	β	J'	J''
64	0.97	0.97	0.156×10^{-1}	0.594	0.151
32	0.97	0.97	0.78×10^{-2}	0.451	0.094
16	1.02	0.97	0.39×10^{-2}	0.334	0.063
8	1.03	0.96	0.195×10^{-2}	0.239	0.0405
4	1.06	0.94	0.97×10^{-3}	0.171	0.0245
2	1.12	0.90	0.48×10^{-3}	0.119	0.0167
1	1.17	0.861	0.24×10^{-3}	0.0824	0.0103
0.5	1.20	0.746	0.12×10^{-3}	0.056	0.0065
0.25	1.15	0.604	0.61×10^{-4}	0.038	0.004
0.125	1.09	0.455	0.305×10^{-4}	0.024	0.0026
0.625×10^{-1}	0.927	0.326	0.15×10^{-4}	0.017	0.0016
0.312×10^{-1}	0.764	0.223			

only in the case when, for one or several perturbing levels, the parameter β is of the order of or less than unity. If $\beta \gg 1$ for all perturbing levels, then the perturbation is adiabatic and γ and Δ are expressed by means of (7.1.23) in terms of the quadratic Stark effect constant C_4 for a given line. The magnitude of this constant is determined by the total perturbing effect of all atomic levels.

If for the nearest perturbing levels which give the principal contribution to η and Γ in (7.4.2, 3), the parameters $\beta \ll 1$, then the width γ can be obtained by summing (7.4.4):

$$\gamma = 2N\langle v\rangle \sum_s{}' \sigma'_{ns}(\langle v\rangle)J'(\beta_s) . \qquad (7.4.10)$$

Such an approximation is valid because in this case the broadening is caused by inelastic collisions, and the partial widths corresponding to different collisional transitions are additive.

The shift of a line cannot be calculated by summing (7.4.5) even if for all perturbing levels $\beta_s \ll 1$.

In the general case when both levels n and k (initial and final) are perturbed,

$$\Gamma = \Gamma_n + \Gamma_k, \quad \eta = \eta_n - \eta_k,$$

where Γ_n, Γ_k, η_n and η_k must be calculated by means of (7.4.2, 3). An extensive bibliography on numerical calculations of shapes of the nonhydrogenlike spectral lines in plasma can be found in [7.6].

Results of numerical calculations of widths and shifts for a large number of nonhydrogenlike spectral lines of different atoms and ions also are given in [7.6]. The experimental data on Stark broadening in nonhydrogenlike spectra may be found in [7.12.13]. The very accurate experimental data on the lines of neutral He are given in [7.83].

7.5 Broadening by Uncharged Particles

7.5.1 Perturbation by Foreign Gas Atoms (Van der Waals Interaction)

The interaction of neutral atoms at large distances has the form

$$V(R) \propto R^{-6} .$$

Therefore usually one assumes

$$\kappa = C_6 R^{-6} . \qquad (7.5.1)$$

The crude estimate of the constant C_6 is given by

$$C_6 \simeq \frac{e^2\langle r^2\rangle\alpha}{\hbar}, \; \alpha \simeq \frac{4}{3} m \left(\frac{\mathrm{Ry}}{I}\right)^2 a_0 \langle r_{\mathrm{p}}^2\rangle , \qquad (7.5.2)$$

where $\langle r^2\rangle \simeq 5n^{*4}a_0^2/2$ is the mean value of r^2 for the excited state of the radiative atom, $\langle r_p^2 \rangle$ is that for the perturber, n^* is the effective principal quantum number, α is the polarizability of the perturbing atom, I is its ionization potential, and m is the number of equivalent electrons. The constant C_6 has the order of magnitude $10^{-30} n^{*4}$. Thus for $v \sim 10^5\,\mathrm{cm\,s^{-1}}$, we have

$$h = \left(\frac{3\pi}{8}\frac{C_6}{v}\right)^{3/5} N \simeq 10^{-21} N . \tag{7.5.3}$$

This indicates that at not very high pressures, of the order of few atmospheres or less, line broadening can be described in the impact approximation. We shall also compare the quantities $\Omega = v^{6/5} C_6^{-1/5}$ and $\Delta\omega_D$. As $\Omega \simeq 10^{12}\,\mathrm{s^{-1}}$ and $\Delta\omega_D \simeq 10^{10}\,\mathrm{s^{-1}}$, we have $\Omega \gg \Delta\omega_D$. Consequently the region of impact broadening extends far beyond the limits of the Doppler width. In accordance with (7.1.23) the width and shift of a line can be estimated using the relations

$$\gamma = 8.16\, C_6^{2/5} v^{3/5} N, \quad \Delta \simeq \gamma/2.8 . \tag{7.5.4}$$

The typical values of γ are $\gamma \sim 10^{-8} N$. In order to treat impact broadening by uncharged particles more accurately it is necessary to take into consideration that at small distances the interaction $V(R)$ has a more complicated form than $V(R) \propto R^{-6}$. Depending on the type and states of interacting atoms both attraction and repulsion can take place at large distances R. At small distances the potential $V(R)$ is repulsive. In some cases atom and perturber can form a quasi-stable molecule. Moreover in the general case, the interaction V is dependent not only of R but also of the angular variables. The results of calculations in which a more realistic interaction than $V(R) \propto R^{-6}$ is used cannot be described by a simple Lorentzian distribution with width and shift as in (7.5.4). In particular, the intensity distribution depends on the type of the transition j—j'. A detailed treatment of the foreign gas broadening is given in [7.4, 10,11,14 14–16]. The repulsive part of the interaction is usually taken into account in the form of the Lennard–Jones potential $V(R) = C_{12}R^{-12}$-$C_6 R^{-6}$. The line shift Δ and the ratio γ/Δ are especially sensitive to the form of the potential $V(R)$.

Experimental data on line broadening in the spectra of alkali atoms obtained at low values of foreign gas pressure, less than 10 atm, are in qualitative agreement with the impact theory. The broadening and shift of the lines are proportional to the concentration of perturbing particles. For the initial members of a principal series perturbed by different foreign atoms (He, Ne, Ar, Kr, Xe, H_2, N_2, and so on), as a rule, a red shift is observed, the ratio γ/Δ being close to 2.8. In some cases (usually for the higher members of a principal series), a blue shift instead of a red one is observed. The sign of the shift of one and the same line can be different for different perturbing particles.

The dimensionless parameter h reaches values of order unity only when $N > 10^{21}$, i.e., at pressures of about tens of atmospheres. In this case the mean distance between atoms has the same order of magnitude as atomic dimensions and consequently the simplest expression $V(R) \propto R^{-6}$ is not valid. The experimental data on line shapes are usually used to obtain information about the form of potential $V(R)$ at small distances.

Specific features of molecular-lines broadening have been described in [7.4, 8, 9].

7.5.2 Self-Broadening

We shall consider now the single-component gas. With an increase of density of such a gas, resonance lines broaden considerably more than on the addition of a foreign gas. This is due to the fact that in the case of collision of two identical atoms, one of which is excited, a resonance transfer of the excitation energy is possible, the effective cross sections of such collisions being extremely large. They can exceed considerably (by several orders) the gas kinetic cross sections.

The effective cross sections of resonance energy transfer σ were calculated in Sect. 4.2. For electric dipole transition, the energy transfer is caused by the dipole–dipole interaction $V \propto R^{-3}$. The cross section σ and corresponding line width γ are of the order of magnitude

$$\sigma \sim \frac{e^2}{\hbar v} \frac{S}{3e^2 a_0^2} a_0^2 \simeq \frac{e^2}{m\omega_0 v} f, \qquad \gamma \sim \frac{e^2}{m\omega_0} fN, \qquad C_3 \sim \frac{e^2}{m\omega_0} f, \qquad (7.5.5)$$

where S is the line strength and f is the oscillator strength. Assuming that $f \simeq 1$ and $\omega_0 \simeq 10^{15}$, we have $\gamma \sim 10^{-7}N$. In the case of foreign-gas broadening, typical values of widths are $\gamma \sim (10^{-9}\text{–}10^{-8})N$. The effective cross section of energy transfer can be relatively large not only under conditions of exact resonance but also in the case of a collision of two atoms with close energy levels. Thus, when calculating the width of the component ${}^2P_{1/2}$—${}^2S_{1/2}$ of the resonance doublet of an alkali atom, it is necessary to take into account not only energy transfer ${}^2P_{1/2} \to {}^2S_{1/2}$ (radiating atom), ${}^2S_{1/2} \to {}^2P_{1/2}$ (perturbing atom), but also the excitation ${}^2S_{1/2} \to {}^2P_{3/2}$ of the perturbing atom.

The cross section of the energy transfer ${}^2P_{1/2} \to {}^2S_{1/2}$, ${}^2S_{1/2} \to {}^2P_{3/2}$ has the same order of magnitude as the cross section of the resonance energy transfer ${}^2P_{1/2} \to {}^2S_{1/2}$, ${}^2S_{1/2} \to {}^2P_{3/2}$, if the following condition is fulfilled (Sect. 4.2):

$$\left(\frac{e^2}{2m\omega_0} f\right)^{2/3} \left(\frac{\Delta E}{\hbar}\right)^{4/3} \frac{1}{v^2} \ll 1,$$

where ΔE is the doublet splitting. Of all alkali atoms, only Li atom satisfies this condition. In the case of Li the doublet splitting of the resonance level is 0.34 cm^{-1}.

In order to calculate the line broadening due to resonance energy transfer more accurately it is necessary to take into consideration the degeneracy of levels and dependence of the interaction on the angular variables. Such calculations using the impact approximation have been carried out by many researchers [7.84–89].

The summary of the results can be given as follows

$$\frac{\gamma}{2} = A(J_0J_1)\sqrt{\frac{2J_0+1}{2J_1+1}}\frac{2\pi e^2}{m\omega_0}f_{01}N \tag{7.5.6}$$

where J_0 and J_1 are the angular momenta of the ground and excited levels, respectively, and f_{01} is the line oscillator strength. The values of the factor $A(J_0J_1)$ for isolated lines calculated numerically in [7.87, 88] are given in Table 7.6 together with the ratios of the width $\gamma/2$ to the shift Δ from [7.89].

As was said above, in the special case of Li atom one has to take into consideration the perturbation of both excited levels $^2P_{1/2}$ and $^2P_{3/2}$. Because of that one can expect almost equal widths for both lines of the resonance doublet. The approximate value of $\gamma/2$ in this case can be obtained using (7.5.6) and assuming $J_0 = 0,\ J_1 = 1$ [7.85].

For the case of interaction $V \propto R^{-3}$ the intensity distribution in quasistatic wings $|\omega-\omega_0| \gg v^{3/2}C_3^{-1/2} = v^{3/2}(e^2 f/m\omega_0)^{-1/2}$ is also proportional to $(\omega-\omega_0)^{-2}$ as in the impact approximation. However, the intensity $I(\omega)$ in the quasistatic wings is somewhat different from that in the impact approximation [7.84, 85, 90]

$$I(\omega) = a\,(J_0J_1)\,\frac{2\pi e^2}{m\omega_0}f_{01}N\frac{1}{(\omega-\omega_0)^2}\,. \tag{7.5.7}$$

The factors $a(J_0J_1)$ are also given in Table 7.6. In the case of Li resonance doublet $J_0 = 0,\ J_1 = 1$ should be assumed in (7.5.7). The calculations taking into account the accurate adiabatic potential curves [7.90] show a slight asymmetry of blue and red wings. The direct experimental studies of the resonance broadening as a rule encounter very serious difficulty connected with extremely large optical depth in the center of the lines [7.91].

The lines corresponding to transitions between the resonance level and other excited levels also undergo broadening due to the resonance interaction. The resonance contribution to the widths γ of such lines can be evaluated to be equal to the linewidth of the resonance line.

Table 7.6. Parameters describing the resonance broadening

J_0	0	1/2	1/2	1
J_1	1	1/2	3/2	1
$A(J_0J_1)$	1.042	0.903	1.039	0.983
$2\,\Delta/\gamma$	0.092	−0.031	0.050	−0.01
$a\,(J_0J_1)$	0.698	1.047	0.805	

7.6 Spectroscopic Methods of Investigating Elastic Scattering of Slow Electrons

7.6.1 Perturbation of Highly Excited States

The broadening of a line corresponding to a transition between the ground state and a state with a large value of the principal quantum number n is completely determined by the perturbation of the upper level. For sufficiently large values of n, the mean distance of the valence electron from the nucleus $\sim a_0 n^2$ is so large that the neutral perturbing particle either interacts with the electron and does not interact with the atomic core, or interacts only with the atomic core. In this case, the broadening is caused by the scattering of the atomic electron by the perturbing particles and by the scattering of the perturbing particles by the atomic core. These two mechanisms of the broadening are statistically independent.

We shall first consider the interactions of the first type. If only one level is perturbed, then in accordance with (7.2.39) we have

$$\sigma' = \frac{2\pi}{k} \operatorname{Im}\{f(0)\}, \; \sigma'' = -\frac{2\pi}{k} \operatorname{Re}\{f(0)\}\,, \tag{7.6.1}$$

where $f(0)$ is the amplitude of forward scattering of the perturbing particle by the atom, and $\hbar \boldsymbol{k}$ is the momentum of the perturbing particle (we assume for simplicity that the mass of the atom is large as compared with the mass of the perturbing particle). If $a_0 n^2 \gg \rho_{\text{eff.}} \simeq (\pi\alpha/4)^{1/3}(e^2/\hbar v_e)^{1/3}$, where $\rho_{\text{eff.}}$ is the effective radius of the interaction between the electron and the perturbing particle and α is the polarizability of the perturbing particle, then in the volume of interaction the field produced by the atomic core and consequently the electron velocity v_e are practically constant. In the state with principal quantum number n, v_e is of the order of magnitude v_0/n, where v_0 is the atomic unit of velocity. If the velocity of the perturbing particle $v_p = \hbar k/M$ is less than $v_e \sim v_0/n$, then the scattering amplitude $f(0)$ in (7.6.1) can be expressed in terms of forward scattering amplitude $f_{\boldsymbol{q}}^{e}(0)$ of a free electron with momentum $\hbar \boldsymbol{q}$ by the perturbing particle [7.92] for the derivation of this, and subsequent formulas of this section):

$$f(0) = \frac{M}{m} \int |\, G(nlm|\boldsymbol{q})|^2 f_{\boldsymbol{q}}^{e}(0)\, d\boldsymbol{q}\,. \tag{7.6.2}$$

Here m is the electron mass, and $G(nlm|\boldsymbol{q})$ are the coefficients of the expansion of the atomic function ψ_{nlm} in plane waves:

$$G(nlm|\boldsymbol{q}) = G_{nl}(q)\, Y_{lm}(\vartheta_{\boldsymbol{q}}, \varphi_{\boldsymbol{q}})\,. \tag{7.6.3}$$

States with large values of the principal quantum number n are hydrogenlike. It is therefore possible to use as the expansion coefficients $G(nlm|\boldsymbol{q})$ the well-known expressions for hydrogen functions in the momentum representation in terms of Gegenbauer polynomials [7.93].

By substituting (7.6.2) in (7.6.1), integrating over the angular variables, and averaging over all possible orientations of the perturbing particle angular momentum, we have

$$\gamma = N\frac{4\pi\hbar}{m}\int dq\, W(q)\,\mathrm{Im}\,\{f_q^{\mathrm{e}}(0)\}\,, \tag{7.6.4}$$

$$\Delta = -N\frac{2\pi\hbar}{m}\int dq\, W(q)\,\mathrm{Re}\,\{f_q^{\mathrm{e}}(0)\}\,, \tag{7.6.5}$$

$$W(q) = q^2|G_{nl}(q)|^2,\ \int W(q)\,dq = 1\,. \tag{7.6.6}$$

As already noted above these formulas describe the width and shift caused by scattering of the atomic electron in the highly excited state by perturbing particle if the following conditions are fulfilled:

$$n^{5/3} \gg \alpha^{1/3} a_0^{-1}\,, \tag{7.6.7}$$

$$n \ll \frac{m}{M}\frac{e^2}{\hbar v_{\mathrm{p}}}\,. \tag{7.6.8}$$

We shall now consider interaction of perturbing particle with the atomic core. As the charge of the atomic core is e, this interaction has the form

$$U = -\frac{\alpha e^2}{2R^4}\,. \tag{7.6.9}$$

It produces the polarization of the perturbing particles by the atomic core and leads to a shift of the frequency of the atomic oscillator,

$$\kappa(t) = -\frac{\alpha}{2\hbar}e^2\frac{1}{[R(t)]^4}\,. \tag{7.6.10}$$

Broadening due to interactions of this type was examined in Sect. 7.4. If

$$h_i = \rho_{\mathrm{i}}^3 N \ll 1,\ \rho_{\mathrm{i}} = \left(\frac{\pi\alpha}{4}\right)^{1/3}\left(\frac{e^2}{\hbar v_{\mathrm{p}}}\right)^{1/3}\,, \tag{7.6.11}$$

where N and v_{p} are the density and velocity of the perturbing particles, then the central part of the line $|\omega-\omega_0| \ll \Omega = (2\hbar v_{\mathrm{p}}^4/\alpha e^2)^{1/3}$ is described by the Lorentzian distribution with width γ' and shift Δ' given by the following formulas:

$$\gamma' = 11.4\left(\frac{\alpha e^2}{2\hbar}\right)^{2/3} v_{\mathrm{p}}^{1/3} N,\ \Delta' = -\frac{\sqrt{3}}{2}\gamma'\,. \tag{7.6.12}$$

It can be shown that when $h_{\mathrm{i}} \gg 1$, the corresponding quasi-static distribution has width of the order of $10N^{4/3}\,\alpha e^2/\hbar$. The case $h_{\mathrm{i}} \ll 1$ corresponding to pressures of the order of or less than an atmosphere is the most interesting. In this case, one can calculate γ' and Δ' with sufficiently good accuracy. Subtracting the calculated values γ' and Δ' from experimental values of the width and shift, one can determine γ and Δ from (7.6.4,5).

Expressing in these equations the amplitude $f_q^e(0)$ in terms of the phase of scattering of the electron by the perturbing particle η_l, we also have

$$\gamma = N\frac{\hbar}{m}\int\left[\frac{4\pi}{q}\sum_l(2l+1)\sin^2\eta_l\right]W(q)\,dq = N\frac{\hbar}{m}\int q\sigma(q)\,W(q)\,dq\,, \quad (7.6.13)$$

$$\Delta = -N\frac{\hbar}{m}\int\left[\frac{\pi}{q}\sum_l(2l+1)\sin 2\eta_l\right]W(q)\,dq\,. \quad (7.6.14)$$

Here $\sigma(q)$ is the effective cross section for elastic scattering of the electron with momentum $\hbar q$ by the perturber.

If exchange interaction is also taken into account, then the following substitution must be made in (7.6.13, 14):

$$\sin^2\eta_l \to C^+\sin^2\eta_l^{(+)} + C^-\sin^2\eta_l^{(-)}\,, \quad (7.6.15)$$

$$\sin 2\eta_l \to C^+\sin 2\eta_l^{(+)} + C^-\sin 2\eta_l^{(-)}\,, \quad (7.6.16)$$

where $\eta_l^{(+)}$ and $\eta_l^{(-)}$ are the scattering phases calculated taking exchange into account for states of the system perturbing particle plus electron with given value of the total spin $S = S_p \pm 1/2, S_p$ being the spin of the perturbing particle; and

$$C^+ = \frac{S_p+1}{2S_p+1}, \quad C^- = \frac{S_p}{2S_p+1}\,. \quad (7.6.17)$$

We shall now consider the resonance transitions $n_0\,s$—$np\,(n \gg 1)$ of the alkali atoms. For np state [7.93]

$$W(q) = \frac{n^2}{2\pi(n^2-1)}\left(\frac{1}{nqa_0}\right)^2[(n+1)\sin(n-1)\varphi$$
$$-(n-1)\sin(n+1)\varphi]^2 a_0\,, \quad (7.6.18)$$

$$\cos\varphi = \frac{n^2q^2a_0^2-1}{n^2q^2a_0^2+1}\,. \quad (7.6.19)$$

The function $W(q)$ has n peaks, $n/2$ peaks being located in the range $0 < q < 1/na_0$. In this range, the envelope behaves approximately as $(1+n^2q^2a_0^2)^{-2}$. For $q \gg 1/na_0$, the function $W(q)$ decreases monotonically:

$$W(q) \simeq \frac{2^7}{3}\frac{(n^2-1)n^3}{\pi}\left(\frac{1}{nqa_0}\right)^8\,. \quad (7.6.20)$$

Thus the principal contribution to the integral over q in (7.6.13, 14) is given by the range $0 < q < 1/na_0$.

If n is so large that the principal contribution to the sum with respect to l is given by the term $l = 0$ (s scattering) and in addition $q^{-1} \sin 2\eta_l$ differs little from its limiting value

$$\lim_{q \to 0} \left(\frac{1}{q} \sin 2\eta_l \right) = \frac{\eta_0}{|\eta_0|} \sqrt{\frac{1}{\pi} \sigma(0)}\,, \tag{7.6.21}$$

then

$$\Delta = -\frac{\hbar}{m} \sqrt{\pi \sigma(0)} \frac{\eta_0}{|\eta_0|} N\,. \tag{7.6.22}$$

Here $\sigma(0)$ is the limiting expression for the elastic scattering cross section at $q \to 0$. By η_0 is understood that part of the phase which after subtraction of $p\pi$ where p is integer, lies in the interval $-\pi/2$, $\pi/2$.

Since as $q \to 0$, $q\sigma(q) \to 0$, in the range of applicability of (7.6.22), $\gamma \ll |\Delta|$. Some additional effects of broadening are discussed in [7.94–96]. Experimental data on broadening of Rydberg levels can be found in [7.97–100].

7.6.2 Fermi Formula

Equation (7.6.22) has been obtained by Fermi [7.4]. In accordance with this equation it is possible using the experimental value of Δ to determine the elastic scattering cross section for extremely slow electrons (in the limit $q \to 0$). Thus the cross section $\sigma(0)$ for the atoms He, Ne, Ar, Kr, and Xe have been found by the shift of the absorption lines of Cs in an atmosphere of noble gases. Some other gases have also been investigated by the same method (see [7.4] and also [7.101]). The Fermi method enables one to obtain information on elastic scattering of electrons at very small energies, i.e., in the range most difficult to investigate by other experimental methods. It must be noted that shift Δ is sensitive not only to the magnitude but also to the sign of the phase η_0.

In the general case, when several terms of the sum over l contribute to γ and Δ, it is not possible to determine the scattering phases from known values of γ and Δ. Knowledge of these quantities, however, enables one to control the quality of approximate calculations of the scattering phases [7.102, 103].

References

Chapter 1

1.1 I.I. Sobel'man: *Atomic Spectra and Radiative Transitions*, 2nd edn., Springer Ser. Atoms Plasmas, Vol.12 (Springer, Berlin, Heidelberg 1992); 1st edn., Springer Ser. Chem. Phys., Vol.1 (Springer, Berlin, Heidelberg 1979)
1.2 M. Venugopalan (ed.): *Reactions under Plasma Conditions* (Wiley-Interscience, New York 1971) Vol.1

Chapter 2

2.1 N.F. Mott, H.S.F. Massey: *The Theory of Atomic Collisions* (Pergamon, Oxford 1965)
2.2 M.L. Goldberger, K.M. Watson: *Collision Theory* (Wiley, New York 1964)
2.3 Ch.J. Joachain: *Quantum Collision Theory* (North-Holland, Amsterdam 1975)
2.4 M.R.H. Rudge: Rev. Mod. Phys. **40**, 564 (1968)
2.5 R.P. Peterkop: *Teoriya Ionizatsii Atomov Elektronnym Udarom* (Theory of Ionization Atoms by Electron Impact, in Russian) (Zinatne, Riga 1975)
2.6 I.I. Sobel'man: *Atomic Spectra and Radiative Transitions*, 2nd edn., Springer Ser. Atoms Plasmas, Vol.12 (Springer, Berlin, Heidelberg 1992); 1st edn., Springer Ser. Chem. Phys., Vol.1 (Springer, Berlin, Heidelberg 1979)
2.7 L.A. Vainshtein, I.I. Sobel'man: Zh. Eksp. Teor. Fiz. **39**, 767 (1960)
2.8 M.J. Seaton: Proc. Phys. Soc. **77**, 184 (1961)
2.9 M.J. Seaton: Adv. Atom. Molec. Phys. **11**, 83 (1975)
2.10 H.S.W. Massey, E.H.S. Burhop, H.B. Gilbody: *Electronic and Ionic Impact Phenomena* Vol.1 (Clarendon, Oxford 1969)
2.11 R. Courant, D. Hilbert: *Methoden der Mathematischen Physik* (Springer, Berlin 1931) Vol.1
2.12 L.A. Vainshtein: Phys. Scripta **33**, 336 (1986)

Chapter 3

3.1 L.A. Vainshtein, I.I. Sobel'man: Zh. Eksp. Teor. Fiz. **39**, 767 (1960)
3.2 B.L. Moiseiwitsch: Rep. Prog. Phys. **40**, 843 (1977)
3.3 K. Smith: *The Calculation of Atomic Collision Processes* (Wiley-Interscience, New York 1971)
3.4 I.I. Sobel'man: *Atomic Spectra and Radiative Transitions*, 2nd edn., Springer Ser. Atoms Plasmas, Vol.12 (Springer, Berlin, Heidelberg 1992); 1st edn., Springer Ser. Chem. Phys., Vol.1 (Springer, Berlin, Heidelberg 1979)
3.5 M. Inokuti: Rev. Mod. Phys. **43**, 297 (1971)
3.6 K. Omidvar: Phys. Rev. **188**, 140 (1969)

3.7 M. Matsuzawa: Phys. Rev. A **9**, 241 (1974)
3.8 M.R.H. Rudge: Rev. Mod. Phys. **40**, 564 (1968)
3.9 R.P. Peterkop: *Teoriya Ionizatskii Atomov Elektronnym Udarom* (Theory of Atom Ionization by Electron Impact, in Russian) (Zinatne, Riga 1975)
3.10 H.S.W. Massey, E.H.S. Burhop, H.B. Gilbody: *Electronic and Ionic Impact Phenomena* (Clarendon, Oxford 1969) Vol.1
3.11 G.H. Wannier: Phys. Rev. **100**, 1180 (1955)
3.12 K.L. Bell, H.B. Gilbody, J.H. Hughes, A.E. Kingston, F.J. Smith: J. Phys. Chem. Ref. Data **212** (1983)
3.13 M.J. Seaton: Adv. Atom. Molec. Phys. **11**, 83 (1975)
3.14 K.T. Dolder, B. Peart: Rep. Prog. Phys. **39**, 693 (1976)
3.15 I.L. Beigman, L.A. Vainshtein: Zh. Eksp. Teor. Fiz. **52**, 185 (1967) [English transl.: Sov. Phys. – JETP **25**, 119 (1967)]
3.16 V.I. Ochkur: Zh. Eksp. Teor. Fiz. **45**, 735 (1963) [English transl.: Sov. Phys. – JETP **18**, 503 (1964)]
3.17 M.J. Seaton: Proc. Phys. Soc. **77**, 184 (1961)
3.18 I.P. Zapesochnyi: Teplofiz. Vys. Temp. **5**, 7 (1967)
3.19 Ch. J. Joachain: *Quantum Collision Theory* (North-Holland, Amsterdam 1975)
3.20 I.I. Sobelman: *Introduction to the Theory of Atomic Spectra* (Pergamon, Oxford 1972)
3.21 R.K. Nesbet: Comput. Phys. Commun. **6**, 265 (1973)
3.22 R. Damburg, E. Karule: Proc. Phys. Soc. London **90**, 637 (1967)
3.23 S. Geltman: Applications of Pseudo-State Expansions, in *Electronic and Atomic Collisions*, invited papers and progress repts. of VII ICPEAC, ed. by T.R. Govers, F.J. de Heer (North-Holland, Amsterdam 1972) p.216
3.24 P.G. Burke, W.D. Robb: Adv. Atom. Molec. Phys. **11**, 144 (1975)
3.25 L.P Presnyakov, A.M. Urnov: Zh. Eksp. Teor. Fiz. **68**, 61 (1975) [English transl.: Sov. Phys. – JETP **41**, 31 (1975)]
3.26 L.P. Presnyakov, A.M. Urnov: J. Phys. B **8**, 1280 (1975)
3.27 A.I. Baz', Ya.B. Zeldovich, A.M. Perelomov: *Rasseyanie, Reaktsii i Raspadny v Nerelyativistkoy Kvantovoy Mekhanike* (Scattering, Reactions and Disintegrations in Nonrelativistic Quantum Mechanics, in Russian) (Nauka, Moscow 1971)
3.28 I.L. Beigman, A.M. Urnov: J. Quant. Spectrosc. Radiat. Transf. **14**, 1009 (1974)
3.29 L. Hostler, R.H. Pratt: Phys. Rev. Lett. **10**, 469 (1963)
3.30 L.A. Bureeva: Astron. Zh. **45**, 1215 (1968)
3.31 R.C. Stabler, Phys. Rev. A **133**, 1268 (1964)
3.32 M. Gryzinski: Phys. Rev. **115**, 374 (1959)
3.33 M. Gryzinski: Phys. Rev. A **138**, 305, 322, 336 (1965)
3.34 I.L. Beigman, L.A. Vainshtein, I.I. Sobleman: Zh. Eksp. Teor. Fiz, **57**, 1703 (1969) [English transl.: Sov. Phys. – JETP **30**, 920 (1969)]
3.35 I.C. Percival, D. Richards: Adv. Atom. Molec. Phys. **11**, 2 (1975)
3.36 M. Born: *Vorlesungen über Atommechanik* (Springer, Berlin 1925)
3.37 M.J. Seaton: Proc. Phys. Soc. **79**, 1105 (1962)
3.38 H.E. Saraph: Proc. Phys. Soc. **83**, 763 (1964)
3.39 L.P. Presnyakov, A.M. Urnov: J. Phys. B **3**, 1267 (1970)
3.40 I.L. Beigman: Zh. Eksp. Teor. Fiz. **73**, 1729 (1977) [English transl.: Sov. Phys. – JETP **46**, 908 (1977)

Chapter 4

4.1 N.F. Mott, H.S.F. Massey: *The Theory of Atomic Collisions* (Pergamon, Oxford 1965)
4.2 H.S.W. Massey, E.H.S. Burhop, H.B. Gilbody: *Electronic and Ionic Impact Phenomena* (Clarendon, Oxford 1969) Vol.1
4.3 H.S.W. Massey, H.B. Gilbody: *Electronic and Ionic Impact Phenomena*, Vol.4 (Pergamon, Oxford 1974)
4.4 E.E. Nikitin, S.Ya. Umanskii: *Theory of Slow Atomic Collisions* (Springer, Berlin, Heidelberg 1984)
4.5 R.K. Janev, L.P. Presnyakov, V.P. Shevelko: *Physics of Highly Charged Ions* (Springer, Berlin, Heidelberg 1985)
4.6 B.N. Bransden, M.R.C. McDowell: *Charge Exchange and the Theory of Ion-Atom Collisions* (Clarenden, Oxford 1992)
4.7 H.B. Gilbody: Adv. Atom. Molec. Phys. **22**, 143 (1986)
4.8 L.D. Landau, E.M. Lifshitz: *Quantum Mechanics* (Pergamon, Oxford 1965)
4.9 L.A. Vainshtein, I.I. Sobelman, L.P. Presnyakov: Zh. Eksp. Teor. Fiz. **43**, 518 (1962) [English transl. Sov. Phys. – JETP **16**, 370 (1962)]
4.10 D.R. Bates: Proc. Phys. Soc. A **73**, 227 (1959)
4.11 M.R. Flannery: Phys. Rev. **183**, 241 (1969)
4.12 M.R. Flannery: J. Phys. B **2**, 909 (1969)
4.13 J.C. Gay, A. Omont: J. de Phys. **35**, 9 (1974)
4.14 I.I. Sobel'man: *Atomic Spectra and Radiative Transitions*, 2nd edn., Springer Ser. Atoms Plasmas, Vol.12 (Springer, Berlin, Heidelberg 1992); 1st edn., Springer Ser. Chem. Phys., Vol.1 (Springer, Berlin, Heidelberg 1979)
4.15 K. Alder, A. Bohr, T. Huus, B. Mottelson, A. Winther: Rev. Mod. Phys. **28**, 432 (1956)
4.16 D.R. Bates, R. McCarrol: Adv. Phys. **11**, 39 (1962)
4.17 M.R.C. McDowell, J.P. Coleman: *Introduction to the Theory of Ion-Atom Collisions* (North-Holland, Amsterdam 1970)
4.18 O.B. Firsov: Zh. Eksp. Teor. Fiz. **21**, 1001 (1951)
4.19 H.C. Brinkman, H.A. Kramers: Proc. Acad. Sci. Amsterdam **33**, 973 (1930)
4.20 B.M. Smirnov: *Asimptoticheskii Metod v Teorii Atomnykh Stolknovenii* (The Asymptotic Method in the Theory of Atomic Collisions, in Russian) (Nauka, Moscow 1973)
4.21 R.A. Mapleton: *Theory of Charge Exchange* (Wiley-Interscience, New York 1972)
4.22 D.S.F. Crothers, N.R. Todd: J. Phys. B **13**, 2277 (1980)
4.23 V.P. Shevelko: Z. Phys. A **287**, 18 (1978)
4.24 A.M. Brodskii, V.S. Potapov, V.V. Tolmachev: Zh. Eksp. Teor. Fiz. **58**, 264 (1970) [English transl.: Sov. Phys. – JETP **31**, 144 (1970)]
4.25 V.V. Afrosimov, R.N. Ilyin, E.S. Solovyev: Zh. Techn. Phys. **30**, 705 (1960)
4.26 V. Schryber: Helv. Phys. Acta **40**, 1023 (1967)
4.27 A.V. Vinogradov, L.P. Presnyakov, V.P. Shevelko: JETP Lett. **8**, 449 (1968)
4.28 H.D. Betz: Rev. Mod. Phys. **44**, 465 (1972)
4.29 N.V. Fedorenko: JTP **15**, 1947 (1972)
4.30 A. Salop, R.E. Olson: Phys. Rev. A **13**, 1312 (1976)
4.31 A. Salop: Phys. Rev. A **13**, 1321 (1976)
4.32 G. Harel, A. Salin: J. Phys. B **10**, 3511 (1977)
4.33 J. Vaaben, J.S. Briggs: J. Phys. B **10**, L521 (1977)

4.34 P.A. Phaneuf, F.W. Meyer, R.H. Knight, R.E. Olson, A. Salop: J. Phys. B **10**, L425 (1977)
4.35 R.A. Phaneuf, R.K. Janev, H.J. Hunter: Nuclear Fusion, Special Supplement 7 (IAEA, Vienna 1987)
4.36 H. Tawara, T. Kato, Y. Nakai: *Cross Sections for Charge Transfer of Highly Ionized Ions in Hydrogen Atoms* (Inst. of Plasma Physics Report IPPJ-AM-30, Nagoya 1983)
4.37 H. Tawara: *Total and Partial Cross Sections of Electron Transfer for* Be^{q+} *and* B^{q+} *Ions in Collisions with H,* H_2 *and He Gas Targets. Status in 1991* (NIFS Data Series, ISSN 0915-6364, Nagoya, June 1991)
4.38 H. Tawara: *Bibliography on Electron Transfer Processes in Ion-Ion/Atom/ Molecule Collisions - updated 1990* (NIFS Data Series, ISSN 0915-6364, Nagoya 1990)

Chapter 5

5.1 I.I. Sobel'man: *Atomic Spectra and Radiative Transitions*, 2nd edn., Springer Ser. Atoms Plasmas, Vol.12 (Springer, Berlin, Heidelberg 1992); 1st edn., Springer Ser. Chem. Phys., Vol.1 (Springer, Berlin, Heidelberg 1979)
5.2 I.L. Beigman: Zh. Eksp. Teor. Fiz. **73** 1729 (1977) [English transl.: Sov. Phys. – JETP **46**, 908 (1977)]
5.3 H. Van Regemorter: Astrophys. J. **132**, 906 (1962)
5.4 H.W. Drawin: Rep. EUP-CEA-FC-383 Assoc. EURATOM-CEA (1967) Fontenay-aux-Roses, France; ibid: in *Plasma Diagnostics*, ed. by W. Lochte-Holtgreven (North-Holland, Amsterdam 1968)
5.5 R. Mewe: Astron. Astrophys. **20**, 215 (1972)
5.6 M. Gryzinski: Phys. Rev. **115**, 374 (1959)
5.7 M. Gryzinski: Phys. Rev. A **138**, 305,322,336 (1965)
5.8 V.P. Shevelko, E.A. Yukov: Phys. Scripta **31**, 265 (1985)
5.9 M.J. Seaton: Planet Space Sci. **12**, 55 (1964)
5.10 W. Lotz: Z. Physik **232**, 101 (1970)
5.11 R.C. Stabler: Phys. Rev. A **133**, 1268 (1964)
5.12 A. Burgess: Astrophys. J. **139**, 776 (1964)
5.13 B. Shore: Rev. Mod. Phys. **39**, 439 (1967)
5.14 E. Trefftz: J. Phys. Atom. Molec. Phys. B **3**, 763 (1970)
5.15 M.J. Seaton, P.J. Storey: Dielectronic Recombination, in *Atomic Processes and Applications*, ed. by P.G. Burke, B. Moiseiwitsch (North-Holland, Amsterdam 1976)
5.16 A. Burgess, M.J. Seaton: Mon. Not. Roy. Astron. Soc. **125**, 355 (1964)
5.17 A. Burgess, H.P. Summers: Astrophys. J. **157**, 1007 (1969)
5.18 A. Gabriel, T.M. Paget: J. Phys. Atom. Molec. Phys. B **5**, 673 (1972)
5.19 Yu.I. Grineva, V.I. Karev, V.V. Koreneev, V.V. Krutov, S.L. Mandelshtam, L.A. Vainshtein, B.N. Vasilyev, I.A. Zhitnik: Solar Phys. **23**, 441 (1973)
5.20 L.M. Biberman, V.S. Vorobyov, N.T. Yakubov: Usp. Fiz. Nauk **107**, 353 (1972) [English transl.: Sov. Phys.Usp. **15**, 375 (1972)]
5.21 H. Griem: *Plasma Spectroscopy* (McGraw-Hill, New York 1964)
5.22 R.H. Huddlestone, S.L. Leonard (eds.): *Plasma Diagnostics Techniques* (Academic, New York 1965)

5.23 W. Lochte-Holtgeven (ed.): *Plasma Diagnostics* (North-Holland, Amsterdam 1968)
5.24 W. Neumann: Spectroscopic methods of plasma diagnostic, in *Progress in Plasmas and Gas Electronics*, Vol.1, ed. by R. Rompe, M. Steenbeck (Akademie, Berlin 1975) p.3
5.25 H.W. Drawin: Validity conditions for local thermodynamic equilibrium, in *Progress in Plasmas and Gas Electronics*, Vol.1, ed. by R. Rompe, M. Steenbeck (Akademie, Berlin 1975) p.593
5.26 M.J. Seaton: Mon. Not. Astron. Soc. **119**, 90 (1959)
5.27 I.L. Beigman, E.D. Mikhalchi: J. Quant. Spectrosc. Radiat. Transl. **9**, 1365 (1969)
5.28 G. Ecker, W. Weizel: Ann. Phys. **17**, 126 (1956/57)
5.29 G. Ecker, W. Kröll: Phys. Fluids **6**, 62 (1963)
5.30 H.R. Griem: Phys. Rev. **128**, 997 (1962)
5.31 H.W. Drawin: Ann. Phys. (Leipzig) *14*, 262 (1964)
5.32 S.T. Belyaev, G.I. Budker: Mnogokvantovaya Rekombinatsiya v Ionizovannom Gaze, in *Fizika Polasmy i Problema Upravlyaemykh Termoaydernykh Reaktsii* (Plasma Physics and the problems of Controlled Nuclear Fusion) Edition of Academy of Sciences, Moscow 1958) Vol.3, p.41
5.33 L.P. Pitaevskii: Zh. Eksp Teor. Fiz. **42**, 1326 (1962) [English transl.: Sov. Phys. – JETP **15**, 919 (1962)]
5.34 A.V. Gurevich, L.P. Pitaevskii: Zh. Eksp Teor. Fiz **46**, 1281 (1964) [English transl.: Sov. Phys. – JETP **19**, 870 (1964)]
5.35 M. Cacciatore, M. Capitelli, H.W. Drawin: Physica C **84**, 267 (1976)
5.36 D.R. Bates, A.E. Kingston, R.W.P. McWhirter: Proc. R. Soc. London A **267**, 297 (1962)
5.37 D.R. Bates, A.E. Kingston, R.W.P. McWhirter: Proc. R. Soc. London A **270**, 155 (1962)
5.38 R.W.P. McWhirter, A.G. Hearn: Proc. Phys. Soc. London **82**, 641 (1963)
5.39 L.C. Johnson, E. Hinnov: J. Quant. Spectrosc. Radiat. Transf. **13**, 333 (1973)
5.40 H.W. Drawin, F. Emard: Physica C **85**, 333 (1977)
5.41 M. Venugopalan (ed.): *Reactions under Plasma Conditions* (Wiley-Interscience, New York 1971) Vol.1
5.42 R. Hess, F. Burrell: J. Quant. Spectrosc. Radiat. Transf. **21**, 23 (1979)
5.43 H.W. Drawin, F. Emard: Physica C **94**, 134 (1978)
5.44 H. Risken: *The Fokker-Planck Equation*, 2nd edn., Springer Ser. Syn., Vol.18 (Springer, Berlin, Heidelberg 1989)

Chapter 6

6.1 L.A. Vainshtein: Trudy FIAN **15**, 3 (1961)
6.2 I.I. Sobel'man: *Atomic Spectra and Radiative Transitions*, 2nd edn., Springer Ser. Atoms Plasmas, Vol.12 (Springer, Berlin, Heidelberg 1992); 1st edn., Springer Ser. Chem. Phys., Vol.1 (Springer, Berlin, Heidelberg 1979)
6.3 D.A. Varshalovich, A.N. Moskalyv, V.K. Khersonske: *Quantum Theory of Angular Moment* (World Scientific, Singapore 1988)
6.4 A.R. Edmonds: *Angular Momentum in Quantum Mechanics* (Princeton Press, Princeton, NJ 1957)

6.5 M. Rotenberg, R. Bivius, N. Metropolis, J.K. Wooten, Jr.: *The 3j and 6j Symbols* (MIT Press, Cambridge, MA 1959)
6.6 H. Appel: *Numerical Tables for 3j, 6j, 9j Symbols*, Landolt-Börnstein (Group I), Vol.3 (Springer, Berlin, Heidelberg 1968)
6.7 A.P. Jucys, A.J. Savukynas: *Mathematical Foundations of the Atomic Theory* (Mintys, Vilnus 1973)

Chapter 7

7.1 R.G. Breene Jr.: *The Shift and Shape of Spectral Lines* (Pergamon, New York 1961)
7.2 M. Baranger: In *Atomic and Molecular Processes*, ed. by D.R. Bates (Academic, New York 1962)
7.3 G. Traving: In *Plasma Diagnostics*, ed. by W. Lochte-Holtgreven (North-Holland, Amsterdam 1968)
7.4 S. Chen, M. Takeo: Rev. Mod. Phys. **29**, 20 (1957)
7.5 H.R. Griem: *Plasma Spectroscopy* (McGraw-Hill, New York 1964)
7.6 H.R. Griem: *Spectral Line Broadening by Plasmas* (Academic, New York 1974)
7.7 J. Cooper: Rev. Mod. Phys. **39**, 167 (1967)
7.8 C.J. Tsao, B. Curnutte: J. Quant. Spectrosc. Radiat. Transf. **2**, 41 (1962)
7.9 H. Rabitz: Ann. Rev. Phys. Chem. **25**, 155 (1974)
7.10 W.R. Hindmarsh, J.M. Farr: in *Progress in Quantum Electronics*, Vol.2 (Pergamon, Oxford 1972) p.141
7.11 F. Schuller, W. Behmenburg: Phys. Rpt. **12**, 273 (1974)
7.12 N. Konjevic, J.D. Roberts: J. Phys. Chem. Ref. Data **5**, 209, 259 (1976)
7.13 N. Ksonjevic, M.S. Dimitrijevic, W.L. Wiese: J. Phys. Chem. Ref. Data **13**, 619 (1984)
7.14 E.L. Lewis: Phys. Repts. **58**, 1 (1980)
7.15 G. Peach: Advances in Physics **30**, 367 (1981)
7.16 N. Allard, J. Kielkopf: Rev. Mod. Phys. **54**, 1103 (1982)
7.17 J.R. Fuhr, L.J. Roszman, W.L. Wiese: *Bibliography on Atomic Line Shapes and Shifts*, NBS Spec. Publ. 366 (1972(); Suppl.1 (1974)
7.18 J.R. Fuhr, G.A. Martin, B.J. Specht: *Bibliography on Atomic Line Shapes and Shifts*, NBS Spec. Publ. 366, Suppl.2 (1975)
7.19 J.R. Fuhr, B.J. Miller, G.A. Martin: *Bibliography on Atomic Line Shapes and Shifts*, NBS Spec. Publ. 366, Suppl.3 (1978)
J.R. Fuhr, A. Lesage: Ibid., Suppl.4 (1993)
7.20 P.R. Berman: Appl. Phys. **6**, 283 (1975)
7.21 P. Anderson: Phys. Rev. **76**, 647 (1949)
7.22 J. Szudy: Acta Phys. Polon. A **40**, 361 (1971)
7.23 J. Szudy, W.E. Baylis: J. Quant. Spectrosc. Radiat. Transf. **15**, 641 (1975)
7.24 S.G. Rautian, I.I. Sobleman: Usp. Fiz. Nauk **90**, 209 (1966) [English transl.: Sov. Phys. – Uspekhi **9**, 701 (1967)]
7.25 S. Chandrasekhar: Rev. Mod. Phys. **15**, 1 (1943)
7.26 R.H. Dicke: Phys. Rev. **89**, 472 (1953)
7.27 D.R.A. McMahon: Austral. J. Phys. **34**, 639 (1981)
7.28 V.N. Faddeyeva, N.M. Terentyev: *Tables of the Probability Integral for Complex Argument* (Pergamon, Oxford 1961)

7.29 M. Abramowitz, I.A. Stegun (eds.): *Handbook of Mathematical Functions* (NBS Math. Ser., Washington 1964)

7.30 I.I. Sobel'man: *Atomic Spectra and Radiative Transitions*, 2nd edn., Springer Ser. Atoms Plasmas, Vol.12 (Springer, Berlin, Heidelberg 1992); 1st edn., Springer Ser. Chem. Phys., Vol.1 (Springer, Berlin, Heidelberg 1979)

7.31 L.D. Landau, E.M. Lifshitz: *Quantum Mechanics* (Pergamon, Oxford 1965)

7.32 I.I. Sobelman: Opt. Spectrosc. **1**, 617 (1956)

7.33 M. Baranger: Phys. Rev. **111**, 481, 494 (1958); ibid. **112**, 855 (1958)

7.34 P.R. Berman, W.E. Lamb, Jr.: Phys. Rev. A **2**, 2435 (1970); ibid. A**4**, 319 (1971)

7.35 E.W. Smith, J. Cooper, W.R. Chappel, T.D. Dillon: J. Quant. Spectrosc. Radiat. Transf. **11**, 1547, 1567 (1971)

7.36 W.R. Chappell, J. Cooper, E.W. Smith, T.D. Dillon: J. Stat. Phys. **3**, 401 (1971)

7.37 V.A. Alekseyev, T.L. Andreyeva, I.I. Sobleman: Zh. Eksp. Teor. Fiz. **62**, 614 (1972) [English transl.: Sov. Phys. – JETP **35**, 325 (1972)]

7.38 V.A. Alekseyev, T.L. Andreyeva, I.I. Sobelman: Zh. Eksp. Teor. Fiz. **64**, 813 (1973) [English transl.: Sov. Phys. – JETP **37**, 413 (1973()

7.39 V.S. Letokhov, V.P. Chebotayev: *Nonlinear Laser Spectroscopy* (Springer, Berlin, Heidelberg 1977)

7.40 V.A. Aleseyev, I.I. Sobelman: Zh. Eksp. Teor. Fiz. **55**, 1874 (1968) [English transl.: Sov. Phys. – JETP **28**, 991 (1968)

7.41 A.B. Underhill, J. Waddell: NBS Circular No.603 (1959)

7.42 H.R. Griem: Astrophys. J. **132**, 883 (1960)

7.43 C.F. Hooper, Jr.: Phys. Rev. **165**, 215 (1968)

7.44 H.K. Wimmel: J. Quant. Spectrosc. Radiat. Transf. **1**, 1 (1961)

7.45 G.V. Sholin, V.S. Lisita, V.I. Kogan: Zh. Eksp. Teor. Fiz. **59**, 1390 (1970) [English transl.: Sov. Phys. – JETP **32**, 758 (1970)]

7.46 W.L. Wiese, D.E. Kelleher, V. Helbig: Phys. Rev. A **11**, 1854 (1975)

7.47 R.L. Green: J. Quant. Spectrosc. Radiat. Transf. **27**, 639 (1982)

7.48 R.L. Green, D.H. Oza, D.E. Kelleher: in *Spectral Line Shapes* Vol.5, ed. by J. Szudy (Ossolineum, Wroclaw 1989) p.127

7.49 J. Seidel; Z. Naturforsch. **32a**, 1195, 1207 (1977)

7.50 J.W. Dufty: in *Spectral Line Shapes*, Vol.1, ed. by B. Wende (de Gryter, New York 1981) p.41

7.51 D.B. Boercker, C.A. Iglesias, J.W. Dufty: Phys. Rev. A **36**, 2254 (1987)

7.52 D.B. Boercker: in *Spectral Line Shapes*, Vol.5, ed. by J. Szudy (Ossolineum, Wroclaw 1989) p.73

7.53 V.S. Lisitsa: Usp. Fiz. Nauk **122**, 449 (1977) [English transl.: Sov. Phys. – Uspekhi **20**, 603 (1977)]

7.54 C. Deutsch, L. Herman, H.-W. Drawin: Phys. Rev. **178**, 261 (1968)

7.55 R.L. Green, J. Cooper, E.W. Smith: J. Quant. Spectrosc. Radiat. Transf. **15**, 1025, 1037, 1045 (1975)

7.56 H.R. Griem, A.C. Kolb, K.Y. Shen: Phys. Rev. **116**, 4 (1959)

7.57 H. Pfennig: Z. Naturforsch. **26a**, 1071 (1971); ibid. J. Quant. Spectrosc. Radiat. Transf. **12**, 821 (1972)

7.58 R.L. Green, J. Cooper: J. Quant. Spectrosc. Radiat. Transf. **15**, 1490 (1975)

7.59 M. Lewis: Phys. Rev. **121**, 501 (1961)

7.60 Nguen-Hoe, H.-W. Drawin, L. Herman: J. Quant. Spectrosc. Radiat. Transf. **4**, 847 (1964)

7.61 M. Caby-Eyrand, G. Gouland, Nguen-Hoe: J. Quant. Spectrosc. Radiat. Transf. **15**, 593 (1975)

7.62 D. Voslamber: Z. Naturforsch. **21a**, 1458 (1969); ibid. **27a**, 1783 (1972); ibid Phys. Lett. A **42**, 469 (1973)

7.63 H. Van Regemorter: Phys. Lett. A **30**, 365 (1969)

7.64 N. Tran-Minh, H. Van Regemorter: J. Phys. B **5**, 903 (1972)

7.65 V.S. Lisitsa, G.V. Sholin: Zh. Eksp. Teor. Fiz. **61**, 912 (1971) [English transl.: Sov. Phys. – JETP **34**, 484 (1971)]

7.66 P. Kepple, H.R. Griem: Phys. Rev. **173**, 317 (1968)

7.67 C.R. Vidal, J. Cooper, E.W. Smith: J. Quant. Spectrosc. Radiat. Transf. **10**, 1011 (1970); ibid. **11**, 263 (1971)

7.68 C.R. Vidal, J. Cooper, E.W. Smith: Astrophys. J. **214**, Suppl.25, 37 (1973)

7.69 N. Tran-Minh, N. Feautrier, H. Van Regemorter: J. Phys. B **8**, 1810 (1975); ibid. B **9**, 1871 (1976); J. Quant. Spectrosc. Radiat. Transf. **16**, 849 (1976)

7.70 W.L. Wiese, D.E. Kelleher, D.R. Paquette: Phys. Rev. A **6**, 1132 (1972)

7.71 G. Boldt, W.B. Cooper: Z. Naturforsch. **19a**, 968 (1964)

7.72 R.C. Elton, H.R. Griem: Phys. Rev. **135**, 1550 (1964)

7.73 D.E. Kelleher, W.L. Wiese: Phys. Rev. Lett. **31**, 1431 (1973)

7.74 K. Grutzmacher, B. Wende: Phys. Rev. A **16**, 243 (1977)

7.75 K. Grutzmacher, B. Wende: Phys. Rev. A **18**, 2140 (1978)

7.76 R. Stamm, E.W. Smith, B. Talin: Phys. Rev. A **30**, 2039 (1984)

7.77 R. Stamm, B. Talin, E. Pollock, C. Iglesias: Phys. Rev. A **34**, 4144 (1986)

7.78 U. Frisch, A. Brissaud: J. Quant. Spectrosc. Radiat. Transf. **11**, 1753 (1971)

7.79 A. Brissaud, U. Frisch: J. Quant. Spectrosc. Radiat. Transf. **11**, 1767 (1971)

7.80 H.R. Griem: Astrophys. J. **148**, 547 (1967)

7.81 L.A. Minaeva, I.I. Sobelman: J. Quant. Spectrosc. Radiat. Transf. **8**, 783 (1968)

7.82 H. Van Regemorter, D. Hoang-Binh: Astron. Astrophys. **277**, 623 (1993)

7.83 D.E. Kelleher: J. Quant. Spectrosc. Radiat. Transf. **25**, 191 (1981)

7.84 Yu.A. Vdovin, V.M. Galitskii: Zh. Eksp. Teor. Fiz. **52**, 1345 (1967) [English transl.: Sov. Phys. – JETP **25**, 894 (1967)]

7.85 Yu.A. Vdovin, N.N. Dobrodeyev: Zh. Eksp. Teor. Fiz. **55**, 1047 (1968) [English transl.: Sov. Phys. – JETP **28**, 544 (1968)]

7.86 A.W. Ali, H.R. Griem: Phys. Rev. A **140**, 1044 (1965); ibid. A **144**, 366 (1966)

7.87 D.N. Stacey, J. Cooper: Phys. Lett. A **30**, 49 (1969)

7.88 C.G. Carrington, D.N. Stacey, J. Cooper: J. Phys. B **6**, 417 (1973)

7.89 J. Cooper, D.N. Stacey: Phys. Lett. A **46**, 299 (1973)

7.90 M. Movre, G. Pichler: J. Phys. B **13**, 697 (1980)

7.91 R.J. Exton: J. Quant. Spectrosc. Radiat. Transf. **15**, 1141 (1975)

7.92 V.A. Alekseyev, I.I. Sobel'man: Zh. Eksp. Teor. Fiz. **49**, 1274 (1965) [English transl.: Sov. Phys. – JETP **22**, 882 (1965)]

7.93 H. Bethe, E. Salpeter: *Quantum Mechanics of One- and Two-Electron Atoms* (Springer, Berlin, Göttingen 1957)

7.94 L.P. Presnyakov: Phys. Rev. A **2**, 1720 (1970)

7.95 A. Omont: J. Physique **38**, 1343 (1977)

7.96 M. Matsuzawa: J. Phys. B **10**, 1543 (1977)

7.97 B.P. Stoicheff, E. Weinberger: Phys. Rev. Lett. **44**, 733 (1980)

7.98 R. Kachru, T.W. Mossberg, S.R. Hartmann: Phys. Rev. A **21**, 1124 (1980)

7.99 R. Kachru, T.J. Chen, T.W. Mossberg, S.R. Hartmann: Phys. Rev. A **25**, 1546 (1982)

7.100 H. Heinke, J. Lawrenz, K. Niemax, K.H. Weber: Z. Phys. A **312**, 329 (1983)

7.101 H. Massey, E. Burhop: *Electronic and Ionic Impact Phenomena* (Clarendon, Oxford 1952)
7.102 M.A. Mazing, M.A. Vrublevskaya: Zh. Eksp. Teor. Fiz. **50**, 343 (1966) [English transl.: Sov. Phys. – JETP **23**, 228 (1966)]
7.103 M.A. Mazing, P.D. Serapinas: Zh. Eksp. Teor. Fiz. **60**, 541 (1971) [English transl.: Sov. Phys. – JETP **33**, 294 (1971)]

List of Symbols

Constants

$a_0 = \hbar/me^2$ Bohr radius
c Velocity of light
e Elementary charge
$\hbar = h/2\pi$ Planck's constant divided by 2π
m Mass of electron
$\mathrm{Ry} = me^4/2\hbar^2$ Rydberg unit of energy

Quantum numbers

j Electron angular momentum
J Atomic angular momentum
J_T Total angular momentum of a system including atom and outer (scattered) electron
l Electron orbital momentum
L Atomic orbital momentum
L_p Orbital momentum of atomic core (of parent ion)
L_T Total orbital momentum of a system including atom and scattered electron
m, M Magnetic quantum numbers
n Principal quantum number
s Electron spin momentum
S Atomic spin momentum
S_p Spin momentum of atomic core (of parent ion)
S_T Total spin momentum of a system including atom and scattered electron
λ Orbital momentum of outer (scattered) electron

Basic Notations

a_0 Set of quantum numbers for initial state of an atom
a, a_1 Set of quantum numbers for final state
A Fitting parameter for approximation of rate coefficients $\langle v\sigma \rangle$
A_{ij} Einstein coefficient for spontaneous emission (radiative transition probability) [s^{-1}]
C Fitting parameter for approximation of cross sections σ
D Fitting parameter for analytical approximation of calculated cross sections and rate coefficients
DE Energy scaling parameter
E_a Energy of bound electron in state a
$\mathscr{E}_0, \mathscr{E}$ Initial and final energies of free electron
f_{ij} Oscillator strength
$f(\vartheta)$ Scattering amplitude
$F_\lambda, F^{\gamma}_{\gamma'}, F^{\Gamma}_{\Gamma'}$ Radial functions of scattered electron in various representations
$g(a)$ Statistical weight of level a
$G^{SL}_{S_p L_p} = (l^{n-1}[S_p L_p] lSL\} l^n SL)$ Coefficient of fractional parentage [see Ref. 1.1]
$G(\beta)$ Function of analytical approximation for rate coefficients $\langle v\sigma \rangle$
$\mathrm{G}_\Gamma(\boldsymbol{r}, \boldsymbol{r}')$ Green's function
$j_x(z)$ Spherical Bessel function
$\mathscr{H}(\beta)$ Holtsmark distribution function
$n^* = \sqrt{z^2 \mathrm{Ry}/|E|}$ Effective principal quantum number
N Number density of particles [cm^{-3}]
$P_l(\cos\vartheta)$ Legendre polynomials
$P_l, P_{nl}, P_{nl}(r) = rR_{nl}(r)$, where $R_{nl}(r)$ is radial function for bound electron
$\mathrm{P}\int$ Principal value of integral
Q, Q_κ Angular factor defining the dependence of cross sections on angular momenta for transitions with no change of spin
Q'', Q''_κ Angular factor for exchange cross sections
S, S_{ik} Scattering matrix
T Temperature in energy units
T_{ik} Transition matrix
$u = \mathscr{E}/\Delta E = (\mathscr{E}_0 - \Delta E)/\Delta E$ Electron energy in threshold units
v Velocity of particles
$\langle v\sigma \rangle$ Rate coefficient averaged over Maxwellian velocity distribution [$\mathrm{cm}^3\,\mathrm{s}^{-1}$]
$W_{a_0 a}$ Dimensionless transition probability, frequency of collisional transitions [s^{-1}]

$W_a(c)$ Autoionization probability for atomic state c [s^{-1}]
z Charge of atomic core (of parent ion)
$Z = z - 1$ Charge of ion
$\mathscr{Z}$ Charge of nucleus
γ Line width (full width) in Chapter 7, set of quantum numbers for atomic term in Sect. 6.2
$\gamma = aM\lambda mm^3$ Set of quantum numbers for a system including atom and scattered electron
$\Gamma = a\lambda 1/2 L_T S_T$ Set of quantum numbers for a system including atom and scattered electron in representation of total momenta
Δ Line shift in Chapter 7
$\Delta E = E_i - E_k$ Energy difference for levels i and k
κ Multipole order
κ_d Rate coefficient of dielectronic recombination [$cm^3 s^{-1}$]
κ_r Rate coefficient for three-body recombination [$cm^6 s^{-1}$]
κ_v Rate coefficient of radiative recombination [$cm^3 s^{-1}$]
O Solid angle
ρ Impact parameter, density matrix
$\rho_{s\sigma}$ Spherical components of density matrix
$\sigma_{a_0 a}, \sigma(a_0, a)$ Effective cross section
$\sigma(a_0 \lambda_0, a\lambda)$ Partial cross section
φ Fitting parameter for approximation of cross sections
$\Phi(u)$ Functions of analytical approximation for cross sections
χ Fitting parameter for approximation of rate coefficients
$[j_1 j_2 \ldots j_n] = \sqrt{(2j_1+1)(2j_2+1)\ldots(2j_n+1)}$
$(m_1 m_2 | s\sigma)$ Klebsch-Gordan coefficients (abbreviated notation)
$\begin{pmatrix} j_1 j_2 j_3 \\ m_1 m_2 m_3 \end{pmatrix}$ Wigner's $3j$ symbol
$\begin{Bmatrix} a_1 a_2 a_3 \\ b_1 b_2 b_3 \end{Bmatrix}$ $6j$ symbol
$\begin{Bmatrix} abc \\ def \\ pqr \end{Bmatrix}$ $9j$ symbol
$(a_0 \| T \| a)$ Reduced matrix element

Subject Index

Springer-Verlag and the Environment

We at Springer-Verlag firmly believe that an international science publisher has a special obligation to the environment, and our corporate policies consistently reflect this conviction.

We also expect our business partners – paper mills, printers, packaging manufacturers, etc. – to commit themselves to using environmentally friendly materials and production processes.

The paper in this book is made from low- or no-chlorine pulp and is acid free, in conformance with international standards for paper permanency.